Cordless Telecommunications Worldwide

Springer
London
Berlin
Heidelberg
New York
Barcelona
Budapest
Hong Kong
Milan
Paris
Santa Clara
Singapore
Tokyo

Walter H.W. Tuttlebee (Ed.)

Cordless Telecommunications Worldwide

The Evolution of Unlicensed PCS

With 153 Figures

 Springer

Walter H.W. Tuttlebee, CEng
Roke Manor Research Ltd, Roke Manor, Romsey,
Hampshire SO51 0ZN, England
e-mail: walter.tuttlebee@iee.org

ISBN 3-540-19970-5 Springer-Verlag Berlin Heidelberg New York

British Library Cataloguing in Publication Data
Cordless telecommunications worldwide: the evolution of unlicensed PCS
   1. Wireless communication systems
   I. Tuttlebee, Wally H. W., 1953–   II. Harold, William
   621.3′8235
ISBN 3540199705

Library of Congress Cataloging-in-Publication Data
Cordless telecommunications worldwide: the evolution of unlicensed
   PCS / Walter H.W. Tuttlebee, ed. – 1st ed.
      p.   cm.
   Includes bibliographical references and index.
   ISBN 3-540-19970-5 (hardback: alk. paper)
   1. Personal communication service systems.   2. Telephone supplies industry.
   I. Tuttlebee, Wally H. W., 1953–
   TK5103.2.C67   1996   96-30964
   384.5–dc20   CIP

Typeset by T&A Typesetting Services, Rochdale, England
Printed and bound by Cambridge University Press
69/3830-543210 Printed on acid-free paper

Contents

Part III: Technology

Contributors

Dag Åkerberg
Ericsson Radio Systems, S-16480 Stockholm, Sweden

Gary Boudreau
Northern Telecom, PO Box 3511, Station C, Ottawa, Ontario, Canada
K1Y 4H7

Andrew Bud
Olivetti Telemedia, Via Jervis 77, I-10015 Ivrea, Italy

Herman Bustamante
Stanford Telecom, 1221 Crossman Avenue, PO Box 3733, Sunnyvale,
CA 94088-3733, USA

Ed Candy
Hutchison Personal Communications, John Tate Road, Foxholes
Business Park, Hertford, Herts SG13 7NN, UK

Horen Chen
Stanford Telecom, 1221 Crossman Avenue, PO Box 3733, Sunnyvale,
CA 94088-3733, USA

Dominic Clancy
Philips Semiconductors, Binzstrasse 44, CH-8045 Zürich, Switzerland

Graham Crisp
GPT Ltd, Technology Drive, Beeston, Nottingham NG9 1LA, UK

Phil Crookes
AT Kearney Ltd, Lansdowne House, Berkeley Square, London W1X
5DH, UK

Robert Harrison
PA Consulting Group, 123 Buckingham Palace Road, London SW1W
9SR, UK

A. Peter Hulbert
Roke Manor Research Ltd, Romsey, Hants SO51 0ZN, UK

Javier Magaña
Advanced Micro Devices Inc., MS 526, 5204 East Ben White Blvd, Austin, TX 78741, USA

Neil Montefiore
MobileOne (Asia) Pte Ltd, 456 Alexandra Road, 12-01 NOL Building, Singapore 119962
   (Previously with Chevalier Telepoint, Hong Kong)

Yasuaki Mori
Advanced Micro Devices, World Trade Centre, PO Box 11, Rue de l'Aeroport 10, 1215 Geneva 15, Switzerland

Anthony Noerpel
Hughes Network Systems, 11717 Exploration Lane, Germantown, MD 20876, USA

Heinz Ochsner
Ochsner MTC Mobile Telecommunications Consulting, Gibelin-strasse 25, CH-4500 Solothurn, Switzerland

Peter Olanders
Telia Research, PO Box 85, S-201 20 Malmoe, Sweden

Annette Ottolini
Greenpoint, PTT Telecom, PO Box 30150, 2500GD The Hague, The Netherlands

Frank Owen
Philips Österreich, 10 Gutheilschodergasse, A-1102 Wien, Austria

Marc Pauwels
Belgacom, 22nd Floor Room 73, Gacqmaen Laan 166, Brussels 1020, Belgium

William H. Scales Jr
Panasonic, 2001 Westside Drive, Alpharetta, GA, USA

J. Neal Smith
Omnipoint Technologies, 1365 Garden of the Gods Road, Colorado Springs, CO 80907, USA

Richard Steedman
PMC-Sierra Inc., 105–8555 Baxter Place, Burnaby, British Columbia, V5A 4V7, Canada

Bob Swain
RSS Telecom, 8 Birchwood Drive, Rushmere St Andrew, Suffolk IP5 7EB, UK

Yuichiro Takagawa
Multimedia Business Development Dept, NTT Corporation Headquarters, Nishi-Shinjuku, Shinjuku-ku, Tokyo 163-19, Japan

Julian Trinder
Roke Manor Research Ltd, Romsey, Hants SO51 0ZN, UK

Diane Trivett
Dataquest Europe, Tamesis, The Glanty, Egham, Surrey TW20 9AW,
UK

Walter Tuttlebee
Roke Manor Research Ltd, Romsey, Hants SO51 0ZN, UK

Margareta Zanichelli
Telia Research, PO Box 85, S-201 20 Malmoe, Sweden

Foreword

A View from America

Cordless telephony in North America has evolved along a quite different path than in Europe or Asia. US cordless telephones have evolved as handset/base unit sets essentially independent of network functionality, except for the base units providing a wireline interface that is indistinguishable from that of a wireline telephone. Early sets operated at 1.6/49 MHz, then evolved to 46/49 MHz and more recently to the 902–928 MHz ISM (industrial–scientific–medical) band. The speech quality of early US cordless telephones was poor – people bought sets by the millions, but also discarded them by the millions because of the poor quality. When manufacturers finally started producing cordless telephones with speech quality comparable with wireline telephones, the market exploded. There are, today, perhaps 100 million cordless phones in use in the USA.

Again, unlike elsewhere, in the USA there has been a wireless research and development activity completely separate from either the cordless or the cellular industries. Originally sponsored by the local exchange (telephone) companies at Bellcore, this activity was aimed at providing widespread, cordless telephone-like, highly mobile services through wireless access to large-scale intelligent fixed infrastructure networks, e.g. local exchange networks or, perhaps, the combination of cable TV and intelligence in inter-exchange (long-distance) networks. It was pursued under various identities such as wireless local loops (WLL), universal digital portable communications (UDPC) and, more recently, Wireless Access Communication System (WACS). Hughes Network Systems (HNS) and Motorola developed WACS prototypes, which, along with Bellcore prototypes, have been used in many successful technology trials.

With the new opportunities for PCS service, derivatives of WACS, e.g. Personal Access Communications System (PACS), have emerged as serious contenders for both high- and low-tier, cellular- and cordless-like PCS. PACS-UB was thus created to provide a compatible unlicensed companion technology to PACS; it is akin to other digital cordless technologies, but was derived by an entirely different route.

Thus, US evolution to a high-capability digital cordless technology, although not starting from cordless telephone optimised standards like CT2, DECT or PHS, has resulted in technical solutions such as PACS and PACS-UB that offer potential for dual-mode, cordless/ cellular, service and the capability to support wireless loop access. The evidence of the recent spectrum auctions demonstrates the major level of commercial interest in such opportunities.

Given this background, this book is to be welcomed, providing a timely overview of the field, comprehensive coverage of the engineering principles of low-power cordless telecommunications and descriptions of digital cordless technologies. It would appear that in the USA, simple cordless telephones and complex cellular will continue to evolve independently for some time; however, digital cordless technologies have significant potential to supplant existing cordless telephones, cellular telephones and wireline telephones. *Cordless Telecommunications Worldwide* provides a perspective of the low-complexity, low-power side of wireless communications that is quite different from that of high-complexity, high-power cellular communications. Thus, the book is a valuable resource for the reader to explore how different regions of the world are developing, applying and commercialising such technology.

May 1996

Donald Cox
Stanford University
California

A View from Asia–Pacific

Like many parts of the world, the Asia–Pacific region is experiencing a mobile service boom brought about by its recent economic successes (especially in its newly industrialised economies), increasingly liberal regulatory environments and the availability of new technologies. Although the major demand for mobile service is in the cellular and paging services, cordless access services (CAS) such as in the telepoint application still have an important role to play in many markets in this region. In places with high population densities, like Hong Kong, "low-mobility" mobile services based on CAS technology can be set up with good coverage at relatively low cost and are suitable for public-transport- or pedestrian-oriented communities. In countries where the fixed telecommunications infrastructure is inadequate, CAS-based wireless local loop is increasingly becoming a viable means for implementing fast rollout of fixed networks. Over the past few years, cordless telecommunications technology has evolved from that used in simple analogue home cordless phones to complex digital public network access systems capable of handling both inbound and outbound traffic with "handoff".

To policy-makers and telecommunications regulators, the evolutionary changes of CAS pose a challenge because they continue to blur the traditional boundaries between mobile and fixed network services. Difficulties also arise in keeping abreast of the various new and often incompatible standards in cordless telecommunications and in providing spectrum for competing technologies. However, the changing CAS technology may open up new market opportunities and applications and is definitely of interest to the telecommunications industry. Our experiences in Hong Kong, where the early success of CT2 telepoint services has dissipated in the face of rapid developments in digital cellular services, indicate that commercial CAS operators will need to position themselves more as adjuncts to the fixed networks rather than as direct competitors to cellular services. This we attempt to facilitate in Hong Kong with our policies of technology-neutrality, spectrum allocation and open licensing processes.

This book, *Cordless Telecommunications Worldwide*, edited by Walter H.W. Tuttlebee, contains invaluable contributions from distinguished experts in the associated disciplines and is a timely work for telecommunications practitioners. I congratulate the editor and his contributors for producing such a clear and comprehensive volume in this exciting field.

May 1996   Alexander A. Arena
                                                   Director-General
                          Office of the Telecommunications Authority
                                                        Hong Kong

A View from Europe

Without frequencies, no services. Without high-technology radio platforms, no future-proof terminals. Without open standards, no multi-vendor market and no intense competition. These three cornerstones form the basis for the initiatives launched by the European Commission since the 1980s. Regarding frequencies, a Directive was adopted. Regarding the radio platform, DECT (Digital Enhanced Cordless Telecommunications) technology was preferred. Regarding competition, ETSI was mandated to elaborate open European Telecommunication Standards.

So where has this brought us now?

It is a pleasure to note that DECT has proven to provide extremely effective communications (more than $10\,000\,\text{Erlang/km}^2$), high speech quality and short- as well as long-range coverage (the latter by using the DECT repeater concept).

A number of core standards have been turned into the European Regulatory Type Approval regime, so the free-market forces can work to the benefit of users. Cordless telecommunication has been turned into a quality consumer product commodity, and indeed a minimum quality has been guaranteed owing to the obligation laid down by the civil protection codes that emergency calls must always be successful.

The European re-regulation of the telecommunication sector has been carried out with the objective of getting the right balance between free competition in the market, effective use of the spectrum, imposing a minimum regulatory regime and still allowing for innovation regarding new supplementary services and early intro-duction of new equipment.

The European cordless policy is therefore now being seen by many as offering the right solution at the right time: DECT is therefore today an acronym for digital enhanced cordless telephony.

Decision-makers throughout the world can indeed benefit from this concise yet comprehensive book, *Cordless Telecommunications Worldwide*, edited by Dr W.H.W. Tuttlebee. The book will contribute to increased transparency in the cordless market. It provides for a platform upon which issues like economics, tariffs and operating costs can be addressed, and thereby the different technologies can be compared.

The initiative taken to publish the state of the art on cordless telecommunications in a global context matches ideally the globalisation trend in telecommunications.

May 1996   Jørgen A. Richter
                                                       Directorate General XIII
                                       Commission of the European Communities
                                                                      Brussels

Acknowledgements

This book owes its origins to Springer-Verlag's UK Engineering Series editor, Nicholas Pinfield, who over a sustained period sought gently to persuade me to update my earlier book, *Cordless Teleccmmunications in Europe*. When I finally agreed, it was with a recognition that this volume would in reality be a completely new book – cordless telecommunications has evolved from its early European genesis to become today a global industry.

I acknowledge with grateful thanks the support of all of the contributors, experts in their own different aspects of cordless telecommunications, who have so willingly shared their knowledge, especially those who contributed to the earlier book and yet still came back for more – without the willing inputs of these people, this book would not have come to fruition. I wish to thank those who provided their contributions early, and then had the frustration of waiting for the project to reach its conclusion, and also those who willingly contributed, at a late stage and to tight timescales, chapters that have helped to make the book comprehensive. I also wish to acknowledge those many other industry colleagues who do not appear as overt contributors but who have supported and encouraged this project in other ways. It has been a privilege both to renew old friendships and to make new ones in the global cordless industry as this project has developed.

I wish to thank the management of Roke Manor Research, a Siemens company, for allowing me to undertake this project, reflecting as it does more than a decade of involvement in the research and development of cordless telecommunications.

I also thank my wife, Helen, and our three children – David, Joy and Stephen – for their ongoing encouragement and understanding throughout this project, but particularly during the busy latter stages of the process, as the book came to completion.

Finally, I dedicate this book to the one who is the author of all knowledge and the giver of all good gifts – to Jesus Christ, lover and saviour of my soul. He is wonderful, He is alive and I love Him.

September 1996   Walter H.W. Tuttlebee
                                                 Romsey, Hants, UK

Introduction

Walter Tuttlebee

Why a Book on Cordless Telecommunications?

The significance of developments in cordless telecommunications over the past decade has been ably stated in the Forewords to this book, with perspectives from America, Asia and Europe, reflecting the global recognition and importance of this market, the dynamic nature of the technology and the pace of regulatory change, which continues to create new opportunities. From initial concepts in the mid-1980s, cordless – unlicensed personal communications services (UPCS) – has become a global industry in the mid-1990s, addressing not simply domestic cordless telephones, but business systems and public access, with significant opportunities for offering new means of access to both fixed networks and cellular radio systems.

Despite this, cordless telecommunications, based on low-power, short-range, microcellular communications, unlike cellular radio, has still received scant attention in books. This book seeks to rectify this omission, drawing on contributions from experts in the field from across the world. We have attempted to present a foundation of information covering the field from four different perspectives – markets and applications, standards and industry development, technology, and the various regional technical standards.

Who is This Book For?

This book is intended to act as a training resource for those already in the industry, in commercial and engineering roles, and for engineering and business students. The book presents the commercial background to the current scene as well as describing the technological issues underpinning the operation of cordless telecommunications systems. This blend has been sought in order to provide those engaged in the commercial aspects of the industry with a solid foundation for their roles as well as to provide engineers with a commercial context to their development efforts. This interrelationship between the technical and commercial issues may be of particular interest to the undergraduate or graduate reader, to whom these two fields can sometimes appear as separate.

The Structure of the Book

Part I discusses markets and applications. Chapter 1 provides an overview of how cordless telecommunications markets and applications have developed, with particular emphasis on their evolution during the 1990s. The different basic market segments and requirements are described alongside application segments and the interdependence between them. The next few chapters examine each of these applications and markets in more depth, addressing the business wireless PABX application (Chapter 2), the mixed experiences of public telepoint in Europe and Asia (Chapters 3 and 4) and wireless local loop (Chapter 5). The potential for integrated usage of cordless handsets operating across all these environments is addressed in Chapter 6, where new developments in cordless terminal mobility are described.

Part II describes the processes by which the new digital cordless telecommunications standards have been developed across the globe. The standardisation processes are very different in Europe, North America and Japan. Thus Chapters 7–9 describe the different environments in which CT2 and DECT, the various unlicensed PCS systems and PHS have come to birth. The development of the industry around these standards is the topic of Chapter 10, which summarises their origins, the current range of equipments on the market and the types of product. The global competition between the different systems is also addressed in this chapter.

Part III focuses on the technology of cordless systems, covering audio, radio, network, data and multimedia, and terminal implementation aspects in turn. The audio chapter (Chapter 11) discusses issues of digital speech coding, transmission plans, delay, echo control, privacy and security, providing an introduction to the terminology of these fields and the technical approaches employed. Radio propagation, choice of spectrum, micro- and picocellular techniques are explored in Chapter 12, together with the different options for multiple access – FDMA, TDMA and CDMA – modulation, error control and communications protocols. Cordless systems provide new means of accessing existing networks, but may be used in many different ways. Chapter 13 addresses issues of network functionality, such as call control, mobility management, supplementary services and traffic interworking, considering domestic, business and public access networks. Cordless data and multimedia developments are reviewed in Chapter 14. As well as basic requirements and benefits of cordless data, recent developments in the field of cordless LANs are described – including IEEE 802.11, HIPERLAN and DECT. The DECT cordless LAN technology is reviewed; examples of current cordless multimedia developments are also described. Chapter 15 pulls together many of the issues described in the preceding chapters to discuss the practical implementation of cordless handsets and other cordless terminals. Practical hardware architectures and approaches to embedded software are described. Initially the principles and architectural options for radio and baseband design are described, followed by specific examples of architectures for CT2 and DECT equipments. Chapter 16 attempts to pull together in a similar way a view of near-term cordless system evolution, describing developments in cordless access in public networks and cellular networks. Technology development, dual-mode handsets and the associated infrastructure evolution are the focal points of this chapter.

Technical standards that have emerged in Europe, America and Japan are the subject of Part IV. In this part of the book, all of the main standards or emerging standards are described, to levels of detail reflecting their maturity. All the major standards are covered – CT2, PCI, DECT and its associated access profiles (GAP, GIP, etc.), PWT, PACS-UB, PHS, PACS-UA, orthogonal CDMA and IS-661 (CCT).

A comprehensive glossary of acronyms and biographical details of the contributors conclude the book.

Many diverse concepts are involved in digital cordless communications. It would be impossible in a book of this size to do justice fully to each of these different areas. Certain topics may already be familiar, perhaps in a different context, to some readers. An attempt has been made to achieve a reasonably consistent level of treatment, although in some areas specific issues have been explored in more depth. For those new to the field, and also for those wishing to probe a little deeper on particular topics, a selection of references has been included in each chapter, some providing general background and others addressing specific detail.

The contributors to the book, as will be seen from their biographies, are individuals who have significantly shaped, in some cases one could almost say "created", the development of cordless telecommunications over the past few years. There is always a healthy tension in preparing an edited work of this type between the desire to produce a cohesive and fully consistent work and the wish to maintain the full diversity of expression and opinion representative of the individual contributors and their own unique perspectives. As Editor, I have sought to preserve a balance in this respect, but this inevitably means that the views expressed in any particular chapter do not necessarily reflect those of the Editor nor indeed those of some contributors. The contributors, the Editor and the publishers cannot accept liability for any errors or omissions which, in any work of this nature, may occur. It would be appreciated, however, if any errors could be brought to the attention of the Editor for correction in future editions.

Finally, I trust the readers of this book will find it informative, interesting and useful. It is a book for reference, for learning and, I hope, for enjoyment. I trust the reader learns and enjoys as much reading this book as I have learnt and enjoyed in its creation.

Part I

Markets and Applications

1 The Cordless Market

Dominic Clancy

This chapter explores the digital cordless telephony market, beginning by examining how the digital cordless telephony market has been, and continues to be, influenced by the cellular radio market, providing a context in the mobile communications marketplace. It then briefly describes the historical development of digital cordless in Europe, North America and Asia and considers the global aspects of digital cordless – common requirements, yet different standards. It discusses market positioning of cordless products and how perceptions in this area have changed since the start of the 1990s, charting the reasons for these changes.

Market segmentation, by application, for cordless telecommunications is summarised, as an introduction to the more detailed coverage of these applications in subsequent chapters, together with aspects of market sector interdependence. Finally, future market evolution and the potential integration of cordless and cellular terminals and services is discussed.

1.1 The Cordless Telephone and the Cellular Radio

The mobile telephony market has developed significantly during the 1990s. At the end of 1989 there were 2 million users of mobile phones in Europe; market forecasts reported in [1] suggested growth to 6.6 million by 1994. Even this forecast proved conservative, with this figure being easily exceeded by 1993 – see Table 1.1. In 1995, almost 2 million users were added to the cellular user base in the UK alone, and in Italy and Germany over 1 million new users were also added. The total number of cellular users in Europe increased by over 8 million. The cellular market has clearly moved on in terms of marketing, distribution, pricing and user expectations.

Table 1.1 shows how this cellular growth is expected to be mirrored across most of the world. This is indicative of the demand for mobility in all aspects of people's lives across all sectors of society.

The expanding use of cellular radio awakened users to the inadequacies for business use of the cordless equipment that existed in some geographical markets in the 1980s. At the same time increased awareness of the possibilities of mobility in communication has stimulated demand for mobility within the office and the home. The development of digital cellular and digital cordless offered

Table 1.1
Cellular subscriber forecasts ('000s)

Region	1993	1994	1995	1996	1997	1998	Annual growth 1994–98 (%)
Asia	5 730	10 785	15 265	20 290	25 320	30 700	57
Europe	8 740	14 140	21 500	26 630	32 200	38 000	54
Latin America	1 365	2 565	3 840	5 000	6 340	7 940	62
North America[a]	16 700	22 580	29 660	38 200	47 740	58 800	52
Totals	32 535	50 070	70 265	90 120	111 600	135 440	54

[a]Does *not* include any forecast for PCS.
Source: "Regional Cellular Subscriber Forecasts and World Cellular Network Equipment Markets", provided by Northern Business Information, January 1996.

new opportunities for cordless telephones to move away from the basic residential or domestic cordless telephone to a cordless terminal that could be used in a number of different environments. These environments were believed to offer a new opportunity for telephone manufacturers to develop new markets for terminals with higher added value.

As far as the cordless market is concerned, figures from Dataquest (Table 1.2) show the expected increase in production of digital cordless sets in the next few years. The figures in Japan are mainly PHS, with an increasing proportion of DECT from 1997 onwards. The figures in Asia–Pacific are almost all DECT, as CT2 production has not been counted because of an expected decline in this product. European production is predominantly DECT and CT0, and Americas is a combination of DECT, PWT and proprietary digital cordless products for the residential market. These figures illustrate the tremendous surge in interest in digital cordless technology from an ever-growing number of manufacturers.

The number of analogue cordless users is impossible to assess, but is clearly enormous in spite of the low audio quality of these products. The basic CT0 cordless telephone has become the street-child of the terminal market, being used even where it is not officially sanctioned.

Further figures from Dataquest (Figure 1.1) show more detail of the European market. Cordless telephony is the fastest-growing segment, and the study

Table 1.2
Digital cordless telephone handset and basestation production[a]

Region		1993	1994	1995	1996	1997	1998	1999	2000
Americas	units	50	100	200	450	900	1 700	2 600	4 000
	$ASP	228	215	190	165	150	135	120	105
Europe	units	70	190	650	1 260	1 850	3 000	3 800	4 700
	$ASP	105	80	75	75	75	75	70	65
Japan	units	35	270	2 300	4 600	8 300	10 500	11 500	13 500
	$ASP	230	215	250	205	145	120	115	110
Asia–Pacific	units				225	450	2 000	4 600	8 700
	$ASP				165	150	130	115	105

[a] Handset with or without basestation equals one unit.
Source: Dataquest 1996 High Volume Electronic Equipment Unit Production Forecast.

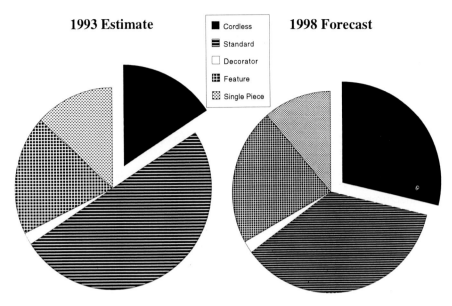

Figure 1.1 The European residential telephone market, showing market share by product type. (*Source*: "Voice Telephones Service", Diane Trivett, Dataquest European Telecommunications Group, August 1995.)

comments that, although the market is dominated by analogue products, digital cordless is starting to take a significant share. This view is shared by a number of cordless manufacturers, who believe that digital cordless will replace CT1 in the market. In addition, the prospects of PWT systems in the USA based on DECT offer substantial opportunities for increasing the volumes of terminals and hence benefiting from further economies of scale.

The cordless market has grown in a similar way to the cellular market. Cordless telephones have become part of mainstream life for the residential user, just as cellular has become part of mainstream life for the business community. Figures from Giga Information Group for the cordless PBX (CPBX) market in Europe indicate strong growth as DECT CPBX systems become available (Table 1.3).

Table 1.3
European CPBX market forecast ('000s)

Type	Year						
	1994	1995	1996	1997	1998	1999	2000
CT2	18	70	195	430	770	1190	1680
DECT	12	67	250	700	1600	3100	5170
Totals	30	137	445	1130	. 2370	4290	6850

Source: Giga Information Group, January 1996.

There are some areas where the cellular service products overlap into what has often been seen as a cordless opportunity, particularly in the areas of on-site use, and where cellular is targeted at the non-business user; in the cordless telephone market, these equate to wireless PBX and telepoint. The market positioning of cordless telephony relative to cellular radio and other systems is considered further in section 1.3.

1.2   The Global Emergence of Cordless

Digital cordless telephony is still establishing its role in the global marketplace, as early pioneers are now being followed by a wider range of telephone manufacturers. Thus, a broader, competitive, product range will become available in the next few years, which will initially vary from region to region, but perhaps will see increasing convergence.

In 1990 there was still much to be done in the area of standardisation to recognise the commercial needs; by 1995 this had been largely achieved. DECT products are now available at realistic prices in Europe and early indications are that DECT will realise its promise of a large-volume market across the whole of Europe. CT2 has found some success in Asia–Pacific and indications are that it will continue to be important in these markets, whilst new cordless standards, such as PWT and PACS derivatives, are emerging in North America. In Japan, after extensive earlier trials, the PHS system was commercially launched to an eager and receptive home market, which now appears to be having second thoughts.

This section briefly reviews the history of developments in these three world regions. The following section explores the position of digital cordless with respect to other competing wireless technologies – a comparison that is independent of geographical market.

1.2.1   The Evolution of Digital Cordless in Europe

European manufacturers and regulators have been the driving force for the world of digital cordless telephony since its origins in the mid-1980s [1]. The story has been complicated, but is summarised by Datapro as follows [2]:

In 1984, several British manufacturers and the British Department of Trade and Industry (DTI) began to develop CT2. In 1985, the European Conference of Telecommunication and Post Administrations (CEPT) commenced its own research initiative into the digital cordless telephony market, spurred on by research undertaken by ESPA[1] and parallel work carried out on the CT2 standard in the UK.

The CT2 standard was put forward to CEPT as a proposal for a pan-European standard. One member of the CEPT, Televerket (now Telia), responded with a vision of an improved cordless technology that was first proposed by Ericsson in 1984. With the company's help, Televerket added to Ericsson's TDMA technology better techniques for echo control and facilities for non-synchronised adjacent systems. In 1987, Ericsson announced a test system based on this improved technology, which subsequently became known as CT3.

The CT2 products being developed operated in the same 900 MHz area as the CT3 products. Some markets in Europe had seen the introduction of what are now called

CT1 telephones, which conformed to a general specification from CEPT. A number of other countries used "CT0" products, which operated around the 1/47 MHz bands, and provided very cheap cordless telephones.

Within the CEPT, the two approaches adopted by the British and the Swedish were very hotly debated. In January 1988, CEPT accepted the ESPA recommended TDMA/TDD approach and gave it the name DECT, Digital European Cordless Telephone. After much haggling, the frequency allocated to this product was agreed, 1.88 to 1.90 GHz. CEPT did not agree on a timetable for the introduction of DECT, and because of continuing disagreement over standards, frequency allocations and industrialisation, the UK continued with the development of CT2. The first CT2 systems were launched in the UK in late 1989, and the CT3[2] product from Ericsson was launched in 1990.

By this time, the CEC had decided that the standardisation activities of the European PTTs should be separated from their regulatory and operational activities, and ETSI was established in 1988. ETSI was given strong support by the European Commission, which used directives to allocate pan-European radio spectrum for the new GSM and DECT standards, effectively ensuring that they would become the dominant standards in Europe.

The debate within the European standardisation arena continued, however, with the opposing camps broadly categorised as follows. One camp believed that DECT should be the main standard, and that all competition to DECT should be restricted by whatever means possible. The other camp adopted the position that it was not the job of a standardisation group to make commercial judgements that would determine the fate of a particular technology or standard in the marketplace. This group also maintained that there was no real competition between the DECT and CT2 standards, because they were targeted at different markets.

The latter camp believed that CT2 would always be as good as DECT, and would always be cheaper, thus winning a higher market share. The more protective of the DECT camp were afraid that the CT2 proponents might be right in this argument, and wanted to preserve their potential market position. There were, of course, other positions, but the regulatory and standardisation meetings at the time were often polarised between these two different camps.

1.2.2 In North America Things Were Different . . .

The battle over the technologies in Europe had been largely within the relatively private confines of the standards bodies. However, in late 1989, the attention of the manufacturers started to focus on North America. To the bewilderment of those unfamiliar with the European debate, the digital cordless discussion in North America was public, extremely fierce, at times acrimonious, and certainly more polarised than had been the case in Europe.

Throughout the 1980s cordless telephones in the North American market were all CT0, low performance, low quality and low range [3]. They were typically called garden phones or yard phones, because that was the limit of their performance.

A peculiarity of the US regulatory system is that there are two bodies that are responsible for regulating the spectrum, one agency, the Federal Communications Commission (the FCC), for civil usage, and another, the National

Telecommunications Information Administration (the NTIA), for military/state usage [4]. The agencies cooperate, because some areas of the spectrum are shared by government and non-government users, but the NTIA does not disclose which parts of the radio spectrum are used by the state agencies, so any potential user has to apply for access to a particular band and wait for permission or refusal to be given. The law forbids the FCC from showing any preference to companies or individuals, and there are no federally endorsed or mandated standards in the radio arena. Thus, any new "standard" in the USA is achieved *de facto*, rather than as a result of a formalised standardisation process. In any given frequency band, there may be a number of different technologies or quasi-standards operating to offer the same service. An example is the large number of different platforms that are endorsed for use in PCS systems in the USA.

The concept is one of coexistence, and easy access to the radio spectrum is only available if equipment conforms to "Part 15", which is the set of regulations outlining power limitations, etc., designed to ensure coexistence. Effectively, manufacturers have to cause no interference, and to tolerate any interference that they suffer. In addition the 900 MHz band is popular for cordless telephony in the USA, because it is already available for telephony applications, so there are no licensing issues or regulatory hurdles for terminal manufacturers to address[3].

Against this backdrop, a number of the Regional Bell Operating Companies (RBOCs) and independent companies saw a large market potential for CT2-based telepoint operations in the USA. The CT2 manufacturers eagerly supported these potential customers as lobbying of the US FCC started, and as the Canadian Department of Communications started to investigate the relative merits of the rival technology candidates. There was immediate intense lobbying by a large number of interested parties, who broadly fell into the following categories:

• Companies proposing solutions based on non-US-originated open standards.
• Companies proposing proprietary solutions.
• Existing users who were concerned that their current spectrum allocations would be threatened.
• Companies who had ideas for different applications and who would use any spectrum that might become available.

Faced with a tidal wave of submissions, in what had initially been a somewhat *ad hoc* activity, in the first half of 1990 the FCC issued a Notice of Inquiry, to formalise the process. The lobbying intensified, and the result some years later was the series of proposals for PCS services in the USA and the clearing of spectrum in the 2 GHz region for licensed and unlicensed PCS. These issues are discussed more fully in Chapter 8 of this book, which is devoted to the North American standards environment.

The result of the regulatory structure in the USA is that there is not and will not be a single preferred US-originated standard in the cordless arena. If standards such as DECT, CT2, CT3, PHS, etc., can be modified to work within the Part 15 regulations, then they may be sold and operated legally, with the same status as any of the other proprietary devices that are sold in very large numbers in the USA. In addition, a variety of cordless standards are being developed for operation in the unlicensed PCS band, most of which are described later in this book.

1.2.3 Whilst in Asia . . .

The Asia–Pacific region is home to many of the traditional major suppliers of analogue cordless telephones, with some 200 or more companies in the region supplying this market, some very small. Digital cordless developments in Asia–Pacific can really be considered in four distinct aspects:

- Acceptance by many of the (mainly smaller) countries of CT2 technology to meet an existing market need.
- Development by Japan of its own standard (PHS).
- Development of proprietary digital cordless solutions by larger terminal manufacturers, which were targeted at the higher-end cordless markets in North America and the Pacific region.
- Growing interest in DECT by larger Asia–Pacific manufacturers seeking to enter the market in Europe, with an eye on emerging opportunities for PWT in the USA.

The impact of digital cordless technology was initially low, with most countries in the region quietly unaware of the European activities until a major worldwide marketing push for CT2 in the late 1980s. This served to alert many Asia–Pacific countries to the potential for digital cordless, particularly for the telepoint application, and prepared the way for its initial success in this area, described in Chapter 4, in places such as Hong Kong, Singapore and China.

In Japan, the response was to acknowledge a need for a new national standard, to compete with the European activity. Work on the Personal Handyphone System (PHS) began in 1989, aiming at a cordless-based system that could provide public access, as well as domestic and business usage. The specifications were to include network aspects, not just air interfaces, in this respect resembling the European CTM activity described in Chapter 6 (which has built upon DECT, CT2 and other areas of telecommunications standardisation). Trials of the PHS began in 1993 and many companies were involved in equipment development and the trials.

PHS has now matured and commercial service was launched in 1995 to a hungry local market, faced with a high-cost cellular alternative. Liberalisation of the cellular market in Japan has only just begun and PHS has been launched into a different environment than was the case with telepoint in Europe. Although the early signs were that PHS would be eagerly accepted by consumers in Japan, by the start of 1996 it appears that they were having second thoughts, and growth in the PHS user base slowed dramatically before again exploding (see Table 1.2). In other parts of Asia it remains to be seen whether the market environments are suitable for the success of cordless public access technologies.

1.2.4 Global Markets – Regional Standards?

The basic market requirements of the different cordless market segments are comparable across geographic regions. However, the market for digital cordless is today fragmented regionally by standards; at present there is no worldwide digital cordless standard, nor any clear prospect for one. Whilst a single global standard has a certain inherent attraction to the industry, it is not yet a market requirement for cordless products, and is not likely to become one for some time to come. Thus whilst the different technologies compete to address emerging markets, it is likely that instead we will see continuing competition between them to secure presence and ownership of different geographical markets.

In Europe, as already described, two European digital standards, DECT and CT2, have emerged with products already commercially available in high volume from several manufacturers. It was recognised in the late 1980s that CT2 had a window of opportunity which would begin to close when DECT products were launched. CT2 has failed to make the depth of penetration in Europe that was expected and DECT equipment is indeed now competing with and complementing CT2, as was predicted [1]. In addition, CT2 has not been allocated pan-European spectrum, unlike DECT. Given that frequency access is the ultimate market enabler in such markets, it is to be expected that DECT will emerge as the dominant European standard. CT2 is still likely to remain strong in some markets – the acceptance of CT2Plus in Canada, a modified version of CT2, PCI, being promoted in the USA for PCS, and the success of CT2 for telepoint services in some Asia–Pacific territories and in France all point to the fact that CT2 is an established competitor in important niche markets.

In the USA, it is not at all clear that a dominant standard will emerge; rather, a few strong contenders are likely to emerge over the next five years, with others becoming casualties en route. As part of the US PCS initiatives that are being commercialised at the time of writing, new systems such as PACS and PWT are being developed as PCS platforms, for which both licensed and unlicensed spectrum will be made available, i.e. variants of the standards will be able to address cellular and cordless-type applications. Such dual-use potential enhances the likelihood of these becoming important cordless standards in North America.

It is too early to say whether PHS will remain solely a Japanese standard or whether it will succeed in other Asia–Pacific markets; certainly the early local response has encouraged promotion of PHS more widely. Clearly Japan would be delighted to see the PHS standard exported, and a number of manufacturers, including some global, non-Japanese, companies, have agreed to collaborate to promote the standard outside of Japan.

Perhaps it is in Asia that the largest market prizes are to be found, and it is here that it is hardest to anticipate how the situation will develop. CT2 has an existing and proven track record. The maturity and potential of DECT products are being increasingly recognised and explored. As identified above, more Asia–Pacific companies are now developing DECT, some as a natural progression of their CT2 activities. The launch of PHS represents a wild card in the pack. Perhaps, as in many other markets, developments may be a stronger reflection of commercial and personal relationships rather than technological merit.

1.3 Market Positioning

In 1990, in [1], the positioning of cordless technologies in the market in Europe, relative to other wireless alternatives, was described as in Figure 1.2. Comparison of this with Figure 1.3, which represents a current view, helps to explain some of the changes in market thinking and target marketing that have occurred in the intervening five years.

The most remarkable development has been the growth of analogue cellular into a mass-market product, and the corresponding penetration of digital GSM

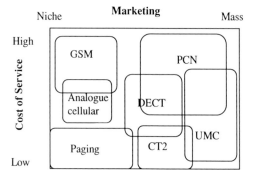

Figure 1.2 Positioning of mobile technologies – a 1989 view. (*Source*: reproduced from [1]. *Original source*: K Edmonds and D Clancy, GEC Plessey Telecommunications presentation to the ETSI Strategic Review of Mobile Communications, Mobile Experts Group, October 1989.)

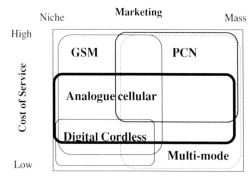

Figure 1.3 Positioning of mobile technologies – a 1996 view. (*Source*: author.)

into a much broader market than was anticipated in 1989. At that time, many of the GSM licensees were indicating that they would charge GSM at a premium to analogue services, although many analysts (including the author) believed that this was an untenable position. In the interval, cellular costs have fallen in real terms – in many countries the fall has been significant. It is this change that accounts for the remarkable cellular market growth figures presented in Table 1.1.

The proposed differentiation between analogue cellular and PCN, which only existed as a concept at that time, was never realised, for a number of reasons. The first of these was that the analogue cellular operators, faced with PCN competition, made pre-emptive moves into the PCN target market. In retrospect, this should have been easily predicted, but at the time many were convinced that the analogue cellular operators would be too constrained by the desire to maintain the profitability of their primary user base, the business community, to be able to make such a move. In practice, these operators have walked this tightrope without major stumbles, but the inherent conflict of the two target markets will eventually become impossible to reconcile, and some convergence is inevitable as tariffs come under pressure with increasing competition.

Partly because of developments in the cellular market, CT2 has failed to make anything like the market penetration that was expected five years ago. This is a result of the failure of the telepoint market in Europe, which was perceived as the main driver of, and was chosen as the initial target for, the CT2 market. In terms of installed base, CT2 has been dominated by its telepoint facet, with its main market in telepoint systems and handsets for the Asia–Pacific territories. CT2 has also made some progress in the cordless PBX area, notably with Motorola/Peacock, GPT and Nortel, but its market potential will be curtailed by DECT once these cordless PBX systems become more widely available.

The concept of universal mobile communications (UMC) of Figure 1.2 is replaced in Figure 1.3 by multi-mode systems. This represents a recognition that the investments required in mobile infrastructures, and the demand for radio spectrum with favourable propagation characteristics, will encourage re-use of as much of the existing systems and technologies as is feasible.

Multi-mode systems encompass a wide range of potential products, which will vary between the different regions of the world. They will clearly include cordless, cellular and satellite mobile communications and will utilise DECT, GSM, one or two of the various US digital standards, as dominant ones emerge, as well as a suitable satellite-based technology. It is of course clear that the multi-mode solution that will be most popular will be the one that combines greatest service coverage with lowest handset price, which will of course imply a minimum number of supported systems. Whilst standards remain implemented on only a regional rather than worldwide basis, a global-coverage multi-mode terminal remains only a dream.

Even if standards could be coordinated there is still a need for roaming agreements for terrestrial services, and it is this niche that is addressed by the various satellite mobile phone enterprises initiated in the early 1990s.

In any market of the future it is important to restate one important consideration: the cordless market is no more homogeneous than the automobile market, and market requirements in each sector have to be addressed with as much care and attention to detail as is the case in the automobile market, and given appropriate offerings. This leads on to market segmentation.

1.4   Market Segments and Requirements

1.4.1   Segmentation by Application

The cordless market is not a single sector of mobile telephony – it comprises a number of segments, which are summarised in Figure 1.4, based on a segmentation by end-user application.

The relative importance of these various segments varies according to the geographical market. For example, telepoint services in Europe have been a failure, largely because the service product offered was totally inadequate for the perceived need in a competitive market. In Asia–Pacific, by contrast, telepoint services in the early 1990s experienced considerable success, mainly because there was no low-cost mobile service in competition with such services at that time.

Similarly, in Europe, cordless telephones are starting to be treated seriously as a business tool, yet this is not the case elsewhere; in the USA, for example, cordless terminals for the business market are still relatively undeveloped, which is surprising since AT&T launched its first business cordless phone in 1989. This is changing with the introduction of digital cordless telephones in the USA.

As time goes on, the user base will segment further, between those who want a basic cordless telephone and those who want their cordless phone functionality

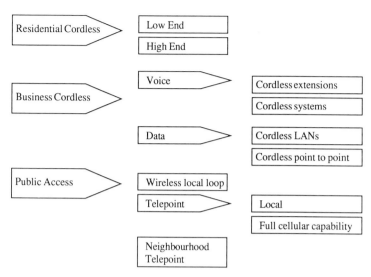

Figure 1.4 Cordless market segmentation. (*Source*: "Examining the Current Market for Cordless Communications", P White, PA Consulting, presented at the IIR DECT CT2 and Cordless Communications Conference, September 1994.)

integrated within another device, to maximise its usability. This approach is being considered by a number of the personal computer (PC) and personal digital assistant (PDA) manufacturers such as Apple, Sony, Sharp and Motorola, and is covered later in this chapter – see section 1.6.2.

Segmentation of the cordless market by technology was attempted in the early 1990s. Like all such technology-based attempts, this approach failed; users buy functionality, not technology, and they make their choices from the products and services that are offered to them, only considering technology when all other things are equal. Thus, the debate over the relative benefits to users of CT2 or DECT was largely a sterile one from a user perspective – all the user is worried about is the price of the equipment and whether it meets the application needs that are being specified. Look and feel are also important.

In the light of these comments, it is appropriate to set out the needs of the market in the broad segments that appear to be most universally acknowledged and agreed.

1.4.2   Generic Requirements

To understand the potential for cordless telephones, we begin by presenting some generic requirements and basic statements. These may appear obvious to the cordless cognoscenti, but had they been fully acknowledged, it is possible that the outcome for telepoint in Europe might have been somewhat different:

- Cordless handsets must be lightweight with long battery life.
- The user interface of any cordless telephone or cordless service must be easy to understand and operate.

- Common air interfaces are essential for products to succeed in any application that is not primarily residential or domestic.
- Coverage areas must be complete and comprehensive, whether indoors or outdoors.
- Usage costs have to be low, and compare favourably with fixed telephone running costs. In a public or mobile service, airtime costs must be well below that of the appropriate competitive cellular service or trunked mobile radio products. Thus, telepoint cannot compete with low-cost cellular (similarly, cellular cannot address the PBX market cost-effectively).
- Given good coverage, a cordless user will not pay a premium to walk past functioning fixed telephones to make a cordless call.

1.5   Application Segments

Any examination of the market for cordless telephones clearly indicates a number of criteria that are important if a product is to succeed in its target market segments. If a simple single-line cordless telephone is considered (the majority of the market to date), it has to offer certain attributes for the domestic market, and others for the business market. These attributes are not exclusive, and, from a manufacturer's perspective, the combination of all these in a handset that can be targeted at all market sectors and configurations has overwhelming attractions, particularly if they can be packaged and differentiated by their housings alone.

This section describes the different market segments and their specific requirements.

1.5.1   The Domestic Residential Market

The domestic cordless market varies across the world, but is dominated by CT0, which offers low-cost analogue terminals. The performance of these terminals is relatively poor, as anyone who has used one will testify. The handsets are often very light, but they suffer from interference from mains electricity, from poor range and from lack of security in very basic phones. Their popularity primarily derives from the flexibility and portability that they offer at a very low cost.

In some parts of Europe the market is primarily served by telephones to the CEPT CT1 standard. These are higher-specification analogue cordless phones that operate around 900 MHz. The disadvantage of these phones is that they are significantly more expensive than CT0, and the specification makes it difficult to reduce the cost of the product. In addition, the phones are often heavy, and are much more expensive than the CT0 products available elsewhere.

In Japan, there is a further specification for cordless phones with a range of only 10–20 m. This is popular for the relatively low cost of the equipment, and serves a market where the required density of usage is very high, even though the real need for a cordless telephone in most Japanese apartments is minimal.

For the domestic market, the requirements are simple, and have become essential for a cordless telephone to be accepted as a serious product on the shelves of the electrical retailers. They are:

- Low cost
- Good voice quality
- Security against unauthorised access to the basestation and therefore to the telephone line
- Lightweight handset

The residential market was not originally seen by many as a prime target market for digital cordless, but the launch of highly integrated semiconductors for DECT now means that many terminal manufacturers believe that they can launch products at price points that allow them to address the small office/home (SoHo) market in the same bracket as the higher-end analogue cordless telephones. The success of this strategy is evident in Germany where DECT cordless telephones are beginning to outsell analogue products.

1.5.2 The Business Market – Cordless in the Office

The growth in the use of mobile telephony served to awaken many manufacturers to the opportunities that are available for portable telephones in the office, and also to the potential of cordless LAN access for desktop computers.

Early marketing concepts were that cordless could be sold easily into any application, as the avoidance of the costs of changing wiring in office reorganisations would easily exceed the price premium for a cordless installation. In reality, developments in PBXs and wiring approaches weakened such arguments and the "reason to buy" became much more application specific, requiring real and concrete business and user benefits to justify cordless implementations. In 1996, there are still only relatively few, although major, manufacturers offering business cordless systems and the cordless LAN market has still not yet developed beyond a niche market. Most of the digital systems are available only in Europe, although some are now being promoted in North America.

The following attributes are required for a business cordless telephone:

- Good voice quality
- Lightweight compact handsets
- Security against unauthorised access or eavesdropping
- Ability to provide high density of usage
- Ability to operate in conjunction with existing PBX and key system equipment
- Ability to work without interference in office environments
- Compatibility of equipment from different vendors
- Availability of handover between basestations
- Availability of cordless Centrex

1.5.3 The Public Access Mobile Market

The third main target market for digital cordless is the mobile public access market, originally conceived as telepoint. However, the growth in cellular, in part stimulated by the rapid decline in cellular pricing, increasing liberalisation and the launch of PCN services, prematurely closed the window of opportunity for telepoint in Europe. This however has not precluded a role for cordless in the public access market segment – paradoxically it may even have increased its potential importance, as operators look to use cordless in microcellular networks.

Even in the late 1980s there was a recognition that there could be a prominent future role for digital cordless as part of a high-capacity personal communications infrastructure, in combination with other technologies. The idea was conceived that networks would be configured with cellular umbrella cells, with cordless providing high-capacity microcells within these umbrellas. This concept faded from public view as the cellular operators developed new ways of sectoring cells and decreasing cell sizes. However, as the cellular user base has continued to grow dramatically, the potential of this concept, to relieve pressure on spectrum and hence increase subscriber capacity in high-density urban areas, has received renewed attention. Some operators and infrastructure/handset manufacturers have implemented dual-mode cordless/cellular solutions to begin to explore and realise this vision.

For such an evolution to be viable, the key market segment requirements for dual-standard terminals are:

- Low premium in terms of size, weight and cost, over a single-mode terminal.
- Simple, easy-to-use, user interface – i.e. not unnecessarily complicated by being dual-mode.

Recent technology developments have meant that it should be possible to meet such requirements in the near future. Indeed, in the USA, dual-mode cellular products already exist, as operators introduce systems and handsets that can operate on both analogue and digital cellular standards (AMPS, IS54, IS95).

In Europe, Ericsson has started trials with both Telia and DeTeMobil (the mobile operator that was formerly part of Deutsche Telekom) of a dual-mode DECT and GSM system. This is a strong indicator of the future direction of DECT, as Ericsson is a leader in the provision of GSM terminals and infrastructure, as well as being a major pioneer of DECT alongside Philips and Siemens.

In such a system, the DECT elements are not a network, but simply provide an access medium to the existing cellular switching infrastructure. This does not require an expensive parallel network, so the cost is lower for the operator and the user alike. For the cellular operators, this is an excellent way of providing high-density coverage without having to provide costly cellular capacity. One drawback is that terminals will be more expensive, but will have to retail for the same price as a single-mode handset. Subsidies of these handsets will therefore have to be greater than for single-mode handsets.

As it continues to grow, the cellular market is expected to change radically over the next decade. In a maturing market, revenues per customer are expected to fall, as the mobile phone further penetrates the consumer market, and the cost of using a mobile phone will have to be reduced to compare more favourably with the fixed network alternatives. (It is notable that the highest penetration of mobile telephony is in those countries where the price premium attached to a mobile service is smallest.) The way that services are marketed will also have to change if growth is to reach the ambitious targets set by the operators and terminal manufacturers.

In this type of changing market, there are two approaches that the operators will take to maintain their profitability. The first is that they will increasingly deal directly with end-users, and dealers and service providers will become less important. The second is that they will look for ways of increasing the density of coverage in their systems to provide high capacity in cities and towns, and to enable them to offer services to all segments of the market that are currently served by the fixed telephone networks. This is the route towards an integrated wireless telephony service, or "personal communications service". Some operators are already attempting to introduce this service concept using solely cellular standards, and are in some cases really bridging the telepoint and wireless local loop market segments.

This change in the perceived potential role of cordless for public access represents a major shift over the past five years. In 1990, the perception was of cordless as a stand-alone, telepoint, service alongside the other mobile services, as the main way that low-cost mobile telephony would be provided into a mass market [1]. The emergence of dual- or multi-mode terminals in the next few years could have a significant impact upon long-term market evolution, as discussed elsewhere in this book.

1.5.4   The Wireless Local Loop Market

Early in the development of digital cordless, the potential application of DECT and CT2 for provision of wireless local loop was recognised; indeed, such an application was explicitly written into the early DECT specifications by ETSI. During the early 1990s several trials took place to evaluate the effectiveness of such systems, as described in Chapter 5. Various factors, such as the need to deploy improved telecommunications infrastructure rapidly in the newly open Eastern Europe and increasing regulatory liberalisation, have been instrumental in stimulating strong interest in this application.

The initial limited availability of cordless equipment, the relatively short range of the first digital cordless products, and the high cost of basestation equipment encouraged operators to look in other directions for wireless local loop technologies. A large number of these now exist, some cellular in origin, others based on fixed microwave solutions, and others on proprietary approaches designed for low-density wide-area coverage of developing areas.

Increasing liberalisation of the services market around the world will encourage many more independent operators to look at wireless and cordless systems as a means of entering the market. Even established operators now believe that radio offers the most cost-effective way of installing new capacity in areas that do not have ducting for copper cable or optical-fibre-based systems. One example of this is in Spain, where Telefonica used its analogue NMT cellular system to install telephones in remote regions where no telephone service had been available previously. Even though the per-line infrastructure cost was high, this was balanced by the elimination of overhead lines and their associated civil engineering costs. The costs of this system then compared more favourably with a traditional fixed wired system. The speed of installation also allowed a much more rapid extension of the telephone system than would have been the case with a landline network. The use of cellular systems to meet this need is unlikely to be widespread, as purpose-made systems are usually cheaper.

For the wireless local loop market a number of additional requirements arise, largely relating to accessing the fixed network and the deployment scenario, viz.:

- Low cost, relative to copper or other fixed alternatives
- High-quality speech capability
- Low delays in network dialtone access
- Range performance appropriate to the required deployment
- Suitable interfacing to the network infrastructure
- Environmental performance

In addition, possible integration with, or evolution towards, a neighbourhood mobility capability may be considered as a requirement, a requirement that clearly would favour use of a cellular or cordless technology.

In this market segment, cordless approaches clearly compete with other technologies. Their low cost, deriving from their complementary high-volume consumer markets, places them in a favourable position, particularly for deployment in high-density, relatively short-range, urban/suburban scenarios.

It is clear that the restricted range of cordless-based systems means that they can never be an appropriate solution for rural wireless local loop, but in towns and cities they will become a valuable tool for all operators looking to install new or additional capacity at low cost in the shortest possible time.

1.6   Interdependence and Evolution

1.6.1   Market Segment Interdependence

The cordless applications that are detailed in the previous sections have some relationships, and operators will exploit these in the future to gain the maximum traffic and revenue possible.

Previously, the relationship was illustrated by three equally sized overlapping circles, but it is here revised. Figure 1.5 attempts to show the relative sizes of the long-term market opportunity as now perceived, given that cordless telephones seem likely to become integrated with cellular handsets to provide integrated dual-mode solutions. Such a development is clearly dependent upon the

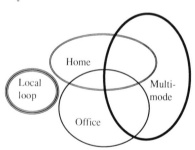

Figure 1.5   Market segment interdependence.

continuing liberalisation of the regulatory environment for telecommunications provision for basic voice services around the world.

All the research that has been done on the cordless market to date has indicated that the main demand for cordless is in the home and business sectors. The wireless local loop is driven not by end-user demand, but by operators seeking to use cordless as a medium to provide capacity and services – given this factor, the range of alternative technologies and the embryonic nature of the market, the development of this market is perhaps harder to predict.

Perhaps the most striking aspect of Figure 1.5 is the absence of telepoint. This is explained by the earlier analysis of the integration of cordless into cellular systems (for dual-mode microcellular networks), and by the continuing downward trend in cellular tariffs, which undermines the economics of telepoint as a stand-alone service. It is acknowledged that telepoint may continue to be popular in some high-density population areas where cellular service remains premium priced, such as parts of Asia–Pacific.

1.6.2   Market Evolution

The longer-term market evolution of digital cordless is not straightforward. There are a number of directions that this could take. Digital cordless could:

- Become fragmented regionally according to the different standards permitted or used in each country.
- Become dominated by a single worldwide standard.
- Remain a stand-alone product.
- Become integrated with cellular as part of different multi-mode systems throughout the world.

The first of these options would appear to be unlikely, as the constant trend is for standards to become more common, and proprietary solutions are less popular in anything except a domestic cordless market. Increasingly, even residential users will expect more than one handset, and may want to buy these from more than one supplier, so even here open standards will become a requirement.

The evolution of the market will first be driven by applications. The first market evolution will be driven by products that are application specific, but conform to a core technology (CT2, DECT, PHS, etc.) in order to provide compatibility and minimise the development costs. Such systems are increasingly likely to come from vendors who are not traditional cordless suppliers, but who buy their radios from module suppliers and merely integrate these in new product packages. Commercial viability will be the watchword of these new products in a world that is increasingly populated by wireless devices of many varieties. In this case it is unlikely that cordless can remain a stand-alone product, and as different standards are optimal for different applications, and as we now have several entrenched regional digital standards, it is unlikely that a single standard will emerge as the dominant one over the next five years.

The path is clearly towards integration, albeit not overnight but steadily, especially as users increasingly expect to be able to use cordless or remote control of many different devices that are not simply telephones. Cordless has a very important part to play, not only in the home or in the office, but in the whole of the mobile or personal communications world. It is also clear that DECT will become a communications medium for many other applications outside the scope of voice telephony.

In the longer term, an all-embracing mass-market mobile telecommunications product, universal mobile communications (UMC[4]), may emerge. To succeed, the concept of UMC has to be driven by the market; this requires that manufacturers, operators and regulators continually strive for the goal of truly low-cost networks and terminals as they jointly determine the characteristics, specifications and functionality of UMC. They will always struggle in their task against the constantly growing expectations of the users, who will demand increased functions and services from an ever-cheaper terminal, and will always expect to pay less and less in real terms for the services that they use. This suggests an evolutionary route based on operators integrating multi-mode systems and evolving from today's technology.

An indication of the increasing functionality that will be expected is shown by the illustration of the new user interface approach taken by General Magic Inc. in its Magic Cap software (Figure 1.6).

General Magic is an alliance of operators and manufacturers which is developing software for use in personal organisers or personal digital assistants (PDAs) and communicators/mobile telephones. It is the integrated approach and simplicity of use that such interfaces permit that will be the main driver of the value-added services market, which in turn is the main driver for the increasing sophistication of the mobile terminals of the future. Plug-and-play approaches will increasingly be perceived as essential for the success of advanced products in the cordless market, whether they are intended for personal or business use, for domestic or public environments. Other similar and competing solutions will be available from Pegasus, Geos and Microsoft.

Five years ago it was clear that the world would move increasingly to radio-based communications equipment; the only surprise has been the continuing acceleration of the rate at which this has occurred. The opportunity available for

Figure 1.6 Example intuitive user interface. (*Source*: General Magic Inc.)

cordless products in this market is immense, particularly now that the political and regulatory battles are coming to a conclusion, and manufacturers are able to focus on designing equipment and selling it to an eager user population. As methods emerge of allowing all users to operate advanced features such as intelligent messaging and fax from their handsets without needing a diploma in computer science, so the popularity of such value-added services will grow.

1.7   Summary

The digital cordless market continues to remain radically different from the cellular radio market, and from the market for analogue cordless phones. These differences will be exploited by manufacturers to develop new applications for a range of cordless terminals and products and by operators to offer enhanced services to users. In both cases the manufacturers and operators will expect reasonable returns from their investments and innovations. In turn the products and services will be accepted by users because they offer useful tools and facilities that are relevant to their everyday needs at an appropriate price.

In the long term, the distinction between cellular and cordless in the market will become less clear to the user. A key step to this, the integration of cordless and cellular into single handsets, is already in development and the operators will develop service packages that take advantage of the facilities offered by such products.

It is important to consider the application requirements of the users when developing new cordless terminals and applications, to ensure that the resulting product is appropriate for its intended target market. Users will choose according to their needs. Any change in the perceived technology requirements will arise from intensive marketing effort directed at raising the consciousness of a mass user population to the potential benefits of new services. In this area, engineers and marketing groups will have to work closely together to drive the evolution of the cordless telecommunications market.

References

[1]   "Cordless Telecommunications in Europe", WHW Tuttlebee (ed.), Springer-Verlag, 1990
[2]   "Mobile and Wireless Analyst: A Global Perspective: DECT", D Trivett, Datapro, March 1994
[3]   "Mobile Communications in the US and Europe", M Paetsch, Artech House, 1993
[4]   "Wireless, The Revolution in Personal Telecommunications", I Brodsky, Datacomm Research Co., April 1995

Notes

1   The European industry association for on-site paging manufacturers, who saw cordless telephony as both a threat and growth opportunity.
2   Further details of the Swedish CT3 system and the Ericsson product to this standard, the DCT900, may be found in [1], in the appendix entitled "Operation of the DCT900 System" and the chapter "Proprietary Digital Cordless Products".
3   In the later part of 1995, this whole unwieldy structure for spectrum allocation was under threat from Republican proposals to break up the FCC and pass spectrum administration to commercially oriented companies. The laissez faire attitude to standards is unlikely to change.
4   For UMC also read UMTS, FPLMTS, etc.

2 The Wireless Private Automatic Branch Exchange

Phil Crookes and Robert Harrison

Market research conducted in 1990 by PA Consulting suggested that the market potential for wireless PBX extensions could be as much as 25% of all extensions. The demand for communications for on-site mobile staff is clear. It has been estimated that European employees spend up to 60% of their time working away from their desks but on-site; this can result in time-consuming and irritating delays and the high cost of returning calls. However, by 1995 the market was running at little more than 1% of shipments of PBX lines.

This chapter examines why the uptake of wireless PBXs has proved to be slower than anticipated, what suppliers can do to overcome the barriers to market development, and how the benefits of wirelessness can be recognised and quantified. Section 2.1 reviews the current market for wireless PBX in Europe and section 2.2 describes the potential benefits, users and applications for wireless office systems. The barriers that have hindered market development are outlined in section 2.3 whilst section 2.4 examines how these barriers can be overcome. Finally, section 2.5 provides a number real-life case studies showing how wireless office systems have been used in practice.

The scope of this chapter is limited to the use of wireless technology to meet basic voice communications needs in the business environment. Wireless technologies also have applications in the business environment for data communications; these are described in a later chapter.

Whilst this chapter is written from a European perspective, where cordless PBX products have been available now for some years, many of the lessons drawn will have equal applicability in the near-future to the North American and Asia–Pacific markets, where such products have yet to be seriously marketed. Many of the barriers to market growth are generic rather than geographic, as are the proposed solutions.

2.1 The Market in the Mid-1990s

The market for cordless PBXs has grown slowly since their introduction in the early 1990s, as can be seen from Figure 2.1.

The level of demand has varied considerably from country to country and has not been in proportion to the size of the national PBX markets. Germany has proved to be the market with the strongest demand, and more recently

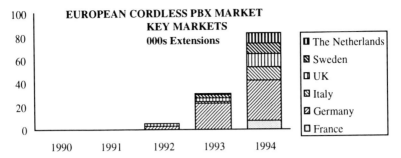

Figure 2.1   Growth of European wireless PBX markets. (*Source*: ICC.)

significant demand has been evident in the Netherlands and Sweden. By contrast, demand in France, Italy and the UK has been relatively subdued.

This varied level of demand in part reflects early experience of cordless telephone systems. In Germany, for example, the first cordless products available during the 1980s used the CEPT CT1 standard, which provided high-quality communications at a high cost. These products were too expensive for the consumer market, but proved attractive for business users, who thus became aware of the potential of the technology. The same standard was adopted in Scandinavia and Italy.

However, the UK adopted the inexpensive CT0 standard developed in the USA. Here the low cost proved attractive to the consumer market, but problems of poor-quality reception, including interference and lack of security, made products using this standard unattractive to business users. The subsequent development of analogue cellular radio in the UK tended to confirm to users the idea that radio reception was by its very nature variable and unreliable, and not necessarily an acceptable alternative to a wired phone unless there were clear benefits from mobility. It is partly as a result of this experience that site surveys are seen as necessary to ensure that complete coverage can be guaranteed. France chose to introduce a variant of the US/UK standard, which provided poor quality at a higher price.

The results of this experience are well illustrated by the pie chart in Figure 2.2, which shows the relative sizes of the six largest European country markets by 1994.

The technologies employed vary from country to country, depending upon the technologies adopted by the major suppliers to those markets, and are shown in Figure 2.3; the early dominance of DECT in the European wireless PBX market so soon after product launch is striking. Sales in Germany and Italy have been largely of systems using CEPT CT1, whilst sales in France and the UK have been dominated by CT2, and in the Netherlands and Sweden by CT2 and DECT. During 1994–95, sales of DECT overtook CT1 sales in Germany, signalling the demise of CT1. Although DECT products have also been introduced in the UK, CT2 can be expected to last far longer than CT1 owing to the support of major manufacturers such as AT&T, Nortel and Motorola. Most users will perceive little difference between the offerings.

EUROPEAN CORDLESS PBX MARKET 1994-KEY MARKET$
83,000 Extensions

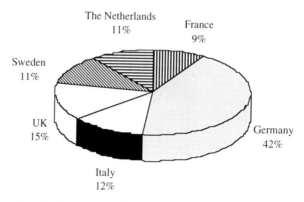

Figure 2.2   Key European wireless PBX markets in 1994. (*Source*: ICC.)

CURRENT CORDLESS PBX TECHNOLOGY
1994

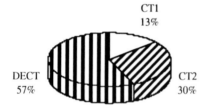

Figure 2.3   Wireless PBX technologies in 1994. (*Source*: ICC.)

Future growth of the market will depend upon a number of factors, which are dealt with in the following pages, and the industry's ability to address them satisfactorily.

2.2   Users, Applications and Benefits

The limited growth of the wireless PBX market has resulted from the fact that use is currently restricted to applications where the benefits are easily quantifiable and clearly justify the premium costs that early wireless PBX products have commanded. This section identifies the "early adopters" and their applications, and describes the benefits that they have found to arise from wirelessness in the office. We begin by outlining the potential benefits of wireless PBX extensions and then go on to describe the types of users and applications where these benefits have been most clearly recognised so far. Section 2.5 also presents a number of case studies, which describe real users and their applications in more detail.

2.2.1 The Benefits of the Wireless PBX

In today's competitive business environment, organisations must strive to maximise customer satisfaction, maximise productivity and reduce costs wherever possible. The wireless PBX can help to meet all these fundamental business needs. The business case for wireless in most organisations is based on four basic benefits, which are described in more detail in the following paragraphs:

- Increased competitiveness through improved customer service
- Reduced costs because of improved contactability
- Reduced costs or increased revenues through productivity improvements
- Reduced costs and improving staff satisfaction because of the convenience and flexibility

Improved Customer Service

Improved customer service is increasingly being recognised as fundamental to competitiveness and profitability. In many organisations, the telephone represents a major route for customer contact, and wirelessness can help to ensure that the right person is available, with the right information, to turn the customer contact into a "magic moment". Examples of organisations for whom this has proved to be a major benefit include: the Netherlands Tax Administration (see section 2.5); large retailers, such as DIY stores, which have given wireless extensions to departmental managers to enable them to deal promptly with customer queries; garages, which are using wireless extensions to enable customers to contact sales and service staff wherever they may be; and hotels, to ensure that maintenance and customer service staff can be contacted immediately to respond to guests' needs.

Improved Contactability

A second benefit that comes from this improvement of contactability of key staff is the cost saving of not having to return calls. Typically, around 40% of telephone traffic is from incoming external calls. It has been estimated that 70% of all these calls fail to reach their destination first time, so almost 30% of an organisation's phone bill could be in returning failed incoming calls. Generally organisations that make extensive use of wireless PBX extensions have found that call costs can be reduced by 12–15%, giving a payback on the investment of 2–3 years.

Another often-quoted source of cost savings can be through reductions in PBX wiring costs. Each customer must carefully evaluate the scale of any possible savings in his or her particular situation. Significant reductions can be made in some cases, especially where the nature of the building makes wiring very difficult or even impracticable; for example, one organisation was faced with the cost of lifting and repairing a marble floor in order to install a small number of extensions in its prestigious headquarters building. However, for most organisations the savings from a reduction in wiring to individual extensions must be offset against the cost of wiring to individual radio basestations around the site. The equation can still be positive, particularly in open areas where a small number of basestations can serve a large number of extension users.

Improved Productivity

Wireless extensions can be an important tool in helping to improve productivity both of on-site mobile staff, and of the organisation as a whole. For example, a German toy manufacturer equipped the staff who maintain production line equipment with wireless extensions and was able to quantify the benefits from reduced downtime such that the payback on the investment in the wireless PBX was less than one year. Other users have been able to improve productivity by equipping PC support staff and facilities maintenance staff with wireless extensions.

Convenience and Flexibility

Further benefits come from the convenience and flexibility that wireless PBX extensions can offer. For example, organisations that regularly need to move phone extensions or make use of temporary installations can make significant cost savings; it has been estimated that most companies spend 10–20% of the initial cost of their PBX every year on reconfiguring the system because of moves and changes. More difficult to quantify, but no less important, are the "soft benefits" that come from improved staff motivation and internal "networking" that wirelessness can bring. For example, DEC in Stockholm is using a wireless PBX to support "hot desking" where office staff are free to sit wherever they please as work needs dictate changing working relationships with various colleagues.

2.2.2   Users and Applications

Early adopters of wireless PBX solutions have been those for whom the benefits outlined above are most easily visible and quantifiable. Wireless PBX extensions are not a high-technology toy for the chief executive, but a working tool for support staff and customer service staff. Early uptake has been led by the following types of organisation:

- *Manufacturing*   Large sites, mobile technical support staff and the high costs of any delays in the production line combine to create a genuine need for wirelessness in this sector.
- *Warehousing and retail*   Anyone who counts the number of staff calls put out on the public address system during a visit to their local supermarket or DIY store will understand how wireless PBX extensions can improve productivity and customer service in these sectors. Garages and builders' merchants are other examples of retailers where on-site mobile staff can be more easily contactable to increase customer satisfaction and reduce the costs of returned calls.

- *Hotels* In addition to administrative and maintenance providing a faster and more responsive service, hotel owners are attracted by the opportunities for guests to make premium charged calls not just from their rooms but also from the bar, restaurant and poolside.
- *Hospitals* Wireless extensions can be used to improve the effectiveness and productivity of emergency teams, porters and other support staff, as well as to improve patient service by providing payphone facilities at the bedside. Such systems also, perhaps surprisingly, offer potential for reduced electromagnetic interference with clinical equipment over current hospital radio systems[1].
- *Specific general office applications* There are a number of applications for work environments where nobody has a desk, such as the Stock Exchange, or where people are often away from their desks but need to be contacted immediately, such as a Social Work department.

The pattern of use in terms of size of company has been, to a large extent, affected by suppliers' offerings; most of the systems available today are designed for small organisations or for use as "piggyback" systems behind a larger PBX to be used by a particular workgroup or to provide wireless coverage for a certain area. The per-extension cost premium of cordless is not as large for a small organisation because few basestations are required to provide coverage and the cost of radio coverage planning is likely to be lower. Early uptake among smaller companies may also be a consequence of the fact that these organisations tend to be more dependent on a small number of key staff who must be contactable to respond to customer enquiries. In fact the majority of business premises in Europe are small enough to be comfortably served by a single basestation using either DECT or CT2 technology, and in almost all these cases suppliers could confidently offer systems without the need for radio surveys.

2.3   Barriers to Market Development

So far the market for cordless PBXs has failed to develop as rapidly as had been expected in the early 1990s. This sluggish development of the market can be attributed to a number of factors including:

- High cost of early products
- Low levels of awareness in the market
- Relatively complex sales process

Whilst these factors are faced by most new products at the time of their introduction, in the case of cordless PBXs they have combined to form a vicious circle that has been impeding further market development, as shown in Figure 2.4; the elements of this diagram are discussed below.

The high costs of the technology, coupled with uncertainty about its capabilities, means that potential purchasers need to identify quantifiable benefits clearly in order to justify the purchase. In addition, the uncertainty about the performance of radio systems in places of work means that users and suppliers need the reassurance of a radio survey before making a commitment. As a result, the costs of sale of cordless PBX products are high, adding to the total costs to be borne by the purchaser, and products are largely sold into niche

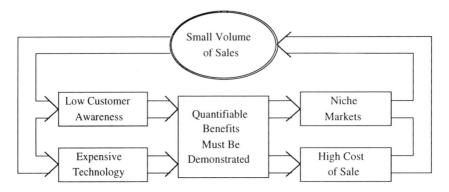

Figure 2.4 Factors inhibiting market growth.

markets, resulting in small sales volumes. The low level of sales in turn perpetuates the high costs and low level of awareness in the market.

The low level of awareness among potential users of the capabilities of cordless PBX products is not just a simple matter of ignorance of the existence of the technology. It also includes perceptions based on past experience of the nature and quality of cordless technologies and their relevance to the applications the potential user has in mind. These factors, which vary from country to country, depending on the history of cordless telephony and, to a lesser extent, mobile telephony in the country, have been described above in section 2.1.

The second factor is the very high cost of the technology. At present, cordless extensions cost up to three to five times the price of a wired phone, at around £800 to £1000 per user. From the user's point of view, this is very expensive for a telephone, and is significantly more than many mobile phones. At these levels it is not surprising that purchasers are looking for tangible benefits that will justify the additional cost. A price more in line with other feature phones, at well under £200, would be likely to be far more acceptable, and could also be expected to reduce the requirement for cost justification.

Replacement of a PBX is usually a fairly major item of capital expenditure, and with the current mature state of the technology, and the increased level of reliability of modern equipment, users are tending to extend the life of their systems. As a result, users are understandably reluctant to consider purchase of a new cordless system unless the PBX is due for replacement.

Taken together, these two factors increase the potential purchaser's desire to seek quantifiable benefits in order to justify the expenditure on a new system. Such benefits are most clearly identifiable and quantifiable only in a small number of specific applications, with the result that most products are sold into niche markets.

The relative difficulty of selling cordless PBXs is made more significant by the use of the traditional distribution channels used for PBX sales. With a mature technology, and in a highly competitive market, PBXs are becoming commoditised, and the costs of sale have to be kept to a minimum. For smaller systems, margins only allow for a single sales call. If, as is the case with cordless PBXs, detailed cost justification and a radio survey are required, the sales costs are significantly higher, and the time to sale is increased. In these circumstances,

EUROPEAN CORDLESS PBX
KEY MARKETS

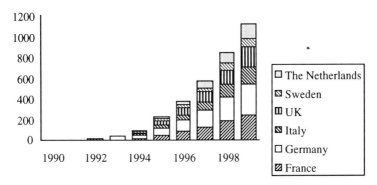

Figure 2.5 Forecast growth of European wireless PBX markets. (*Source*: ICC.)

the salesperson, with an eye on his or her sales targets and commission payments, is driven towards straightforward sales that can be achieved in the minimum time.

Predictably these factors combine to give small sales volumes and perpetuate low awareness. Future development of the market will depend upon the industry's ability to address these issues, and Figure 2.5 illustrates the likely growth that could be achieved if prices were to be reduced to levels nearer to £100 per handset and £100 per line in 1996, representing a penetration of 10% of all PBX extensions by the end of the decade.

2.4   Addressing the Market Potential

If the barriers to market development outlined in the previous section can be successfully addressed, the market can be developed to its full potential. To do so, manufacturers will need to look closely at the issues discussed in the previous section. In addition, they must deal with a number of competing technological solutions.

Cordless PBX systems satisfy the customer needs for contactability on-site. This need can also be met by a number of other technological solutions, including paging, public address systems, voice-mail and even cellular radio. Any of these, either individually or in combination, may appear to the user as a more cost-effective alternative to cordless telephony. Undoubtedly use of cordless telephony provides the best guarantee of reaching the desired person immediately, but the person being called may not wish to be so readily contactable at all times, or may fear that inadequate radio coverage may prevent that one urgent call from getting through. In practice, use of a cordless PBX may be complementary to the other systems, and careful consideration of the application may well identify a role for cordless telephony in conjunction with other technologies.

If the user wishes to be reached most of the time, but wants to avoid intrusion at others, combination with paging or voice-mail could offer an effective solution. If the user needs instant notification of a call, but the option to accept it, then use of a paging system in conjunction with a cordless phone would provide the required capability. If immediate knowledge of the call is not required, then voice-mail would be an effective alternative. In both cases the use of cordless phones would allow the user to be contacted at once when desired, and the freedom to make calls from any location, without having to seek the nearest fixed connection.

If the user is worried about coverage, the voice-mail will ensure that any missed calls are captured. Again, paging can be used as a complement if, as is likely, it offers superior coverage. An effective public address system could also fill this need, although of course such a solution can be intrusive.

Products that combine the benefits of cellular radio and cordless telephony are being developed, typically offering a GSM/DECT interface. These are intended to allow users to gain the best of both worlds with high-density coverage of the local site combined with wide-area service, with users able to take advantage of the choice of technologies to obtain the cheapest tariffs.

2.5   Case Studies

This section presents a number of real-life case studies that describe users' experiences of wireless PBXs. The case studies from Coca-Cola and Dickens are edited versions of articles that first appeared in Telecomeuropa's "Advanced Cordless Communications" [1]. The examples from the Netherlands Tax Authority and Axis Holdings were first presented at the 1994 Mobile Conference at TMA27 organised by the publishers of MTN.

2.5.1   A Production Facility: Coca-Cola, New Zealand

Coca-Cola Amatil (NZ) Ltd (CCANZ) uses Ericsson's Freeset wireless PBX to improve communications to and from the production floor. This case study looks at the installation of the Freeset system and the reaction of the staff.

CCANZ's Auckland plant is the largest of the company's three facilities. It produces, packages and distributes famous drinks brands such as Coca-Cola, Fanta, Sprite and Schweppes mixers. There are four filling lines and in total the plant has an annual production of 125 million litres. Eighty people work two shifts on the production lines; this figure rises to over 100 during the summer months, when the plant operates for 21 hours a day.

A machine breakdown, often due to a crushed can or bottle, can put an entire production line out of action. An hour's lost production can cost NZ$2000 in high season, so getting production restarted quickly is essential – that means getting maintenance technicians to the right place, at the right time, with the right tools and spare parts. CCANZ found that paging could not provide a direct link between the production foremen and the maintenance technician, leading to delays in contacting the right staff. "We lost five to ten minutes every time a machine stopped working; often it took longer to get the technician to the machine than it did to make the repair", says Joseph Utama, production manager at CCANZ.

CCANZ solved its problem with a wireless PBX system. "We've been able to build a system that covers virtually the whole site – not just the production area, but our offices and warehouses as well – so that people can be contacted wherever they are. The phones themselves are small and light and the system is very flexible and easy to use."

CCANZ's system has six radio basestations serving the production area of 11 000 square metres, plus the administration and warehouse buildings. At present the cordless phones are used only by production and maintenance staff, who are continually on the move in the production areas. Joseph Utama reports strong demand for cordless phones from all parts of the organisation; over the next two years, he expects the number of cordless users at the Auckland plant to expand by around 50%.

"Dealing with production stoppages is much easier now", he says. "Foremen can get directly in touch with maintenance technicians, and talk through the problem in advance. That means the technician can bring the right tools and spare parts with him when he comes to fix the problem. It's certainly helped to reduce lost production time."

Production time is not the only improvement that Utama has noticed. "Being able to make contact immediately, wherever you are, is a great benefit", he says. "It means you can act on ideas as soon as you have them – rather than waiting until you get back to your desk and your phone. Like any manufacturing business today, flexible working and fast responses to customer requirements are among our main goals. We aim to slim down stocks, and gear production closely to demand", says Utama. "That means we must be able to react quickly to changes in demand for products, by switching lines from producing one drink to another, for instance. Every drink production run requires the syrup concentrate to be made up four hours in advance of starting the filling line. Anything that can speed up the switch-over between production runs is welcome and Freeset has definitely had an effect here."

"It's hard to put a dollar value on what we have achieved with Freeset", Joseph Utama concludes. "Undoubtedly it has improved productivity and efficiency, but it's gone further than that. One of the biggest improvements we've seen is in job satisfaction. Better communications seems to eliminate much of the stress and frustration of people's jobs and makes them work better in teams."

2.5.2   A General Office Environment: the Netherlands Tax Administration

The Netherlands Tax Administration (NTA), which employs 30 000 staff spread over 300 offices, is responsible for implementing tax legislation as efficiently and effectively as possible, whilst ensuring fair and reasonable treatment of its customers. NTA's goals of improving quality of service to customers relies extensively on effective communications. Through the application of a wireless PBX solution, NTA is achieving demonstrable benefits that support this goal.

Fred van den Bosch is responsible for strategic planning and investment in a range of services, including communications, across NTA's offices. "Research revealed that the general public viewed the administration's image as fair and reliable, but also slow and cumbersome. With customer workload increasing rapidly – telephone enquiries alone were increasing at more than 20% per year – it was evident that changes would have to take place. We realised that efficiency gains could only be achieved through substantial investment in support services and facilities, such as automation, communications and training."

Starting in 1990 and continuing over the next two years, the NTA underwent a wide-ranging reorganisation involving restructuring into target groups, focusing on customer needs and providing improved services and counselling to its client base. For the new organisation and its associated working methods to be successful, improvements to the voice communications facilities within the NTA had to be implemented.

NTA invited tenders for the provision of a cordless system and an in-house paging system for a tax office near Utrecht. "The cost difference between the two systems was not so great and the functional advantages of cordless were obvious", says Fred van den Bosch.

Ten wireless PBX systems, based on the GPT iSDX, have now been installed in offices of the NTA. Users of cordless handsets include staff who previously relied on other mobile services. These include managers, bailiffs, inspection officials, agents handling customer enquiries and security staff. These total around 10% of the NTA's workforce.

"The functions of these staff do not generally tie them to a specific location but they must be accessible. In-house paging and a fixed telephone in the office plus, in some cases, a cellular phone would have been provided to ensure that accessibility. All of these are now replaced by a single wireless handset."

Integrated cordless telephony is making a direct positive impact on the day-to-day working of staff answering calls from clients. By having cordless handsets, staff are able to make and receive calls away from their desks and consult client files that are held in separate filing cabinets. Among the benefits being realised are improved telephone access, rapid turnaround of telephone enquiries, and dual usage of handsets in the office and on the public access "Greenpoint" telepoint service available in the Netherlands.

2.5.3 A Retail Store: Dickens Specialist Home Improvement Retail Group

The leading home improvement retailer in north-east England, Dickens Specialist Home Improvement Retail Group, is using the Multitone CS500 cordless telephone system to provide senior staff at its Stockton site with a means of always being in touch. The Dickens store in Stockton has over 175 000 square feet of retail space and each of the nine departments has a turnover of around £2 million a year.

The chief executive, his assistant, the systems manager and all departmental managers have been equipped with cordless phones linked into the store's PABX. Each manager is given a direct-dial number that allows customers to contact them without going through the switchboard.

"All our business managers spend most of their time moving around their department in the store and it's vital to our business that they are instantly contactable by other staff and our customers", said Dickens' computer services manager, Jim Small. "Customer service is our top priority and making this as efficient as possible is very important. If our business managers and other key staff can be in touch with people as cost-effectively and as well as if they were sitting at a desk that has a fixed telephone, that's obviously better for our business and will help to ensure that we keep ahead of our competitors."

2.5.4 A Manufacturing Company: Axis Holdings

Axis Holdings occupies a four-storey building where it carries out its business of producing extruded polythene film and PVC profiles, and stockholding and merchandising raw materials for the plastics industry. Axis operates in a highly competitive business sector where "winning" is often dependent on the level of customer service offered to the customer base. In addition, the company had dramatically expanded its customer base and was becoming increasingly aware that its existing telephony system was not able to meet the demands being placed upon it by the business.

The four-storey factory building, a former tannery, is large, with a complex network of walls and corridors that often make it difficult to obtain quick access to a telephone whilst out of the factory. The directors of the company are far from office-bound, spending considerable periods "walking the job". Warehouse and distribution staff are constantly on the move as they deal with stock and expedite customer orders. Notice of incoming calls for staff moving around the site or away from their desks was given by a receptionist over a public address system. Direct contact between the customer and the key contact was not maintained, with the possibility of considerable delay in responding to enquiries.

Finance director, Steve Turner, who has direct responsibility for communications, explains: "The greatest need for effective communications is in the warehouse and distribution department. When staff were called away to the telephone, mostly to take messages from haulage companies or to answer customers' enquiries, it meant interruptions and delays for callers. Management were concerned that customers would have difficulty reaching them with the distinct possibility they might well be losing business. We did not go out specifically to examine a cordless system but we were impressed by what the handsets could achieve. We also saw an opportunity to reduce our telephone bill by not having to ring back so many incoming callers."

Having identified radio coverage requirements with Axis, the GPT distributor carried out a site survey, which confirmed that two radio basestations would provide coverage of 90% of the premises – a large production area, three floors, a cellar and many corridors, including a separate warehouse as an extra.

Axis originally chose to have four cordless handsets interworking with the integrated cordless facility – two handsets for use by warehouse and distribution staff and two for directors and managers. The cordless handset users have access to all system features available to users of wired handsets. The most-invoked feature has become "call forward", diverting incoming calls to the cordless handsets. With the large expansion of the customer base, direct evaluation of actual cost savings due to fewer returned calls is difficult, but Axis is sure that savings are being made and may well increase the number of cordless handsets in time.

2.6 Summary

In this chapter we have looked at the development of the market for cordless PBX to date, and found that it has fallen far short of initial expectations. A number of factors, including price, awareness of the capabilities of the technology, and the relative complexity of the selling process, have contributed to this situation, and these have contributed to the desire on the part of potential users for clear, quantifiable financial benefits to be identified prior to purchase.

However, we have identified several actual benefits that have been realised by users, including improved customer service, direct cost savings, improved productivity and increased convenience and flexibility of use. We have also identified a number of applications for which cordless technologies are well suited, and illustrated these with case studies.

Our analysis suggests that many of the barriers that have hindered market growth can be overcome if the suppliers use their experience to address the market with greater confidence, and if this is done then the market can be expected to grow to meet its potential. These lessons from the early markets in Europe are likely to be mirrored in North America and Asia–Pacific as wireless PBX products begin to be launched into these markets in the coming years.

Acknowledgements

Permission to reproduce the studies reported in section 2.5 was given by the publishers there mentioned, and is gratefully acknowledged.

Reference

[1]   Advanced Cordless Communications, Vol. 2, Issue 7, Telecomeuropa, July 1995

Note

1   A UK Health Service working group established to investigate possible electromagnetic interference between clinical equipment and mobile phones recently concluded that "whilst much [Health Service] equipment was susceptible to interference from VHF handportables . . . very few pieces of equipment were susceptible to interference from cordless PBX handsets". Their report concluded that "the introduction of cordless PBX handsets may well enable most VHF handportable systems to be phased out" [1].

3 Telepoint in Europe

Ed Candy, Annette Ottolini, Marc Pauwels and Walter Tuttlebee

The first chapter of this book outlined the emergence of the digital cordless telephone market, its various application segments and the development of the market environment during the first half of the 1990s. Telepoint was seen as an integral part, indeed a driver, of the early cordless market, as recently as 1990.

By the middle of the 1990s telepoint in Europe had failed to make a commercial impact. The window of opportunity for telepoint in the UK was foreclosed, in part, ironically, by the initiative that led to personal communications networks (PCN). In some other European countries, telepoint never emerged past the trial stage. Meanwhile, in France and the Netherlands, commercial systems have enjoyed a measure of success, albeit not living up to early expectations. The situation of commercial European telepoint networks in the mid-1990s is summarised in Table 3.1.

Despite the lack of widespread commercial success, telepoint has been an important conceptual step towards bringing low-cost mobile communications services to the mass market – it was inextricably bound up with the launch of the PCN concept, which has driven cellular telephony into the consumer market. This chapter thus describes the history of telepoint in various European countries, from the perspective of the relevant telepoint operator organisations. Clear lessons are apparent for those with eyes to see and ears to hear. As new forms of public access cordless systems emerge – such as PHS in Japan, cordless complementing cellular and wireless local loop – operators have the opportunity to benefit from the lessons of the past.

The first section in this chapter reviews the events that led to the award of licences in January 1989 to operate the world's first commercial telepoint networks, in the UK. The nature and implementation of these networks are described. The changes in environment that led to the closure of the UK networks are then summarised and a brief analysis given of the factors that precluded commercial success.

A French perspective is then presented, outlining the adoption of CT2 telepoint in France, the requirements and implementation, market issues and a view of anticipated future developments. The first French service, Bi-Bop, has seen reasonable success, albeit below initial expectations. The experience in France provides an interesting contrast to the UK situation.

In the Netherlands the Greenpoint system has found a place within PTT Telecom's portfolio of mobile services and, like France, gradual expansion of the network from its initial deployment has occurred. The positioning of the Greenpoint proposition, its implementation and experience to date are discussed along with some directions for the future.

The CITEL telepoint service was introduced in Belgium in late 1992. Several important lessons were learnt from the initial trial service, which are described, along with the changes in the local situation that resulted in a decision to close the network in 1995.

Finally, a brief summary of developments in Germany and Finland is included for completeness.

Table 3.1
Telepoint networks in Europe

Country	Operator and *name of service*	Network supplier	Number of subscribers (as at June 1996)
France (Paris, Lille, Ile-de-France, Strasbourg)	France Télécom *Bi-Bop*	Dassault Cap Sesa Electronique Médélec Monétel	93 000
France (Acquitaine)	Prologos *Kapt*	Dassault	Launched March 1996 in Bordeaux
The Netherlands	PTT Telecom *Greenpoint*	Motorola	56 000
Belgium	Belgacom *CITEL*		Closed in 1995
Greece (Rhodes town)	OTE	Dassault	Launched July 1995

Source: adapted from [1].

3.1   UK – Rabbit and its Competitors

The concept of public cordless telephony (or telepoint) was perhaps the first example of a low-cost personal communicator destined for use by the general public. The cost of implementing a telepoint basestation network is very low, compared to a cellular network, providing the opportunity to establish a public network with a comparatively low-cost infrastructure and thereby providing a low-cost personal public telephone service. Telepoint can offer a unique capability – a handset user is by definition able to use the handset in the home, in the office or in public places with call charges inherently (significantly) lower than if such a utility were provided by cellular radio. It is these points that make telepoint different from cellular radio.

3.1.1   Licence Awards

By the late 1980s, in the UK, manufacturers and prospective network operators were in a position to request the Government to issue licences for the operation of public telepoint services. Recognising the growth of the UK mobile communications industry during the 1980s, and confident of the maturity of the CT2 technology, the Government invited applications in July 1988, the closing date for submissions being October 1988. The invitation resulted in eleven high-quality applications for up to four network licences. The UK Department of Trade and Industry (DTI) and Office of Telecommunications (OFTEL) assessed the applicants for their ability to create a successful business enterprise and quality telepoint service. The outcome was announced in January 1989 that four licences would be granted to: Ferranti Creditphone; the Philips, Shell and Barclays consortium; a consortium comprising STC, British Telecom, France Télécom and Nynex; and a fourth consortium involving Motorola, Shaye and Mercury. Significantly, at the same time as announcing awards of telepoint licences, the launch of the UK's "Phones on the Move" initiative was also announced. The latter led to personal communications networks (PCN), the success of which was later to reduce the window of opportunity for telepoint in the UK.

In formulating the policy to allocate telepoint licences, the DTI believed the success of the service would depend on two factors:

- A large geographic distribution of basestations
- Users being able to access the basestations provided by any of the network operators, i.e. internetwork roaming

Under the early national CT2 specifications (BS 6833), different proprietary and incompatible interface standards had emerged, which, whilst adequately providing the necessary technical specifications for telepoint, would ultimately lead to separate networks, with users unable to roam between basestations of the various networks. Thus, a group of manufacturers and operators were brought together in October 1988 under the guidance of the DTI to produce what became the Common Air Interface (CAI) standard. The objective was to provide a uniform standard for public telepoint that would enable handsets to roam between networks and operators. This standard was later to be adopted by ETSI (European Telecommunications Standards Institute) as the I-ETS 300 131 digital cordless standard.

The UK Government statement in January 1989 formally set out the requirement of the licensees to adopt CAI by the end of 1990 and to form roaming agreements with other network operators so that telepoint users had improved service, with roaming between networks and freedom of choice of equipment. To encourage the telepoint service to start as quickly as possible, however, all the licensees were given the option of commencing the initial service with proprietary equipment provided they met the requirement to support the CAI standard from 1990 onwards.

3.1.2 Implementation Plans

Following the award of licences, establishing the service itself required a combination of equipment and potential basestation locations, as well as the foundation of companies and their manpower resources to implement the system. Two of the four licensees – the Ferranti Creditphone consortium operating the Zonephone service, and the Phonepoint consortium (STC, British

Telecom, France Télécom and Nynex) – commenced operation in the latter part of 1989.

The equipment and call charges for both services were broadly similar. The Zonephone service was launched using proprietary equipment supplied by Ferranti. Phonepoint launched using a proprietary standard originally developed by Shaye Communications, using Shaye handset equipment and basestation equipment modified by British Telecom from Shaye submodules. Shaye proprietary equipment was also adopted by the third consortium, Callpoint (Motorola, Shaye and Mercury), who announced its intention to launch in December 1989.

Typically equipment prices were around £200 for a handset[1] and charger, an additional £200 for a domestic basestation, connection fees of £20 to £30 and monthly access charges of £8 to £10. The telepoint networks employing them used synthesised voice messages (generated in the basestation) to prompt and inform the user of the progress of a call.

In October 1989, the Barclays–Philips–Shell consortium (BYPS) announced that it would commence service in 1990 using handsets, basestations and network equipment manufactured by GEC Plessey Telecommunications to the CAI standard. This was subsequently delayed to April 1992 following the acquisition of BYPS by Hutchison in February 1991. By choosing to launch immediately with CAI, the BYPS consortium argued that this would guarantee freedom of choice for the customer as well as guaranteeing that equipment purchased would not become obsolete when the proprietary services were required, by the terms of their licence, to convert to CAI. It was further argued that a CAI network would provide customers with an opportunity for roaming between networks. At that time the CAI standard was being adopted by various international and European operators because of the stability of the standard, the availability of CAI infrastructure from several manufacturers, and the potential source of low-cost handsets in significant quantities. The establishment of a European telepoint operators' Memorandum of Understanding, based around the adoption of the CAI and described in [2], also seemed to validate such an approach.

Two of the CAI product manufacturers, GPT and Orbitel, announced details of their CAI products in late 1989, with product delivery during 1991. Both handsets included liquid-crystal displays (LCDs) and volume controls, and used standard penlight or rechargeable cells. The LCDs were to be used for displaying the call status and for recalling telephone numbers previously stored. In September 1990, Motorola announced their "Silver Link" CT2 handset, a compact flip phone, which could easily be carried in a shirt or suit pocket, and which was powered by a rechargeable pack or replaceable miniature AAA cell. Some early products are shown in Figure 3.1.

The successful implementation of telepoint required the acquisition of sites for the installation of basestations. All four UK operators were engaged in commercial discussions with a range of property owners to provide the necessary distribution of basestations throughout the UK – all the operators found this to be a resource-consuming and slow activity, which delayed network rollout.

Whilst telepoint offers a low-cost personal communications service, continuous coverage is generally only available in city centres and at key travel points. In other areas, coverage is more fragmented and limited to identifiable locations. In the absence of complete coverage, a telepoint user needs to be able to identify the location of a telepoint site easily and quickly to make a call; thus, each of the operators developed distinctive signage or site acquisition policies, which, to a lesser or greater extent, simplified this task.

3.1.3   Network Operation and Enhancement

Basic Network Operation

The telepoint system itself is a network of a large number of public cordless telephone basestations. Each basestation is a small, easily installed piece of equipment, about the size of a briefcase, and connected by ordinary telephone lines to the public switched telephone network (PSTN). Handsets generally can be used up to 200 metres away from the public basestations. While each telepoint basestation is quite simple, it is the sheer number and spread of all the basestations that can allow telepoint systems (in conjunction with office, domestic, paging and messaging systems) to provide an effective and affordable service.

When a telepoint call is made, the accessed basestation records the details of its duration and destination. Each call is individually authorised before it is allowed to pro :eed. ɪhe handset sends the user's account and personal identity number (PIN) and the system verifies that they match correctly, also checking that the user is creditworthy. These two checks can be performed either locally in each basestation or in one central computer. Performing these functions locally requires that the basestation be updated each night with the latest information, but also avoids delays whilst the central computer is contacted. Performing the functions centrally is a secure and easily managed method, but involves a call to the central computer for every customer call made.

Two-Way Calling

While office and domestic cordless telephones can both make and receive telephone calls, early UK telepoint users were limited to making only outgoing calls – this was a restriction of the initial licences, not the technology. The basic telepoint system can be enhanced in a variety of different ways to allow two-way calling, as has been done in France and Asia.

In the UK, the Government indicated in early 1990 its intention to allow the network operators to adapt their systems to permit two-way calling. By supplementing telepoint with paging and voice-mail-boxes, a quasi-two-way service can be provided, with users never out of touch but spared the irritation of incoming calls at inconvenient moments. For example, when a person is called at the office but happens to be out, the call can be automatically diverted to his or her voice message box; the caller can be asked to leave a message, the relevant person can be paged and the call can be returned from the nearest telepoint if appropriate.

A further enhancement of telepoint can be provided by a so-called "meet me" switch. This facility, in conjunction with paging and a telepoint network, provides a two-way calling facility. Telepoint subscribers are allocated a unique telephone number. Callers wishing to contact a telepoint subscriber ring this number and the "meet me" switch initiates a paging call to the telepoint subscriber. The telepoint subscriber can respond to this page by dialling a silent

Figure 3.1   (*This page and opposite*) Early CT2 handset products.

number, which, when recognised in the "meet me" switch, connects the telepoint subscriber to the incoming caller. The "meet me" switch includes a series of voice messages to the incoming caller, advising progress of the call and giving the caller the option to wait until the telepoint subscriber is able to answer or to divert to a message service. With appropriate care, very low call completion delays can be achieved.

3.1.4   The Changing Environment

Before the launch of the Hutchison (BYPS) network in April 1992, under the heavily promoted "Rabbit" brand name, all three other licensees withdrew from the market, for different reasons:

- *Zonephone* had deployed some 1700 basestations before it was forced to close down its network owing to the well publicised financial difficulties suffered by its parent organisation in October 1990.
- *Mercury Callpoint* had installed a network of 1500 basestations in London but, following reluctant take-up of a proprietary system, and faced with the costs of converting to CAI and conflicts with the then-new Mercury One2one PCN digital cellular licence recently secured by its parent, it ceased operation in mid-1990.
- *Phonepoint* had commenced with a proprietary system but quickly announced a relaunch of a CAI network for October 1991. Within weeks of the proposed relaunch, with over 1700 CAI basestations deployed in central London, British Telecom announced that Phonepoint would be closed "due to lack of market interest". Declining interest in the UK by the overseas investors in Phonepoint

could have left British Telecom with a majority shareholding and in breach of their licence conditions on majority ownership.

These events left Hutchison with a dilemma if they were to launch on their own: on the one hand, they had a *de facto* monopoly; on the other, they would be unable to benefit from roaming agreements with other operators, a key element in providing the basestation density necessary for a predominantly suburban population distribution. Thus it was that, following successful pilot testing, Rabbit announced details of a soft launch in the Manchester area in April 1992, followed by a national launch in October 1992.

By January 1993 Rabbit had over 11 000 basestations in operation throughout the UK, with 1700 in central London and 3000 inside the M25 London orbital motorway. The remainder were spread throughout roadside restaurant chains on the motorways and principal arteries. Basestations were located in all the high-street shopping centres throughout the country, and in many local shopping arcades.

Although an extensive quality national network had been developed, Rabbit's fortunes were thus constrained by external events:

- The premature demise of Rabbit's competitors removed the two factors identified as key to success when it was licensed – a large geographic distribution of basestations and internetwork roaming – thus limiting the capability and reducing the attractiveness of the service.
- The launch of Rabbit occurred in the midst of the UK's deepest recession in recent times.
- The cost of calls was distorted by a PSTN access charging regime that resulted in a cost above the standard retail rates for exchange line usage. The regulator seemed unable to resolve the consequences of a BT pricing policy that resulted in retail access rates below the so-called wholesale rate, and effectively held Rabbit subscribers' call rates at or above payphone levels.

Finally, in October 1992, Mercury One2one launched its new PCN digital cellular service, with an offer of free local calls in London after 7 p.m. This initiative effectively removed one of telepoint's unique advantages, namely the ability to make low-cost PSTN calls through a cordless basestation when at home. This free-call policy had a major impact on opportunities for Rabbit, with potential subscribers being tempted instead to go for a low-cost cellular service.

By June 1993, Rabbit had sold around 15 000 connections and 20 000 cordless handset/home basestation combinations, but the growth rate was considerably short of the level needed to sustain a long-term business. Hutchison, with a close eye on the highly competitive mobile market, following an extensive review, announced in November 1993 that it would concentrate on its own PCN business, Orange, due to be launched in April 1994, and close down the Rabbit network by the end of 1993.

3.1.5 Analysis

Some would maintain that the failure of telepoint in the UK, its place of origin, signals the failure of the concept. Whilst such a simplistic analysis might sound attractive, reality is more complex.

Strengths of the Telepoint Concept

It is generally accepted in the mobile communications industry that cordless telephones and handheld cellular telephones can provide, in essence, the same service. Moreover, despite the withdrawal of Rabbit from service in the UK, cordless and cellular technologies will still, in the long term, eventually converge into a unified form of service – cf. experiments with combinations of DECT and GSM in Germany, Scandinavia and elsewhere. For a mass-market service, cordless systems, with their high spectrum efficiency and low cost, undoubtedly offer the best starting point for two scenarios:

- High-density urban residential situations
- Local cooperation in developing economies

As a result of microcell and spectrum re-use technologies, cordless systems are able to deliver a very high density of service. It is generally believed that they will be needed to achieve the capacity and density as mobile communications move into the lower end of the mass market. Cordless systems are much more spectrum efficient than a pure cellular system using cells of the same size, since the great mass of customers, encouraged by tariff structures, can use private and domestic office systems as well as the public networks; the use of dynamic channel allocation procedures further improves spectrum efficiency. Cordless telephones can also be used inside cars, buses, trains, etc., which are connected to the public network by cellular or other radio systems.

The arrival of multimedia applications will give those cordless systems which have the potential for higher bandwidths in microcellular environments, such as offices, further unique opportunities to deliver new-generation services. When data communication standards are agreed, public access cordless users will be able to access a variety of data or ISDN-type services such as fax, e-mail and remote logging on to computers.

Critical Success Factors

Early public access cordless systems have had mixed success, and it is helpful to consider the differences between those systems whose services have continued to develop and those where service has been withdrawn. Telepoint appears to have been most successful in those situations where there is coexistence between the commercial centre and high-density residential population, e.g. in Hong Kong and central Paris. Population distribution in the UK is more suburban in nature and there is very little high-density residential living in the major cities. There is no doubt that telepoint is far more attractive in those societies where people at all levels in the community often have multiple jobs and interests, and effective communication is needed to manage the multiplicity of lifestyle activities. Whilst Hutchison (Rabbit) was able to achieve reasonable densities of telepoint basestations in city centres and in transport centres such as bus stations, railway stations and motorway service stations, it was uneconomic to build a telepoint network that provided a similar density and continuous coverage necessary to serve the population in suburban areas. So whilst the service was effective in commercial centres, it could not, as a result of the population distribution in the UK and with the demise of its competitors (and in another sense necessary collaborators), provide the level of coverage needed to meet fully the three elements of the telepoint proposal – namely to provide service in the home and the office and dense coverage in the street throughout residential districts.

Telepoint systems in conjunction with domestic and office cordless systems can provide low-cost and hence mass-market personal communications. Installation of extensive telepoint networks in popular places and complementing these with paging and messaging services can ensure a high-quality and effective service to users. The relatively low cost of infrastructure means that call charges can be very modest, much closer to fixed-network than mobile-network call charges.

In developed regions, it is clear that telepoint is best suited for applications in which low-cost public access is required in cities that are characterised by high-density residential living, intertwined with business and commercial activity centres. Recent initiatives to establish a DECT Operators' Group (DOG) [3] suggest that, despite the experience of telepoint, there is still keen interest in using cordless systems to provide public access portable telephony. Similarly, it is likely that, if the growth in demand for digital personal communication systems is to continue, then cordless systems with their very high capacities and efficient spectrum re-use may be the only way in which the long-term mass-market needs can be fulfilled.

3.2   France – Bi-Bop *et al.*

The concept of telepoint ("Pointel", *en Français*) emerged in France even before the specification of the CT2 Common Air Interface was begun. In late 1985 engineers at the Centre National d'Études des Télécommunications (CNET) were considering the microcellular "communication area" concept, which had been proposed by Matra Communication to France Télécom. In November 1986, various papers were presented at the IDATE Conference in Montpellier [4] on that subject, and the concept was officially presented at the France Télécom stand at Telecom '87 in Geneva.

In early 1989, press releases were issued by various companies interested in manufacturing telepoint equipment, as well as by France Télécom. A request for proposals for equipment supply was launched in July 1989 by France Télécom, and around ten offers, some from Franco-British consortia, were made, with a view to implementing a pilot Pointel network during 1990.

3.2.1   Trial and Service Launch

Thus, a pilot Pointel network consisting of some 300 basestations was constructed, initially in Strasbourg, which was commissioned in October 1991. Thorough testing of the Bi-Bop service ensued throughout 1992, involving some 2000 users. This trial period resulted in improvements to the quality of service, development of engineering rules for radio coverage planning relating to traffic profiles, and improvements to the handset man–machine interface and subscription mechanisms.

The two-way calling service (i.e. the ability to receive incoming calls at a handset) – Bi-Bop Réponse – was also introduced and successfully tested during this Strasbourg trial phase. Bi-Bop Réponse offers the users the ability to receive incoming calls on their handsets under certain conditions. To do this, in public areas, a user must notify his or her presence to a nearby basestation and remain within the coverage area of that basestation or its nearest neighbours. In semi-public areas, such as airports or railway stations, users register with a basestation but their locations are monitored as they move and calls are routed to them as they move within range of any basestation. Voice-mail is also incorporated as an essential part of the Bi-Bop Réponse service, so that all calls to a subscriber can be delivered, even when the user is not in range of a basestation. A paging service, Bi-Bop Alphapage, was also developed, offering customers a single combined bill as well as the use of a combined CT2 and pager handset.

Commercial Bi-Bop service was introduced in Strasbourg in January 1993, closely followed, in April of that year, with service launch in Paris. Since late 1994 service has been available in Ile-de-France and the Greater Paris region, as well as the Strasbourg and Lille metropolitan areas. Bi-Bop basestations have been principally located in areas of potentially high telephony traffic – e.g. city centres, motorways, shopping malls, railway stations, airports, etc. – with some 7000 basestations in Ile-de-France and 900 in Lille. Areas of coverage are indicated by the use of distinctive markings – blue, white and green stripes around lampposts at a height of 3.5 m or stickers in shop windows. Such markings appear obvious to the user, but unobtrusive to the uninitiated.

3.2.2   Network Architecture

The Bi-Bop network architecture is illustrated in Figure 3.2 and can be seen as comprising three key elements – the basestation subsystem, the management subsystem and the subscriber system – all interconnected via the X.25 packet-switched (Transpac) network [5, 6].

The Basestation Subsystem

The basestation subsystem comprises the basestations and basestation controllers (BSCs). These equipments in France Télécom's Pointel network have been supplied by SAT and Dassault Automatisme et Télécommunications (DAT) and by Ascom/Monétel. The basestation is housed in a small cabinet suitable for wall or pole mounting and is primarily mains powered, although with a one-hour battery backup capability. The basestation is modular and can be expanded to support two, four or six telephone lines.

The type of BSC provided, for example, by SAT, housed in a 19 inch rack mounting or stand-alone cabinet, operates from a 48 V DC power source and supervises the signalling interchange with subscribers' handsets located within the coverage areas of the basestations connected to the BSC, as well as establishing calls through the PSTN via digital PCM ports. An ISDN 2B+D U interface is employed to connect the BSC to the basestations, whilst X.25 links provide communication with the management subsystem.

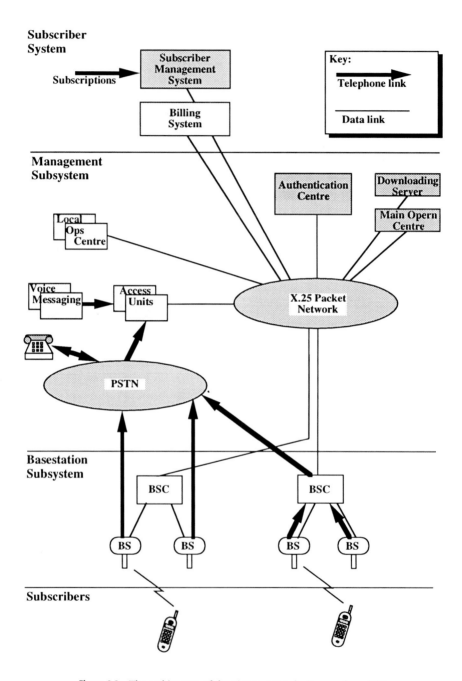

Figure 3.2   The architecture of the Bi-Bop network. (*Source*: from [5].)

The Management Subsystem

The management subsystem comprises a national authentication centre, a main and regional operations centres, the downloading servers and access units. For the France Télécom network, this equipment has been supplied by Cap Sesa, DAT, Electronique Médélec and Ascom/Monétel.

The authentication centre (AC) stores information on all Bi-Bop users and their subscription details; this information is used to validate and centrally manage authorised access to the various service offerings. Subscription management issues require communication between the AC and the subscriber system. Likewise, the AC communicates with the BSCs to manage call establishment and with Pointel access units receiving incoming calls for Bi-Bop Réponse subscribers.

The AC provides three key functions – data management, authentication and location, and incoming call management. Data management relates to initialisation of the subscriber database for new or modified subscriptions. Authentication and location relate to authenticating cordless handsets and confirming subscribers' access rights. On granting service, the AC updates its database on the location of the cordless handset. Incoming call management is handled as follows. When an incoming call for a Pointel subscriber arrives at an access unit, it signals the AC with the called handset number and the access unit number to be called back. The AC then searches for the required cordless handset in the location table and, assuming it locates the handset, signals the appropriate BSC and further on the appropriate basestation. The basestation then searches for the requested cordless handset and finally calls back the access unit, completing the call.

The main operations centre collects all billing data and provides an overall nationwide network management facility, presenting statistical traffic data to the network managers. It is essentially a database, collecting data from the regional operations centres, together with a set of analysis and expert software tools. It interacts with the authentication centre in order to monitor usage and the billing system to generate call ticket data. The localised operations centres manage the installed base, monitoring hardware and software status and statistics on network operation and usage.

The downloading server provides access to basestation and BSC software packages, enabling a national management capability of the software status of these parts of the network.

The Subscriber System

The subscriber system comprises the subscriber management system, which handles customer and subscription data management, and the billing system, which processes call tickets received from the main operations centre. A Minitel-based service allows remote access to the billing system, enabling subscribers to this service the ability to access a detailed up-to-date listing of calls made. The subscriber system used by France Télécom has been supplied by Sema.

Network Functionality

The Pointel network provides a number of functions essential to offering a service suitable for the consumer, notably security, over-the-air registration (OTAR), handset location management, call processing, call data management, and operations and maintenance (O&M).

Network security functions are located at the authentication centre, which contains the subscriber database. Subscriber authentication confirms that the identity claimed on the radio access corresponds to the person registered as the subscriber. As such, it provides protection to the user against unauthorised use of their identity and provides the operator with confirmation that the subscriber should be billed for the service provided. Subscriber access control functions provide further protection, by checking that the subscriber is entitled to the service he or she has requested – e.g. national or international call restrictions.

The Bi-Bop network supports over-the-air registration of handsets to allow quick and secure initial access to the network for a new subscriber. The new user enters an identification code via the keypad on the cordless telephone, which the network recognises as an OTAR request for public Bi-Bop service. Encrypted subscription data is then downloaded through the radio link and stored in the handset for use for subsequent call access.

Location management is necessary to support the two-way calling service offered by Bi-Bop. Thus the network performs handset location registration and deregistration upon subscriber request or, in the case of deregistration time-outs, as determined by the network. Discrete automatic polling of handset location is also undertaken to track handsets as they move within basestation clusters.

Call processing functions include simple outgoing call interconnect to the PSTN and more complex processes for incoming calls. For the latter, the presence of an incoming call at a Pointel access unit causes the authentication centre to activate the incoming call processing function. This triggers a call to the called handset via the relevant basestation controller (BSC) and basestation(s). When the handset answers, the BSC alerts the authentication centre and calls the access unit via the PSTN, which then interconnects the two parties to complete the call. If the handset does not respond, the call is routed instead to a voice-mail facility.

Call data management is localised in the BSCs, where call duration and metering pulses from the PSTN are measured and stored, with the billing data downloaded to the main operations centre daily.

A wide range of O&M functions are supported. These include defining and supervising the installed base, and collecting, collating, processing and forwarding of call data statistics for normal and other (e.g. toll-free) calls. Data and query tools are provided to the operators to manage the installed base, in addition to tools for management of the other, localised, operations centres. Maintenance functions embrace generation and processing of local and remote alarms and self-test. At the localised level, provision also exists for simple test of basestations and basestation controllers.

3.2.3 Market Positioning and Subscriber Base

Early predictions for the French CT2 market anticipated an installed base of 2.5 million users, of which 1 million were expected in Ile-de-France, after five years.

Those figures included residential, small office and street telepoint applications. As with other European telepoint markets, such predictions were wildly over-optimistic. Despite this, however, telepoint in France has indeed secured an important niche in the personal communications marketplace, partly helped by the favourable position enjoyed by France Télécom and the relatively high prices of cellular telephony.

Telepoint was always seen as addressing a very different market segment than cellular. This has been borne out in practice, as shown in Table 3.2. As can be seen, the much lower monthly cost of Bi-Bop compared with cellular has attracted a substantial residential customer segment, largely untapped by the cellular market. Analysis of the Bi-Bop subscriber base has indicated a high satisfaction rate of some 86%. Subscribers include executives, middle-class employees and significant numbers of young people. On average, users are making some nine calls per week, of two minutes per call.

Table 3.2
Differences between Bi-Bop and cellular in France, as at December 1995

	Bi-Bop	Cellular (GSM)
Number of subscribers	95 000	700 000 (France Télécom) 300 000 (SFR)
Coverage	Regional – three cities	National
Average monthly bill	£16 (incl. tax)	£60 (incl. tax)
Split of customer type	20% professional 80% residential	95% professional 5% residential

Source: from [7].

The success of Bi-Bop in France has been attributed to the low penetration of cellular telephones at the launch of Bi-Bop, the dense city environment and street culture of Paris, and innovative consumer marketing, with a clear differentiation from cellular.

As a consumer product, branding was seen as an important aspect to address – hence the adoption of the "Bi-Bop" name. Two options were made available to subscribers: either to purchase a handset at a price of FFr 590 or to rent one at a price of FFr 35 per month. Tariff structures were revised in 1996 as follows:

- *Bi-Bop Complice* for intensive users, with a per-minute charge of FFr 0.5/1.5 (local/national) and a monthly charge of FFr 85.
- *Bi-Bop Malin* for low users, with a per-minute charge of FFr 2.0 and no monthly fee.

The Bi-Bop Réponse service, supporting incoming calls, originally an optional service at additional cost, is now included in these fees.

A variety of new sales methods have been developed to promote Bi-Bop, including partnerships, co-branding and mail-order distribution channels. Becoming a new subscriber has been intentionally kept simple, with immediate handset activation. Developing a strong customer relationship was seen as important, and to this end France Télécom has established a customer care centre, regular newsletters and a customer loyalty programme.

The relative success of Bi-Bop must be seen in the context of its initial and ongoing growth – whilst initially very successful, growth of the subscriber base has been steady but not huge. From a base of some 60 000 in October 1994, subscriber numbers have grown by some 50% to 95 000 by the start of 1996 – this compares with a stated target user base of 150 000 by the end of 1995. The introduction of the Bi-Bop Malin tariff contributed to this, and further new initiatives are anticipated. Thus, whilst subscriber numbers have fallen below target, the network has nonetheless continued to see respectable growth.

3.2.4 Future Developments

The experience of France Télécom has been that a telepoint service is cheap to implement, can be extended proportionately as the subscriber base grows, and, once installed, has very low maintenance costs. For France Télécom, telepoint is seen as serving two complementary objectives – one being provision of a low-cost limited-mobility telephony service in itself and the other being a means of educating the consumer market with a view to subscribers upgrading to full cellular service with France Télécom. Thus it is seen as having a strategic role, as well as being a commercial service in its own right.

In the short term, the goals of Bi-Bop are to develop the numbers of subscribers in the Ile-de-France region, to educate them with Bi-Bop and to encourage them to switch to cellular with the same operator – clearly, providing an experience of good customer service is key to such strategy. New tariff structures to get closer to the price of the fixed telephone are envisaged, along with other innovations such as hot billing. In the medium term, a second objective is the development of a new offering to address the residential market. The use of the same CT2 CAI technology for wireless local loop has obvious operational synergies [8, 9].

Alongside the France Télécom Bi-Bop service, a second operator has now entered the field of play. Prologos, a consortium including Olivetti and a number of French financial groups, applied during 1994 for a licence to operate CT2 telepoint services in France. The French regulator, the Direction Général des Postes et Télécommunications (DGPT), awarded a licence to Prologos in May 1995, which allows the company to offer regional telepoint service in the départements of Gironde, Charente-Maritime, Landes and Pyrénées-Atlantiques. The licence terms also permit the establishment of a roaming agreement with Bi-Bop – of benefit to both parties. Thus, rather than seeking to compete directly with France Télécom, Prologos plan a complementary strategy, aiming to offer services in regions not covered by Bi-Bop, with applications to operate telepoint services in other regions of France anticipated. Initial service was launched in March 1996 in Bordeaux, with a monthly subscription of FFr 73 and offering handsets, supplied by Matra, at FFr 995. How well the new operator can perform, at a time when other new personal communications services, such as DCS1800 offered by Bouygues Télécom, are being introduced and forcing down prices, remains to be seen; unlike France Télécom, Prologos have no opportunity to encourage subscribers to migrate upward to a cellular service provided by themselves.

3.3   The Netherlands – Greenpoint

The Greenpoint service was introduced in 1992 in the Netherlands by the national PTT Telecom organisation. It was a very logical step in the whole development of the mobile communications portfolio; PTT Telecom felt that Greenpoint was, and still is, one of the important pillars of that portfolio. This section describes the implementation, the developments, the success factors that have contributed to its growth and why, in the Netherlands, telepoint has succeeded as a service in contrast to the UK experience.

Rather than primarily offering a telepoint service, Greenpoint was positioned as a premium cordless phone for the home, which had an extra functionality, namely a capability of cordless calling when in the neighbourhood of a Greenpoint. It was therefore seen as a two-in-one concept: the only phone that could be used in the street as well as at home.

3.3.1   Technology

Greenpoint is based upon the international CT2 CAI technology, a technology for cordless phones that is approved by the European Telecommunication Standards Institute (ETSI). The characteristic feature of this standard is that digital information transmission is used over radio, in contrast to the previous CT1 analogue standard. This technology supports improved speech quality and enables new functionalities, e.g. a handset can be used on multiple basestations and multiple handsets can be used at one basestation. In parallel to the technology development, thoughts about a new network emerged, which could enable new applications – the telepoint concept, described more fully in the earlier sections of this chapter. The service was introduced in the Netherlands when the above-mentioned standard had been developed to the point that it was transparent and compatible.

3.3.2   Network Deployment

The network had 2000 Greenpoints at launch in 1992. The whole philosophy from the start was to have Greenpoint at places that were:

- Easily recognised by the caller – e.g. preferably chain stores.
- Logical for people to make a call from – e.g. main shopping streets, bus and railway stations, outskirts of cities, M-roads.

In 1993 the decision was taken to extend the Greenpoints to 5000 in number. This would ensure that a Greenpoint could be found within a ten-minute drive on M-roads and within a five-minute walk within inner cities.

All the Greenpoints are easily recognisable by stickers and signs. By 1995 Greenpoints could be found near every railway station, post offices, many chain stores and restaurants, e.g. McDonalds. The Greenpoint network is not a network in the pure sense of the word, in that the basestations are actually connected to the fixed network. Every Greenpoint needs to be powered by electricity; in rural areas, where it is difficult to get electricity, Greenpoints are running on solar and wind energy.

3.3.3 Target User Groups

A few very important trends are clearly seen for the coming years:

- Customer needs are becoming individual needs.
- Professionalism of the service industry is increasing in importance.
- Cost to the consumer is an important factor.

Mobile communications can contribute to the individualisation and the independence of consumers. With a Greenpoint handset everybody can have a real personal phone that can be used in multiple settings – at home, in the office and on the road. The target groups are people who are mobile (business and non-business) and who want to have contact with their home base on a regular basis. Greenpoint attracts people who make a trade-off between functionality, price and quality.

Amongst current users are many business people for whom it is necessary to be able to place a call and be reached by their office – this has been facilitated by incorporating an inbuilt pager into the Greenpoint telephone. By analysing the business process of potential customers and their communication needs, it is thus very easy to identify whether they could be interested in the Greenpoint service – business users who work within a large organisation and are organised in such a way that the contact with the end-customer always takes place via the office are prime candidates for the Greenpoint service.

The non-business users are mostly young or elderly people who have some money to spend. Although the Greenpoint service is still cheap compared to other mobile services, mobile services as such attract people who have some disposable income. The trade-offs that people make are every time within the frame of the mobile portfolio.

3.3.4 Positioning – the Greenhopper

The positioning is chosen from the angle of the hardware; in effect the product on offer is the telephone – the Greenhopper – for which the telepoint service is offered as an additional facility. The reason for this approach is that customers get a feeling for a mobile service when it is marketing-wise connected to the hardware necessary to unlock the service. The Greenhopper is a telephone that is cordless and digital for use indoors (at home and in the office) and at Greenpoints. Greenpoint is an extra facility for the Greenhopper and is an extra unique selling proposition (USP). Greenpoint is positioned as the cheapest method of cordless calling at calling points and where it is also possible to be reached (using the inbuilt pager).

3.3.5 Service Development

By 1995 more than 58 000 subscribers[2] were using the Greenpoint service – each year the number of subscribers has doubled. An important driver has been the introduction of a Greenhopper with a pager in it; from market research it became clear that, especially in the business segment, the growth could be increased significantly by adding a functionality to overcome the drawback of being unable to be reached. At the moment more than 40% of the users are business people.

In late 1994 the service capabilities of the network were expanded with the arrival on the market of a PCMCIA CT2 CAI modem, jointly developed by PTT Telecom, Digicom and Motorola. Plugging into a standard PCMCIA 2.0 slot on a modern personal computer, this modem enables users to use the Greenpoint network for transmission of PC data; the modem can also operate via a normal PSTN line.

It is interesting to note that, since the introduction of the GSM cellular telephone network, Greenpoint has grown significantly as well. The reason can be found in the explanation that the market has started to become more aware of all the choices that can be made in the mobile portfolio.

In the future it will become more important to add extra functionality in terms of the service and to stay competitive on the triangle of functionality, price and quality – the customer will become more and more critical about offered services. Owing to the flexible organisation of Greenpoint and the possibilities that are still not used, it must be possible to maintain a place in the mobile portfolio and to grow into a pre-PCS service.

3.4 Belgium – CITEL

3.4.1 Origins and Service Concept

The decision to start a small-scale telepoint trial in Belgium was taken by the Board of Directors at the end of 1989, following the establishment of the CT2 MoU agreement, signed by the Belgian operator RTT. The aim of the trial was mainly to investigate the ability of the CT2 technology to implement a telepoint service, whilst trying to estimate the potential of the limited-mobility concept as a low-end, mass-market service.

The telepoint trial infrastructure consisted of 120 basestations directly connected to the PSTN and with a modem connection to one central maintenance and registration centre. In order to keep network costs as low as possible, authentication and authorisation were done locally, off-line, by the basestations, based on a stored blacklist, which was updated over the network every night.

In its first phase CITEL was mainly a technical test; thus, given the small number of basestations, the CITEL trial could not be exploited as a fully premium mobile pilot service. Instead, it had to be positioned as a service offering some added value to owners of CT2 equipment for domestic use. CITEL could be seen as an added-value public payphone service – the "payphone in your pocket" concept. Basestations were mainly located in telephone booths and call charges were based on public payphone tariffing, with an extra charge per call for the offered mobility.

At the beginning of 1990, RTT launched a tender for domestic CT2 equipment and decided to distribute Orbitel equipment for residential use on a nationwide scale. Whilst only first-generation handsets from Orbitel and GPT were available at that time, it was clear that very soon other CT2 equipment manufacturers would launch their equipment in the liberalised Belgian market for cordless terminal equipment. The commercial success of the CITEL service was heavily dependent on the success of CT2 equipment for home use together with the perceived added value of telepoint for the owner; for this reason CITEL had to ensure access to different types of CT2 handsets from the very beginning. The price of CT2 telephones was slightly higher than that of CT1 cordless telephones in use at the time.

It was economically impossible to offer a telepoint service in all the residential areas of Belgium; thus CITEL was to be implemented in Brussels, where it could best offer added value to cordless users coming from all over the country. For users using CITEL as a premium service only, RTT offered the possibility to rent a handset when subscribing to the service. Several upgrades of both CITEL equipment and first-generation terminal handsets were needed to ensure interworking and sufficient grade of service and integration in RTT billing and client systems.

When CITEL was finally launched at the end of 1992, after a delay of more than a year, the picture was completely different. The results of telepoint networks in the neighbouring countries were not encouraging – Deutsche Telekom was giving up its "Birdie" trial and most of the UK operators had closed down their networks. All this and the need to give priority to the introduction of GSM put the RTT in an indecisive state concerning CITEL. No further actions were taken. CITEL suddenly had to prove itself a viable commercial service, whilst support and extension to the service was completely abandoned, focusing commercial effort on the new GSM network.

3.4.2   Lessons

Despite the small number of subscribers, the trial led to some interesting indications about the necessary evolution of CT2 CITEL to fulfil the conditions for a very profitable service.

Firstly, almost 30% of the subscribers rented a handset (at nearly £18 per month) and used CITEL as a premium service only. Their first concern was to have good coverage in their working area only, not needing the full coverage of cellular, and some means of two-way calling (most had a pager), for which they were apparently willing a pay a surcharge nearly half of that of GSM.

Secondly, cordless equipment prices needed to fall to an acceptable level (£300) compared to CT1 to reach a mass market. Nearly 50% of the few domestic sets sold on a nationwide scale were registered to CITEL, indicating the importance of telepoint to support the CT2 terminal market. Most of these users were not resident in Brussels and wanted coverage in city access points: railway stations and parking places on the main routes accessing the city.

As a result of this evaluation, an upgrade of CITEL in Brussels and Antwerp – with more residents in the city centre – was proposed, offering the following features:

- Nearly full coverage in the city centre and limited coverage in surrounding suburban areas.
- Evolution to two-way calling combining a "meet me" service, joining paging and one-way telepoint, and network routing based on a manual localisation call by the user.
- Replacement of the first-generation terminal equipment.
- Adapted tariffing dependent on the offered service level.

These proposals led to an in-depth study of the needs and possibilities of a mobile urban telecommunication service. This study indicated a priority for public mobile service with full coverage and handover. Belgacom considered that there was sufficient need for an upgraded telepoint service, coexisting with cellular-based services. However, it decided to postpone the launch of such a service until it could benefit from integration in the fixed network via intelligent networking, and the CITEL network was closed down in 1995.

3.5   Elsewhere in Europe

In the late 1980s, in addition to the countries already discussed, a number of other European countries were considering the possibility of deploying telepoint networks but have not proceeded to do so. Of these, perhaps the most significant was Germany. The German decision not to proceed with telepoint, following swiftly after the demise of telepoint in the UK, was undoubtedly a major blow to the chances of further telepoint networks being launched in Europe. The background to the German decision is briefly summarised below.

In addition, we also summarise below the early telepoint activities in Finland. Whilst telepoint did not proceed to a successful commercial service, the impact of those early activities continues to be felt today, as Finland pioneers cordless wireless local loop activities described later in this book.

3.5.1   Germany

Until 1985 customers looking for cordless telephones in Germany could only buy imported, non-approved, equipment and connect it, illegally, to their access lines. As these phones caused problems, for example with respect to frequency band, privacy, security and for other reasons, and as the demand for cordless telephony was steadily growing, Deutsche Bundespost, being the sole (legal) supplier of telephone sets at that time, asked for cordless telephones that were designed according to CEPT, and CCITT, recommendations. These phones were to provide 40 channels in the frequency range from 914–915 to 959–960 MHz. Delivery of this CEPT CT1 apparatus started in 1985–86. Based on technical experiences and on user requirements – one of which was the lack of channel capacity on the "Black Friday" of 1987 in the Stock Exchanges – an enhanced version (CT1+) was developed providing for 80 channels and using the frequency bands 885–887/930–932 MHz. The features of these cordless phones comprised memory functions for abbreviated dialling, last number redial facility and (with some) LCD display. Type-approved cordless phones were excepted from the Deutsche Bundespost monopoly from the very beginning, i.e. they were available from Bundespost or from any retailer.

Deutsche Bundespost became aware of the telepoint licensing discussions occurring in the UK in 1988. This led to their decision in early 1989 to carry out telepoint field trials in Germany – but the tricky question was: "Which technology could best meet the requirements of a telepoint application?" Deutsche Bundespost Telekom was of the opinion that neither CT2 nor CT1+ technologies was so far ahead of the other that they should focus on one of them only, and thus they decided to run field trials with both systems. Thus, a CT2 system was implemented in Munich, a CT1+ system was deployed in Dortmund/Meschede, and both systems (CT2 and CT1+) operated side by side in Münster. This gave the opportunity to test the performance of two products of both techniques. Each of the four trial systems included around 150–200 telepoint channels and up to 2000 handsets.

The aim of the field trials, apart from testing the reliability of the technical system, was to establish customer reaction to the kind of service offered. Deutsche Bundespost Telekom implemented the telepoint service as it stood at the time, with no plans for enhancements or add-on features like incoming calls or even handover facilities. Although handsets were, of course, capable of receiving calls when used on home basestations or wireless PABXs, this capability was not envisaged as a feature of the public telepoint service. It was envisaged that this would allow a reasonable market size to be secured, for the level of equipment price and service fees (subscription, call charges). To draw the service features of telepoint close to those of the cellular telephone service would have required a considerable increase of costs.

Thus it was that the German telepoint trial service, branded "Birdie", was launched in late 1990. The trial ran until early 1993, in both Munich and Münster, with a total of around 4500 customers.

By 1992, of course, three of the four UK telepoint operators had closed down their operations and the new low-cost cellular, PCN, concept had been proposed. By the end of the Birdie trial, Mercury One2one in the UK had actually launched their PCN service, complete with the free-call offering. It was within the context of this markedly changed commercial environment that, at the end of the Birdie trial, a decision was made not to proceed to implement a full commercial service – essentially it was believed that in Germany at that time a telepoint service could not be operated on a profitable basis. The impact of this decision in such a key European telecommunications market, on top of the experience of telepoint in the UK, was far-reaching in other European countries.

3.5.2   Finland

The second country in the world to implement a telepoint network, after the UK, was in fact Finland, where a trial CT2 system based upon proprietary Shaye equipment was established in September 1989 by the Helsinki Telephone Company (HTC) in collaboration with Nokia Mobile Phones [2].

The regulatory situation in Finland has for many years been much more liberalised than most other European countries. This encouraged, at a very early stage, both competition and the usage of radio technology for telephone access. Liberalisation of the market has led to a situation where in the late 1980s there existed over 50 private regional operating companies providing telecommunications infrastructure within mainly the urban areas of the country. In addition, the Finnish PTT also provides the national long-distance network, international connection, some local operation and the NMT mobile cellular radio telephone network. Unlike some countries, regulatory restrictions do not preclude the regional companies from using new technologies for connection of customers to the network. This framework encouraged early experimentation with the telepoint concept, and indeed with CT2 PABX applications.

The initial trial telepoint network established by the Helsinki Telephone Company used proprietary CT2 technology. The project began in 1987 with the intention of starting a pilot telepoint service and also a wireless PABX trial. Having proven the telepoint concept, it was anticipated that during 1990 larger telepoint networks would be established using CAI-compatible products. In practice, however, reflecting the loss of confidence and market weakness of CT2 telepoint in other European countries, progress in Finland faltered, as in Germany.

Despite, or perhaps because of, these experiences with telepoint, continued experimentation with cordless public access technologies has continued in Finland throughout the early 1990s, based upon the maturing DECT technology. Wireless local loop trials based on DECT have been particularly successful and have led to the introduction of commercial service in the Helsinki area, as described in Chapter 5. As with other Scandinavian countries, interest has also grown in dual-mode cellular/cordless applications, and these may also yet find commercial application in this country.

3.6 Summary

Telepoint, based on CT2 CAI technology, may be said to have enjoyed, at best, mixed success in Europe. The advent of personal communications networks and accompanying dramatic price reductions in cellular telephony in the UK, along with other factors, contributed to the early closure of the window of opportunity for telepoint in that country. In France and the Netherlands, telepoint service continues to fulfil a niche in the portfolio of mobile services offered by the traditional PTT Telecom suppliers. Even in these countries, however, market growth has not met initial expectations.

Across Europe, the enormous growth of cellular telephony has made it hard to differentiate and market the concept of a reduced capability service such as telepoint, even at reduced prices. It remains to be seen in the future what will be the outcome of operators' experiments currently under way into combined dual-mode, cordless and cellular, services.

Acknowledgements

The UK contribution to this chapter was provided by Ed Candy, formerly Technical Director of Rabbit. Annette Ottolini kindly provided the description of the Netherlands' Greenpoint network, whilst Marc Pauwels provided the section on CITEL in Belgium. The remaining material has been compiled by the editor from various sources. In this respect, the help of Alain Charbonnier, Catherine Gargouïl and Jaqueline Madier in providing information for and reviewing the description of France Télécom's Bi-Bop network and activities is particularly acknowledged.

References

[1]　"Telepoint Networks: Europe", Advanced Cordless Communications, Vol. 3, Issue 6, Telecomeuropa, June 1996
[2]　"Cordless Telecommunications in Europe", WHW Tuttlebee (ed.), Springer-Verlag, 1990
[3]　"The DECT Operators' Group", Advanced Cordless Communications, Vol. 3, Issue 2, Telecomeuropa, February 1996 (see also Chapter 5 in this volume on Cordless Local Loop)
[4]　Proceedings of IDATE Conference, Montpellier, France, November 1986, documented in Bulletin No. 25, Les Services de Futures, IDATE
[5]　"Telepoint in France", M Brussol, D Bolus, H Thomas and P Cazein, Commutation et Transmission (Spécial), pp. 21–30, 1993
[6]　"The Common Air Interface (CAI): The Experience Gained by France Télécom on Three Application Fields of This European Wireless Network Access Standard", O Blondeau, presented at the International Symposium on Subscriber Loops and Services, ISSLS, Vancouver, Canada, 1993
[7]　"Telepoint and Cellular: Bi-Bop en France", J Madier (France Télécom), presentation at the IIR Cordless Communications Conference, London, January 1996
[8]　"Existing and Short Term Solutions for a Public and Private Telephone Service in Fast Developing Urban Areas", O Blondeau, presented at TELKOM 95, Johannesburg, South Africa, 1995
[9]　"From Bi-Bop to Cordless Access", J Le Bastard, presented at the IIR Cordless Communications Conference, London, November 1995
[10]　"Greenpoint to Offer Data Services", Advanced Cordless Communications, Vol. 1, Issue 8, Telecomeuropa, October 1994

Notes

1　Proprietary handset products on offer at the end of 1989, from Shaye and Ferranti, were described in [2].
2　To set this figure in context, by late 1994 there were less than 100 000 cordless telephones (both analogue and digital) in use in the Netherlands [10].

4 Telepoint in Asia

Neil Montefiore

Telepoint in Asia developed in stark contrast to Europe – whilst networks languished with slow growth or even closure in Europe, in Asia in the early 1990s new networks opened and flourished. In the mid-1990s, new competitive pressures from the adjacent cellular market began to squeeze telepoint in Hong Kong, with severe erosion of the customer base during 1995 leading to a first announcement of network closures in March 1996.

In this chapter we describe the early success and subsequent demise of telepoint in Hong Kong, we survey the telepoint scene in other Asian countries and we briefly consider future developments in the region.

4.1 Cordless Evolution and Revolution – Europe Versus Asia

4.1.1 The Cordless Evolution in Europe

Although the first application of cordless telephony (CT1) only offered mobility in the home, it generated enough interest for a confusing number of standards to appear. Europe, and particularly the UK, sought consolidation and enhanced capabilities. With the aim of providing additional capacity and a higher quality of service to the residential market, new digital standards were introduced, in particular, initially, CT2. It also appeared that a public or outdoor application was a viable proposition, and from this developed the telepoint concept.

Although early telepoint took cordless telephony into the street, early implementations still did not have the features of true mobility. The major drawback was undoubtedly their inability to receive calls, but there were others, including the fact that it was impossible to have a continuous conversation on the move. The future looked bleak. The disadvantages to mobility, combined with a market positioning that was totally wrong for the product, resulted in a very poor start for CT2 in its fledgling European market. It seemed that CT2 had nowhere to go – it took Asia to put it back on the map.

4.1.2 The Cordless Revolution in Asia

With the exception of France – where the Parisian street culture proved ideal for France Télécom's Bi-Bop service and 30 000 subscribers signed up in the first six months – Europe has largely been a telepoint graveyard. This contrasts starkly with Asia's digital cordless services, where, for example, the market grew 50% in a single year, from 196 000 subscribers at the end of 1993 to 290 000 at the end of 1994.

The operators in Asia were, of course, fortunate in being able to learn from the mistakes that the UK made when it launched telepoint services in 1989. One of the main reasons the service failed was poor product positioning, and Asia was quick to recognise this weakness. In the UK, telepoint was at first positioned too close to cellular; this was compounded by targeting car owners as a primary market and attempting to provide coverage by installing basestations in motorway service areas, filling stations and roadside cafes. The service was also positioned as an alternative to public payphones but, because the availability and reliability of these had greatly improved in the UK, telepoint found itself squeezed between the payphone and the success of mobile cellular.

Clear product positioning was undoubtedly one of the major factors in the Asian telepoint market. Being clearly differentiated from cellular from the start, telepoint was targeted at the young, street-wise person on the move and the large number of existing pager users.

However, despite clear price and image differentiation, there was still confusion in the Hong Kong market because cellular telephony was so well established. Telepoint was initially perceived as a poor man's cellular phone, and it became clear that advertising would have to play an educational role, rather than simply one of raising brand awareness. The fact that there were three competitors in the market increased the overall advertising media spend, helped to educate the consumer further and underscored the proposition that telepoint was "a phone box in your pocket".

4.1.3   The Paging Phenomenon – Asia Versus Europe

When examining the early success of cordless communications in Asia, it is important not to overlook the role of paging. The paging market has a direct bearing on the use of telepoint. Pager users enjoy a different style of communication from mobile cellular users. They can be contacted in a controlled manner and can choose how, when and where they respond to messages. Telepoint complements that communication pattern. In addition, paging, and particularly paging plus telepoint, is a cheap alternative in countries where fixed-line networks are scarce or non-existent.

Throughout Asia, paging is continuing to grow dramatically. Between 1993 and 1994 the subscriber base increased by 50% and revenues increased by 60%, representing almost US $5.2 billion. By contrast, the European base grew just 4% and revenues actually fell. Today, having almost doubled its number of subscribers to 10 million in 1994, China constitutes the largest base of pager users and the most significant growth area in the region.

4.1.4   The Telepoint Squeeze in Hong Kong

The Asian marketplace is a dynamic one, where technology and markets move in unexpected directions with surprising speed. An example of this was the impact of the decision of Hong Kong's Office of the Telecommunications Authority (OFTA) during 1995 to issue up to ten new licences for Cordless Access Services (CAS) and Personal Communications Services (PCS). Subsequent developments mirrored those in the UK, when its Government announced its PCN initiative, i.e. an immediate increase in competitive pressures in the existing mobile communications marketplace that squeezed telepoint. One analyst's view of the impact of this is summarised in [1].

All of the existing Hong Kong cellular operators slashed costs or introduced incentive packages in the first three months of 1996, in an effort to build subscriber base pre-emptively, prior to the advent of new PCS competitors. Heavily subsidised, and even free, cellular handsets were distorting the market even prior to these initiatives and encouraged telepoint subscribers to migrate to these higher-functionality, but somewhat higher-subscription, networks during 1995.

The result was a slump in telepoint subscribers from a peak of 170 000 in March 1995 to only 100 000 a year later. In March 1996, Chevalier announced its intention to close its telepoint network, planning to transfer its existing 28 500 subscribers at that time across to an independent cellular operator, Smartone. Pacific Link adopted a similar strategy, having been migrating its telepoint subscribers across to its own analogue cellular network since 1995. The intentions of Hutchison had not been announced at the time of writing, although, with their own cellular network to which to migrate their customer base, it is likely that they will follow suit.

An uncertain issue is how the new CAS services will play out in Hong Kong. Both DECT and PHS technologies are strong contenders in this area, and both of these offer the potential for comprehensive public access services. Whilst such offerings will not be "telepoint", it remains to be seen how services based on these technologies will be positioned and priced and how they will compete or cooperate with the established cellular, and fixed, operators. Certainly they have the opportunity to learn from the experience of CT2 telepoint.

4.2 The Asian Marketplace

This section includes a short description of the digital cordless services in operation around the Asia–Pacific region in 1996. For purposes of brevity and to keep repetition to the minimum, the early experience of CT2 in Hong Kong has been used as the model; many of the early experiences of Hong Kong have been paralleled in these other countries.

Digital cordless services in Asia are in various stages of maturity (Table 4.1). During the early 1990s, Hong Kong with its three operators was the front-runner, with Singapore coming in a somewhat distant second with its CallZone service. Mainland China now has 12 city networks in four of its provinces. Malaysia has Smartfon on both the peninsula and in East Malaysia (Borneo); Thailand had Fonepoint in Bangkok; and Vietnam has CityNet in Ho Chi Minh City. Taiwan is likely to develop as the next major telepoint market in the region; installation of networks in South Korea is also planned. Elsewhere nearby, in Australia, Telstra launched its Talkabout service in Brisbane, which subsequently closed.

Regardless of its success in certain Asian markets, however, it is worth noting that the average revenue per subscriber across the region is only US $250, which

Table 4.1
Telepoint networks in Asia

Country	Operator and name of service	Network supplier	Number of subscribers (as at March 1995)
Hong Kong	Chevalier Telepoint[a] Bo Bo Tung	GPT	52 000
Hong Kong	Hutchison Tien Dey Seen	Motorola	93 000
Hong Kong	Pacific Link[a] TeleLink	Orbitel	25 000
Malaysia	Telekom Malaysia Smartfon	Motorola	18 000
Singapore	Singapore Telecom CallZone	Motorola	44 000
Thailand	Fonepoint[b] Fonepoint	Motorola	14 000
Vietnam	Steamers SMRC CityNet	GPT	10 000

[a]Closure of the Chevalier and Pacific Link networks in Hong Kong was announced in early March 1996 (see section 4.1.4).
[b] Subsequently closed.

is significantly lower than that of cellular – illustrative of similar differences in network investment costs and market positioning.

4.2.1   China

China has seen significant growth in CT2 usage, albeit for different socio-economic reasons than its neighbouring countries. In China, CT2 is a cheap, quick alternative to a fixed-line service. It is predominantly being used as a means of accessing and extending the local telecommunications service where there are currently too few lines – i.e. a means of providing a wireless local loop service.

Whereas in Europe the cost of installing fixed infrastructure has long since been amortised, in Asia it is perfectly viable to compare the costs of installing a fixed service with those of mobile. In such cost and ease-of-installation comparisons, mobile is a clear winner. As a result, some Asian countries – and mainland China is an obvious example – have accepted digital cordless technology as a serious contender as the primary means of communications.

The number of CT2 subscribers in mainland China had risen to approximately 40 000 by the end of 1994. This figure represents network implementation in 12 cities: Chengdu and Chongquing, where Dassault is the network provider; Dalian, Fengxian, Hangzhou, Panyu, Shaoxing, Shenzhen, Shenyang, Wenzhou and Wuzhou, which are supplied by Motorola; and Ningbo, which is supplied by GPT.

With demand for fixed-wireline and cellular services likely to outstrip supply for some considerable time, the climate would appear to be right for CT2 to continue its growth. However, the proximity of Hong Kong, where CT2 has been squeezed by rapid and highly competitive cellular growth and where the role of new cordless access technologies is yet to be defined, may well affect the long-term picture.

4.2.2 Malaysia

The Malaysian Government has issued three CT2 licences – to Telekom Malaysia, to Malaysia Resources Corporation Berhad and to Electronic and Telematique. To date only Smartfon from Telekom Malaysia is operational.

Based on Motorola equipment, Smartfon was launched in 1992 with the operator predicting a subscriber base of 20 000 in two years. However, growth has been far slower than anticipated. The number of subscribers reported at the end of September 1994 was approximately 9000 – under half the projected figure.

Although Smartfon's 4000 basestations currently cover transport hubs, restaurants and shopping malls in Malaysia's main urban centres, the poor initial take-up of the service has caused the operator to rethink its strategy, and it will now concentrate on city-centre business districts.

Telekom Malaysia's monopoly over the country's telecommunications industry has declined over the past four years. This, combined with the fact that the Government granted PCN licences to three operators in 1994, raises the question of whether Malaysia's CT2 market has now stalled.

4.2.3 Singapore

With the privatisation of telecommunications in Singapore not due for completion until 1997, the Government is currently still the sole provider of most mobile telephony. However, Singapore Telecom does not have exclusive operating rights in the digital cordless area, where the regulatory body – the Telecommunications Authority of Singapore – deals with licence applications on a case-by-case basis.

That said, in February 1992, Singapore Telecom became Singapore's first and thus-far only telepoint operator with a service called CallZone. After an extremely positive initial take-up, with 28 000 subscribers signing up in the first year, the market slowed, with users ending 1994 at around 50 000.

Singapore Telecom was the first non-European signatory to the Memorandum of Understanding on the Common Air Interface standard for telepoint. CallZone is based on a Motorola infrastructure and Motorola had a monopoly on handset supply at the outset, although this has since changed, with GPT entering the arena.

4.2.4 Thailand

Thailand's only telepoint service – Fonepoint – met with only limited success. The service was first launched in 1991. It was operated by a consortium called Fonepoint (Thailand), in which Motorola had a stake, and the service underwent a number of relaunches in a bid to gain acceptance.

The service covered the metropolitan area of Bangkok, but subscriber numbers at the end of 1994 were only in the region of 12 000. It is generally believed that lower tariffs and reductions in handset prices in the cellular market, combined with the true mobility cellular offers, caused Fonepoint to stagnate and led to closure.

4.2.5   Taiwan

Currently Taiwan has no telepoint service but, under the first signs of deregulation, the Directorate General of Telecommunications (DGT) signalled its intention to award a number of licences in 1995. As a result, there has been a great deal of interest from potential operators, and several Taiwan consortia submitted applications by the deadline at the end of March 1995. Many interested parties are already deploying trial networks and generating advanced sales of handsets. In addition, in an unusual reversal of roles, Taiwan site owners are paying operators to install basestations – rather than operators renting the sites – for a share of future revenues.

Orders for substantial amounts of network equipment have been made in advance of licences being issued. One such order, announced by GPT in 1994, was hailed as the world's largest for telepoint equipment. Leading Taiwan company Excelview International placed the order under which GPT would supply more than 12 000 basestations and associated network management systems initially to serve the 6 million people living in and around Taipei.

This level of activity suggests that the market will ultimately be very competitive, with two or three operators in a number of regions covering all the major cities. Whatever the outcome, this first attempt at deregulation is certain to be monitored carefully by countries around the region.

4.2.6   Vietnam

Vietnam has also taken up CT2 – although so far only in Ho Chi Minh City – with a system called CityNet. GPT Australia supplied the 200-basestation two-way system to a joint venture company formed by Steamers of Singapore and Ho Chi Minh City Posts and Telecommunications. The network was deployed in 1993, but subscriber growth has been slow. The total number of subscribers is not believed to exceed 10 000 as at mid-1995.

4.2.7   Korea

In Korea, deployment of a CT2 telepoint system supplied by Dassault is anticipated in Seoul during 1996; Dassault will be working in cooperation with a local Korean manufacturer, Samwoo. Korea Telecommunications, the national operator, undertook a successful closed trial, providing one-way coverage, in Seoul during 1995; it is anticipated that initial commercial service will be in the business district of Seoul. One national and ten regional licences were awarded in June 1996.

4.3 Hong Kong – the Pioneer

Hong Kong pioneered the Asian telepoint trend. After just three years, the total number of users in the territory topped the 170 000 mark, making telepoint the fastest-growing service in Hong Kong's telecommunications history. By March 1996, however, this had slumped to 100 000, with users migrating to cellular as prices fell in this adjacent market and as telepoint operators with cellular networks focused on growth of these higher-value businesses.

Table 4.2 provides a historical and marketplace context for telepoint in Hong Kong, summarising the development of cellular telephony within the country. Whilst telepoint was not positioned to compete directly with cellular telephony, clearly these adjacent markets impacted upon each other; indeed, certain conditions in the cellular market initially assisted the growth of telepoint, and latterly eroded it.

Table 4.2
Evolution of mobile telephone services in Hong Kong

Year	Network operator	Event	Network standard	Cumulative number of cellular networks	
				Analogue	Digital
1984	CSL	Introduction of analogue mobile telephone service	NEC (proprietary)	1	0
1987	Hutchison	Second analogue network	AMPS	2	0
1987	CSL	Switch to TACS standard	TACS	2	0
1989	Hutchison	Third analogue network	TACS	3	0
1989	Pacific Link	Fourth analogue network	E-TACS	4	0
1990 to 1992		Analogue network capacity problems. High demand results in customer base reaching and exceeding the network capacity			
1992	Pacific Link	First digital network	D-AMPS	4	1
1993	SmarTone	Second digital network	GSM	4	2
1993	CSL	Third digital network	GSM	4	3
1995		Digital networks experience capacity pressure. SmarTone issues sales quotas to manage excess customer demand for service. CSL approaches capacity ceiling. Pacific Link suffers from shortage of D-AMPS handsets			
1995	Hutchison	Fourth digital network	GSM	4	4
1996		Six PCS licences to be issued	DCS 1800		
1996		Anticipated launch of PCS services		4	up to 10

Source: adapted from "Mobile Communications in Asia and the Pacific 1995", CIT Research.

4.3.1 Telepoint Introduction and Growth

In March 1991, as the analogue mobile phone networks were saturating and prior to the advent of digital cellular radio networks, the Hong Kong Government granted four CT2 telepoint network licences. In 1992, Chevalier Telepoint and Hutchison became the first service providers; Pacific Link began operations a year later. The fourth licensee – Hong Kong Telecom – did not take up its option. By the end of 1994, Hutchison reported approximately 93 000 subscribers, Chevalier 52 000 and Pacific Link 25 000.

To set this early success in context, it must be recognised that in its first year telepoint growth outstripped cellular – *amassing more subscribers than cellular telephony had secured in its first four years*. The fact that the service subsequently matured and diversified was certainly one reason for this initial success. Subsequent cellular growth, however, has been dramatic, with over 700 000 cellular subscribers by 1996; undoubtedly the success of telepoint itself contributed to awareness of personal communications and helped this cellular growth.

There was, however, really no single reason for this early telepoint growth. It was attributable to a complex mix of issues ranging from aspects of geography and demographics to product positioning, price and value-added services. Hong Kong is probably the most receptive city in the world for technological innovation; in 1992, when the first telepoint licences were granted, it was uniquely well placed to develop the service. The demographics enabled telepoint to be aimed squarely at approximately 1 million pager users and 1.3 million street users who are away from their desks 30% of their time. In addition, the territory's small, densely populated urban areas meant that operators could establish rapid coverage in key business and residential districts with minimum infrastructure investment. Also, of course, at the time of launch the price window between fixed phones and cellular was substantial.

4.3.2 Coverage

The early days were not without their problems – the first major hurdle was coverage. With basestation sites often controlled by property companies – some of which were themselves aspiring operators – restrictive practices came into play. For example, Chevalier launched with the predicted 1000 basestations, which it subsequently more than trebled, but Hong Kong's tightly knit business community with its established business practices unquestionably hampered the rollout.

In Hong Kong, blanket coverage is essential and, in a race to get sites, operators often had to take what was available whilst bidding high to obtain prime sites. Both Hutchison and Chevalier provided blanket coverage in main streets, business and commercial centres, bank automated teller machines (ATMs), subway and overground railway stations, major shopping arcades, restaurants, markets, hiking trails, theatres and cinemas.

Restrictive practices were not confined to real estate, however. Exclusive distribution agreements also worked to suppress competition between suppliers. This made it difficult for operators to negotiate supply contracts. And exclusive supplier arrangements restricted the use of the technology – whilst the handset-basestation interface was standard, the basestation–network management systems are proprietary, which limits the operators' choice of basestation supplier.

4.3.3   Marketing – Positioning, Pricing and Programmes

From the outset, correct product positioning was of paramount importance. Hong Kong's telepoint providers recognised that they must not make the same errors as their counterparts in the UK's fledgling market. The message was one of *utility* not *mobility* – it would compete at the fixed network end of the market and not with mobile cellular.

Telepoint is linked closely to the fixed network in price, utility and service, but it is actually more useful than the fixed network. It has everything the fixed network can offer plus a bit more – and with the same handset. It was positioned as a convenience phone or a "pocket phone booth" and it was priced 70% cheaper than cellular.

Lying between the fixed service and cellular, however, the telepoint niche is highly price-sensitive. Hong Kong's fixed network costs are fairly stable, but there must be pressure to reduce price as competition increases and, as cellular capacity increased, the niche was inevitably squeezed.

By 1995 telepoint in Hong Kong had reached consumer product status. Advertising, sponsorship and all other promotional activities targeted the young, active, predominantly Cantonese-speaking community, who respond well, for example, to the use of their favourite pop-star endorsing the product in television advertising. Concert sponsorship, discount shopping and restaurant deals, competitions and loyalty awards are typical of the promotional marketing mix.

Aggressive innovation was the overall name of the marketing game and Chevalier in particular developed a reputation for extending the frontiers of technology. This strategy included delivering the widest choice of the best handsets and making the most feature-rich, value-added network services available to Hong Kong's telepoint customers. This included the introduction of a flat-rate free tariff allowing subscribers to pay a fixed charge for their telepoint service regardless of how many calls they make. This approach removed a psychological purchasing barrier in the consumers' minds – they were no longer frightened by an unknown cost of ownership. It also highlighted Hong Kong's dynamic marketplace; user demographics changed and broadened to encompass females and younger groups.

This also led towards a new marketing trend – that of telepoint as a "people's phone". Success with this as a marketing goal is driven by discovering how people purchase. The less formal the purchase becomes – perhaps by removing contracts and paperwork – the easier it becomes for the consumer to make the decision to buy.

Functionality continued to increase, with all three Hong Kong operators having announced a two-way service by the end of 1994. In keeping with Chevalier's leading-edge image, it was the first company to offer a district-based two-way service offering far more mobility than the competitive services that were tied to a single basestation.

4.3.4   Paging as a Complement to Telepoint

Interestingly, the apparent inability of telepoint to receive calls was not a barrier to its early success. One of the reasons for this was almost certainly the maturity of Hong Kong's enormous pager market, which is quite independent of cellular. Radio paging services and telepoint services are highly complementary.

Hong Kong's pager base is approximately 1.2 million and this market itself has segmented. Paging combined with CT2 included secretarial paging, operator paging and autopaging. Chevalier's Page-Link service provided direct access to incoming calls at the touch of a button, offering two-way functionality at a fraction of the cost of a cellular service. Hutchison's Meet Me service had similar attributes.

CT2 has showed itself to be highly successful when used in conjunction with paging; 85–90% of CT2 owners used them with a pager. It will be interesting to monitor developments in other significant paging markets, for example South Korea with 5.5 million and Taiwan with 1.7 million.

4.3.5   Usage in Multiple Environments

The concept of a handset that would be used in multiple environments – at home with a single-line basestation, at the office with a cordless PABX, and in the street with the telepoint service – has not yet been realised. The reason? Price, price and price.

The domestic consumer will not pay much more for the "added value" of being digital. To penetrate the existing analogue cordless market, manufacturers must take a long-term view and not seek to recover their development costs in initially low-volume production. The prices for a CT2 handset and home base may reach acceptable levels in the near-future if consumer manufacturers enter the market. However, the inability of telepoint to compete against subsidised cellular handsets sends a clear message.

Although businesses may perceive some value in office mobility, they will not pay four to five times the cost of a wired extension for the privilege. At current prices, the market is limited; however, this may change as the reality of achieved sales figures against projections force reviews of manufacturers' margins and entry strategies.

4.3.6   Cordless Access in the PCS Era

From a consumer perspective, Hong Kong's mobile telephony marketplace is already crowded and fiercely competitive; with deregulation, it is becoming even more so. In 1995, the Hong Kong Government announced its intention to award up to ten new licences – four for Cordless Access Services (CAS) and six for Personal Communications Services (PCS). The Office of the Telecommunications Authority (OFTA) defines CAS as a low-mobility service for public access to fixed-line networks. Under PCS, it includes the US PCS, European PCN or any similar service in the 1.7–2.0 GHz band.

Unfortunately, lack of agreement between the British and Chinese members of Hong Kong's Joint Liaison Group on this issue delayed the awards, and as a consequence denied the existing CT2 operators an upgrade path and the opportunity to retain their customers whilst repositioning their cordless services. This certainly hastened CT2's demise in Hong Kong.

The technologies vying for pole position as the basis on which these services will be deployed are cordless and cellular-based: Europe's Digital Enhanced Cordless Telecommunications (DECT); Japan's Personal Handyphone Service (PHS); the Personal Access Communications System (PACS) from the USA; Qualcomm's Code Division Multiple Access (CDMA); and the Digital Communications System DCS1800 – the GSM derivative that has already been deployed in Europe with some success. The cordless technologies are described in the later chapters of this book.

Considering the cordless technologies, the general belief was that DECT's robust technology would almost certainly find application in Hong Kong as an access service to existing GSM cellular technologies. However, both DECT and PHS – which offer increased functionality over CT2 with advanced paging and handover techniques – would require a sophisticated and extensive network structure, the estimated cost per subscriber being in the order of 2–2.5 times the current CT2 cost. The cordless technologies – DECT, PHS and PACS – certainly have the potential to become the delivery platforms for fixed network capability, as an alternative to in-building wiring. Once the infrastructure is installed for such a purpose, clearly they could also support mobile users. Microcellular services based on DECT or PHS could be implemented with infrastructure and functionality to mimic those of cellular. To do this in Hong Kong, however, they would have to deliver 95–98% coverage of the entire 1000 square kilometres of the territory, provide two-way services with handover, and offer roaming with compatible networks in neighbouring countries.

As described earlier, the immediate impact of increased competition had a profound impact upon telepoint in Hong Kong. It remains to be seen at what pace other countries in the region will follow Hong Kong down the same path of deregulation and increased competition. Certainly the likely impact upon telepoint can perhaps now be clearly seen.

The low radio power and high traffic capacity characteristics of CT2-based telepoint make it an ideal reduced-functionality, low-cost access service, and as such it may have a place in the services of tomorrow. On this basis, and as long as it continues to position itself – particularly in price – very closely to the fixed network, and is able to maintain a differentiation from cellular telephony, there should be a space in the market for telepoint; however, these may not be easy conditions to meet in some markets. As new cordless access services are deployed in Hong Kong, so such systems and technology may be adopted elsewhere in Asia.

4.4 Summary

This chapter has described the development and status of telepoint in Asia. The national situation in several Asian countries has been summarised, with particular emphasis upon the Hong Kong market. Key issues that contributed to the early success of telepoint in Asia have been described – notably issues of marketing and synergies with paging – together with the increased competitive pressures from cellular that led to the slump in the Hong Kong market.

Cordless public access based on CT2 telepoint served to pioneer low-cost, low-functionality mobility; indeed, it was this very success that encouraged the issue of new licences and the technologies that increased the competitive pressures it faced in Hong Kong. These new technologies and systems will now compete for role and market position. Whilst the future of the cordless technology on the Asian scene is still fluid, the relative costs and maturities of the various technologies suggest that, in Asia at least, cordless access of some form will continue to play an important role into the twenty-first century.

Reference

[1]   "Cheaper Mobiles Kill Market for Telepoint in Hong Kong", Asia–Pacific Telecoms Analyst, Financial Times, Issue 42, 18 March 1996

5 Cordless in the Local Loop

Margareta Zanichelli

Since its introduction, and until very recently, the telephone network has used wired analogue circuits to connect the user's telephone to the local exchange of the network; this interconnection is referred to as the "local loop". In the 1980s the arrival of cheap analogue cordless telephones enabled the wire to be broken within the home, over the short distance between the handset and the telephone socket on the wall. Whilst previously radio has been used for isolated long-range rural links, it is only within the last decade that the idea of using radio to provide the link from an ordinary telephone subscriber, perhaps in a city, to his or her local exchange – the concept of "wireless local loop" – has emerged, as the costs of radio technology have fallen and wireless local loop has become an economically viable proposition.

In this chapter we examine the role of digital cordless technology for wireless local loop application, a potential role that was recognised in the early specification phase of such technology. We first consider the benefits offered by wireless local loop in general and then consider two specific approaches to radio access – fixed wireless access and wireless access with local mobility. Whilst these are currently separate approaches, such simplistic distinctions between fixed and mobile services will increasingly disappear as we enter the new century. Next, we examine more specifically the role of cordless technologies – mainly CT2 and DECT – from the viewpoints of services, network issues, radio issues and environmental issues. We then describe a number of wireless access trials undertaken in various countries during the early 1990s. Finally we examine the economics of wireless local loop; from this viewpoint, in particular, cordless technologies have great attractions, given the economy-of-scale benefits that arise from use of a technology in very high-volume production for consumer cordless telephones.

5.1 Why Wireless Local Loop?

There are many aspects that make the wireless local loop (or "radio access") concept dissimilar from other local loop access alternatives such as wire or fibre. These features are presented here to give an idea of the many viewpoints that must be considered in answering this question.

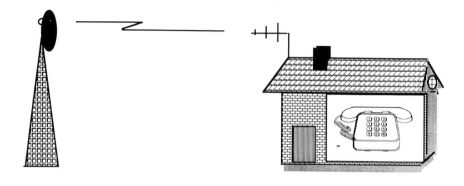

Figure 5.1 Fixed local loop replacement.

Introduction of a new technology in the local loop (Figure 5.1) may be justified for one of two different basic reasons. Firstly, a new technology may substitute old components of the existing network and give an improved cost/performance ratio. Secondly, the technology may enable totally new services and applications to be implemented, offering a network operator a competitive advantage. Both of these arguments are applicable to radio access. For introduction of new services, there may be push from the technology side or pull from the market. In either case, the service and quality requirements of the customers must be fulfilled at a price that the customers can afford and that is profitable to the operator. The decision on the use of radio access is an operator option based on various factors such as cost-effectiveness and speed of deployment.

Possible reasons for an operator wishing to use radio in the local loop include to:

- Provide terminal mobility
- Replace obsolete copper cabling
- Provide access in a competitive environment
- Provide services to new areas
- Enhance capacity of an existing network
- Provide backup for old deteriorating lines
- Facilitate fast deployment

During the 1990s two major factors are coinciding to create major interest in wireless local loop – the (worldwide) liberalisation of telecommunications provision, and the major cost reductions of radio technology, which have been realised largely as a result of the enormous growth in the personal communications market over the last decade. The fall of Communism and the resultant need to provide improved telecommunications services rapidly in Eastern Europe has also been an important factor accelerating such interest. The fact that a radio-based solution also potentially allows mobility to be supported is a further important factor contributing to this interest.

5.2   Approaches to Radio Access

Two quite obvious approaches exist to radio access – simple fixed wireless local loop, and wireless access with ubiquitous coverage that supports mobility. These alternatives represent the two extremes of radio access applications. For the user the main differences are in the level of mobility that the systems provide, the set of services available and the quality of service that an operator can guarantee. For an operator these two types of service are quite different to run and they both require specialist expertise. This chapter aims to explain the potential of available cordless technologies for such roles or, if deemed more appropriate by the operator, for a solution between these extremes. Just as cordless and cellular technologies are beginning to converge, so also are fixed and mobile services; whilst telepoint *per se* may not have seen great success in Europe, today's trials seem to indicate considerable interest in mobility as a service associated with wireless local loop.

5.2.1   Fixed Wireless Access

In a wireless local loop (WLL) application, the service provided to the user is essentially equivalent to that of the copper lines; however, WLL can also be used to provide new services to some customers. For instance, the wired access network in an area may not have the capacity or capability to support ISDN services, and there may be too few customers interested in these services to justify laying new cables (at least for the moment); a WLL system could be used to provide such services until the wired network is upgraded. Alternatively, in a liberalised situation, a newly licensed operator may wish to use WLL to provide a competing network to the established operator's copper network.

The remote radio terminal is in most cases owned by the network operator, who is also in charge of operation and maintenance. It is necessary to provide, at the operator's expense, a terminal radio part located at the user's premises; the latter remains the property of, and will be maintained by, the operator.
Depending on range, such basestations may be externally mounted on the outside of the house or may comprise indoor equipment. The rest of the installation, from a standard PSTN/ISDN interface socket, will in most countries be the responsibility of the user. An example basestation and subscriber terminal unit are shown in Figure 5.2.

5.2.2   Wireless Access With Mobility

In this second scenario, users may be offered the attractive value-added service of terminal mobility – i.e. instead of a fixed wired telephone they can use a portable cordless telephone throughout the residential area, and preferably also in the local area nearby (Figure 5.3). With the introduction of CTM (Cordless Terminal Mobility) – described in the next chapter – the user will be able to roam into other cordless clusters as well.

The user will normally own the terminal and will be responsible for its maintenance. Requirements, in this case, are more similar to those of the mobile network operator. This approach can also be evolutionary. Users may initially

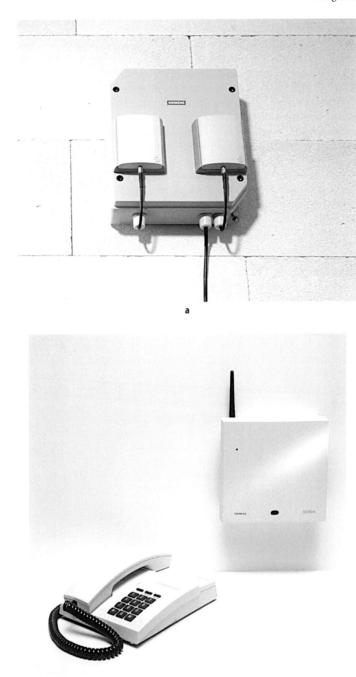

Figure 5.2 WLL basestation and subscriber terminal – the Siemens DECT*link* product. (Photographs courtesy of Siemens.)

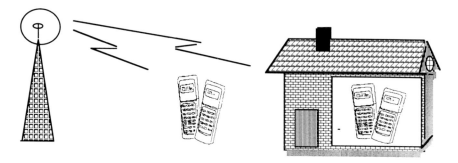

Figure 5.3 Neighbourhood access.

have very limited mobility and then be upgraded. In the process, service features will have to be modified as well as the tariff structures.

5.3 Cordless Technologies for the Local Loop

In the early standardisation of cordless technologies, the requirements of the wireless local loop application were included as important constraints upon the specifications. It is therefore not surprising that such technologies are in fact very well suited to the shorter-range (up to several kilometres) wireless local loop applications. In this section we examine some of these constraints and see how cordless technologies match these needs. Whilst the emphasis of the discussion is upon DECT[1] and CT2, since these two technologies are the most mature, with products already implemented, many of the arguments will also apply to the other cordless technologies, such as PHS, PACS and PWT, which have been designed in Japan and North America with similar constraints in mind. Spread-spectrum CDMA technologies have also been proposed for wireless local loop applications and equipment developments are under way. However, whilst in the longer term CDMA may offer some additional benefits, at present cordless technologies are further advanced, in terms of both standardisation and product development. Another European proprietary technology, developed by Nortel for the new UK operator Ionica, has also begun to be deployed [1]; however, this technology has not been developed for cordless telephony as such and is not covered in this chapter.

5.3.1 Service Aspects

Speech Telephony

It is an essential requirement that speech telephony can be provided with a high quality to the local loop user. The speech quality provided by DECT and CT2 is comparable to the wireline network quality owing to the fact that $32\,\text{kb s}^{-1}$ ADPCM speech codecs are used. Speech delay can also be an important consideration; both systems have short delays – of order 2 ms for CT2 and 10 ms for DECT, both one-way delay figures.

It is also important that a wireless local loop service be indistinguishable, from the user's viewpoint, from the wired service. An important aspect of this is dialtone access. When the user picks up the telephone, he or she expects to hear a dialtone from the local exchange immediately. Thus it is important that the WLL technology can deliver low perceived dialtone access delay.

Voice-Band Data and Fax

Data at 14.4 kb s^{-1} and fax at 9.6 kb s^{-1} are very common nowadays in developed regions. These requirements should therefore be met in the local loop. DECT and CT2 both pass voice-band data and fax up to 4.8 kb s^{-1} (9.6 kb s^{-1} is achievable in some deployments), which may be insufficient in the local loop even today. The complete operation of all voice-band modem equipment can, however, be assured using DECT either by providing a 64 kb s^{-1} (2 × 32 kb s^{-1}) DECT channel or by using appropriate terminal adapters included in the single-channel (1 × 32 kb s^{-1}) DECT equipment.

Higher-Rate Data Services

The high data transmission capability in DECT is an important advantage also in the local loop, where the requirement for data services is increasing. In DECT data rates up to 552 kb s^{-1} may be supported, using multiple time slots, whilst in CT2, only 32 kb s^{-1} can be provided from a single radio unit.

ISDN Service

The provision of ISDN will become increasingly important. In a few years' time, about 10% of residential users in Europe are expected to require basic rate (BRA) ISDN 144 kb s^{-1} service. The expected penetration will, however, vary from country to country. In this respect DECT has a great advantage, having been designed to be fully BRA ISDN transparent.

Broad-Band Services

The need for broad-band services is expected to increase also for residential users. By the end of the century, 5% of residential users in Europe are expected to require some kind of broad-band services, such as video on demand, telegames, etc. Neither DECT nor CT2, however, is capable of providing broad-band services – for such applications, future systems will be needed with greater spectrum allocations.

Mobility

Mobility can be one very strong reason for using wireless access instead of wireline in the local loop. Mobility should at least be provided within the home and preferably also be extended to the local area nearby where the user is expected to move frequently. Depending on which part of the local loop the cordless system is designed to cover, different grades of mobility can be provided. An important economic trade-off exists here – to provide coverage to a portable terminal inside the user's house directly from the operator's wireless basestation will require increased infrastructure density, and hence higher costs, compared to providing simply a link from the operator's wireless basestation to a fixed wireless access unit elevated on the exterior of the subscriber's premises. Provision of mobility as a service has its inherent associated costs.

Security

Security is here only discussed in terms of logical security, which implies that a set of information processing functions is added to the system's original operation; physical security is not considered. In the cordless local loop application, the network must be able to authenticate the user terminal, so that only the authorised subscriber can access his or her exchange line. In the case of mobile users, it is also necessary that the cordless terminal can authenticate the network or the basestation. Further, an encryption mode should be available, at least as an option, for the users.

Both DECT and CT2 have a standardised authentication algorithm. DECT also has a standardised encryption algorithm; in CT2 encryption is implementation dependent.

5.3.2   Network Aspects

Network Integration

The fixed network interface is not standardised for DECT or CT2 but is dependent on implementation. Both can support digital $2\,\mathrm{Mb\,s}^{-1}$ V5.x or analogue. The cordless system may be connected to different parts of the existing local loop network, as is illustrated in Figure 5.4, showing alternative connection points for the cordless system.

Generally, only the drop segment is substituted by wireless in the case of mobile terminals, providing a basestation to terminal range of perhaps 50–100 m. In the case of fixed subscriber units, the cordless control unit is preferably placed at the local exchange, since a much longer range can be achieved in this case, perhaps 1–5 km. Where to connect the cordless system is dependent upon the area in which the system is to be installed, the required grade of mobility, the required range and capacity.

Operation and Maintenance

Remote operation and maintenance can be obtained, but today only proprietary solutions are available, as no O&M system is specified for either DECT or CT2. In general, however, all the manufacturers do provide some kind of centralised and

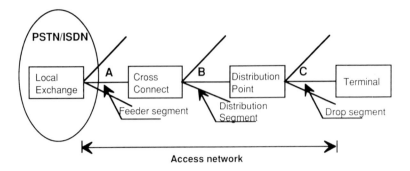

Figure 5.4 Components and segments of the standard wireline access.

remote operation and maintenance. Clearly this is an important requirement, as an operator needs to be able rapidly to locate, identify and diagnose any faults on the access network, preferably prior to the subscriber noticing them.

5.3.3   Radio Aspects

Range and Capacity

The range is highly dependent on the environment as well as on the type of application, as noted above. In a cordless local loop system with fixed subscriber units, much longer range can be achieved than in the case of a cordless local loop system with mobile user terminals. To extend the range, repeaters may be used; DECT repeaters are currently specified within ETSI RES 3, and products are under development by several manufacturers. The use of repeaters in the access network will decrease the system capacity, which has to be considered when the system is planned. Nonetheless, the capacities for which DECT has been designed mean that this need not be a significant factor in most situations.

Time Dispersion and Equalisation

In longer-range or highly cluttered deployments, multiple reflections of the radio waves can result in multipath echoes – time dispersion of the received signal (see Chapter 12). This can impair reception of the transmission. In the cellular radio environment, such a situation is the norm, and, for TDMA systems, this is overcome by the use of equalisation, a method of constructively recombining the time-spread signals using digital signal processing. Despite a number of theoretical papers suggesting that significant time dispersion should occur, in practice field trials have failed to identify this as a major problem in most cases; indeed, the implementation of diversity reception in some of the equipment used for these trials will have mitigated such effects. Some manufacturers have considered the incorporation of simple equalisation in DECT equipments designed for WLL applications, to provide enhanced performance in those situations where time dispersion may be significant. This can in fact be done without the need for any changes to the DECT specification, as it simply relates to the receiver processing.

Radio Planning and Installation

DECT and CT2 do not require any frequency planning, owing to the dynamic channel allocation techniques employed. This feature makes it also easy to modify existing systems if, for example, more capacity is needed or if there are environmental changes. This is a great advantage, especially in newly built areas, as these most likely will change during the coming years: trees will grow high, new houses may be built that may lead to increasing capacity requirements, and so on.

Line of sight is not a requirement for planning of the DECT and CT2 radio path, which makes it very flexible. Typically many base units, and/or repeaters, generally are required to cover an area; these have to be installed on already existing lampposts, on other street furniture or on the buildings in the locality. Given this, the fact that the basestation units for DECT and CT2 can be small, and therefore easy to install, is thus an important advantage. The external appearance can be easily made satisfactory, which makes cordless technology attractive also from an aesthetic environmental viewpoint. Permission from the city architect to do the installations may, however, be needed in some countries.

Most of the cordless systems available on the market today have not yet been adapted for outdoor use. Many manufacturers are, however, currently developing cordless local loop equipment adapted to the harsher outdoor environment.

5.4   Cordless Local Loop Trials

Field trials with different cordless systems have been performed in many countries around the world in the early 1990s. Fixed local loop replacement (Figure 5.1) as well as neighbourhood access (Figure 5.3) have been tested. This section gives an overview of some of these trials and early commercial service offerings. Commercial DECT WLL systems are also being implemented in Asia and South America.

5.4.1   Norway

The Norwegian operator Telenor has established a DECT trial network in the small community of Førde [2]. Førde is located in a mountainous area in western Norway and has 9000 inhabitants. The main reason for choosing Førde for the field trial was that residential, down-town, business and industrial areas were all concentrated within a small local area, giving the possibility of covering the entire area with a reasonable number of basestations and thereby allowing roaming between different types of environments. Seen from the market point of view, this trial gives a special opportunity for studying the customers' reaction to this new service. Seen from a technical point of view, the trial, covering both indoor and outdoor locations, as well as different types of environments, gives valuable information on how well the DECT technology can operate in such a multi-operating environment.

The DECT network went into operation in mid-1994, and trial customers were connected to the network in December 1994. Market tests were conducted throughout 1995. The DECT trial system used in Førde was delivered by Ericsson and consisted of 160 radio basestations controlled by a radio controller. Some 240 pocket-sized telephone handsets are connected to the system.

The trial area comprised both detached and terraced housing, mainly of wooden construction with a density of some 800 dwellings per square kilometre, in hilly terrain with some vegetation. The average distance between a basestation and a house was some 80 m. With this scenario, the outdoor basestations gave around 95% indoor coverage to cordless handsets. Thus the system supported not only simple fixed wireless access, but also wireless access with mobility. An increased traffic level was found with personal handsets. A comparative cost analysis indicated that cordless access could reduce both the installation investment cost and also the operations and maintenance costs, since access to customer premises is not required for service installation. A potential for increased profits based on wireless access with local mobility is foreseen by Telenor, who operated the trial [2].

5.4.2   Finland

The Finnish Helsinki Telephone Company (HTC) began a commercial DECT service, called Porvoon Linja (Porvoo Line), in the city of Porvoo, 50 km east of Helsinki, in 1994. This was mainly a mobile DECT service intended for residential, public and business use. The entire city centre was covered by the DECT system as well as some business campus areas outside the central city area. A traditional copper network was built to interconnect the basestations.

The DECT system was delivered by Ericsson and has been in commercial operation since the beginning of 1994. The Porvoo application consists of an AXE switching system, a DECT radio control switch, some 80 basestations and handsets. Again, as with the Norwegian trial, the provision of local mobility alongside simple telephony provision has proved to be important. Prior to the launch of Porvoon Linja, Telecom Finland had been the only telecommunications operator in the town; the new service offered local mobility as an additional service but for only a marginal increase in cost compared to the wired telephone offered by the established monopoly operator.

The positive results of the Porvoon Linja encouraged HTC to launch a similar service called CityPhone in Helsinki in November 1994. The service offers limited mobility in the home and on the streets in the local neighbourhood – in the case of Helsinki all the main city-centre streets, a region of area two square kilometres. The customer sees a conventional telephone connection (POTS) with standard services at his or her disposal; in addition he or she has local mobility limited to the town area. The low infrastructure cost allows HTC to offer call charges much closer to those of a fixed phone than those of cellular. In some ways the service can be considered to be similar to the PHS service in Japan, but with pricing and positioning closer to wireline service. The CityPhone service offers customers transparent access to all the service features of the ordinary wired network, such as call forwarding, call barring, voice-mail, etc., and the ability to stay within the existing numbering schemes. Centrex capability is also available.

The experience of HTC has been that digital cordless technology has been an effective means of rapidly penetrating a competing incumbent operator's market. The speed of network rollout has been an especially notable factor, arising from the small size and quick installation associated with DECT basestations. Radio coverage has been good and cost-effective. The added benefit of mobility, the ease with which an additional line for data services can be installed and the ability to retain ordinary numbering schemes have all been important factors influencing customer acceptance.

HTC [3] have concluded very positively that:

> DECT is a very viable, powerful and flexible platform for offering customer services in the competitive local loop market. DECT will undoubtedly enrich customer choice and enhance the dynamism of the newly liberalised telecommunications market and represents a common on-going technical and economic opportunity well into the future. In addition DECT has interfaces to GSM/DCS1800, which makes it future proof and a potential candidate for the PCS concept and a means to tap into the mass market with local mobility too. Altogether from the investments and services viewpoint DECT is an attractive option as one of the wireless mainstreams for the established network operators.

Perhaps not surprisingly HTC are continuing to roll out the CityPhone service in new areas, with service launched in the town of Järvenpää in November 1995.

5.4.3   Sweden

The Swedish operator Telia carried out a local loop field trial during 1994 with both DECT and CT2 equipment. The trial was carried out in a neighbourhood in Halmstad, a city located in the southern part of Sweden. The purpose of this trial was to evaluate the technologies and to estimate the production cost of a neighbourhood local loop service. The CT2 trial focused on the fixed local loop service, whilst both fixed and mobile services were evaluated in the DECT trial.

The CT2 equipment used was delivered by Dassault and the DECT equipment by Ericsson. The number of CT2 users was very small, only about 10, whilst there were about 30 DECT users. To cover the pilot area of one square kilometre, 15 DECT base units were needed. The trial gave Telia important experience of planning and operation of cordless local loop networks. Telia also experienced that the customers were mainly interested in the mobile service.

In 1995 Telia started another DECT field trial in Lund, a small university town also in the south of Sweden. In this trial the customers were offered DECT-based mobility in public and business environments. This is a market trial with a focus on understanding coverage needs, acceptance levels and offered traffic. The main purpose is to evaluate the profitability of a DECT service in the addressed environments.

The DECT system in Lund, which is delivered by Ericsson, covers the centre of the town, parts of a business campus and parts of the university. The number of DECT base units used totals 60, of which 20 are placed in the city centre and 40 in the business and university areas. The number of customers in the trial is in total 270, of whom 260 are business customers and 10 are private customers. The trial has shown that there is a very low public usage of the system in the city centre but that the usage is very high in the business campus area. The private users do not have any coverage at home, which they see as a major drawback. The business customers are quite satisfied with the system, but they would like to have functional integration with the company PABX, which is missing in the initial trial.

During 1996 Telia is planning to extend the trial in Lund with tests of DECT/ GSM services with dual-mode terminals. There are also plans to phase in IN-based functionality in the trial.

5.4.4 Denmark

Plans for a cordless local loop trial in Denmark were begun in 1993 and involved TeleDanmark, Ericsson and Aalborg University Centre for Personal Communications [4]. The trial was located in the region of Aalborg University, where there was a large selection of both private and business customers within the expected range of more than 1 km.

The selection of trial participants was made on the basis of received signal strength measurements, made from mobile measurements on roads in the area, and a maximum path loss, determined from available antennas, of 120 dB. Other criteria for selection of trial participants included the following:

- Participants did not use fax, modem or public alarm systems (out-of-band signalling).
- They did not have outstanding telephone bills.
- Their monthly bill was not too low (i.e. they were reasonably heavy telephone users).

Business users were not initially included until the data and fax capability had been verified. Of 70 eligible triallists, 65 responded positively to the opportunity to participate.

The equipment used was made by Ericsson, with analogue interconnection used, as $2 \, \text{Mb s}^{-1}$ compatible equipment was not initially available. The radio basestation antenna arrangement comprised three sectors each with 15 dBi directional antennas, supplemented by a 10 dBi omnidirectional antenna providing redundancy and capacity. Subscriber installations employed either 8 dBi directional or 2 dBi omnidirectional antennas.

A number of technical and market lessons have emerged from the trial. Firstly, it was found that mobile signal strength measurements provided a good basis for coverage estimations. Installation times of some two to three hours were sufficient – in the future this could be reduced. Fax transmissions at data rates of $4800 \, \text{b s}^{-1}$ were comparable in speed and quality to the wired network; at $7200 \, \text{b s}^{-1}$ the quality was found to be slightly degraded. Data transmissions at 4800 and $7200 \, \text{b s}^{-1}$ had nearly identical quality as on the wired network, whilst at $9600 \, \text{b s}^{-1}$ quality was degraded; thus users were advised to force the modem to operate at 4800 or $7200 \, \text{b s}^{-1}$. Despite these apparent restrictions, a market survey conducted in October 1995 indicated a high degree of customer satisfaction.

5.4.5 France

In France the Compagnie Général des Eaux (CGE) has been running a DECT field trial in Saint-Maur-des-Fossés, a small town of 90 000 inhabitants, located in the suburbs of Paris. The trial offers a mobile cordless service to the residential users living in the area. The technical objective of the trial was to test the DECT technology for indoor and outdoor use, whilst the marketing objective was to test the usage by people at home of a cordless telephone, which is a first step towards the personal telephone.

The trial started at the end of 1994 with a few users and the goal is to have about 400 users by the end of the trial. The DECT system is delivered by Alcatel. The network is composed of a standard PBX connected to the public network and to a number of concentrators. About 450 basestation units were needed to offer a complete indoor service to handsets within the pilot zone, which is about two square kilometres.

During 1995 the trial was further extended in scope when CGE were given permission by the French regulator to interconnect the DECT network to the digital cellular GSM network of SFR (of which CGE is a major shareholder). This has primarily allowed technical aspects of such interworking to be addressed, paving the way for an integrated DECT/GSM service offering to be evaluated.

5.4.6   Hungary

The Hungarian trials were undertaken during 1993–94 and were some of the earliest DECT wireless access experiments to be undertaken within a public forum; results from the trial were presented to ETSI RES 3 in February 1994 and to a wider audience in April that year at an open demonstration in Budapest. Their purpose was to investigate the technical feasibility of DECT for fixed local loop, telepoint and neighbourhood access applications, to demonstrate the public access potential to Eastern European operators, and to validate the economic viability of DECT for such applications. Participants in the trial consortium included Scientific Generics, Siemens, Cordless Technologies A/S, Hungarian Telecommunications Company, Swedish Telecom (now Telia), British Telecom, DBP Telekom (now Deutsche Telekom) and Quotient Communications.

The trials used first-generation Siemens Gigaset DECT equipment employing diversity in the basestation and a data-logging system to enable detailed monitoring of the link performance. A variety of antennas were used to demonstrate performance over various ranges. Preliminary trials undertaken in the UK using directional antennas demonstrated ranges from 1 km to in excess of 3 km for the fixed local loop application; the demonstration in Budapest included a 0.5 km link across the Danube river. Ranges of up to 0.3–0.4 km were demonstrated using standard omnidirectional antennas in line-of-site scenarios. Perhaps one of the most significant aspects of this trial at the time was the absence of any significant problems associated with delay spread or intersymbol interference, which, until this time, had been anticipated. Full details of the trials results were published widely and are available from the European Commission [5].

5.4.7   Japan

In Japan a number of field trials with the Japanese Personal Handyphone System (PHS) have been carried out at various locations beginning in 1993. The objective of the trials was to determine how the PHS service would be received by the public and to identify what types of functions could be provided by the PHS technology. They also served to validate and demonstrate interoperability between equipment from a variety of vendors.

The field trials consisted of two main types – service verification tests and interoperability tests. The service verification tests were conducted by installing the test equipment in certain specified areas and by selecting local residents to participate in the tests as monitors using the PHS. The interoperability tests were carried out by public demonstrations using equipment from a large number of manufacturers. These trials are described further in Chapter 9 on developments in Japan.

5.5   Economics

A key factor for the operator is the payback period for the investment in any new technology such as a wireless local loop deployment. The economic viability of the system is determined by the revenues from the provided services offset by the total costs, i.e. initial, installation, and operation and maintenance costs.

5.5.1   Initial Costs

The number of basestation sites needed is a very important factor when considering the initial costs, as early European telepoint operators have found. Each basestation needs powering, some kind of cabinet for the equipment and maybe a mast for the antenna. The amount of basestation equipment needed does not say very much about the initial costs of the system, because the price of basestation equipment can vary significantly. One major advantage associated with radio access systems compared to cable-based systems is that it is not necessary to invest in an individual subscriber before he or she is connected to the network and is starting to use services.

5.5.2   Installation Costs

Usually the installation costs of the basestation can be divided between many subscribers, and thus these costs are not a major concern. However, the powering of the basestations is one cost issue that should be considered carefully.

The customer terminals of cordless systems with handheld phones do not need any installation, only programming of the user's information into the network database and coding of the phone number into the phone or some kind of identification card. Therefore these systems achieve substantial savings with respect to installation costs compared with fixed systems. On the other hand, the basestation densities required to support mobility are much higher than to support fixed radio access.

In the case of fixed cordless systems, it would be preferable that the antenna is integrated in the terminal and that the terminal is located in an elevated position on the wall of the customer premises or some other place, so that no separate mast is needed for the antenna. A wire can then be run from the terminal to provide a standard telephone socket for the user.

5.5.3   Operation and Maintenance Costs

As explained earlier, it should be possible for all network functions to be administered in the central network operations room to minimise costs, i.e. it is essential that every basestation and customer terminal can be tested and monitored remotely. In the wired network, which does not have a centralised administration system, the operation and maintenance costs can be divided so that operation is 60% and maintenance is 40% of the total O&M costs [6]. A centralised administration system offers significant savings of the administration part of the O&M costs.

Maintenance of the drop line is a major cost factor in a wireline system. A very rough estimation of such factors has been presented [7], according to which the maintenance cost per year is 5% of embedded initial costs for wireline network and 2.5% for a radio system. However, reliable results of the maintenance costs relating to the existing wireless networks are not available.

In a fixed cordless system an important cost issue to be considered is vandalism against parts of the system located outdoors, which could potentially produce higher maintenance costs than might initially be expected. Having said this, this has not been found to be a major problem by telepoint operators, who are faced with similar deployment situations.

5.5.4   Salvage Value

Masts and the basestation premises can probably be used again or shared with other systems. Radio systems can also be used as a first-stage solution, i.e. when the number of subscribers is high enough, the radio solution can be replaced by a wireline network and the radio system moved to another place.

5.5.5   Cost Evolution

In a radio system the radius of a cell can be changed easily as a function of the subscriber density. Therefore it is obvious that the costs of the basestation can be divided among a sufficient number of customers. The most critical factor in both the fixed and mobile radio solutions will be the costs of the individual subscriber terminal.

In the case of the radio systems, the evolution of prices should be considered. The technical solutions used in radio systems are still today quite often proprietary, i.e. manufacturer dependent, have a low production volume and are therefore expensive. Increase in volume of production, standardisation and competition in the latter half of the 1990s will result in lower prices in the future. Wireless local loop systems based on standard cordless technology are thus more promising in this respect than other proprietary solutions. At least the critical parts – subscriber radio terminals and basestations, which have the lowest number of subscribers sharing the costs – should be standardised.

5.5.6   Cost Estimates

The economic issues associated with cordless local loop have been evaluated by many operators as part of their internal business activities and, as such, are not in the public domain. However, economic analysis of the viability of cordless local loop comprised an essential element of the Hungarian validation activities described above, a report on which was made publicly available [5].

Three environments were considered – city centre, urban and suburban. For each of these, three options for local loop interconnect were costed – underground cabling, overhead drop wire and DECT wireless. The costs of basestation interconnections and radio exchanges were excluded from the analysis in order to allow a true "last drop" comparison – i.e. it was assumed that the cost of reaching the DECT basestations is the same as that of reaching the last distribution point in the conventional local network. For more detail of the assumptions underlying the analysis, the reader is referred to the original referenced report. The resulting per-subscriber costs, in "ecu", are shown in Table 5.1.

Table 5.1
Cost-per-subscriber (in "ecu") comparison of DECT and wired access options

	Underground cabling	Overhead drop wire	DECT wireless
City centre	160	Not used	124
Urban	160	70–88 (not often used)	113
Suburban	245–521	105–140	110

Source: from [5].

This analysis clearly indicated the cost competitiveness of cordless access in all three scenarios, when compared with conventional practice – although overhead drop wires can be cheaper in high-density deployments, environmental reasons dictate that they are not often used in such situations. In suburban areas the cost differential can be particularly significant.

Today, even though DECT and CT2 systems are still primarily targeted at indoor applications and as yet limited equipment is available for the wireless local loop application, the fact that many operators are developing such solutions as an alternative to copper is a clear response to such economics. Once volume production of WLL products to these standard technologies is under way, the economics will no doubt speak for themselves.

5.6   Summary

In this chapter we have explored the requirements of wireless local loop and the match of cordless technologies to these needs, from the service, network, radio and environmental viewpoints. The main trials that have occurred to date and their outcomes have been described and the economic issues also reviewed. As radio technology and product costs continue to fall, and as the liberalisation of telecommunications service provision accelerates, we can expect to see wireless local loop based around cordless technologies become a significant factor in local access provision in the latter half of the decade.

References

[1]   "Where There's a WILL", R Dettmer, IEE Review, pp. 145–8, July 1995
[2]   "DECT and Residential Access", J Loevsletten, presented at the DECT '96 Conference, London, January 1996
[3]   "Wireless Local Loop Residential Segment and Citywide DECT", P Vepsäläinen, presented at the DECT '96 Conference, London, January 1996
[4]   "DECT and the Wireless Local Loop: The Aalborg Trial", P Faergeman, presented at the DECT '96 Conference, London, January 1996
[5]   "Validation of DECT Standard in Public Access Applications", Management Summary and Report, Scientific Generics, May 1994 (Report on a study undertaken for the European Commission, from whom copies are available)
[6]   "Access Network Alternatives from RACE 2087/TITAN.EFOC & N'94", Olsen et al., Heidelberg, Germany, June 1994
[7]   "Wireless Access and the Local Telephone Network", G Calhoun, Artech House, Boston, 1992
[8]   "Services, Facilities and Configurations for DECT in the Local Loop", ETR 308, available from ETSI, 1996
[9]   "Radio in the Local Loop", ETR 139, available from ETSI, 1994
[10] "DECT RLL Access Profile", ETS 300 765, available from ETSI, 1996

Note

1   For more detail specifically related to DECT, the reader is referred to references [8–10].

6 Cordless Terminal Mobility

Graham Crisp

The terminology "cordless terminal mobility" (CTM) may be new to some readers, particularly those from outside Europe, although not necessarily the concept. CTM is in essence the exploitation, in combination, of three new telecommunications technologies, to create a low-cost, mass-market, mobile telecommunications offering, complementing the higher-cost cellular telephony offering. As such, CTM is simple, whilst its implications are profound.

In 1992, the European Telecommunications Standards Institute's (ETSI) Strategic Review Committee on Public Networks (SRC4) recommended that a feasibility study be conducted into the support of terminal mobility on fixed public and private networks using cordless (CT2 [1] and DECT [2]) and network access standards combined with mobility management standards. The feasibility study, which looked into providing users with the possibility of roaming with their cordless terminals between home, work and public areas, was conducted between October 1992 and April 1993. The conclusion of the task group responsible for the study was that network support of cordless terminal mobility within and between public and private networks was feasible and that standardisation and study activities should be initiated. The task group also supported the SRC4 view that "there is a market for mobility in and between the different environments".

Following the feasibility study, ETSI created the Cordless Terminal Mobility (CTM) Project and appointed a Project Manager (the author of this chapter). The CTM Project Specification has been refined, a Joint Management Group (JMG), with representatives from the different ETSI and ECMA (the European Association for Standardising Information and Communication Systems) committees, has been formed and standardisation is progressing.

In this chapter, the CTM concept, service, potential market, enabling technologies, functional architecture and the role of CTM in the evolution to future mobile telecommunications are presented.

6.1 The CTM Concept

First-generation (analogue) cordless telephones, used within individual homes and workplaces, have been very successful in providing cordless terminal mobility within the radio coverage area of individual basestations. In 1994, 20 million cordless telephones were sold in the USA; approximately 50% of homes in that market now have a cordless telephone.

Second-generation (digital) cordless technology, with its improved radio range, security, traffic handling capability, etc., offers greater possibilities than simply being a technical improvement on first-generation analogue cordless telephones. Furthermore, the price advantage of first-generation cordless systems is already disappearing as the technology for second-generation cordless systems matures. The improved facilities offered by second-generation cordless systems, together with the possibility of embedding mobility management functions in the network, provides the basis for Cordless Terminal Mobility. CTM will allow personal cordless terminals with a single service registration to be used to communicate via any air-interface-compatible and CTM-capable basestation connected to any CTM-supporting network that has a roaming agreement with the user's home network.

The improved signalling, processing and database handling capabilities provided by integrated services digital networks (ISDNs) [3] with intelligent network (IN) [4] facilities provide an ideal basis for the network-based mobility management functions required to support CTM.

The combination of digital cordless technology and network-based terminal mobility management functions allows the network to associate users' service registrations and directory numbers with their cordless terminals, rather than with the fixed network addresses of their cordless basestations. A business or household can then have a number of cordless terminals, each with a separate service registration, which allows the individual cordless terminal users to originate and receive calls independently of each other. When cordless terminal users leave their home location, they can take their cordless terminal and hence their calls with them. If this is not convenient, they can forward their calls to an answering machine, a voice-mail-box or another person, or they can simply switch their terminal off and be unavailable. When the cordless terminal users are away from their home location, with their terminal enabled, the network-based mobility management enables them to be tracked as they roam between residential, business and public basestations, thus allowing them to continue to originate and receive calls related to their own personal service registration.

CTM is not intended to provide the degree of mobility that is provided by cellular mobile telecommunications networks. Instead, it is conceived that CTM will provide a good-quality and attractive service that could be afforded by a greater percentage of business and private subscribers. In addition, CTM offers the possibility of providing a mobile telecommunications service, suitable for a mass market, without requiring the large amount of radio spectrum that would be required to provide cellular telecommunications to an equivalent market. CTM should therefore be seen as being a basic complementary service, rather than as a competitor, to the enhanced mobile service offered by cellular mobile telecommunication systems.

6.2 The CTM Service

In principle, CTM should be capable of supporting all of the telecommunication services that are available on the supporting ISDNs, subject to the services being appropriate to CTM and feasible via the chosen digital cordless access system. In addition, CTM will allow users of cordless terminals to move between basestations in between calls (roaming), and within the radio coverage of groups of basestations during calls (handover).

6.2.1   CTM Services

European CTM standardisation will be phased. In the first phase, standards to support the following services will be developed:

- Telephony teleservice – incoming and outgoing calls
- Roaming – residential, business and public access
- Handover – intra-cell and intra-cluster (see section 6.5)

In later phases, standards will be added to support the following categories of services:

- Bearer services
- Other teleservices
- Supplementary services
- Handover – intra-exchange(?), other levels(?)

Further details of the European standardisation processes and committee structures are included in Chapter 8.

6.2.2   CTM Service Deployment

With any mobile telecommunication system, the main question from a user point of view is: "How much mobility does the system provide?" In the case of the CTM service, the degree of mobility provided by the system is dependent on two factors, namely the radio coverage provided by the cordless basestations and the distribution, or deployment, of those cordless basestations. These two issues are discussed below.

Cordless Basestation Radio Coverage

The radio coverage provided by a basestation is dependent on the RF transmit power used by both the basestation and the terminals and the radio environment in which they operate. In the case of CT2 and DECT, the RF powers are quite different owing to the different radio frequencies, multiplexing schemes, etc. However, in an open environment, both CT2 and DECT basestations can typically serve terminals within a range of about 200 metres. This range, or cell radius, is considerably reduced in cluttered environments, e.g. office complexes, where 50 metres may be a more reasonable figure to assume. Given these typical figures, then a basestation in an open environment may be expected to serve an area of around 125 000 square metres and in a cluttered environment an area of around 8000 square metres. These figures lead to a requirement for anything between 8 and 125 basestations per square kilometre to provide complete coverage of a given area. Obviously, such figures would lead to a requirement for very large numbers of basestations to provide complete coverage of very large areas. For this reason, CTM is not intended to provide such coverage. Therefore, the key to CTM's success must depend on finding basestation deployment strategies that can offer an attractive service to the end-users but do not result in exorbitant costs for either the CTM users or the CTM service providers.

Deployment of Cordless Basestations

There is no single answer to the question of how basestations should be deployed to provide a CTM service. Indeed, different service providers or network operators may adopt quite different strategies depending upon the resources that they own, their position in the marketplace and their target user population. This section describes some possible deployment strategies. This set of strategies is not intended to be exhaustive and the strategies may, or may not, apply to particular service providers, network operators, territories or user groups. In addition, the deployment strategies are not intended to be mutually exclusive – indeed, combinations of the strategies can be used to complement each other.

CTM access in residential and small businesses

This scenario is concerned with the deployment of individual basestations in residential premises (houses, apartments, etc.) and small business premises (shops, small office units, small industrial units, etc.). In this scenario ISDN-compatible and CTM-capable basestations are installed in the subscribers' premises to serve both the resident and regular users or visitors.

These basestations could be owned by a service provider, the network operator or the subscribers. The users' terminals are registered on the database in the public network. A given terminal's location is updated on the database when the terminal, with its location registration capability enabled, enters the radio coverage area of a basestation.

The location registration process enables cordless terminals to access the public network to make and receive calls whilst the terminal remains within the vicinity of the basestation. The cordless terminal users can roam between different such basestations and originate and receive calls within the coverage of those basestations. Each user's outgoing calls are charged to their individual service registration, and calls made to their individual number are automatically redirected to their terminal.

The big question related to this scenario is: "Should all terminals registered on the system have equal access via all such basestations?" There are a number of possible answers to this question: basestations could limit access to a known set of terminals; basestations could limit the traffic capacity used by visitors (e.g. to one of the two ISDN access B channels); basestations could provide a priority mechanism for a known set of terminals; etc. In addition, the network operator could provide some inducement to encourage subscribers to allow open access via their basestations, e.g. the ISDN access line rental could be reduced according to the quantity of traffic carried.

CTM access in medium and large businesses

This scenario is concerned with the deployment of basestations in medium and large administrative, commercial and industrial premises. In this case, the CTM-compatible basestations may be connected to a private integrated services network exchange (PINX) or directly to the public network, which may provide cordless Centrex or virtual private network (VPN) facilities[1]. In these cases, the enterprise system (PINX, Centrex or VPN facility) will provide internal and external telecommunication facilities for terminals registered within the enterprise system's database. This will allow those registered terminals to roam between the enterprise's basestations and to make and receive calls in a similar manner to those described in the first scenario, but with roaming limited to the coverage of the enterprise system. In addition, the enterprise system may have roaming agreements with public networks and/or other enterprise systems. These roaming agreements will allow internal users to roam outside, and/or external users to roam inside, the enterprise system.

CTM access in semi-public places

This scenario is concerned with the deployment of basestations in semi-public places, e.g. department stores, shopping malls, airports, bus and railway stations, hotels, exhibition and conference centres, etc. In principle, this case is very similar to the previous scenario. The main difference is that, in the case of the enterprise system, the dominant traffic would be generated by internal users and roaming agreements would only be used to provide a convenient service for business visitors' and/or employees' personal use. In the case of semi-public systems, the majority of the traffic may be generated by visitors and, therefore, the majority of the cost of providing and running the system may need to be recovered from the usage charges levied on visiting users.

CTM access in public places

This scenario is concerned with the deployment of basestations in public places, e.g. streets, town squares, parks, etc. These public basestations may be provided as a complement, or an alternative, to the residential, enterprise or semi-public basestations described above. In the case of a complementary system, then the public basestations will be installed in public locations (i.e. locations not served by individual, enterprise or semi-public basestations). These public basestations will serve CTM users in transit between locations that may be served by the other classes of basestations. In the case of an alternative system, then the public basestations will be installed in public locations, which could serve residential and small business users at their normal locations as well as in transit between those locations. In this case the most important question is related to the extent to which public basestations can be used to provide indoor coverage. This obviously depends on issues like housing densities, construction methods, etc., and will, therefore, vary from area to area and from country to country.

6.3   The CTM Market

With a few exceptions, most telecommunication users are potentially mobile telecommunication users – it is usually only the constraints of the telecommunication facilities that make users immobile. In order to understand the market for CTM, it is therefore necessary to understand the mobile nature of telecommunication users. The big question is: "How mobile are potential mobile telecommunication users?" In order to answer this, we need to consider the movement patterns of typical telecommunication users. There are certainly users who spend a significant part of their life on the move, over very wide areas, and spend a lot of their time communicating. At the other extreme there are those people who are desk-bound, house-bound, or genuinely immobile for other reasons. However, these two extremes tend to represent a small percentage, possibly of the order of 20%, of the total telecommunications user population. In between these two extremes there is the vast majority of private and business users who tend to be mobile, but only within relatively restricted areas, e.g. business users confined to one or a few business sites, or urban areas, private users who spend the majority of their lives at, or moving between, home, work and social activities, all within populated areas. In addition, the vast majority of these users only have a limited need to communicate. It is at this vast majority of people, with limited mobility and limited communication need, and for whom cellular telephony would be an expensive overkill, that CTM is aimed.

The above assumes that telecommunication users have to make a simple decision between subscribing to CTM or cellular telecommunication services. If we consider today's cellular mobile telecommunication users, very few of them are pure cellular users, i.e. they use fixed, often cordless telephones, when they are convenient and use a cellular telephone when that is more appropriate. Users often make a choice between telecommunication options on the basis of:

- Cost
- Quality
- Convenience
- Need to communicate

In other words, each time they want to communicate, users consider several questions: How much will it cost? How good will the communication be? How convenient are the alternative facilities? And how much do I need to communicate at this particular time? For example, if a user's life is in danger, it will not matter what the cost, quality or convenience is – the need to communicate will be overwhelming. In other circumstances, most users will give more consideration to cost, quality and convenience.

Ideally all users should have access to a selection of facilities and be able to select the most appropriate option each time they need to communicate. This leads to a need to provide a single service registration/subscription and personal number, and the freedom of the user to be able to choose between, for example, CTM and cellular communications on a call-by-call basis, i.e. the integration of fixed and cellular services to provide true personal communications. Dual-mode cordless/cellular terminals are already being developed. Therefore, the possibility of providing users with dual-mode CTM/cellular terminals, which, with service integration in the networks, will give them greater control over their communications, cannot be far away. Having said that, this chapter concentrates on pure CTM, which will be positioned by comparison as a lower-functionality, lower-cost service. However, the possibility of integrating CTM with other services should not be overlooked.

6.3.1   Service Benefits

The indication from the above, and from the success of cellular mobile telecommunications and first-generation cordless telephones (single-cell cordless terminal mobility), is that mobility should be, and is indeed becoming, the norm rather than the exception.

Let us consider the financial aspects of the market for mobile telecommunications. The highly mobile, highly communicative and affluent subscribers perceive mobile telecommunications to have a high value, as the benefits outweigh the cost. However, as the market penetration increases, the perceived value of mobile telecommunications decreases, and hence the potential revenue per subscriber decreases. This has been seen already in Europe, as reductions in cellular handset prices and network subscription charges have been reflected in lower usage revenues from new users. In order to serve the vast majority of users who are attracted by mobile telecommunication services, but still cannot afford cellular prices, or do not have such demanding mobility requirements to justify the cost, a lower-cost but attractive service is required. CTM is expected to satisfy that requirement.

6.3.2   Regulatory Questions

CTM raises some interesting regulatory questions. For example: "Is a network supporting CTM between residential, business and privately owned public access basestations, a mobile or a fixed network?" Currently the regulatory situation, with respect to mobile telecommunication services based on cordless technology, varies from country to country. It should, therefore, be noted that the CTM standards must cater for a variety of network access configurations to allow for the ownership of basestations by service providers, network operators or subscribers, and that regulations may dictate different commercial solutions in various countries.

6.4   Enabling Technologies

As has been indicated earlier, Cordless Terminal Mobility is dependent on three key new technologies, namely:

- Digital cordless technology
- Integrated services digital network (ISDN)
- Intelligent networking (IN)

The relevance of these technologies is discussed in the following subsections. Before discussing the technologies, it is worth noting that these technologies have emerged independently of CTM, i.e. they are all justified independently of CTM;

CTM should be seen as an end-user service based on the application of a combination of these new technologies.

6.4.1   Digital Cordless Technologies

Europe

In Europe, a great deal of effort has been expended in the development of standards for digital cordless systems, i.e. CT2 and DECT. It is not the intention of the CTM Project to develop new cordless access systems. The main emphasis is on the identification of network interface standards and the specification of the protocols required on those interfaces to allow users to roam with their cordless terminals between different basestations and networks. However, in order that users can roam with their cordless terminals between residential, business and public basestations, standard access profiles, defining the mandatory and optional features of the existing CT2 and DECT fixed and portable equipment, are required. In the case of CTM, the network architecture and services will define the subsets of the CT2 and DECT standards that will be used.

Outside Europe

The primary cordless technologies being considered for CTM are the European CT2 and DECT systems. However, as has been stated earlier, the primary consideration with CTM is not with the cordless access technologies but with the network interfaces and protocols required to allow users to roam with their cordless terminals between different basestations and networks. In doing this, it has been a major requirement to make the network interfaces and protocols for CTM as generic as possible. In this respect, the network services and protocols tend to have a greater influence than do the cordless access standards on the CTM standards. Thus, CTM standards will have wider applicability than CT2 and DECT, and could be used to support other existing or proposed cordless access systems outside of Europe.

6.4.2   Integrated Services Digital Network (ISDN)

In Europe, and indeed across the world in general, extensive development of standards and equipment for ISDNs occurred during the 1980s. Many network operators have implemented the digital transmission, digital switching, common channel signalling and computer-controlled exchanges required for ISDN. However, the facilities offered by ISDN, of faster call establishment, clearer transmission, higher-bit-rate data transmission and a range of supplementary services, need to be packaged into new added-value services to excite the end-user.

At the same time, significant advances have been made in customer premises equipment (CPE) with portable computers, cordless telephones, etc., which are selling in large numbers. What is required is a new range of ISDN services that provide significant benefits to end-users who want more than a fixed telephone.

An increasing emphasis on home working, networked PCs and electronic mail (e-mail) is expected to increase demand for ISDN access to provide higher-speed modem-less e-mail access – even children's television call-in programmes provide e-mail addresses these days. CTM may be used to complement this by providing advanced telephony services to these users and later allowing them to use their portable PCs in conjunction with CTM to provide, for example, mobile e-mail access.

One of the most important considerations in the development of digital cordless access technology in general has been ISDN compatibility. With respect to CTM in particular, ISDN provides three significant capabilities:

- Open network access interfaces
- Message-based signalling, particularly at the network access
- Digital transmission techniques

The open network access interfaces allow products with standard interfaces and mobility management protocols to be developed, rather than a multitude of country-specific variants. The use of message-based signalling provides a natural means of transporting the mobility management protocols required for CTM. The use of digital transmission techniques not only provides a high-quality digital speech transmission medium, but also allows digital data bearer services to be added in later phases.

6.4.3 Intelligent Networking (IN)

The single most important feature of CTM is the introduction of the network-based mobility management. The heart of this mobility management is the set of databases and the related protocols that are used to keep track of the terminals, to allow authentication of the terminals, and to provide the information to allow the processing and routing of incoming and outgoing calls.

One of the major decisions that had to be made in this area was whether to base the mobility management within public networks on the GSM Location Registers and GSM Mobile Application Part (MAP) [5] of ITU-T Signalling System Number 7, or on the Intelligent Network (IN) architecture and Intelligent Network Application Part (INAP) [4] of ITU-T Signalling System Number 7. The final decision was influenced by the following:

- The primary networks under consideration being ISDN.
- The planned deployment of IN in ISDN networks.
- The planned use of IN for terminal and personal mobility management in future mobile telecommunication systems – ETSI Universal Mobile Tele-communication Service (UMTS) and ITU Future Public Land Mobile Telecommunication System (FPLMTS).

As a result of these considerations, it was decided to base the mobility management in the public networks on the IN standards. As public telecommunication networks are large, complex and difficult to evolve, it is important that the mobility management protocols should be made as generic as possible. To this end it is expected that an early subset of the mobility management protocols being developed for UMTS, and FPLMTS, will be suitable, or adaptable with careful design, for managing the terminal mobility required for both CT2 and DECT cordless terminals.

It should also be noted that there is a general trend in the telecommunication industry to consider new telecommunication services within the context of IN. Therefore, we can expect in the coming years a general migration of ISDN services from the existing exchange platforms to the emerging IN platforms. The introduction of CTM, and the need for location-independent access, necessitates the movement of service logic from the exchanges to the IN service control points. IN provides the ideal platform for the development, testing and deployment of such new and innovative fixed and mobile telecommunication services.

6.5 The CTM Functional Architecture

The CTM functional architecture is shown in Figure 6.1. In the following subsections, the CTM functional entities, reference points and interfaces are described.

6.5.1 CTM Functional Entities

The cordless telecommunications network (CTN) consists of one fixed radio termination (FT) and its associated portable radio terminations (PT).

Two important functional entities within the FT are the cluster control function (CCF) and the cell site function (CSF). The CCF is responsible for the overall control of the cluster, i.e. the cells of the FT, whilst each CSF controls one single cell.

The portable radio termination (PT) contains all cordless technology-specific processes and procedures on the portable side of the air interface. All other functions are located in the portable application(s) (PA), associated with the PT. Collectively the PT and its associated PAs are referred to as a portable part (PP).

The fixed radio termination (FT) may be connected to a public network or a private integrated services network (PISN). In the public network the FT will be connected to a local exchange, and in the PISN to a private integrated services network exchange (PINX). The public networks and/or PISNs also contain location databases. These are used to hold the data associated with the PPs.

6.5.2 CTM Reference Points

Reference points are the conceptual points between two functional entities. In the CTM functional architecture, four reference points are defined – α, β, γ and δ.

- The α reference point divides the FT and the LE, as well as the FT and the PINX.
- The β reference point divides public networks from private networks. It is thereby also dividing location databases in the public and private networks. This identifies an important difference between the α and the β reference

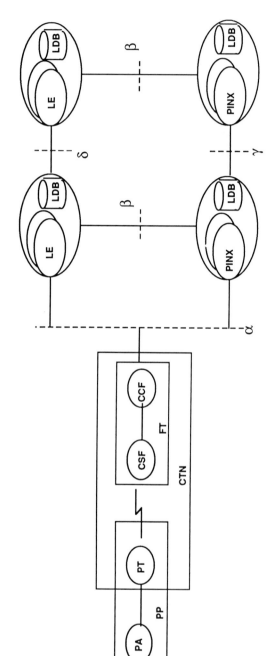

Figure 6.1 The CTM functional architecture. *Key:* CCF, cluster control functions; LDB, location database; CSF, cell site function; PA, portable application; CTN, cordless telecommunication network; PP, portable part; PT, portable termination; FT, fixed termination; LE, local exchange; PISN, private integrated services network; PINX, private integrated services network exchange.

points – no location database to location database information will be transferred over the α reference point.
- The γ and δ reference points divide different private and public networks respectively. Internetwork roaming involves information transfer over these and the β reference point.

6.5.3   CTM Interfaces

The interface at the α reference point, hereafter referred to as the α interface, is proposed to be either an ISDN basic access (2B+D) or a primary rate access (30B+D) with an access mobility management protocol. The traffic capacity of the CTN will determine which interface to use in a particular implementation.

The interface at the β reference point, hereafter referred to as the β interface, will be an ISDN primary rate access or a basic access with a network mobility management protocol.

6.6   The Role of CTM in the Evolution to Future Mobile Telecommunication Systems

This section looks at the relationship between CTM and future mobile telecommunication systems. It aims to show that CTM is not a transient concept that will disappear as other systems emerge. It is based on firm foundations and will, therefore, play an important role alongside cellular mobile telecommunications in the evolution of the telecommunications market.

There are a number of features related to CTM that are expected to be significant in the evolution of mobile telecommunications. These include:

- Use of microcells
- Public/private system interworking
- Use of the fixed network infrastructure
- Use of IN for service control and mobility management

6.6.1   The Use of Microcells

The use of microcells offers a number of advantages to designers of a mobile telecommunications system. Among the advantages are:

- Increased spectrum efficiency, providing higher traffic capacity and/or higher bandwidth per channel, to support higher-bandwidth services and/or higher-quality speech transmission.
- Lower RF power for longer battery life, smaller and lighter terminals and safer RF power levels.

Offset against these advantages is the increased dependence on the fixed access network and the consequential cost of larger numbers of basestations. However, as mobile telecommunications market penetration, user traffic, variety of services and service bandwidth all increase, the use of microcellular techniques will become essential, in order to better utilise the limited radio spectrum.

Microcells are, therefore, expected to become an important feature of future telecommunication systems. CTM provides an early opportunity to gain experience with the large-scale use of microcells.

6.6.2 Public and Private System Interworking

The past trend for business owners to provide their own internal telecommunication facilities can be expected to continue in the future. The increasing interest in the use of microcellular radio techniques for telecommunications within business premises, and the necessary use of microcells for future mobile telecommunication systems (see above), enables private mobile telecommunication systems for corporate networks to be developed. These systems can be used to provide the benefits of mobile telecommunications to the corporate user within their organisation's premises. By using compatible standards for corporate and public mobile telecommunication networks and by developing network interworking standards, it is possible to allow corporate users to roam into the public domain and vice versa. Interworking between public and private systems is expected to become an important feature of future telecommunication systems. CTM provides an early opportunity to gain experience with public and private system interworking, in terms of the practical and commercial issues, not simply the technical ones.

6.6.3 The Use of the Fixed Network Infrastructure

As the market penetration of mobile telecommunications increases, and the use of microcells increases, the cost of the network infrastructure will increase, owing to the larger number of basestations and the increased cost of connecting those basestations into the network. In addition, as the market penetration increases and mobile telecommunication services move down-market, the potential revenue per subscriber reduces. Therefore, a method of increasing the market penetration without increasing the cost of provision out of all proportion must be found. The obvious route is to build upon existing fixed network infrastructure, rather than to deploy special overlay networks. The use of fixed network infrastructures is expected to be an important feature of future telecommunication systems. CTM provides an early opportunity to develop mobile telecommunication services based on fixed network infrastructures.

6.6.4 The Use of IN for Service Control and Mobility Management

Intelligent network (IN) techniques and standards are expected to be used to provide the service control and mobility management for future mobile telecommunication systems (ITU FPLMTS and ETSI UMTS). The IN capabilities to support future mobile telecommunication systems are being introduced in IN Capability Sets 2 and 3 (CS2 and CS3). The IN features required to support CTM are expected to be included in CS2, which is in the closing stages of standardisation within the ITU-T. CTM will, therefore, provide an early opportunity to gain experience with the use of IN for mobile applications and to contribute to the IN standardisation process.

6.7 Summary

This chapter has introduced the reader to the concepts of Cordless Terminal Mobility (CTM), providing an understanding of how CTM relates to other cordless and cellular mobile telecommunication systems of the past, present and future.

The reader may wish to reflect on how early attempts at providing wide-area Cordless Terminal Mobility consisted of adding functionality to the periphery of networks, e.g. telepoint; in contrast it is perhaps a sign of the growing maturity of cordless technology that greater consideration is now being given to provision of support for Cordless Terminal Mobility within the heart of both public and private networks.

References

[1]   "Common Air Interface Specification to be Used for the Interworking Between Cordless Telephone Apparatus in the Frequency Band 864.1 MHz to 868.1 MHz, Including Public Access Services", ETSI I-ETS 300 131: 1994-11, Edition 2, Radio Equipment and Systems (RES)

[2]   "Digital European Cordless Telecommunications (DECT) Common Interface Part 1: Overview (to Part 9)", ETSI ETS 300 175-1 (to -9): 1992-10, Radio Equipment and Systems (RES)

[3]   "Integrated Services Digital Network (ISDN); Standards Guide", ETSI ETR 076: 1995-02, Edition 3

[4]   "Intelligent Network Capability Set 1 (CS1) Core Intelligent Network Application Protocol (INAP) Part 1: Protocol Specification", ETSI ETS 300 374-1: 1994-09, Intelligent Network (IN)

[5]   "European Digital Cellular Telecommunications System (Phase 2); Mobile Application Part (MAP) Specification (GSM 09.02)", ETSI ETS 300 599, Edition 2

Note

1   The International Standards Organisation (ISO) terms PINX (private integrated services network exchange) and PISN (private integrated services network) were adopted during 1995 by ETSI and ECMA to replace the terms previously used, namely PTNX (private telecommunications network exchange) and PTN (private telecommunications network).

Part II

Standardisation and Industry Development

7 Cordless Standards in Europe

Heinz Ochsner

In the 1990s cordless personal telecommunications has begun to emerge in Europe as a mass market, attracting a large number of customers and suppliers of equipment and services. It is therefore not surprising that on the international playgrounds the creation of binding standards has been one of the prime activities in preparation for this growth.

The need for standards reflects market importance. As soon as the application of a technology begins to promise a market that is big enough to attract the interest of many customers and several manufacturers, the question of standards becomes extremely important. Today's mass markets such as television, compact discs or personal computers simply could not exist without standards. In the field of personal computers, for example, the problems of incompatibility or quasi-compatibility of equipment from different vendors still cost a lot of money and time; such issues create a demand for increasing but rapid, and in some cases *de facto*, standardisation. In the field of telecommunications, the European approach has been to encourage rapid but formal, rather than *de facto*, standards.

To understand a telecommunications standard, it is helpful to understand its creation and its regulatory context. For this reason we outline the procedures that finally led to a mandatory regulation to which the manufacturer of a cordless telephone product has to comply, within the context of the European market. The environment in which telecommunications standards are created in Europe has fundamentally changed over the past decade – this chapter attempts to describe this new environment. Whilst the topic of this book is cordless telecommunications, the explanations given in this section apply to most telecommunications standardisation projects that currently are and in future will be of public interest.

7.1 European Telecommunications Standards

It is difficult to explain what a standard is without knowing who actually creates it and vice versa. Thus, Figure 7.1 provides an overview of the different bodies involved. This diagram shows the worldwide context within which standardisation bodies in Europe and, indeed, in other world regions operate.

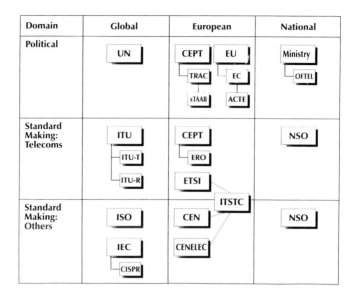

Domain	Global	European		National
Political	UN	CEPT EU		Ministry
		TRAC EC		OFTEL
		xTAAB ACTE		
Standard Making: Telecoms	ITU	CEPT		NSO
	ITU-T	ERO		
	ITU-R	ETSI	ITSTC	
Standard Making: Others	ISO	CEN		NSO
	IEC	CENELEC		
	CISPR			

Figure 7.1 Overview of standards bodies and their roles.

7.1.1 The World Context

Worldwide telecommunications standards have historically grown out of the International Telecommunications Union (ITU) of the United Nations, the structure of which is outlined in Figure 7.2. Founded in 1866 as the International Telegraph Union, its main purpose is to establish cooperation between member states in the field of telecommunications [1]. The Union's Plenipotentiary Conference is held once every six years.

In recent years the ITU has been in the process of restructuring. Originally formed as a union of national monopoly PTTs, which at the same time were administrations, it must now accommodate a rapidly changing situation. What makes it even more complicated is the fact that many countries have completed the process of telecommunications liberalisation, implying separation of the regulatory and operational functions and the introduction of private operating companies (some of them operating worldwide), whilst others still remain in the traditional fully regulated monopoly mould.

ITU standards are created within the Telecommunications Standardisation Sector (abbreviated ITU-T). This Sector corresponds mostly to the former CCITT (Comité Consultatif International des Télégraphes et Téléphones). Its plenary assembly meets normally every three years, whilst study groups are set up to deal with particular questions. As with the other Sectors, ITU-T has a Director supported by a technical and administrative bureau. Laboratories and other technical installations can be made available to the study groups. ITU-T creates Recommendations and Reports, which are not binding upon the member countries of the ITU. However, some of their work may later become part of international treaties.

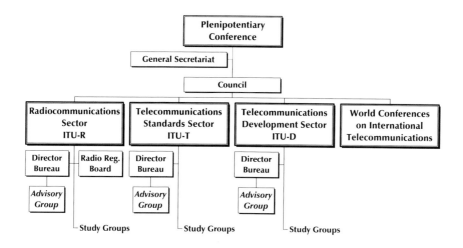

Figure 7.2   Structure of the International Telecommunications Union (ITU).

The Radiocommunication Sector embraces both the former CCIR (Comité Consultatif International des Radiocommunications) and the International Frequency Registration Board (IFRB). It is responsible for advising on and recording the worldwide usage of radio spectrum. For the allocation of frequencies the world is divided into three regions: Region 1 with Europe, the Middle East, the former USSR and Africa; Region 2 with the North and South American continents plus Greenland; and Region 3 with the Far East, Australia and New Zealand. The current allocation table comprises all the radio frequencies between 9 kHz and 275 GHz. The last World Administrative Radio Conference (WARC) was held in 1992 and discussed the allocation of frequencies to mobile and personal services. WARCs have been replaced by World Radio Conferences (WRCs), which are held every two years with limited agendas that are elaborated over a four-year period and ratified by the previous WRC. WRC-95 was mainly concerned with mobile satellite issues.

A special unit organises world conferences on international telecommunications; in particular it organises the World Radio Conferences. Frequency allocations defined by the WRC become mandatory for the Union's member states. However, changes of frequency allocations may result in the need for existing services to change their operating frequencies. Therefore, a reallocation of a frequency band to other services may take many years from the time of the decision to final implementation. The timescales associated with the release of a common frequency band for widespread geographical usage for a common application obviously pose a major constraint in standardisation of products.

The third Sector reflects the new way ITU understands its existence. The Development Sector aims at creating the environment to make telecommunications (and its driving function for development in general) available and affordable to all countries in the world.

In addition to the three Sectors the Union maintains a General Secretariat.

With the growth of computer telecommunication the standards created within the International Organisation for Standardisation (ISO) are gaining more and more importance and beginning to impinge upon those of the ITU. Other international organisations are also increasingly emerging to accommodate the rapid developments in these fields – examples being the ATM Forum and DAVIC – and these increasingly are setting the international standardisation agenda in some fields.

Finally, the International Electrotechnical Commission (IEC) has to be mentioned. Its CISPR (Comité International Spécial Perturbations Radio), dealing with radio interference, impacts substantially on all systems and technologies using radio spectrum.

In many cases the Recommendations and Reports created by the above-mentioned committees serve as prime inputs for European or national standards.

7.1.2 The European Organisations

In the past decade Europe has begun a phase of transition within which individual nations are becoming integrated into a "Europe" that is far more than just the geographical description of a continent. As a result, the situation continues to evolve.

The European Union (EU)

The European Union (EU) must be mentioned first in the European context. The Union's decision to implement a Single Market in Europe by 1992 and the recognition of the vital political, social as well as economic role of telecommunications have resulted in a major change of the telecommunications scene since 1987. The way for the Common Market to develop in the field of telecommunications was outlined in the famous "Green Paper on . . . Telecommunications Services and Equipments" [2]. The Green Paper was published in 1987, and its proposed action lines endorsed by the Council of Ministers for telecommunications one year later. Among other things, the Green Paper proposed:

- Reinforcement of the rapid development of standards . . . , supported by the creation of a European Telecommunications Standards Institute.
- Definition of the conditions needed for open network provision to service providers and users.
- Stimulation of the development of new services and the setting up of an information market.

To do so, the Green Paper requested three major changes of the Community's telecommunications scene:

- Complete opening of the terminal equipment market to competition, including the subscriber's first telephone set.
- Opening of national networks to service providers of other Member States, with the possible exception of a few basic services, such as voice telephony.
- Clear separation of the regulatory and operational functions of Telecommunications Administrations.

This Green Paper set the scene for a number of subsequent EU initiatives and resultant development of the European telecommunications environment.

In addition to its political steps, the European Commission is initiating, controlling and financing a variety of research programmes. Within the field of radio communications, projects of the RACE, ACTS, COST and ESPRIT programmes are important. Aside from the purely technical objectives, these programmes aim at the promotion of collaboration between manufacturers across national boundaries. Furthermore, many of these projects in the past have led and in the future will lead to important knowledge finding its way directly into the standardisation process.

For the issue of standards, the European Commission is assisted by the Approval Committee for Telecommunications Equipment (ACTE). ACTE approves the creation of a CTR (see section 7.2) together with TRAC (see below).

The Conférence Européenne des Postes et Télécommunications (CEPT)

The Conférence Européenne des Postes et Télécommunications (CEPT) has as members the European PTT administrations. Provision of telephony services as well as of equipment was a State monopoly in all European countries for a long time. Therefore, standardisation of telephony was treated uniquely within the CEPT Comité de Coordination et Harmonisation (CCH). Use of radio spectrum, because administered by CEPT, was standardised by CCH as well. However, after the creation in 1988 of the European Telecommunications Standards Institute (ETSI), with wider participation than just the Telecommunications Administrations (as proposed by the Green Paper), all standardisation issues were moved from CEPT to ETSI and CEPT/CCH was subsequently dissolved in early 1989.

Nevertheless, two committees within the CEPT are still of prime importance to the creation of European standards: the ERC (see below) and TRAC. The Technical Recommendations Applications Committee (TRAC) like ACTE is approving the creation of CTR. It also maintains Advisory Boards, such as the DECT Type Approval Advisory Board (DTAAB), which try to anticipate possible specific problems within the context of the current type approval regime and to make recommendations, for example for mechanisms to avoid potential type approval delays such as were previously experienced with GSM.

The European Radiocommunications Committee (ERC)

The European Radiocommunications Committee (ERC) is responsible for developing strategies for the use of radio frequencies and where possible agrees common frequency allocations for Europe. Detailed studies are undertaken in three Working Groups - Frequency Management (FM), Radio Regulatory (RR) and Spectrum Engineering (SE). The European Radiocommunications Office (ERO) was established to provide a permanent resource of expertise to the ERC and also undertakes studies under mandate from the European Commission.

The ERC publishes Reports on various aspects of its work. It issues Recommendations, which are not mandatory, but in regard of each, Members put on record whether or not they have adopted or intend to adopt them. The most binding output is the ERC Decision, which is intended to deal with issues on which there is broad European consensus and which when adopted by a Member is binding on that Member. Decisions may typically be adopted for frequency allocations and regulatory provisions for the use of equipment or systems defined in ETSI standards.

The European Telecommunications Standards Institute (ETSI)

The European Telecommunication Standards Institute (ETSI) was established in 1988 [3]. Within this Institute many future standards relevant to European telecommunications are being created. The Institute consists of Members, which are classified into five categories: administrations, network operators, manufacturers, user groups, and research bodies. Members pay a membership fee, which gives ETSI more possibilities in accelerating the creation of standards, e.g. by hiring technical experts. The structure of ETSI is shown in Figure 7.3.

The ETSI General Assembly (GA) is the highest committee in ETSI. The GA's main tasks are to determine policy, to elect new Members and to manage the resources, in particular the budget. The GA is, however, not involved in technical activities. The administrative tasks of ETSI are performed by a permanent secretariat located in Sophia-Antipolis near Nice in southern France. Headed by a Director, it also supervises the Project Teams (see below).

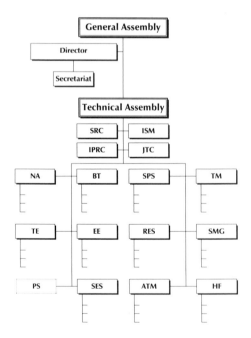

Figure 7.3 Structure of the European Telecommunications Standards Institute (ETSI) in 1995.

The Technical Assembly (TA) approves and issues European Telecommunications Standards (ETS). Approval is gained by voting, after a period of Public Enquiry (PE) has been completed and any feedback assimilated, as deemed appropriate, into the standard document. In addition the TA creates the Technical Committees (TC) and elects their chairmen. TCs create the standards and undertake the related drafting work. They may divide themselves further into Sub-Technical Committees (STC). The TC Radio Equipment and Systems (RES) deals with all low-power radio applications. Its STC RES03 is the body responsible for creating the standards for CT1 and CT2, as well as for the Digital Enhanced Cordless Telecommunications (DECT).

Under the TA, committees other than TCs do exist. For example, the Strategic Review Committee (SRC) studies and defines future possible tasks, the Intellectual Property Rights Committee (IPRC) deals with problems arising from patents that apply to certain standards, and the ISDN Standards Management Committee (ISM) controls and coordinates the creation of the ISDN standards that are the responsibility of many TCs.

As mentioned earlier, ETSI may hire experts to join Project Teams (PT). Such PTs are allocated to TCs or STCs and support the creation of the standard. The work on DECT, for example has been supported by such a team, PT10, originally formed in autumn 1989. More recently, the concept of voluntary project teams has been introduced. These teams are financed by sources other than ETSI, for example the European Commission or industry. RES03 has been supported by voluntary project teams financed by the Commission to create the test-case libraries.

The European Telecommunication Standards (ETS) are still voluntary standards and not binding on national Telecommunications Administrations. Nevertheless, administration members of ETSI will not introduce competing standards once ETSI has started to work on a new ETS. Furthermore, once an ETS is established, existing national standards will eventually be withdrawn.

Sometimes ETSI publishes a standard for application during a limited time. The reasons for such an interim ETS (I-ETS) may be several. The standard may not yet have been fully completed but may still be of some use and hence worthy of publication. Thus the early DECT Conformance Test Specification was published in this way, as I-ETS 300 175. Another reason to declare a standard "interim" is to show that the standard is expected to be withdrawn after a few years and to be replaced by another one. This was the case, for example, for CT1 and CT2. An interim ETS can be upgraded to a full ETS should circumstances change; thus, for example, many manufacturers have argued for the CT2 I-ETS to be upgraded to ETS status, in view of its wide acceptance outside of Europe since its original publication.

ISO and IEC have their European counterparts in CEN (Comité Européen de Normalisation) and CENELEC (Comité Européen de Normalisation Electrotechnique). Since the work done in ETSI, CEN and CENELEC may conflict, the Information Technology Steering Committee (ITSTC) has been established to coordinate the work of the three European standardisation bodies.

What has just been described reflects the situation of ETSI at the start of 1996. At this time, several activities are in hand to restructure the ETSI organisation as well as the way ETSI works to further improve and speed up the creation of European standards. Notably, the GA and TA are to merge, and TCs will approve their own outputs; increasing use will also be made of electronic communications.

7.1.3   The National Aspect

Even though there is a mandatory European type approval regime through the CTRs or European regulations to apply standards (Directives), implementation and enforcement of standards is a purely national matter. Therefore, the mechanisms of standardisation are different in each country. National Governments and Ministries, their Offices for Telecommunications, sometimes network operators and national standards organisations share the responsibility for telecommunications standards and their implementation.

7.2   Standards – What Are They?

There exist a vast variety of terms for documents that describe a standardised telecommunication system or, maybe more exactly, a telecommunication technology. There are two different types of documents that must be distinguished – the technical description of the standard and those documents, or Acts, which legally enforce the use of the standard.

7.2.1   Standards, Specifications, Recommendations and Reports

The terms "Standard", "Specification", "Recommendation" and "Report" usually denote a technical description. It is nearly impossible to assign different meanings to the different terms since they depend mostly on the organisation that created the documents. For the rest of this chapter the term "Standard" will be used for the complete system description whilst "Specification" may be used for different parts of it. The International Organisation for Standardisation (ISO) defines a standard as follows:

> *Standard*: A technical specification or other document available to the public, drawn up with cooperation and consensus or general approval of all interests affected by it, based on the consolidated results of science, technology and experience, aimed at the promotion of optimum community benefits and approved by a body recognised on the national, regional or international level.

Standards generally define interfaces between different modules of a system, e.g. the radio interface specification between the cordless handset and its associated basestation. Standards must, however, avoid describing (or even implying) possible technical realisations of modules. A standard usually will also contain rules on how to verify the compliance of a particular product to a mandatory standard (type approval). It must, however, be noted that for most standards in effect today the interface descriptions and the type approval procedures have been created by different bodies. As an example, the CEPT CT1 interface specification for analogue cordless telephones (see below) has been created by CEPT, whilst each country having introduced CEPT CT1 has its own national type approval specification.

7.2.2 Regulations, Directives, CTRs and MoUs

Having great impact not only commercially, but also socially and politically, telecommunications has been regulated by the national authorities from the very beginning. Therefore, in addition to the technical recommendations, there are "Regulations" that enforce the introduction of the recommendations. Quoting again from the ISO:

> Regulation: A binding document which contains legislative, regulatory or administrative rules and which is adopted and published by an authority legally vested with the necessary power.

Within the EU, a special kind of "Regulation" is the "Directive". A Directive is usually proposed by the European Commission but finally issued by the Council of Ministers; more recently, the Commission has been given new power to issue Directives in some cases directly, without the formal approval of the Council, which however retains a right of veto. A Directive may, for example, instruct the Member States to perform certain actions, usually by issuing regulations themselves, or to notify the EU on certain national actions. Member States are free to choose the way in which they implement the Directive. It is noted that Directives are issued not only to implement standards but to perform actions in any area in which the EU can direct its Member States. Some actions thus fall outside the powers of the EU. In this situation the EU may issue a "Recommendation". (Such an EU Recommendation has nothing to do with a recommendation for a standard, in the sense of a technical description, just described.)

As an example, in the area of cordless telecommunications, the EU issued a Recommendation proposing actions for a coordinated introduction of DECT in the Community and a Directive requesting Member States to free the necessary frequencies by 1992. In particular, the Directive requested from the Member States:

- That the frequency band 1880–1900 MHz be assigned to DECT by ERC (or CEPT/CR at that time).
- Recognition that the band may be extended in the future.

Another important tool of regulatory character is the Common Technical Regulation (CTR). A CTR is an interface standard for connection of telecommunication terminals to public networks. It usually has the form of a Type Approval Specification[1]. Once the TRAC and ACTE Committee have decided to turn a standard into a CTR, it becomes obligatory for all nations that are Members of CEPT. The legal basis for these procedures has been issued by the EU in the Telecommunications Terminal Equipment (TTE) Directive.

Originally, any standard was eligible to become a CTR. Today, the CTR is drafted by ETSI: ETSI creates the Technical Basis for a Regulation (TBR) – ACTE makes it a CTR. Ideally ACTE would do it just by changing the title, but there have been cases where ACTE have considerably reworked the TBR before releasing it as a CTR. Once approved by ACTE and TRAC, the CTR comes into force by publication in the EU's *Official Journal* (OJ) stating the base standard. In practice this means that there exists no physical document bearing a title "CTR xx: . . .". Rather, the CTR comprises its base standard together with the OJ publication.

An agreement to undertake a concerted course of action is called a Memorandum of Understanding (MoU). Operators of public telecommunication networks may sign such an MoU and agree, for example, to introduce a telepoint service by a certain date. Since the MoU is a contract between the signatories, it becomes binding for them. An example of an MoU is that between telepoint operators, PTTs and PTOs, which was signed in March 1990. Another group of operators established during 1995 was the DECT Operators' Group (DOG); DOG has been established as a voluntary group, that is they have no binding contractual relationship. DOG, however, may become an MoU Group in the future.

7.3 Standards – Why Do We Need Them?

The Greek mythical footpad Procrustes invited exhausted travellers to take a rest in his bed. For some travellers this bed was too long, for others it was too short. If the bed was too long, Procrustes used to stretch the traveller to a size fitting the bed; if the traveller was too long, Procrustes took an axe and cut the traveller's legs to the appropriate size. In all cases, the travellers were equally long and equally dead.

Standards that are too rigorous limit the application for which they were designed. The standard and its makers live in permanent danger of being the modern Procrustes by killing the application through tight and inflexible standards. In order not to create a "Procrustean bed" a standards maker has to make sure, firstly, that only those areas are standardised for which there is a need to standardise and, secondly, that standards are flexible enough such that the standard fits the application and does not force the application to fit the standard (Figure 7.4).

The need for standards has already been mentioned in the introduction to this chapter. In the case of public telecommunication networks, the need for standards is most obvious. In addition, however, big markets, in particular the European common market, need standards to allow even small manufacturers to create equipment that may be used with other manufacturers' systems. The standard thus allows a system to be assembled from modules produced by

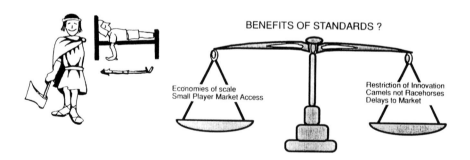

Figure 7.4 The Procrustean bed.

different manufacturers. Such a manufacturer independence gives small manufacturers access to big markets. Finally, standards may lead to large volumes of components and hence to cheaper equipment.

On the other hand, standards always restrict developers of equipment because they have to follow given rules. The reduced freedom of implementation of new features into products may reduce the potential of new developments and in the worst case result in nearly identical products from all manufacturers. As a further disadvantage of standardisation, very generally, the long creation time of standards has to be mentioned. Since standards have to be verified after their creation, several years may elapse between the first ideas of a new system to be standardised and its commercial introduction. The result can be a system that is no longer state-of-the-art. To ameliorate the latter problem, standards makers often attempt to forecast technological progress in order to define a technology intercept at the anticipated introduction date of a new standard – that is, the standard definition is based upon the expected technology that will be available at the time of its completion, rather than simply that already available. Such approaches have been adopted in the context of the CT2 and DECT standards.

Seeing the benefits as well as the dangers of standardisation, a concept of two planes of standards has been developed: coexistence specifications and common interface specifications.

7.3.1 Coexistence Specifications

In the context of a radio interface, a coexistence specification ensures that the equipment of different manufacturers may operate and coexist in the same environment. Coexistence means that if, for example, one system uses, say, a couple of radio channels, then the number of channels denied to a second system from a different manufacturer would be about the same as if the first system were another of his own. A coexistence specification, however, does not treat the communication between the two peers in detail, the communication protocols being proprietary to the manufacturer of the equipment. Coexistence specifications are common in those situations where equipment is used only privately.

7.3.2 Common Interface Specifications

If equipment of different manufacturers has to interoperate, then an exact description of the radio communication is needed. Interoperability is needed in the case of public services such as the telepoint service. As explained above, interoperability also allows small manufacturers to participate in a market by not needing to produce a whole system but only parts of it. A large number of manufacturers are believed to result in a big market with a large variety of different products. A common interface may further allow high volumes of standard components and eventually cheap equipment. However, this is only true if the standard is wisely designed and no "Procrustean bed".

A common interface standard for cordless telecommunications, of course, needs to define the exact format of all signals being transmitted. Furthermore, the protocols for some procedures need to be defined. Such defined protocols make sure that one side knows what answer to expect once he has sent a message to his peer. Nevertheless, in order to allow freedom of implementation, the standard shall not say how these functions should be implemented. In DECT, for example, the standard defines the type of modulation, but not the demodulation technique to be employed. Furthermore, the standard shall not say how certain decisions are to be taken. It should, for example, be within the product designer's freedom to choose a preferred algorithm for handover or for call curtailment in case of bad propagation conditions, although the standard may place requirements upon the performance of such algorithms. Today, ETSI issues coexistence radio standards only as interim standards to allow rapid introduction of a new technology, which later would become covered by a common interface specification.

7.4   Analogue Standards

This section presents the analogue standards for cordless telephones introduced in Europe during the 1980s. The intention is to present the standards and their application rather than the technical details of the systems they describe. The only system parameters being presented here are the frequency bands and the number of channels, because these parameters touch upon regulatory issues.

7.4.1   Low-Band Systems

Early products, introduced during the early 1980s, used short-wave frequencies with only a few channels. In the UK and France, relatively inexpensive low-band systems were introduced, rather than the CEPT CT1 standard.

The UK Analogue Standard UK/CT1

In the UK it was felt that CT1 in the CEPT – and later ETSI – version was not suitable for a free telephone equipment market because the standard resulted in expensive products as compared to, for example, the units sold in the USA. Therefore, CEPT/CT1 was not introduced in the UK, but instead a UK/CT1 standard (sometimes also referred to as CT0) was published, MPT 1322 [4], a standard very similar to that used in the USA.

UK/CT1 offers eight channels: from 47.456 25 MHz to 47.543 75 MHz (portable transmitting, spacing 12.5 kHz) and from 1.642 MHz to 1.782 MHz (base transmitting, spacing 20 kHz). Any apparatus is either tuned to one of the channels or allows switching between two channels. MPT 1322 is a coexistence rather than a common interface specification.

In the Netherlands the market was opened to similar products in the early years of the 1990s.

The French Analogue Standard F/CT1

A system similar to UK/CT1 was also introduced in France. The specification NF C 98-220 allocates 15 channels spaced at 12.5 kHz from 41.3125 MHz to

41.4875 MHz (portable transmitting) and from 26.3125 MHz to 26.4875 MHz (base transmitting). However, only 10 channels are really in use today. Products conforming to this standard have proven to be expensive and, considering the limited capabilities, the market take-up has been correspondingly low.

7.4.2   CEPT CT1 and CT1+

Generally, the term CT1 is used to refer to the standard for cordless telephones originally designed by CEPT and first released in 1983. Later it was taken by ETSI and converted into an Interim ETS. I-ETS 300 235 describes the technical characteristics, test conditions and methods for radio aspects of cordless telephones. This I-ETS is a Recommendation only. Hence, each of the national versions is slightly different. Many of them have been derived from the original CEPT standard and not been changed upon the availability of the ETSI standard. This was only a small problem in times when PTTs were the only buyers of telephone terminal equipment and used to purchase equipment uniquely from national manufacturers. Today, however, it means that a manufacturer has to design a slightly different product for each country in which he intends to sell it. Nevertheless, CT1 cordless telephones are in use in 11 European countries: Austria, Belgium, Denmark, Finland, Germany, Italy, Luxembourg, The Netherlands, Norway, Sweden and Switzerland.

CT1 uses 40 channels spaced at 25 kHz from 914.0125 MHz to 914.9875 MHz (portable transmitting) and from 959.0125 MHz to 959.9875 MHz (base transmitting). Channels are assigned dynamically; that is, one out of the 40 channels is chosen at call set-up.

CT1 is a coexistence specification; it does not allow the cordless handsets to communicate with basestations of other manufacturers. However, like any other telephone, a CT1 set has to be type approved against national specifications of the line interface; this latter specification is a common interface specification.

The 40 channels do not provide enough capacity for applications in the business area. In addition, the frequencies are being released for use by the GSM cellular radio telephone service. Therefore, Belgium, Germany and Switzerland decided to open a new band from 885.0125 MHz to 886.9875 MHz and from 931.0125 MHz to 931.9875 MHz offering 80 channels. This system is sometimes called CT1+. CT1+ specifications are, however, national; the ETSI standard specifies the original frequencies only.

7.5   Digital Standards

Mirroring developments in cellular radio, cordless telephones have made the transition from analogue to digital technology. Such systems offer a wide area of applications described in earlier chapters. The first national digital standards came into effect in the UK and Sweden during the late 1980s (predecessors to CT2 CAI and DECT) and, in the early 1990s, pan-European digital cordless standards have begun to take significant market share. Fuller technical descriptions of these standards are given later in this book.

7.5.1   CT2 CAI

In the UK the poor capacity of the eight-channel UK/CT1 was recognised very early. Consequently a so-called "second-generation cordless telephone" or CT2 system was developed by a collaboration amongst several UK-based manufacturers in the mid-1980s. In order to allow a telepoint service based on CT2, a common interface specification, a refinement of the existing specifications had to be created; this was necessary to allow handsets from different manufacturers to operate with telepoint basestations. This specification became known as CAI (Common Air Interface).

By the time that other European countries opened their markets for CT2 around 1990, notably Germany, France, and Finland, ETSI had taken over responsibility for the CT2 standard and created I-ETS 300 131.

CT2 uses 40 channels separated by 100 kHz from 864.15 MHz to 868.05 MHz accessing the radio medium by a frequency-division multiple access (FDMA). Unlike all the analogue systems, it separates the two directions of communication using the time-division duplex (TDD) technique rather than frequency-division duplex (FDD). In this way duplex separation is achieved with cheap switches rather than with expensive filters. The CT2 specification, and its North American derivative PCI, are described in greater detail in Chapters 17 and 18.

7.5.2   DECT: the Digital Enhanced Cordless Telecommunications Standard

DECT is the first standard for cordless telecommunications that has been fully designed by ETSI. Nevertheless, many of its concepts have been taken from a Swedish system known as DCT, Digital Cordless Telephone, as well as from CT2. DCT was designed after it was recognised that CT1, which is introduced in Sweden, is not particularly suitable for high-capacity applications, in particular for cordless PABX.

The development of DECT started within CEPT, on the one hand, and in industry associations such as ESPA and ECTEL, on the other. Responsibility for the standard was then assumed by ETSI shortly after its creation in 1988.

DECT offers services that go significantly beyond those available with earlier systems. In particular the DECT radio environment has been designed for:

- Cordless telephones for use in private residences
- Telepoint and wireless local loop services
- Cordless PABX applications
- On-site cordless data services

The standard allows design of a handset to be used at home, in the office and in the street.

The system uses frequencies on 10 carriers between 1880 MHz and 1900 MHz. Each carrier may bear 12 telephone conversations by using a time-division multiple access (TDMA) and time-division duplex (TDD) approach. A high degree of flexibility allows DECT to support everything from simple cordless telephones to complex wireless PABX networks integrating voice and data services. The DECT specification, and its North American derivative PWT, are described in greater detail in Chapters 19, 20 and 21.

7.6 Summary

This chapter has described the standards creation process and summarised the existing standards in Europe. Firstly the environment in which standards are created in Europe has been covered. The important bodies – ranging from the ITU-T and ITU-R committees to the newer, but, for Europe, most important, ETSI – have been presented. Next, the need for standardisation as well as its dangers have been considered. The concepts of coexistence and common interface specifications were shown to give manufacturers the possibility of compromise between these benefits and dangers. Finally, the existing standards for cordless telephones have been briefly presented. As seen in market figures in previous chapters, the variety of different national systems is beginning to be replaced by the comprehensive and unique standard, DECT.

References

[1] "International Telecommunications Union: Celebrating 130 Years 1865–1995", International Systems and Communications Ltd, London, 1995

[2] "Towards a Dynamic European Economy – Green Paper on the Development of the Common Market for Telecommunications Services and Equipments", Commission of the European Communities, COM(87) 290, Brussels, 1987

[3] "The Development of Standards", D Gagliardi, presented at the Pan-European Digital Cellular Radio Conference, Munich, February 1989

[4] "Performance Specification – Angle Modulated MF and VHF Radio Equipment for Use at Base and Handportable Stations in the Cordless Telephone Service", UK Department of Trade and Industry, MPT 1322, London, August 1982

Note

1 Note the European usage of the term "Type Approval Specification": A Type Approval Specification is a binding specification applicable uniquely for the purpose of approving a piece of equipment. All other test specifications describing tests to verify the compliance to a standard are called Conformance Test Specifications.

8   UPCS Standards in the USA

Tony Noerpel

This chapter explores the PCS standards environment in the USA insofar as it affects cordless telecommunications – or unlicensed PCS as it is termed. The free-market approach of the US Administration to personal communications has resulted in a far more flexible approach to standards than elsewhere, with an approach based upon coexistence, leading to the proliferation of potential systems. In this chapter the standards bodies and activities are described, along with the process being introduced to clear spectrum for PCS usage. The ingenious "Etiquette Rules" that have been devised to enable a diverse collection of different standards to coexist within the same spectrum are described, followed by an overview of the main UPCS technology contenders. Existing cordless telephone technologies, as well as new ones, are also briefly presented. Whilst 'not yet certain, it is likely that standards and products for the US marketplace will find themselves adopted in the remainder of the NAFTA region, i.e. Canada and Mexico. Which systems will emerge as the dominant standards in years to come remain to be seen.

8.1   Who Defines Standards?

8.1.1   US Standardisation Bodies

Telecommunications standards in the USA are developed principally by two American National Standards Institute (ANSI) accredited organisations. These are the Telecommunications Industry Association (TIA) and the Alliance for Telecommunications Industry Solutions (ATIS), sponsoring organisation of Committee T1.

TIA is organised into four divisions responsible for standards development. In particular, the Mobile and Personal Communications Division develops interim standards for wireless information networks, mobility management and radio air interfaces for, amongst others, 800 MHz cellular, the Advanced Mobile Phone Service (AMPS), US Digital Cellular, IS-95 (CDMA) and IS-54 (TDMA), 900 MHz specialised mobile radio (SMR), 46–49 MHz and 902–928 MHz cordless telephony and network signalling such as IS-41.

The ATIS-sponsored Committee T1 comprises six Technical Subcommittees (TSC). These are: T1P1, which has responsibility for overall programme management for PCS standards development; T1S1, which develops signalling protocols; T1M1, which develops standards for operations, administration, maintenance and provisioning (OAM&P); T1A1, which develops performance guidelines for speech and voice-band data; T1E1, which develops standards for physical layer transmission; and T1X1, which deals with inter-exchange signalling. Committee T1 has developed standards for the PSTN and ISDN wireline network including primary and basic rate ISDN, SS7 and SONET.

Most telecommunications standards are typically developed by a single standards forum having primary responsibility. This works well when other forums have only a casual interest or interest relating to compatibility with the proposed standard. Occasionally, a single forum would have primary responsibility but other forums would provide contributions. However, in the case of the standardisation of common air interfaces for personal communications systems (PCS), member organisations of both TIA and T1 felt critical need for developing suitable Common Air Interface standards [1].

TIA was interested in developing wireless interfaces to the cellular network. Within TIA, engineering committee TR 46 was formed in 1993 to develop 2 GHz PCS standards including services, network architecture, signalling, air interfaces, and privacy and authentication. In addition, engineering subcommittee TR 41.6, "Wireless User Premises Equipment", was formed to develop 2 GHz standards for operation in the unlicensed PCS spectrum.

Committee T1 was interested in developing interfaces to support wireless access to the public network. In 1990, T1P1 was formed and charged with the responsibility for PCS service description, privacy and authentication, network architectures and internetworking interface standards, and air interface standards.

8.1.2 Origins of the Joint Technical Committee

Originally, cordless telephony was considered a wireless adjunct to wireline access to the PSTN separate and independent of other emerging wireless access technologies such as mobile satellite communications and cellular telephony. There was no apparent need for a Common Air Interface from the perspective of wireline service providers, who viewed the cordless telephone as customer premises equipment (CPE) and not part of their networks. Also wireless service providers (cellular telephone companies) did not at that time consider the integration of cellular telephony with wireline access. Finally, manufacturers of cordless telephones did not consider the need for a Common Air Interface or the need for roaming between private systems, since a cordless telephone handset would only be expected to operate with its companion basestation.

As personal communication service (PCS) began to be defined, it became apparent that wireless and wireline network integration and interoperability between public and private access radio systems would be important. Such interoperability is graphically depicted in Figure 8.1. In this figure, a user would have a single portable handset, which could automatically register and deregister the user's location at a home location register (HLR). Calls could then be routed over wireline facilities to a residential cordless basestation when the user is home, to the wireless keyset, PBX, or Centrex service when the user is at work, or to the public PCS network to which the user might subscribe when that user is in

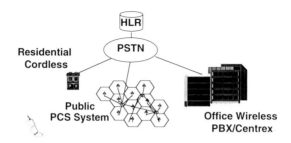

Figure 8.1 A US view of wireless and wireline integration and interoperability.

transit. Such location and routing capabilities could also be extended to public or private systems in shopping centres, hotels and airports. Manufacturers and both wireline and wireless access service providers feel the need for licensed and unlicensed PCS air interface standards.

To meet the needs of the various stakeholders, the formation of a Joint Technical Committee (JTC) [2] between TIA and T1 was proposed by S. Engelman, Chairman of T1P1 [3]. This relationship was finalised in a Statement of Cooperation [4], signed by J. Russell, Chair of TIA Mobile and Personal Communications Division, and A. Reilly, Chair of Committee T1. The JTC is chartered with defining Common Air Interface standards for both public and private systems and for both licensed and unlicensed PCS spectrum.

In addition to the Common Air Interface standards work being done in the JTC, separate work is being carried out in TIA subcommittee TR 41.6, which is concerned with CPE standards, to define air interface standards for use as wireless access adjuncts for private systems.

Several informative articles appeared in the *IEEE Personal Communications Magazine* (Fourth Quarter 1994 issue), which describe the PCS standards process for the USA. These articles make interesting reading and discuss, in greater detail, issues relating not only to air interface standards but to PCS network issues such as operations, administration and maintenance of PCS networks, and their standardisation. These articles and their authors are also the principle source for much of the information presented in this chapter.

To realise the vision of universal and ubiquitous wireless access, the evolution of the public switched telephone network, as well as definition of air interfaces, is necessary to effect this vision. Important issues include network architectures, interfaces, database design, protocols, OAM&P and diverse network integration and interoperability. These issues are being addressed in standards bodies, research institutions and development organisations around the world. The approaches emerging in Europe, with CTM, and in Japan with PHS, are described elsewhere in this book.

8.2 The Advent of UPCS

8.2.1 Licensed and Unlicensed Bands

On 22 October 1993, the Federal Communications Commission (FCC) issued its Second Report and Order [5], in which a total of 160 MHz of radio spectrum near

2 GHz was allocated for new wireless personal communications services. In June 1994, the FCC [6] approved changes to the October 1993 Report and Order, in which they redefined the licensed allocations to include three 30 MHz blocks and three 10 MHz blocks, and limited the unlicensed spectrum to the 20 MHz from 1910 MHz to 1930 MHz, 1910–1920 MHz for asynchronous use and 1920–1930 MHz for isochronous use. This new spectrum allocation is shown in Figure 8.2. During 1995 the licensed spectrum, including the A and B bands shown in Figure 8.2, was auctioned by the FCC to potential PCS service providers [7] in 51 major trading areas (MTAs) and 492 basic trading areas (BTAs), raising substantial sums for the US Treasury. The auction of the remaining spectrum, the C, D, E and F bands, should be complete in 1996.

A primary requirement for unlicensed operation is compliance to new FCC Rules in Part 15, Subpart D, which defined an "etiquette" by which unlike systems can make common use of the allocated spectrum. This etiquette as it pertains to operation in the isochronous spectrum is described in greater detail later in this chapter.

The FCC has encouraged technologies that would interoperate in both the licensed and unlicensed spectrum so as to give future PCS customers opportunity for economic, high-quality and flexible communications. Notably, Julius Knapp of the FCC in an address at the WINForum Conference in Dallas, Texas, stated that we should "expect dual-mode devices that operate in either the licensed or unlicensed PCS spectrum" [8].

The unlicensed allocation was based largely on a proposal made by the WINForum industry advocacy group. WINForum began in response to Apple Computer Corporation's Petition for a spectrum allocation for "Data-PCS." Apple Petition RM-7618 of January 1991 proposed "only general objectives and a regulatory framework for meeting them, and [did] not propose precise specifications to be incorporated in FCC rules for Data-PCS". Apple further asserted that "there should be a thorough dialogue, within the industry and between the Commission and industry, both within and without the rule-making process, to refine the details". Originally created as a working group of the computer industry concerned with spectrum for wireless local-area networks

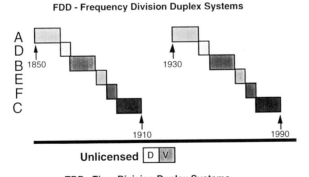

Figure 8.2 FCC allocation of unlicensed spectrum near 1.9 GHz for voice and data communications (June 1994 Memorandum Opinion and Order).

(LANs), WINForum broadened its scope to address user-provided wireless connectivity for both voice and data ("User-PCS"). In its filing with the FCC on 21 June 1992, it stressed seven points in regard to spectrum allocations for emerging technologies:

- User-PCS computing and communications devices will require exclusive *unlicensed* frequency allocations, and cannot share frequencies with fixed point-to-point microwave or with carrier-provided PCS.
- User-PCS should be enabled by a nationwide primary and exclusive spectrum allocation for User-PCS, clear of other uses (existing fixed microwave operations or carrier-provided PCS operations). A fair and equitable process for clearing the corresponding emerging technologies spectrum of point-to-point microwave operations must be implemented.
- The total amount of spectrum needed in the short term for user-PCS is at least 40 MHz.
- The rules for User-PCS spectrum should promote fair access to the spectrum without imposing significant restrictions upon equipment design and implementation. Specifically, the FCC should adopt a minimum set of equipment authorisation rules that provide for equitable sharing of the exclusive User-PCS allocation.
- The Commission should require compliance with FCC Rules solely through the equipment authorisation process, not through licensing of spectrum. The equipment authorisation scheme should not establish exclusive or preclusive access to any portion of the User-PCS allocation by any party.
- The Commission should allocate frequencies for User-PCS that are close to and possibly contiguous with other domestic PCS services, to minimise the cost and complexity of multi-mode equipment.
- The Commission should promote international harmonisation of spectrum allocation and regulation for like services, especially for spectrum around 2 GHz band for development of User-PCS.

WINForum's broad concept of User-PCS includes wireless local- and personal-area communications for portable and desktop computers, for wireless notepad and messaging devices, and for wireless office and home telephone systems. User-provided services are used by individuals and organisations who purchase and operate FCC-authorised equipment for their own benefit, without the need to rent service from a third-party provider or carrier. User-PCS services employ low power and cover limited distances. Many are described as "in-premises" radio services that will not require individual user or station licenses.

The primary motivation for the activities of WINForum was, where possible in User-PCS, to reduce the risk associated with the unlicensed use of the *Industrial, Scientific and Medical* (ISM) bands for two-way wireless communications. Unlicensed communications devices may make use of the ISM band spectrum from 902 to 928 MHz, from 2400 to 2483.5 MHz, and from 5725 to 5850 MHz, in conformance to Parts 15.247 or 15.249 of FCC Rules. Part 15.249 operation has limited utility because transmit power levels are limited to the order of 1 mW, which makes two-way communications over any appreciable distance very difficult. Part 15.247 rules [9] allow systems using either direct-sequence or frequency-hopped spread-spectrum modulations (or hybrids thereof) to transmit at significantly higher power levels (up to 1 W, which must be derated to account for any antenna gain greater than 6 dBi). A number of commercially available products (wireless PBXs, wireless LANs, cordless telephones) make use of this spectrum to avoid the delays of spectrum allocation and licensing processes.

However, devices operating in this spectrum cannot claim protection from interference caused by other approved devices using the same spectrum. The Part 15 rules do not specify that devices must cooperatively share spectrum. If a 902–928 MHz wireless PBX and a 902–928 MHz wireless LAN transmit in the same space, they can render each other inoperable, and neither system operator has recourse except to relocate their systems. Furthermore, Part 15 devices must accept interference caused by other authorised radio services and approved non-communicating ISM devices within this band. Examples of the former include Government radar, amateur radio and vehicle location systems at 902–928 MHz; and of the latter, field disturbance sensors, inventory control devices, industrial plywood dryers, microwave ovens and radiofrequency-stabilised arc welding equipment. WINForum's actions can then be seen as a clear response to these difficulties: firstly, to secure *exclusive* spectrum use for type-accepted unlicensed communications devices; and secondly, to establish a set of rules whereby conforming devices can make cooperative use of shared spectrum, and therefore increase their utility over their ISM band counterparts. However, WINForum was clearly stopping short of advocacy for an air interface protocol, as they were aware of the contentious debate over such protocols in standards activities for licensed spectrum and wanted to avoid possible delays associated with approving standards.

The FCC etiquette adopts major portions of the WINForum Industry Etiquette that was submitted to the commission in June 1993. The FCC received a significant number of petitions for reconsideration with regard to the October Report and Order impacting both the licensed and unlicensed operation. The modifications, introduced by the FCC in the June 1994 Memorandum Opinion and Order, addressed these petitions.

The resultant unlicensed PCS allocation is well suited for provision of primarily indoor wireless access. Possible applications include wireless Centrex, wireless PBX, wireless key sets and coordinated residential cordless access and circuit-switched data in isochronous sub-bands. There is also a potential opportunity for interoperable indoor/outdoor services with licensed spectrum.

8.2.2   Spectrum Coordination

Spectrum coordination in the licensed band is left as the responsibility of the PCS service providers, who acquired this spectrum through the auctions. Several issues are being addressed to consider boundary problems. Boundary problems include microwave paths that either cross or are near two adjacent trading areas, each owned by different service providers, and links whose transmission bandwidth overlaps different licensed bands, again used by different service providers in the same trading area. Such boundary problems are the rule rather than the exception and require mediation, cooperation and sharing of the cost of spectrum clearing.

UTAM Inc. has been charged with the responsibility of clearing UPCS spectrum and coordinating the deployment of unlicensed devices. As part of the equipment authorisation process, all manufacturers will be required to demonstrate procedures to assure that the fixed portion of the system (or ports) will not be activated until installation at its authorised location is verified and that relocation of the system or device will not occur without "recoordination".

UTAM allows only fixed systems to be deployed prior to clearing the spectrum of microwave incumbents nationwide. Owing to the possible costs for coordination with fixed licensed operation, it is possible that only large, geographically bounded systems can be deployed quickly in this spectrum. In particular, wireless Centrex and wireless PBX systems fit this description. The operation of residential cordless units will be difficult to coordinate and control.

There are more than 300 *operational fixed system* (OFS) links in the unlicensed band in the USA that need to be cleared, in addition to many more links in Canada and Mexico. There is no accurate database of all of these 2 GHz links. Links are still currently being installed. Spectrum will be cleared on a county-by-county basis and a database of cleared counties will be maintained by UTAM. There will also be zone 1 and zone 2 counties. Zone 1 counties are those which are partially cleared or are very far from remaining links and which therefore can support some limited UPCS activity. In these zones a total power budget will be calculated, say 50 W of UPCS, and the installed base will be allowed to grow up to this amount. When it is reached, no more unlicensed ports will be approved for installation in that county.

In zone 2 counties, a vendor or customer can pay to have the proposed installation coordinated with existing microwave links just like new microwave links are required to do. If the coordination process is successful, then the UPCS equipment can be installed. All coordinations go through UTAM for a fee. However, an outside contractor could perform the actual coordination provided it gets final approval from UTAM.

Manufacturers will be required to ensure that their systems are able to be coordinated and also develop the method of ensuring this. Manufacturers are also responsible for ensuring that relocated fixed parts cannot operate until they are recoordinated or their location is reverified. Manufacturers are responsible for the inadvertent interference caused by their equipment if it is moved and subsequently operated by a customer without proper recoordination.

Possible engineering innovations encouraged by UTAM and the FCC to comply with these constraints include:

- A device that automatically complies with the "non-movement" constraint described above.
- A device whose location can be easily verified by UTAM.

UTAM estimates that they will require some 67 million dollars to clear a total of 386 links at US $200 000 per link. This cost includes equipment and coordination fees and incidentals. The FCC has specified that incumbents are entitled to reimbursement for the cost of replacement of equipment and not for the cost of upgrading equipment. UTAM is raising funds from manufacturers to begin clearing spectrum. A fee of US $20.00 per transmitter will be required to be paid to UTAM to pay for the cost of spectrum clearing. It is anticipated that spectrum clearing will be complete within 7–12 years.

8.2.3 ANSI Standard Tests

ANSI C.63 Subcommittee 7 is developing standard test procedures for verifying compliance with Subpart D of the FCC Part 15 rules. These test procedures are in the ANSI multistage review and balloting process to become an ANSI standard. These tests will cover physical measurements such as sub-band limits, transmitter RF power, spectral density, emission bandwidth, emissions, carrier stability, frame stability, frame duration, intraburst separation as well as tests to confirm compliance with the channel access protocol prescribed in the etiquette sections of Subpart D.

8.3 Part 15 Subpart D (Etiquette) Rules

The Rules detailed in the June 1994 Memorandum Opinion and Order have two principal goals:

- To enable emerging technologies to be deployed in a manner so as not to interfere with existing fixed microwave systems (until the spectrum is cleared of OFS).
- To enable unlike systems to reasonably share spectrum, in a manner more efficient than is possible in the unlicensed ISM bands [10].

In addition to traditional physical measurements associated with Part 15 compliance, the new rules include a channel access protocol. Aspects of this protocol include the 30-second rule, the listen-before-transmit rule and the packing rule. These and other rules are briefly described below.

8.3.1 The Coordination Rule

The Rules make it clear that radio systems operating in this spectrum must be coordinated with fixed microwave systems. At the heart of any radio system capable of such operation is a suitable marker transmission from fixed, coordinated PCS equipment that advertises its presence to the otherwise uncoordinated equipment (generally PCS portable terminals). This rule is stated in Part III, Section C, Paragraph 91 of the Part 15 rules.

This rule requires that all communications between fixed ports and portables be initiated by the port. In order for portables to originate phone calls, they must first listen for and receive a transmission from a suitable port and be directed to allowable frequencies on which to transmit. This forces ports to transmit some kind of beacon or control signal that advertises their presence and the availability or coordination status of the spectrum. In the presence of such a signal, portable devices can respond or transmit; but in the absence of such a signal, portable devices must be prevented from transmitting.

8.3.2 The Listen-Before-Transmit Rule

The coordinated device must access the spectrum to transmit a marker or beacon in a way that allows shared access to that spectrum by other systems. In order to do so, it must first measure the interference in the time/frequency window in which it intends to transmit, i.e. *listen-before-transmit* (LBT). The channel access procedure is defined in Paragraph 15.323 (c).

Before initiating transmission, devices must monitor the combined time and spectrum windows in which they intend to transmit for a period of at least 10 ms to determine if the access criteria are met. If no signal above a monitoring threshold is detected, transmission may begin and continue in the monitored time and spectrum windows without further monitoring. The monitoring threshold must not be more than 30 dB above the thermal noise power kTB, where k is Boltzmann's constant, T is temperature in kelvins and B is the noise bandwidth equivalent to the emission bandwidth used by the device.

A duplex conversation may be established by allowing a fixed device to monitor both the transmit and receive windows, i.e. the portable does not have to perform the LBT etiquette measurements if the port has already done so for both the forward and return links.

Since duplex channels will be paired, this rule allows a device to measure the received time/frequency window and thus clear the paired transmit window if the transmit window cannot be monitored. This would allow two transceivers to be collocated and to use the same antenna. The second transceiver only has to measure the receive channel of the duplex pair and not the channel in which it is actually going to transmit. This rule does not appear to address the coordination rule requirement that ports or fixed devices must maintain simplex control channels as beacons since portable devices are prevented from transmitting unless they hear such a signal.

8.3.3   The 30-Second Rule

The Rules recognise that, whilst marker transmissions serve an important purpose, the real utility of the spectrum is in carrying two-way information. Therefore, to prevent the marker transmissions from flooding the spectrum, the FCC added a stipulation in Paragraph 15.323 (c). This paragraph states that, once access to a specific combined time and frequency window is obtained, an acknowledgement from a system participant must be received by the initiating transmitter within 1 s or transmission must cease. Periodic acknowledgements must be received at least every 30 s or transmission must cease. Channels used exclusively for control and signalling information may transmit continuously for 30 s without receiving an acknowledgement, at which time the access criteria must be repeated.

We interpret this to mean that, after measuring a channel and finding that it is available, the fixed device can transmit the marker information for a maximum of 30 s, unless a portable terminal or other transmitting device responds with a request for access. If the fixed device does not receive a response, it must release the channel and re-execute the channel access procedure. There is strong likelihood that the port will not be able to access the same frequency the next time for its beacon or control channel when it exercises the etiquette rules, for the following reasons:

- During the port's previous transmission, a lower frequency might have become available and the port must now use this newly vacated channel because of the packing rule (see below).

- There will be churn of frequencies due to power measurement errors or multipath.
- The previously used frequency may be seized by another port during the time that the port ceased transmission and began exercising the etiquette.

8.3.4 The Channelisation Rule

The Rules provide for a maximum channelisation of 1.25 MHz in the isochronous sub-band from 1920 to 1930 MHz; the minimum channelisation is 50 kHz. This is specified in Paragraph 15.323 (a).

8.3.5 The Packing Rule

The "packing rule" requires systems executing LBT to scan frequencies in increasing or decreasing order, and to accept the first frequency channel that meets the interference criterion. This is described by Paragraph 15.323 (b). Furthermore, devices with an intended emission bandwidth less than 625 kHz shall start searching for an available time and spectrum window within 3 MHz of the sub-band edge at 1920 MHz and search upwards from that point. Devices with an intended emission bandwidth greater than 625 kHz shall start searching for an available time and spectrum window within 3 MHz of the sub-band edge at 1930 MHz and search downwards from that point.

8.3.6 The Power Level Rule

The peak transmit power level or the averaged power level measured during a transmission burst or interval is constrained as a function of bandwidth. Paragraph 15.319 (c) describes this limitation.

Peak transmit power shall not exceed 100 mW multiplied by the square root of the emission bandwidth in hertz. Table 8.1 lists the allowable peak transmit power levels for some possible examples of emission bandwidth. The peak power level is based on the emission bandwidth and not on channel spacing, so that whilst two systems might both have 625 kHz channel spacing, their emission bandwidth could be quite different. The emission bandwidth is defined in section 8.3.9.

An important result of these low power levels is that there is not enough power to use complying systems cost-effectively outdoors because of the small port coverage areas. Also, a system with less bandwidth is able to transmit more

Table 8.1
Allowable transmit power levels with respect to emission bandwidth

Bandwidth (kHz)	Transmit power level (dBm)	Number of channels in 10 MHz	Average power spectral density (dBm Hz^{-1})
1250	20.5	4	−40.5
625	19	8	−39.0
312.5	17.5	16	−37.5

spectral energy than a system with more bandwidth. To understand the implications of these power limits, consider the scenario of two nearby systems shown in Figure 8.3.

System W is a wide-band system with a 1.25 MHz channel bandwidth and system N is a narrow-band system with 312.5 kHz channels. Assume that the two systems are separated at a distance corresponding to a path loss of P_1. The wideband *signal-to-interference ratio* (*S/I*) is disadvantaged by 6 dB relative to the narrow-band *S/I*, since there are four 312.5 kHz narrow-band interferers in the 1.25 MHz channel ($17.5 + 10 \log 4 = 23.5$ dB) and only one-quarter of the wideband signal interferes with a narrow-band channel ($20.5 + 10 \log 0.25 = 14.5$ dB). When the LBT rule is considered, the wide-band system can seize 1.25 MHz of spectrum if the narrow-band system is outside a distance corresponding to a path loss of 106.5 dB and prevent a narrow-band system from seizing a channel within a distance corresponding to a path loss of 103.5 dB, since 30 dB above kTB for the wide-band system is –83 dBm and –89 dBm for the narrow-band system.

8.3.7 The Frame Duration Rule

The Report and Order specifies allowable frame duration in Paragraph 15.319 (e). The frame duration of a device shall be 10 ms/N where N is a positive whole number. Therefore, any radio system must have a frame duration of 10 ms, 5 ms, 3.33 ms, 2.5 ms, 2 ms, etc.; frame durations exceeding 10 ms are disallowed. Note that only some frame duration periods allow systems to sub-rate channels whilst maintaining a constant burst duration. For example, a system with a frame duration of 2.5 ms using 32 kb s^{-1} channel information rates could be sub-rated to allow for 16 and 8 kb s^{-1} channel information rates, since an 8 kb s^{-1} channel would then have a frame duration of 10 ms. However, systems with a frame duration of 10 ms could not be sub-rated.

8.3.8 The 20 ms Rule

There are provisions in the rules permitting devices to have a 20 ms frame duration or burst repetition rate. However, such devices must not cause

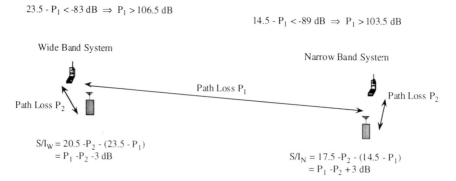

$23.5 - P_1 < -83$ dB $\Rightarrow P_1 > 106.5$ dB

$14.5 - P_1 < -89$ dB $\Rightarrow P_1 > 103.5$ dB

Wide Band System

Narrow Band System

Path Loss P_1

Path Loss P_2

Path Loss P_2

$S/I_W = 20.5 - P_2 - (23.5 - P_1)$
   $= P_1 - P_2 - 3$ dB

$S/I_N = 17.5 - P_2 - (14.5 - P_1)$
   $= P_1 - P_2 + 3$ dB

Figure 8.3 Narrow-band and wide-band systems coexistence.

interference into those devices that adhere to the 10 ms rule. There are no rules incumbent upon devices that adhere to the 10 ms rule preventing them from interfering with devices that have a 20 ms frame duration. At the time of writing, precise behaviour of devices using a 20 ms frame duration has not yet been fully developed.

8.3.9   The Emissions Rule

The emission bandwidth of a device is the width of the transmission between the two furthest points where the level is 26 dB below the level of the peak of the emission. The emission bandwidth is used to determine the monitoring threshold and the allowed peak power level. An emissions mask is also specified by the rules, which governs out-of-channel and out-of-band emissions of the device.

8.3.10   Rule Interpretations

We conclude this section by discussing some of the implications of the above rules, in terms of what is or is not in practice permitted in the unlicensed PCS spectrum.

- The Rules do not explicitly require a conforming system to operate in *time-division duplex* (TDD) mode. However, only TDD systems can feasibly operate in this unlicensed band, because it is not possible to make low-cost terminal components that can provide sufficient isolation between receive and transmitter chains separated by less than 1.25 MHz. A system using a frequency-division duplex technique may also have problems with regard to the packing rules described above.
- Unacknowledged radio port transmissions cannot continue beyond 30 s without exercising the channel acquisition etiquette; thus, dedicated fixed frequency control channels are not permitted.
- The monitoring threshold portion of the LBT channel access procedure is problematic for collocating transceivers. When one transceiver is transmitting at 10 dBm or more, the collocated device must be able to measure received signals on the order of –110 dBm. The rules change allowing collocated transceivers to measure only the received portion of a duplex paired channel ignores the consequences of interference to simplex control and pilot channels by assuming that all communications are duplex.
- Another consequence of the packing and LBT rules is that frequency-hopping spread-spectrum modulations are not feasible.
- The FCC has not specified required spectral efficiency; e.g. a maximum ratio of control to information bandwidth or dead times imposed by synthesiser tuning, etc.
- The Rules allow for sharing of frequencies in time by different systems. But the viability of sharing frequencies in time across systems is questionable because of interruption due to sliding frame structures, since frame structures for different systems will be unsynchronised and unlocked.

- It must be demonstrated that portables cannot transmit outside of their coordinated area, i.e. they must be enabled and disabled via the fixed ports so as not to operate on uncoordinated frequencies. Furthermore, the ports need to be tied to some coordinating arrangement.
- The maximum transmit power level imposed by the Rules limit these systems to primarily indoor use.
- Interoperability with public licensed PCS is highly desirable for successful application of a technology in the UPCS spectrum.

8.4 Technology Proposals

In the TIA/T1 Joint Technical Committee (JTC) on wireless access, the primary emphasis has been on air interface protocol standards for the licensed spectrum allocation. This allocation supports *frequency-division duplex* (FDD) radio systems because of the paired nature of the frequency bands. The JTC has approved, for ANSI publication, four regular Common Air Interface standards intended for public use in the licensed PCS spectrum, including PCS1900 (a derivative of DCS1800 and GSM) [11], IS-136 [12] (an extension of IS-54, the North American digital TDMA cellular standard) and IS-95+ [13] (an extension of IS-95, the North American CDMA digital cellular standard), and at least one Common Air Interface standard, Personal Access Communications System (PACS), intended for operation in both public access licensed PCS spectrum and in private-use unlicensed PCS spectrum. IS-136, IS-95+ and PCS1900 are applicable for "high-tier", i.e. high-power, high-mobility, large-port-coverage, areas, whilst PACS is intended for "low-tier", i.e. low-power, low-mobility, small-port-coverage, areas. Two additional air interface technologies have not been approved for ANSI publication but have been approved as interim standards in TIA and for trial use by Committee T1 including Omnipoint's TDMA/CDMA technology [14] and wide-band CDMA [15]. These systems are all FDD except for the Omnipoint technology.

The JTC is also developing standard air interfaces for the UPCS spectrum. The PACS standard includes two annexes that describe Common Air Interface standards for private unlicensed applications, both of which have been approved for ANSI publication as regular standards. PACS-UA is based on the Personal Handyphone System (PHS) and PACS-UB is based on the Bellcore WACS technology. The application of Omnipoint's radio system to the UPCS spectrum has also been discussed. These systems are all *time-division duplexed* (TDD).

The TIA subdivision TR 41.6 is also considering air interface standards for the UPCS spectrum. These technologies include the Personal Communications Interface (PCI) based on CT2, the Personal Wireless Telecommunications (PWT) based on the Digital Enhanced Cordless Telecommunications (DECT), and Orthogonal CDMA (OCDMA) from Stanford Telecom. These systems are also all TDD.

It should be noted that standardisation is not a requirement for operation in the UPCS spectrum in the USA. A product must have UTAM, C.63.7 and FCC approval, so that it does not cause interference to incumbent microwave users and does comply with the etiquette and emissions rules, but it may use a proprietary air interface. In the future such systems may become available.

8.4.1   PACS-UA

The Personal Handyphone System (PHS) [16] is a standard developed by the Research and Development Centre for Radio Systems (RCR), a private standardisation organisation in Japan. PHS is a digital cordless telephone system that offers telecommunications services in homes, offices and outdoor environments using radio access to the public telephone network, or other digital networks.

In the original PHS standard, a fixed frequency for a control channel that carries system and signalling information is preselected. PACS-UA is a version of PHS wherein the system control channel architecture is modified to reflect FCC LBT rules [17].

The basic radio characteristics of PACS-UA are similar to PACS and PACS-UB including a carrier spacing of 300 kHz, eight slots per frame TDMA, $\pi/4$-shifted QPSK, a transmission rate of $384\,\text{kb s}^{-1}$, and a voice coding rate of $32\,\text{kb s}^{-1}$ ADPCM. Unlike PACS-UB, PACS-UA has a 5 ms frame duration with 2.5 ms for basestation-to-portable transmission and 2.5 ms for portable-to-basestation transmission and can be sub-rated to accommodate $16\,\text{kb s}^{-1}$ channels. PACS-UA supports one-hundred and twenty-eight $32\,\text{kb s}^{-1}$ traffic channels in the 10 MHz UPCS spectrum.

The PACS-UA technology is being standardised within the JTC.

8.4.2   PACS-UB

Personal Access Communications System – Unlicensed B (PACS-UB) [18–20] is a TDD version of the PACS [21] air interface standard for licensed PCS. The combination of PACS and PACS-UB offers a Common Air Interface technology that has both a TDD mode of operation for indoor, private access, unlicensed PCS spectrum and applications and an FDD mode of operation for outdoor, public access, licensed PCS spectrum [22, 23].

PACS-UB employs TDMA/TDM (uplink/downlink) on the radio interface using $\pi/4$-shifted QPSK modulation at $384\,\text{kb s}^{-1}$. The 2.5 ms radio frame comprises eight bursts, each of 312.5 ms duration, per frame. In a burst window of 120 bits, 80 bits (10 octets) are allocated to the *fast channel* (FC). The FC provides a raw data rate of $32\,\text{kb s}^{-1}$, suitable for reasonable-quality speech coders. PACS-UB also supports sub-rate channels of $16\,\text{kb s}^{-1}$ and $8\,\text{kb s}^{-1}$ achieved by using one burst every two frames or one burst every four frames, respectively (with *effective* frame periods of 5 ms and 10 ms, respectively). One-hundred and twenty-eight $32\,\text{kb s}^{-1}$ traffic channels are supported in 10 MHz.

The PACS-UB technology is being standardised within the JTC.

8.4.3   PWT

Digital Enhanced Cordless Telecommunications (DECT) [24] is the digital cordless standard developed by the European Telecommunications Standards Institute (ETSI). DECT is a TDMA-based radio communication system that provides low-power radio access between portables and ports (called fixed parts in ETSI terminology) at ranges up to a few hundred metres using omnidirectional antennas.

The Personal Wireless Telecommunications (PWT) standard is a derivative of DECT as defined in the ETSI Document ETS 300 175 "Digital Cordless Telecommunications Common Interface". PWT was proposed originally for unlicensed operation[1] [25].

The major difference between the PWT and DECT standards is in the physical layer and the related medium access control (MAC) layer.

PWT has a carrier spacing of 1.25 MHz and therefore eight frequency channels are supported in each isochronous sub-band. Each frequency channel supports 12 time-division duplexed channels on 24 time slots per 10 ms frame. Therefore, ninety-six 32 kb s^{-1} time/frequency channels are supported in the 10 MHz.

Each burst is 417 ms long, 49 ms of which is guard time. The gross bit rate is 1152 kb s^{-1} and since the modulation technique is $\pi/4$ QPSK, the gross symbol rate is 576×10^3 symbols per second. Each time slot contains 320 user information bits, 72 bits for system control, 16 bits for synchronisation, 16 bits for preamble and 56-bit guard time. The user information rate is 32 kb s^{-1} per time slot. The net channel rate of the signalling channel is 6.4 kb s^{-1} per time slot. The signalling protocol is based on DECT. The PWT technology is being standardised in TIA subcommittee TR 41.6.

8.4.4   PCI

The PCI radio air interface is based on the CT2Plus standard [26]. CT2Plus is an enhancement to the multinational CT2 standard. A major departure from CT2 is the addition of common signalling channels in CT2Plus. PCI's implementation of this common signalling channel is different from CT2Plus and is called a Marker Channel (MC); the use of the marker channel facilitates that all PCI devices can be coordinated as per the etiquette requirement.

PCI is an FDMA/TDD system, which uses a 100 kHz RF channel to provide one full-duplex circuit. The user information channel is 32 kb s^{-1} with an associated 1 kb s^{-1} signalling channel in a 2 ms TDD frame. The resulting low speech delay is one of the attractive features of PCI. However, the use of frequency-division multiplexing does not allow sub-rating of the channel to multiple lower-rate uses. The average transmit power is 20 mW, with 32 mW peak power on a 50% duty cycle. PCI uses a 32 kb s^{-1} ADPCM voice codec. PCI provides for 96 full-duplex circuits in 10 MHz.

Two-step power control of the cordless portable part (CPP) (portable) output power is used. The low power setting is 16 dB below the normal power setting.

The MC is used for cordless fixed part (CFP) identification, paging, broadcasting system information, polling, registration, deregistration, call set-up, and as a beacon to enable portables. Only one transceiver within a cell transmits the MC. The PCI technology is being standardised in TIA subcommittee TR 41.6 as TIA standard 663.

8.4.5   O-CDMA

The Stanford Telecom technology proposal for the UPCS spectrum is an orthogonal code-division multiple access radio system with 1.25 MHz channel spacing employing TDD. The data modulation is differentially encoded QPSK transmitted at a burst rate 20.8×10^3 symbols per second. This accounts for overhead and guard bands. The data signal is bi-phase modulated with a spreading code at 32 times the burst symbol rate or 665.6×10^3 chips per second. The spreading code is the modulo-2 sum of a length-256 PN sequence and a length-32 Rademacher–Walsh function. The 32 Walsh codes are orthogonal, provided that the portable signals are chip-wise synchronised. The all-ones Walsh function is used as the control- or order-wire channel for each frequency channel.

This system uses an open loop uplink power control with 5 ms delay between the measured downlink burst and the uplink burst to which the power control is applied.

The Stanford Telecom direct sequence spreading technique uses orthogonal spreading codes instead of the non-orthogonal pseudo-random codes used by other CDMA technologies, and this requires tight control on the synchronisation.

The system potentially has one-hundred and twenty-eight $32\,kb\,s^{-1}$ channels per 10 MHz. However, one $16\,kb\,s^{-1}$ channel is required for signalling and control. This channel cannot be given up for a traffic channel because not only does it support access and alerting and system information, but it also carries associated signalling in support of the other traffic channels. Therefore the system supports 120 usable $32\,kb\,s^{-1}$ traffic channels.

The O-CDMA technology began standardisation in TIA subcommittee TR 46.1, although latterly the focus of O-CDMA has shifted onto other applications, notably wireless local loop and satellite PCS.

8.4.6   Omnipoint CDMA

Omnipoint Corporation received a Pioneers Preference Award from the FCC for its pioneering work in early PCS systems. The system utilises a combination of CDMA, TDMA, FDMA and TDD. Direct sequence spread spectrum (DSSS) is used to improve the radio link under severe multipath conditions and to improve the interference rejection of in-band and co-channel interference. TDMA separates users by time and reduces the need for strict power control required by CDMA-only systems. TDD is used in both the licensed and unlicensed versions of the Omnipoint system. RF channelisation is either 5 MHz or 1.875 MHz for the licensed PCS operation and either 5 MHz or 1.25 MHz for the unlicensed version. The frame duration is 20 ms. The Omnipoint system is being standardised as an interim standard by the TIA and for trial use by ATIS Committee T1.

8.5   Other Cordless Technologies

In addition to the wireless technology standards described above for use in the UPCS spectrum, there are also other technologies being trialled, based on AMPS, or already available as commercial cordless telephone products, operating in other bands, which address the same applications and services. The ISM band systems are proprietary and there is no attempt to standardise an air interface for use in this spectrum; a description of two of these systems is also provided below.

8.5.1 AMPS/NAMPS

The applicability of the AMPS and NAMPS analogue cellular radio standard air interfaces as a cordless telephone and office business phone using the 800 MHz cellular spectrum has been trialled by several cellular service providers. From a customer's perspective, these trials have been an overwhelming success, indicating a desire for interoperability between private and public wireless access. From a service provider perspective, the service is difficult to operate and maintain because of hard-to-control interference from private systems into the public system.

The TIA interim standard IS-94 describes the air interface requirements for this application of AMPS, and IS-94A describes the protocol and interface between the cordless basestation and the network both to control the basestation emissions, as necessary to limit interference to the public system, and to register and deregister the location of the handsets to and from the private cordless basestation at the service provider's HLR for the purpose of routing calls. Authentication of the handset is included in this protocol. The networking protocol described by IS-94A is extensible to digital cellular systems as well as effecting the interoperability between public systems using licensed spectrum and private systems using unlicensed spectrum.

8.5.2 ISM Band Technologies

In addition to the UPCS spectrum, unlicensed communications devices may make use of the industrial, scientific and medical (ISM) spectrum as described earlier in this chapter on a non-protected basis, in conformance to Parts 15.247 or 15.249 of FCC Rules. A number of commercially available products (wireless PBXs, wireless LANs, cordless telephones) make use of this spectrum to avoid the delays associated with spectrum allocation and licensing.

The Spectralink radio system wireless PBX product operates in the unlicensed 902–928 MHz ISM band. This product is an FHMA/TDMA hybrid system. It has 50 hopping patterns, or hopping sequence re-uses. There are fifty-two 500 kHz-wide channels in the 26 MHz of the ISM band spectrum, of which 48 are for traffic and four are for signalling. Since two of these 500 kHz channels could fit into the 1.25 MHz channelisation specified for the upper isochronous sub-band, only 16 frequency channels could be provided in 10 MHz. The system uses 24 kb s^{-1} ADPCM speech encoding in a 10 ms frame interval.

The Tadiran system is based on digital radio technology and is intended for use in both public and private environments. This system uses frequency-hopping spread-spectrum technology in the 902–928 MHz and in the 2.4 GHz ISM bands. The radio architecture of the system shares many features of the PACS technology and is in fact based on the original 2 ms frame structure version first proposed by Bellcore [5]. Without frequency hopping, the Tadiran system is very similar to 2 ms WACS, with the layer 2 and 3 protocols derived from the original Bellcore WACS documents.

The basic radio characteristics of the Tadiran system when operating in the 902–928 MHz frequency band are 51 RF carriers with 500 kHz carrier spacing, 30 mW peak transmit power and a frame duration of 2 ms. The multiple access technique is TDM/TDMA/FH – i.e. TDM on the downlink and TDMA on the uplink, with the carrier frequency hopped at a rate of 500 hops per second. The frequency hopping pattern is dynamically allocated. The duplexing technique is TDD with five downlink slots followed by five uplink slots. The channel bit rate is $468 \, \text{kb s}^{-1}$. The modulation is Gaussian minimum phase shift keying with a modulation index of 0.5 and a premodulation filter bandwidth of one-half the bit rate. The voice coding used is $32 \, \text{kb s}^{-1}$.

8.6 Summary

In this chapter we have described the background to the introduction of microcellular UPCS in North America. Unlike the situation in other regions of the world, the FCC has not specified that a particular air interface technology be exclusively used within the isochronous unlicensed frequency allocation, but instead has published a set of rules or an etiquette. Any technology can be deployed provided it adheres to the etiquette. The FCC has not even required a minimum level of spectrum efficiency. There has been a move for open standards, as we have seen, because that makes it easier for manufacturers and users, but there is no requirement for open standards. Thus there may be many closed systems manufactured with proprietary air interfaces. Several of these standard and proprietary systems have been described in overview – more detail on many of these is given elsewhere in this book. Since the intended UPCS applications are largely private in nature, i.e. residential cordless telephony, small wireless keyset systems and large wireless PBX and Centrex systems, the lack of interoperability between private systems may not be viewed as a problem to many customers.

A standard, however, allows a customer to replace broken or old equipment with devices from a different manufacturer, for example, with certainty that the equipment will work. This might be viewed as an advantage to corporate communications directors of large businesses who need to maintain big wireless PBX systems or to Local Exchange Carriers (LECs) who wish to offer wireless Centrex services. Furthermore, interoperability or dual modality between licensed and unlicensed spectrum applications encourages the use of air interface standards in the UPCS spectrum as well as in the licensed spectrum.

Outdoor use of TDD results in a capacity reduction of from 30% to 70% because of basestation-to-basestation interference when the basestation transmissions are not frame synchronised as compared with the capacity of FDD systems. Synchronisation of outdoor ports spread over a large geographic area is more difficult than indoors, where all connect to the same port controller within a building; therefore, PCS radio systems designed for outdoors generally use FDD. TDD, however, does have certain advantages, such as transmission path reciprocity and ease of spectrum allocation, which make it the convenient choice for duplexing for unlicensed, largely indoor, wireless applications.

Macrocellular outdoor technologies (e.g. PCS1900, IS-95, IS-136 and AMPS) tend to use network-controlled handover (NCHO) whereas microcellular technologies tend to use mobile-controlled handover (MCHO). MCHOs take less time and are more reliable than NCHOs and are preferred in general; however, NCHO has an important advantage in high-powered macrocellular systems in that the handover algorithms are under the control of the service provider and can be easily adjusted and modified to account for various local pathological propagation conditions. This is considered to be less of a problem for microcellular systems. In the future, all systems may evolve to MCHO or to a hybrid technique called mobile-assisted handover (MAHO) wherein the network controls the handover but with assistance from the mobile usually in the form of adjacent cell measurements [27, 28].

Issues such as these are the motivation for examining various combinations of outdoor "high-tier" public access technologies such as PCS1900 and IS-136 with indoor "low-tier" technologies such as PACS-UB [29], PACS-UA and PWT within a single handset and for exploring the applicability of various "high-tier" radio standards using licensed spectrum for private use. Interoperability between public and private access is an increasingly important aspect of standardisation work likely to impact all licensed and unlicensed standards.

In North America we see several standards emerging as well as several closed, or proprietary, air interface systems, all with varying degrees of functionality, i.e. handover, call delivery, call origination, call set-up by basestations between handsets, privacy, data capabilities, etc. Over the next decade we will see some of these systems succeeding in the marketplace and will see interoperability and multi-environment operation growing in importance.

Acknowledgements

Many people have contributed to the standards process for PCS in the USA for issues relating to common air interfaces as well as network infrastructure support of wireless access technologies. In writing this chapter the author would like to acknowledge the help and advice of Gary Boudreau, Ken Felix, William Cruz, Gary Jones, Tony Akers, Jay Padgett, Sandy Abramson, Charles Cook and Cliff Halevi; however, the author accepts the responsibility for any errors and inaccuracies.

References

[1]   "Preparing the Way for PCS", B Frison, M Woinsky and A Kripalani, Guest Editorial, IEEE Personal Communications Magazine, Fourth Quarter, 1994
[2]   "Development of Air Interface Standards for PCS", C Cook, IEEE Personal Communications Magazine, Fourth Quarter, 1994
[3]   "Recommendation for Joint Work Between Committee T1 and TIA TR-45 on PCS Standards", S Engelman, Contribution T1P1/92-061
[4]   TIA Mobile and Personal Communications Division – ANSI Accredited Standards Organisation and ANSI Accredited Standards Committee T1, Telecommunications Statement of Cooperation, September 1993
[5]   "Amendment of the Commission's Rules to Establish New Personal Communications Services", FCC Second Report and Order, October 1993, GEN Docket No. 90-314, Vol. 8, Book 22, p. 7700
[6]   "Amendment of the Commission's Rules to Establish New Personal Communications Services", FCC Memorandum Opinion and Order, June 1993, GEN Docket No. 90-314, Vol. 8, Book 14, p. 4461
[7]   "Technico-Economic Methods for Radio Spectrum Assignment", AM Youssef, E Kalman and L Benzoni, IEEE Communications Magazine, June 1995

[8]   "Making User PCS a Reality – A Journey on the Information Highway", J Knapp, First Annual
      WINForum User PCS Workshop, Dallas, October 1994
[9]   FCC Report and Order, adopted 14 June 1990, GEN Docket No. 89-354
[10] "Coexistence and Access Etiquette in the United States Unlicensed PCS Band", DG Steer, IEEE
      Personal Communications Magazine, Fourth Quarter, 1994
[11] "PCS-1900 Air Interface Specification", ANSI J-STD-007, 1995
[12] "PCS-1900 Air Interface Specification", ANSI J-STD-0011, 1995
[13] "Personal Station–Basestation Compatibility Requirements for 1.8–2 GHz Code Division
      Multiple Access (CDMA) Personal Communications Systems", ANSI J-STD-008, 1995
[14] "Composite CDMA/TDMA Air Interface Compatibility Standard for Personal Communica-
      tions in 1.8–2 GHz for Licensed and Unlicensed Applications", TIA Interim Standard PN 3390,
      1995
[15] "W-CDMA Air Interface Compatibility Standard for 1.85–1.99 GHz PCS Applications", TIA
      Interim Standard PN 3502, 1995
[16] "Personal Handy Phone Standard, Version 1", Research and Development Centre for Radio
      Systems (RCR), RCR STD-28, December 1993
[17] "Personal Access Communications System Unlicensed (version A)", ANSI J-STD-014
      supplement A
[18] "Personal Access Communications System Unlicensed (version B)", ANSI J-STD-014
      supplement B
[19] "PACS-UB, A Protocol for the Unlicensed Spectrum", A Noerpel, L Chang and R Ziegler, IEEE
      International Conference on Communications Proceedings, Seattle, June 1995
[20] "PACS-UB for Use in Unlicensed Spectrum", A Noerpel, E Laborde and K Felix, First Annual
      WINForum User PCS Workshop, Dallas, October 1994
[21] "Personal Access Communications System", ANSI J-STD-014, 1995
[22] "A United States Perspective for a Flexible PCS Standard", A Noerpel, Y Lin and H Sherry, ITU
      Telecom 95, Palexpo, Geneva, October 1995
[23] "Low Complexity Hardware Implementation for Interoperable Licensed and Unlicensed
      Personal Communications Services: the PACS-UB System", R Ziegler, A Noerpel, L Chang and
      N Sollenberger, WINLAB Workshop, East Brunswick, NJ, April 1995
[24] The DECT Specification, ETS 300 175; available from ETSI, Sophia-Antipolis, Nice, France
[25] "Petition for Wireless User Premises Equipment Sub-Committee and Project Authorization",
      Hendy, DAmico, Lloyd, Sterkel, Sivitz, Byrne and Willey, TR 441/94.03.07
[26] "CT2Plus Class 2: Specification for the Canadian Common Air Interface for Digital Cordless
      Telephony, Including Public Access Services", Annex 1 to Radio Standards Specification 130,
      Radio Advisory Board of Canada, 1992
[27] "A Handoff Control Process for Microcellular Systems", T Kanai and Y Furuya, IEEE VTC
      Conference Proceedings, 1988
[28] "Analysis of Handover Algorithms", M Gudmundson, IEEE VTC Conference Proceedings,
      1991
[29] "An Interoperable PACS and DCS1900 Subscriber Unit Radio Architecture", R Malkemes, P
      Lukander and P Harrison, IEEE PIMRC 95 Conference Proceedings, Toronto, September 1995

Note

1   *Editor's note*: An enhanced version PWT-E to address public applications has also been proposed
    – see Chapter 21 in this volume.

9 Standards Development in Japan

Yuichiro Takagawa

Soon after the development of early CT2 cordless technology in Europe and the commencement of work on DECT, work began, in 1989, on a Japanese cordless standard, which has become known as the Personal Handyphone System (PHS). Public standards are now available, early field trials successfully completed and commercial service was introduced during 1995. This chapter explains the origins and timescales of PHS, the way the standards were developed and the envisaged developments in Japan. The PHS standard itself is presented later in this book, in Chapter 23.

9.1 Concept and Origins of PHS

The Japanese PHS was first conceived as a system in January 1989, encompassing the idea of a low-cost wireless pocketphone that could be used in both indoor and outdoor environments as an access to network-supported services. In wireless telecommunications services, whether a call is placed indoors or outdoors, the emission power from basestation and the handset, or personal station, must be low, and this has motivated research into ways to reduce power consumption and increase the compactness of the handset unit. Personal handsets also differ from conventional cellular telephones in that they are designed to stand up to long hours of use with a long standby time of days, not hours. Also, their use of a local zone (cell) system makes it possible to make effective use of the available frequency allocation. In these respects, the concepts are similar to those developed in Europe during the late 1980s and early 1990s. Since the early beginnings of PHS, Nippon Telegraph and Telephone Corporation (NTT) and other companies have been aggressively pursuing the research and development of the necessary network and wireless access technologies to support what has become the standardised Personal Handyphone System.

9.2 Standards Development

The Telecommunications Technology Council, a consulting organisation for the Ministry of Posts and Telecommunications (MPT) has had the prime responsibility for coordinating PHS standardisation, which has been addressed at two levels, air interface and network standards, with detailed responsibilities delegated to other bodies, explained below [1]. In this respect PHS standards are more akin, in concept, to the European GSM cellular standards, as opposed to CT2 and DECT, which are primarily air interface standards, and also have similarities to the new CTM standards emerging in Europe, already described in Chapter 6. A very brief outline of the air interface and network standards is given below; for more detail see Chapter 23.

9.2.1 Air Interface Standards

The PHS air interface has been developed under the auspices of the Research and Development Centre for Radio Systems (RCR), based in Tokyo. Early RCR work began in mid-1990, with initial activities on propagation tests in the 1.9 GHz band. This work led to the emergence of an interim draft standard in mid-1991. In July 1992, in Tokyo, RCR carried out a verification experiment of this draft standard. This experiment involved some 47 companies, of which NTT was one. NTT provided the network functions, basestations and terminals, with other companies connecting their terminals into the NTT trial network.

Further refinement of the air interface standard led to the publication in March 1993 of the draft air interface standard STD-28. System verification testing began in the latter part of 1993 and the final version of STD-28 was published at the beginning of 1994 (although further enhancements may still follow).

The essential nature of the PHS air interface is similar to the European standards, in that it employs a TDMA/TDD structure, with four slots – less than DECT, but more than CT2; a more spectrally efficient modulation scheme has been adopted, $\pi/4$ DQPSK, and operation is in the frequency band 1.895–1.918 GHz [2]. An important difference from the European systems is the concept of a higher-power public basestation, or cell station, rated at 500 mW, compared to the 10 mW rating for the personal handset and private basestation. The use of a higher-power and improved-sensitivity cell station results in better ranges and hence a more economically acceptable public access network investment requirement. Furthermore, the system has been designed to allow direct handset-to-handset calls to be established; this mode of communication is possible over a range of about 100 m between handsets registered to the same home basestation.

9.2.2 Network Standards

The network supporting the PHS air interface is based on an ISDN system, making it possible to offer packet mode communication and other non-voice services, together with a variety of value-added services. This is an important feature of PHS, which is seen in Japan as a vehicle for wireless multimedia applications [3].

The network interfaces and signalling system have been developed under the responsibility of the Telecommunication Technology Committee (TTC); this committee is a standardisation organisation for fixed network interface issues. The extent of the standardisation at TTC, and at RCR, is shown in Figure 9.1.

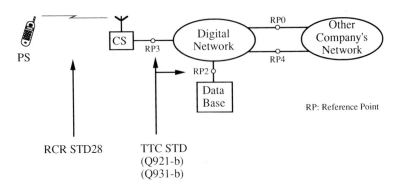

Figure 9.1   Standardisation of PHS: responsibilities of RCR and TTC.

Two types of basestation are employed in PHS – the user station, in the indoor application at home or in the office, and the cell station (CS), for outdoor service. The network interface for the former scenario is the conventional PSTN user-network interface, for which no new standards are required; the latter requires a new CS–telecommunications network interface, which has been defined by the TTC. This network interface is based upon an ISDN interface, with layer 1 (physical layer), layer 2 (data link layer) and layer 3 (network layer) essentially identical to ISDN protocols developed by ITU-T, but with application parts enhanced in order to support PHS-specific functions, such as location registration, handover and authentication. Two standards were published by the TTC in November 1993 relating to this new interface: the Public Cell Station – Digital Network Interface Layer 2 and Layer 3 specifications. The reference points of the interfaces, X1 and X2, are shown in Figure 9.2.

The recommendations do not define items regarding matters between network administrations. The D-channel packet protocol on layer 3 also is not defined. The protocol architecture for PHS signalling is summarised in Table 9.1.

Further standards, including Internetwork Interfaces for PHS Roaming and Public Cell Station–PHS Service Control Procedures, were published at the end of 1994. The internetworking standards (JT-Q761, 762 and 763) relate to the functions required to connect a PHS call between different PHS carrier networks.

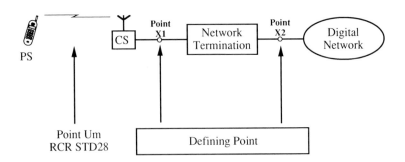

Figure 9.2   Network interface reference points.

Table 9.1
Network standards: layers 1–3

	Reference points	
	X1	X2
Layer 3	TTC Standard JT-Q931b	
	TTC Standard JT-Q932a	
Layer 2	TTC Standard JT-Q921b	
Layer 1	TTC Standard JT-I430	TTC Standard JT-G961

The roaming standard (JT-Q1218a), finalised in November 1994, defines the functionality needed to enable a PHS user to make calls via a PHS network with which the user has no contact.

9.2.3 Standards Evolution

Following successful field trials, described below, PHS was introduced commercially in Japan during 1995, initially as a public access system. A number of other enhanced applications have also been proposed, such as wireless local loop, high-speed data transmission – $32\,\mathrm{kb\,s^{-1}}$ or $64\,\mathrm{kb\,s^{-1}}$ transmission (the latter using two time slots) – various network-based service enhancements, enhanced roaming and introduction of a half-rate codec. Responsibility for standardisation of these aspects resides with RCR.

Further extensions to the basic PHS standards are ongoing. One such is aimed at wireless local loop (WLL) applications and another at PHS over cable TV (CATV) networks. Experiments to verify such operation have already been successfully undertaken.

The PHS has been developed in parallel with the new Japanese PDC (Personal Digital Cellular) system, and both have been conceived within an overall evolutionary framework. Just as dual-mode GSM/DECT terminals have been announced in Europe, it is envisaged that dual-mode PDC/PHS terminals may also be developed in Japan. The future longer-term outlook of land mobile communication systems in Japan is shown in Figure 9.3.

9.3 PHS Products

NTT has already developed complete systems that make it possible to provide application services as a nationwide service, pioneering location registration procedures and the handover feature used during calls, and in the processing of incoming and outgoing calls. NTT was also the first to adopt a new technique that allows the use of an identification procedure in PHS telecommunications. As well as infrastructure products, NTT has developed PHS handsets and personal basestations.

Example specifications for an early basestation and personal station are given below:

- *Basestation* (example) Volume $4800\,\mathrm{cm^3}$, weight $3\,\mathrm{kg}$
- *Personal station* (example) Volume 100–$200\,\mathrm{ml}$, weight 95–$200\,\mathrm{g}$, talk time $4\,\mathrm{h}$, standby time $100\,\mathrm{h}$

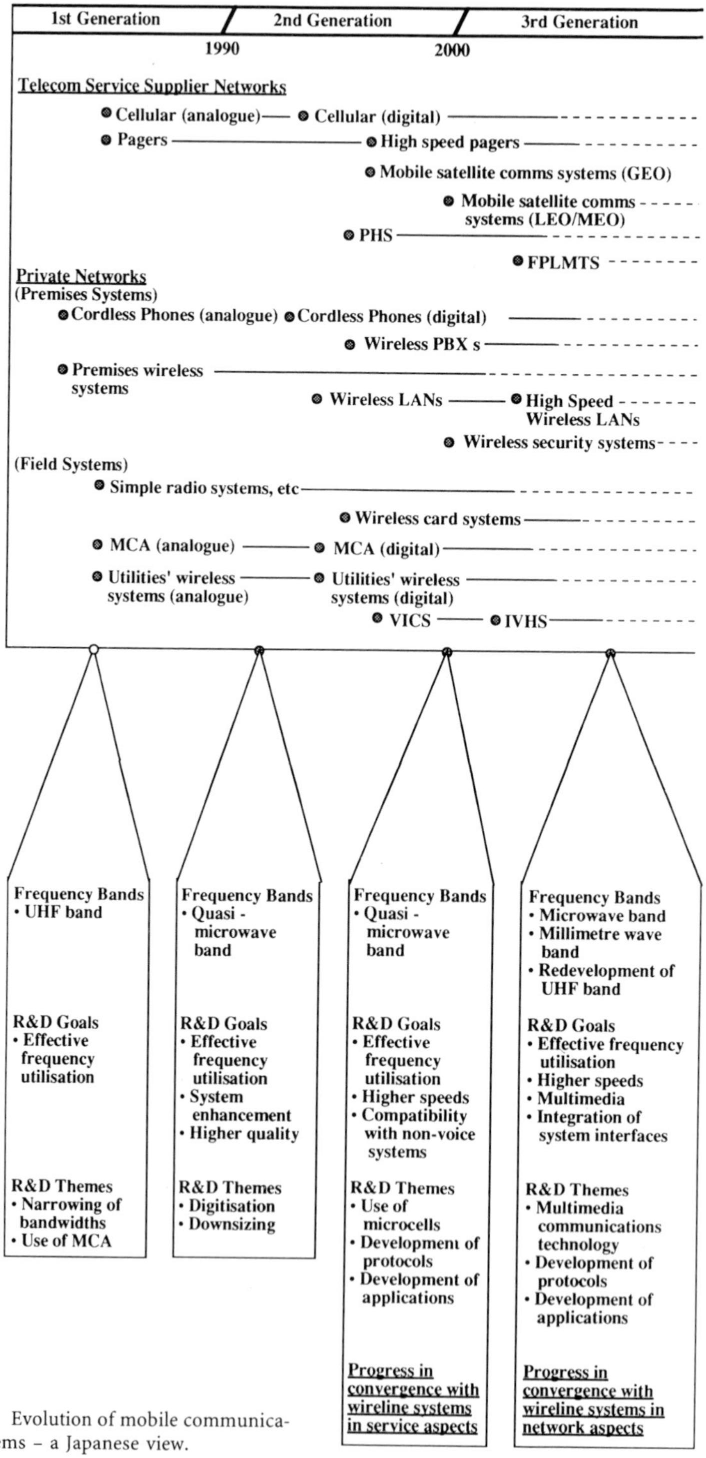

1st Generation	2nd Generation	3rd Generation
	1990	2000

Telecom Service Supplier Networks

- ● Cellular (analogue)——— ● Cellular (digital) ————————- - - - - - - - - -
- ● Pagers ———————————● High speed pagers ———— - - - - - - - - - -
- ● Mobile satellite comms systems (GEO)
- ● Mobile satellite comms - - - - - systems (LEO/MEO) - - - - - - - - -
- ● PHS ——————————— - - - - - - - - - - - -
- ● FPLMTS - - - - - - - -

Private Networks
(Premises Systems)

- ● Cordless Phones (analogue) ● Cordless Phones (digital) ——— - - - - - - - - - -
- ● Wireless PBX s ———— - - - - - - - - - - -
- ● Premises wireless systems ——————————— - - - - - - - - - - - - - - - - - -
- ● Wireless LANs ——— ● High Speed - - - - - - - Wireless LANs
- ● Wireless security systems- - - -

(Field Systems)

- ● Simple radio systems, etc——————————— - - - - - - - - - - - - - - - -
- ● Wireless card systems ——————- - - - - - - - -
- ● MCA (analogue) ——————● MCA (digital) ———— - - - - - - - - - - - - -
- ● Utilities' wireless systems (analogue) ——————● Utilities' wireless systems (digital) ——————— - - - - - - - - - - -
- ● VICS ——— ● IVHS——— - - - - - - - -

| Frequency Bands
• UHF band

R&D Goals
• Effective frequency utilisation

R&D Themes
• Narrowing of bandwidths
• Use of MCA | Frequency Bands
• Quasi - microwave band

R&D Goals
• Effective frequency utilisation
• System enhancement
• Higher quality

R&D Themes
• Digitisation
• Downsizing | Frequency Bands
• Quasi - microwave band

R&D Goals
• Effective frequency utilisation
• Higher speeds
• Compatibility with non-voice systems

R&D Themes
• Use of microcells
• Development of protocols
• Development of applications

Progress in convergence with wireline systems in service aspects | Frequency Bands
• Microwave band
• Millimetre wave band
• Redevelopment of UHF band

R&D Goals
• Effective frequency utilisation
• Higher speeds
• Multimedia
• Integration of system interfaces

R&D Themes
• Multimedia communications technology
• Development of protocols
• Development of applications

Progress in convergence with wireline systems in network aspects |

Figure 9.3 Evolution of mobile communications systems – a Japanese view.

NTT will continue its efforts to make these units even more compact and increase the length of operation time. Figure 9.4 shows an example of an early PHS handset and basestation. Further information on PHS products is included in Chapter 10.

Responsibility for verifying the technical performance of cell stations and personal handsets, in both public and private applications, resides with the Radio Equipment Inspection and Certification Institute of Japan (MKK). Such verification entails both techical standard conformity certification and conformance tests. The purpose of the former is to confirm that the equipment will not interfere with other radio equipment, whilst the latter validate the air interface between the handsets and cell stations. Table 9.2 summarises the equipment parameters tested by MKK to issue Certificate of Conformity with the Technical Standard.

Table 9.2
Equipment parameters tested as part of the MKK conformance testing

Transmitter characteristics	Frequency deviation
	Occupied bandwidth
	Spurious emission values
	Antenna power deviation
	Adjacent channel power
	Leakage power when carrier is off
	Transmission speed of modulation signals
Receiver characteristics	Limit of collateral radiation
Other	Call-name memory device and identification device

Source: from [4].

9.4   Commercial Frameworks

9.4.1   The PHS Field Trials

Commencing in October 1993, an extensive set of field trials of PHS equipment began in various locations across Japan, involving many different organisations. The Conference for Personal Handy Phone System Field Trials was established in July 1993, as a mechanism to facilitate exchange of information between trial participants and to promote their smooth implementation, and produced its final report in October 1994 [5].

The trial objectives were comprehensive. They were to demonstrate technical feasibility, service verification and interoperability at all levels – between different manufacturers' equipment, between cell stations and fixed network, and between local and national networks. The functions explored as part of the service verification activity were outgoing and incoming calling, speech quality, location registration, authentication, billing, operations and maintenance (O&M), and numbering and routing. The trials were also targeted at identifying how PHS service would be perceived and received by the public, including acceptable equipment prices and tariffs. Overall, the trials sought to establish the necessary conditions for rapid commercialisation of PHS. The comprehensive

Figure 9.4   Example PHS basestation and handset from NTT. (*Source*: NTT.)

nature of the trials resulted in clear conclusions that the PHS standards were technically sound and could support effective services and provided clear guidelines for commercial service that has shaped commercial network implementation.

The types of organisation involved in deploying trial networks included national and local telecommunications network operators, cable television operators, utility companies, research bodies, etc., and included the following:

- Nippon Telegraph and Telephone (NTT)
- Daini Denden Inc. (DDI)
- Japan Telecom (JT) and Kokusai Denshin Denwa (KDD)
- Teleway Japan Corp. (TJC)
- Sikoku Information and Telecommunication Network (STNet) and others
- Tokyo Telecommunication Network (TTNet) and others
- Kansai Personal Handy Phone Research Association
- Tokyo Telemessage and others

Amongst these trial networks a full range of potential service capabilities was established, including operation in private, business and public environments. Several private PHS subnetworks were constructed and successfully interconnected to public networks, and thence to the international telephone network. Access connections to cell stations were demonstrated by a range of technologies including fibre, wired line, cable TV lines and infrared transmission. Interoperability was successfully demonstrated, with equipment from some 35 manufacturers used in the field trials programme.

A special task force responsible for evaluating the trials results issued a report that gave their analysis and assessment of the trials. The report concluded that PHS service would be well received by consumers and that the technologies involved were at the practicable service level. Based on the conclusions from the trial, the MPT developed a framework for commercial PHS services. Details of the NTT field trial in Sapporo are summarised in [6].

9.4.2 PHS Operators

The Ministry of Posts and Telecommunications (MPT) published initial PHS licensing guidelines in June 1994 [7]; further additions to these guidelines were published in November 1994 [8]. The objective of the licensing guidelines was rapid promotion of nationwide PHS on a low-cost and good-quality basis.

The MPT decided to issue up to three PHS operating licences in each of the 11 blocks into which the populated area of Japan will be divided. These 11 regional blocks consist of Hokkaido, Tohoku, Kanto, Shinetsu, Tokai, Hokuriku, Kinki, Chugoku, Shikoku, Kyushu and Okinawa. Initially, 12 MHz of the 1.9 GHz frequency bandwidth will be shared between up to three carriers in each block; another 11 MHz will be allocated for private system usage. Each licensed operator has to provide coverage to 50% of the population within five years.

Three groups have emerged as major players in the PHS business as operators – the NTT Personal Group, DDI Pocket Telephone Group and the Astel Group – and service has naturally started from big cities like Tokyo, Osaka and Nagoya, then other cities.

NTT has established nine business planning companies for PHS operation, as joint ventures with its subsidiary NTT Mobile Communications Network Inc. (NTT DoCoMo) and the large trading houses Itochu and Marubeni. Other investors include the international operator Cable and Wireless and other small shareholders. NTT is investing 70% and NTT DoCoMo 30% of the funds required to start these companies. NTT launched PHS service in July 1995 with some 25 000 standard cell stations in Tokyo, using a conventional small cell structure, 150–200 m, with antennas based on telephone boxes.

Daini Denden Inc. (DDI), one of Japan's long-distance carriers, set up a wholly owned subsidiary in July 1994 in preparation for operation of its nine regional DDI Pocket Telephone companies. DDI's investment ratio is expected to reach 75% after it has completed its full investment in each company. DDI has developed the 500 mW cell station to minimise costs. Like NTT Personal, DDI launched PHS service in Tokyo in July 1995, although with a less conventional architecture, using its high-power/improved-sensitivity rooftop cell stations. Whilst potentially enabling lower infrastructure costs, through the use of larger cells, 300–500 m sizes, DDI experienced interference problems in the first two months of network operation. Although this required network reconfiguration, these initial technical problems appear now to have been resolved.

A third PHS operation is being run by a consortium of companies with different areas of concentration across the country. In the Tokyo area, the Telewalker consortium was set up by TTNet (Tokyo Telecommunications Network), Japan Telecom, Mitsui & Co., Mitsubishi and Sumitomo, each with a 14.5% participation. Meanwhile, new carriers affiliated to the local electricity power companies, and others, have set up the Astel Group to cover areas outside

Figure 9.5 Example PHS public basestation in Japan. (Photograph courtesy of P White, PA Consulting.)

Tokyo. Astel launched service in October 1995 with an infrastructure similar to NTT's, with cell stations mounted on power poles along city streets. In addition to the standard 20 mW cell station, Astel has also developed a 100 mW cell station.

All three operators are implementing a very rapid rollout within the first two years of launch. It is anticipated that some 60 000 to 70 000 standard cell stations will be needed to provide coverage of most of Tokyo and its suburban area. Using higher-power cell stations, some 7000 to 10 000 should suffice. Estimates of costs of cell stations range from 100 000 Yen for a standard cell station up to ten times this figure for a high-power cell station [9]. A typical PHS basestation is shown in Figure 9.5.

9.4.3 Network Charges and Product Prices

Access charges to the existing fixed network infrastructure are of vital importance for operator profitability. The PHS operators have negotiated access charges for use of NTT's ISDN network to link their exchange systems and basestations. As NTT will also operate PHS services through its subsidiary, NTT

Personal Communication Network, and the Astel Group are also relying partly on NTT's ISDN lines, the charging basis must be reasonable.

Charges to the PHS subscriber reflect the market research conducted as part of the field trials. Initial network connection charges are around 7200 Yen, with a basic monthly charge of about 2700 Yen and call charges around 40 Yen per three minutes for a local call in the Tokyo area. This level of call charge is only slightly higher than a public payphone charge (30 Yen) and around one-quarter that of the current cellular telephone rate. For longer-distance calls the difference between PHS and cellular call charges decreases, but PHS remains significantly cheaper.

Network operators have negotiated with potential suppliers to establish an early and adequate supply of suitable handsets. Many local Japanese suppliers have entered the PHS market – such as Kenwood, Kyocera, Matsushita, Mitsubishi, NEC, Sanyo, Sharp, Sony and Toshiba – but overseas suppliers, such as ATT and Motorola, are also keen to gain access to the market. It is anticipated that handsets could be priced as cheaply as 20 000 Yen, or even less, depending upon the level of competition; in fact, market take-up has been substantial despite initial product prices of between 38 000 and 48 000 Yen.

9.4.4   Numbering Plans

Every PHS carrier has adopted a numbering system that gives users a ten-digit phone number 050–XX–XXXXX consisting of (PHS service identifier)–(carrier identifier)–(subscriber's number for PHS). This scheme was well received in the earlier field trials.

To make PHS service available nationwide from a single handset, the PHS carrier must provide a roaming capability that goes beyond his own service territory. When two PHS carriers operating in the same territory agree to provide an intra-territory roaming service, they need to ensure the details of the service are known to subscribers.

9.4.5   Market Forecasts

There are strong hopes that the PHS market will be a major business field. An MPT panel has estimated that there could be as many as 38 million PHS subscribers by 2010 – this assumes the provision of seamless coverage service areas; even with spotty service areas, however, a user base exceeding 20 million is forecast.

The PHS market for service and terminals is projected to reach 1.3 trillion Yen in ten years and 1.9 trillion Yen in 15 years. It is expected that the growth of the PHS market will hinge on how attractively the PHS carriers can set their rates from the early stage.

To promote efficient installation of PHS basestations, the relevant authorities are being encouraged to take the necessary measures to make public facilities available to PHS service providers under suitable and impartial conditions – facilities such as traffic-light equipment, utility poles, public telephone boxes, etc. In addition, to maximise the market potential, PHS service providers will need to develop basestations that can be used with a common air interface.

9.5 Summary

This chapter has described the process by which the Japanese Personal Handyphone System was conceived, standardised and introduced. An important part of the process was the comprehensive field trial programme, which proved the technology and established important information with which to define the emerging commercial framework, which has also been described. The systematic and structured approach to the development of a cordless standard in Japan has, as was expected, resulted in rapid early acceptance of this new form of communication; this itself will encourage adoption of PHS in other countries in future years.

References

[1]   "Standardisation of Personal Handy Phone (PHP)", T Habuka and H Sekiguchi, NTT Review, Vol. 5, No. 5, pp. 101–5, September 1993
[2]   "Towards the Personal Communications Era – a Proposal of the 'Radio Access' Concept from Japan", K Ogawa, K Kohiyama and T Kobayashi, International Journal of Wireless Information Networks, Vol. 1, No. 1, pp. 17–27, January 1994
[3]   "Multimedia Wireless Communication System using Personal Handy Phone System", Y Tanaka, J Takeuchi, M Suzuki and M Morozumi, National Technical Report, Vol. 40, No. 6, pp. 56–61, December 1994
[4]   "PHS MoU Group News", Vol. 1, Issue 2, December 1995
[5]   Final Report of the Conference on Personal Handy Phone System Field Trials, October 1994
[6]   "On the Way to Personal Communications: PHS Field Trial in Sapporo", T Kobayashi, NTT Review, Vol. 6, No. 4, pp. 59–64, July 1994
[7]   Press Release from the Ministry of Posts and Telecommunications, dated 24 June 1994
[8]   Press Release from the Ministry of Posts and Telecommunications, dated 1 November 1994
[9]   "PHS Has Got Into Commercial Service", Y Ito, presented at the Digital Cordless Telephone Conference, Hong Kong, September 1995

10 Industry Development and Products

Diane Trivett

As described in the preceding chapters, the first digital cordless telecommunications standards – CT2 and DECT – were pioneered in the late 1980s in Europe. Canada subsequently built upon the CT2 standard, with its evolution to CT2Plus; Japan developed its PHS standard in the early 1990s; whilst the USA is still, in the mid-1990s, developing standards for unlicensed PCS, building on all these standards and more. As one would expect, product developments have reflected these timescales, with products to the CT2 CAI standard having been available since 1989, to the DECT standard since 1992 and to the PHS standard since 1995.

The penetration of cordless working in the residential and business environments in Europe is still restricted by price. In the business market, the cost of a cordless extension is still at least three times that of a standard corded extension, and as a result volumes have remained low. To realise market volume, this price differential needs to fall even further. Distribution is a key factor for success in the cordless market. The PTOs play a significant role in the distribution of cordless telephones and systems, and are a much sought-after distribution channel for manufacturers. Today, business cordless products are mainly distributed through the same channels as their corded counterparts. However, more specialised distribution channels for mobility solutions are expected in the future. Increasing liberalisation of telecommunication provision worldwide and the capability of cordless technology to support wireless local loop applications may both be expected to change this picture.

In this chapter we provide some historical background to the development of the digital cordless industry; more detail on this aspect is given in [1]. We then describe the range of products available in the mid-1990s from the various manufacturers for the different application markets[1]. Finally, we consider the issue of global competition, which continues to grow in importance.

10.1 CT2

10.1.1 Origins

Cordless telephones based on analogue CT0 technology first appeared in Europe in the 1970s as a simple replacement for their corded counterparts. The customer interest generated in some areas meant that they quickly ran into capacity and interference problems; other difficulties encountered included poor security and below-average speech quality. In many countries, where spectrum had not been allocated, such products were made illegal.

To overcome these problems, a new-generation analogue cordless technology was developed, known as CT1. This was released by CEPT (the European Conference of Posts and Telecommunications) in 1983. However, limitations were still in evidence, for example, roaming and handover. In 1985, CEPT commenced its own research into a digital-based, second-generation cordless telephony standard. By this time, work on other concepts had already begun. One such technology was CT2.

The CT2 specification, which originated in the UK in 1984, was developed by several British manufacturers, with encouragement from the UK's Department of Trade and Industry (DTI). The specification was submitted to CEPT as a potential pan-European standard. With competition from the then-emerging Swedish CT3 standard, CT2 was not accepted, but instead a compromise was agreed, which became the DECT standard. CEPT did not provide a firm timetable for the introduction of DECT and, in any case, its decisions were not mandatory, allowing the launch of other digital products including CT2 and CT3 prior to DECT.

CT2 was introduced to the UK in 1989. Unfortunately, it was inextricably linked with the telepoint services licensed at this time, which did not meet customer expectations. The failure of telepoint did affect the early success of CT2 in Europe, but was not its ruin.

10.1.2   Early Products and Services

Early CT2 products were mainly for the public services in operation at the time. In the UK, telepoint began to make its mark in 1989 when four consortia – BYPS, Phonepoint, Zonephone and Callpoint – were awarded licences to construct and operate telepoint networks. All but BYPS launched services in the autumn of that year based on proprietary CT2 products – all three services were withdrawn in 1991. BYPS was bought by Hutchison Telecom in 1991 and launched its Rabbit telepoint service when CAI equipment became available in 1992. Again, optimistic forecasts were not met, and Hutchison closed the service in 1993 to focus on its Orange personal communications service, based on DCS1800 (GSM at 1.8 GHz), which was launched in 1994.

The products for these networks were sourced mainly from the early pioneers of CT2 equipment, which included BT, Ferranti (Libera), GPT, Orbitel, Shaye and STC. In 1989 these companies agreed to share essential intellectual property rights and established a group now known as the Founders Group. BT, GPT, Orbitel, Shaye and STC are still active in the market today, although the company structure and names have changed in most cases, reflecting the increasing globalisation of the telecommunications industry – see Table 10.1. Today, the Founders Group has evolved to include members drawn from Europe, North America and Japan.

Early CT2 adoption in other European markets again focused around the telepoint services on offer in countries such as Belgium, Finland, France, Germany and the Netherlands. Today, all but France and the Netherlands have closed their services.

Table 10.1
Evolution of the Founders Group

1989	1996	Comment
British Telecom	British Telecom	No change
Ferranti (Libera)	Kenwood	Kenwood acquired Libera's patent portfolio in 1993
GPT	GPT	60/40 GEC/Siemens joint venture
Orbitel	Orbitel	Ericsson acquired full ownership of Orbitel in 1996
Shaye	AT&T	AT&T acquired Shaye in 1993[a]
STC	Nortel	Nortel acquired STC in 1991

[a] AT&T changed its name to Lucent Technologies in 1995.
Source: CAI Founders Group.

10.1.3   Common Air Interface (CAI)

As CT2 technology developed, so did its range of applications and the need for a common interface specification. To meet this requirement, British manufacturers and operators developed the Common Air Interface (CAI), with GPT and Orbitel being the early manufacturing players to go via this route; Shaye (now Lucent) also followed down this route, developing a successful range of CAI products.

The CAI standard allows one manufacturer's handset to address all applications and to communicate with other manufacturers' basestations. It was published as DTI specification MPT 1375 in 1989. This approach also proved attractive to other European countries, and CT2 CAI was proposed as a pan-European standard to ETSI (European Telecommunications Standards Institute), backed by an operators' MoU (Memorandum of Understanding).

An I-ETS (Interim European Telecommunications Standard) for CT2 CAI was accepted by ETSI in 1992. A revised version of the standard (I-ETS 300 131 Edn 2) was accepted in August 1994. Discussion as to whether to promote CT2 CAI to full ETS (European Telecommunications Standard) status is still ongoing. At the last ETSI debate on this issue in spring 1995, it was decided to discuss the situation again during the normal course of a standard's review. It will, therefore, be 1997 before the issue is revisited.

10.1.4   CT2 CAI Equipment Availability

CT2 CAI equipment for all applications is available worldwide (see Table 10.2). In Europe, CT2 CAI solutions are sold in all markets where the 864–868 MHz frequency band has been made available for this standard. Countries that have not yet freed the required bandwidth include Denmark, Italy and Spain.

Where CT2 has been adopted outside of Europe (with the exception of Canada, where additional spectrum has been granted), the CAI technical standard has been adopted with minimal change, such as a slightly different frequency allocation. In most countries the standard is referred to by its ETS identity, although in some countries national designations have been given (Table 10.3).

Table 10.2
CAI schedule

Country	UK, Finland, Hong Kong, Australia, Canada, Argentina, Brazil and USA (restricted to PCI)
I-ETS 300 131 Edn 2 extending to	Austria, Belgium, Bulgaria, Czech and Slovak Republics, Cyprus, Denmark, Finland, France, Germany, Greece, Hungary, Iceland, Ireland, Italy, Luxembourg, Malta, Monaco, Netherlands, Norway, Poland, Portugal, Romania, San Marino, Spain, Sweden, Switzerland, Turkey, UK, Vatican City and the former Yugoslavia
I-ETS 300 131 procured by	Singapore, Malaysia, Thailand, China, Taiwan and South Africa
Possibilities	Korea, Venezuela, Vietnam, New Zealand, Zimbabwe, Qatar, Bahrain, Mexico, Ethiopia, China, India, Indonesia, Botswana and Namibia

Source: CAI Founders Secretariat (December 1995) [2].

Table 10.3
Local national references for I-ETS 300 131

Country	National designation
Argentina	3078/CNT 93
Australia	AUSTEL TS019
Canada	RSS 130
Finland	SFS 5604
Hong Kong	HKTA1015

10.1.5   CT2 CAI Products

CT2 CAI products are available for many roles. A brief overview of product availability for the standard's most common applications is provided below.

Personal Communications Services (PCS)

CT2 is currently the standard on which most operators have based their telepoint services. Telepoint networks are available in Europe, Asia–Pacific and China. Worldwide estimates for total subscribers vary from 500 000 to one million.

Europe has not been welcoming to CT2 telepoint services, as described in an earlier chapter. Today, only two of the original telepoint networks are still in service – Bi-Bop in France and Greenpoint in the Netherlands. In Asia, the previously buoyant Hong Kong telepoint market has been squeezed by cellular during 1995–96, with operators closing networks and migrating users to cellular as costs have fallen, as previously occurred in Europe.

France Télécom began trials of its telepoint service, Bi-Bop, in 1991, followed by commercial service in January 1993. The development of the network was delegated to Dassault Automatismes et Télécommunications, which is also involved in the supply of handsets with Matra and Sagem. France Télécom had nearly 100 000 subscribers to its Bi-Bop service by the end of 1995. A second telepoint operator, Prologos, was licensed earlier in 1995 to operate in south-west France.

The Netherlands' telepoint service, Greenpoint, was launched in 1992 by PTT Telecom. Greenpoint is based on Motorola infrastructure equipment and handsets. There were some 60 000 subscribers on the service at the end of October 1995.

Residential Use

The availability of CT2 CAI products for the residential market varies by country. In France, they have been available since 1991 when trials of the Bi-Bop telepoint service commenced. France Télécom offers CT2 CAI cordless telephones as part of its residential portfolio; these are manufactured by Sagem, Matra and Dassault. Sagem's Bi-Bop phone weighs just 150 g and offers up to 4 h talk time with 36 h standby. Motorola, Dassault, Sagem and Sony also supply residential CT2 CAI telephones through retail channels.

In Germany, Sony is currently the only manufacturer supplying CT2 CAI-based residential cordless telephones. It provides a range of cordless handsets and basestations that can be packaged in various combinations through audiovisual retail outlets. Three different handsets and two basestations are available, in three packages: the DCT 200, DCT 201 and DCT 204. One basestation includes a corded handset plus memory. The products also offer Sony's "Multilink" system, which allows one basestation to support up to eight handsets and four basestations to be linked together to provide site coverage.

In the Netherlands, PTT Telecom has led the market for digital cordless telephones by launching residential CT2 CAI telephones under the same brand name as its telepoint service, i.e. Greenpoint. For residential use, these are available with a choice of basestations.

Sales of digital residential cordless telephones in the UK have in the past been in conjunction with the failed CT2 CAI-based telepoint services, but since then this market has disappeared.

In the rest of Europe, suppliers of CT2 CAI-based digital cordless residential telephones are Sony and Motorola, although Motorola has now announced its intention to stop supplying consumer outlets.

Business Use

CT2 CAI-based cordless PBXs (WPBXs) can be found in most European countries, with the exception of Italy, Spain and Denmark, where no frequency allocation has been made for this standard. The availability and success of these systems vary by country. Table 10.4 provides an overview of the products on offer in 1995.

In France, the first manufacturers to make CT2 CAI WPBXs available were Nortel, Matra and AT&T Barphone. Nortel supplies its Companion product range directly and through a distribution agreement with Matra. AT&T Barphone provides integrated cordless working on its Generis PBX and the Domoline 6. Other CT2 CAI systems on the market include the Tangara from Eritelcom (formerly SAT). CT2 CAI cordless telephones are currently mainly sourced from Motorola, Sagem and AT&T Wireless Communication Products (WCP). France Télécom supplies CT2 CAI-based WPBXs within its portfolio as part of its Diatonis range. These products are supplied by Secom and Discofone.

Table 10.4
CT2 CAI WPBX products available in 1995

Product	Maximum capacity	Configuration	Additional features/interfaces
Ascom Tateco CTS 902	100 cordless handsets 50 basestations	Adjunct	
Lucent Domoline 6	10 cordless handsets 2 exchange lines	Integrated	
Lucent Generis	80 cordless handsets 20 basestations	Integrated	
Lucent Forum WBS	32 cordless handsets 49 basestations 2 external lines	Stand-alone, Adjunct	Voice prompts/confirmations, roaming capability, handover
GPT iSDX-Micro	120 handsets 40 basestations	Integrated	Handover, roaming, integrated management, numbering and feature access
GPT iSDX-S	500 handsets 80 basestations	Integrated	Handover, roaming, integrated management, numbering and feature access
GPT iSDX-L	1000 handsets 400 basestations	Integrated	Handover, roaming, integrated management, numbering and feature access
GPT Realitis DX	1000 handsets 400 basestations	Integrated	Handover, roaming, integrated management, numbering and feature access
GPT Wanderer	36 handsets 40 basestations	Adjunct	Identical features to host PBX, handover, roaming
Multitone CS500	152 cordless handsets 128 basestations	Adjunct	Identical features to host PBX, automatic handover, roaming
Nortel Meridian 1 Companion	240 users 128 basestations	Integrated	Meridian 1 feature set
Nortel Companion 10	32 cordless handsets 15 basestations 6 external lines or PBX extensions	Stand-alone, Adjunct	Identical features to host PBX
Nortel Companion 100	80 cordless handsets 80 basestations	Adjunct	Identical features to host PBX
Nortel Companion 200	152 cordless handsets 128 basestations	Adjunct	Identical features to host PBX
Norstar Modular 32	32 handsets 15 basestations	Integrated	Identical features to host PBX
Peacock[a] RTS 12-50	100 cordless handsets 20 external lines	Stand-alone, Adjunct	Enhanced handover, intelligent handset search

[a] Peacock was acquired by Orbitel in 1996.
Source: Dataquest (December 1995) [3].

CT2 CAI will retain a significant proportion of the digital WPBX and cordless telephones markets in France, compared with its share in most other European countries. This is in part due to the number of available CT2 CAI solutions and the continued development of these products. For example, Matra has now introduced its own CT2 CAI basestations for the MC6500 PBX range, and Ascom Tateco has launched a CT2 CAI product. CT2 CAI working is also being added to the new ranges of ISDN PBXs that are now appearing on the market, for example, Discofone's 1208 IS system. Discofone has also enhanced its 1206 system with a low-capacity integrated CT2 cordless facility.

In Germany, the push from DECT manufacturers has ensured the dominance of this standard in the German business sector, although there remains a market for CT2 products. CT2 telephones and systems are currently available from Nortel, Peacock and Multitone, an OEM supplier of Nortel's Companion WPBX. Telephones supplied with CT2 CAI systems are presently mainly sourced from Lucent and Motorola.

In the Netherlands, PTT Telecom is still the dominant supplier in the market for PBXs and from this position has driven the growth in the cordless PBX market. It supplies both the Companion CT2 CAI system from Nortel and the Freeset DECT system from Ericsson. Both products are distributed through this channel on an exclusive basis, as is the case with its corded PBXs. Peacock and GPT are the other manufacturers of CT2 CAI systems in this market, supplied independently through distributors and dealers. CT2 CAI telephones are generally sourced from GPT, Motorola and Lucent.

The leading CT2 CAI PBX manufacturer in Sweden is Nortel, helped by its distribution agreement with Telia. Nortel systems are mainly supplied with cordless telephones from Lucent. The only other CT2 CAI vendor present in the market is Peacock (Orbitel).

In the UK, GPT and Nortel were the first manufacturers to enter the CT2 CAI systems market. GPT has developed a range of WPBXs including integrated cordless extensions on the iSDX, and the recently launched Realitis DX. Nortel manufactures its Companion range providing stand-alone and adjunct solutions. An integrated solution for the Meridian 1 PBX will be launched in the UK in 1996. Other CT2 CAI suppliers are Peacock with the RTS 12-50, Lucent with the Forum WBS, and Multitone. Both Nortel and GPT now supply handsets manufactured by Lucent. Motorola has in the past been one of the biggest manufacturers of CT2 handsets in Europe and has an agreement with Peacock for the development of CT2 technology used in its WPBXs. However, this will change in the future as Motorola moves its focus towards DECT solutions.

In the rest of Europe, the biggest market for CT2 is Finland. In these markets, Nortel is the main supplier along with Peacock and GPT. CT2 CAI cordless telephones are generally supplied by Lucent and Motorola.

10.2   DECT

10.2.1   Origins

The origins of DECT can be traced back to 1987 when ESPA (the European Manufacturers of Pocket Communications Systems Association) finished a four-year study into the various technical approaches to the cordless office. A report was then published recommending TDMA/TDD (time-division multiple access/ time-division duplex) as the best technology for this environment. Following this report, CEPT decided to launch its own initiative into the digital cordless telephony market.

In response to the proposal of CT2 as a pan-European standard, the Swedish PTT put forward an improved version of an original Ericsson technology, now known as CT3, to CEPT. Ericsson announced a test system based on this technology in 1987.

In early 1988, CEPT accepted a compromise based on the Swedish technology, allocated it the frequency band of 1880–1900 MHz and called it DECT. A MoU was signed in 1989 and the basic DECT ETS (standard) was finalised in 1992. An EU Directive came into force in January 1992 which declared that all EU and EFTA countries make the necessary spectrum available for DECT applications.

DECT and CT3 differ primarily in the frequency at which they operate (1.9 GHz versus 900 MHz) and in the number of TDMA time slots (twelve versus eight). DECT is the dominant standard in Europe; outside Europe, CT3 still finds application. A much fuller description of the CT3 standard (there referred to by its earlier acronym of DCT900) is given in [1].

The first DECT product to gain type approval was the Siemens Gigaset in 1992, closely followed by the Olivetti NET[3] cordless LAN product. Early DECT solutions were proprietary, although GAP-compliant DECT cordless telephones are now starting to appear on the market. GAP (generic access profile) is now accepted by ETSI and the common technical regulations are expected in 1996. In the meantime, an interim GAP type approval specification has been established. The introduction of GAP is encouraging more manufacturers to introduce DECT-based cordless solutions for all applications. For example, mobile and telephone specialists, such as Nokia, are introducing handsets that will work with any DECT-based WPBX. Nokia presented its first cordless telephone for the DECT standard at Geneva Telecom in October 1995.

10.2.2   CT3 Products

As already mentioned, CEPT did not provide a firm timetable for the introduction of DECT, and so CT3 and CT2 products appeared on the market ahead of DECT ones.

Ericsson first introduced the CT3 version of its Freeset WPBX in 1991. Today, the company provides three versions of its Freeset cordless system to allow it to compete in markets worldwide: Freeset DECT; Freeset CT3, 4 MHz in the 800 MHz or 900 MHz frequency band; and Freeset CT3, conforming to FCC Part 15 rules. The DECT-based system was launched in 1993.

Freeset is an adjunct cordless solution that is compatible with any PBX or Centrex system. Ericsson is marketing Freeset in over 30 countries worldwide. It is approved in all European markets and several countries in the Far East, Asia–Pacific, South America, Latin America and the USA.

10.2.3   DECT Equipment Availability

DECT equipment for all applications is available worldwide (see Table 10.5).

Table 10.5
DECT approval (December 1995)

Country	Austria, Belgium, Czech Republic, Denmark, France, Finland, Germany, Greece, Hungary, Ireland, Italy, Netherlands, Norway, Portugal, Slovak Republic, Spain, Sweden, Switzerland, Turkey, UK, Australia, Taiwan, Indonesia, Venezuela, Argentina, Brazil, Malaysia, Philippines, Thailand

Source: Dataquest [3].

10.2.4   DECT Products

DECT products are by 1996 available for a wide range of applications. To some degree this relatively rapid product development reflects the fact that some manufacturers have benefited from the learning curve of CT2. A brief overview of product availability for each application area is provided below.

Wireless in the Local Loop

Several vendors have developed a range of solutions for local loop applications based on DECT, and trials of such products have been performed. For example, Philips provides a digital radio subscriber access systems product based on DECT which supports fixed or portable terminals, depending on the application in the public access environment. Siemens provides a product range called DECT*link* which forms part of its Multilink family of access products. It offers all of the services available in the host exchange. From one to four a/b interfaces connect to standard telephone equipment and up to 60 radio channels are supported. Other manufacturers, including Ericsson and Alcatel, are also active in this market.

ETSI is now developing the RAP (RLL Access Profile), a DECT profile for the local loop application. Although DECT faces stiff competition from other wireless options such as cellular, radio, microwave and satellite, there would appear to be a strong operator pull for such products, as evidenced by several papers in DECT '96 [4].

Personal Communications Services (PCS)

DECT already provides two-way working with seamless handover between cells and is developing towards interworking between DECT systems in the public and private sectors. A step towards this is the development of the GSM/DECT Interworking Profile (GIP). GSM/DECT multi-mode handsets are already available from Ericsson and are being used as part of a user trial by Telia in Sweden; other manufacturers and operators are also considering such developments.

In Europe, DECT-based PCS trials have been ongoing since 1994 in several countries, including Norway, Finland and France. In France, CGE is operating an experimental DECT-based Pointel service near Paris. Results of the various trials so far show that DECT has the potential for outdoor local mobility, but is unsuited to wide-area mobility. In this sense it complements GSM.

Wireless Local-Area Networks (WLANs)

Most WLAN solutions available today are based on spread-spectrum technology (SST). However, there is a DECT-based product available from Olivetti, known as NET[3]. NET[3] can run Novell NetWare and Microsoft LAN Manager. The hub supports up to two radio basestations, each supporting 30 terminals. Whether DECT or SST will be the dominant technology in this sector remains to be seen.

Residential Use

DECT cordless telephones for the residential market are now available in most European countries.

CT2 CAI-based residential cordless telephones have been the only products on the French market to date. However, in 1995 Grundig launched the CP-830 DECT cordless telephone and Siemens shipped its Gigaset cordless products to France.

In Germany, Deutsche Telekom continues to play a key role in promoting cordless products. Its current portfolio includes analogue and DECT models. There are now several DECT-based residential cordless telephones on the market, the most popular of which is the Siemens Gigaset (see Figure 10.1). Gigaset is targeted at both the residential and business markets. Two new GAP-compliant products for the Gigaset range were shown for the first time at Telecom Geneva in October 1995 – Gigaset 1030 and Gigaset 1054i. Gigaset 1030 is a basestation with integrated featurephone and digital answering machine, supporting up to six DECT telephones. Gigaset 1054i, which supports up to eight cordless telephones and two analogue extensions, incorporates basic-rate Euro-ISDN access. Other products for this sector are available from Hagenuk, DeTeWe, Philips and Grundig.

Hagenuk DECT residential cordless products are also suitable for the small office. They include the DigiCell A and i. Each basestation supports up to six OfficeHandys, one featurephone and one fax machine. Digicell A supports up to one exchange line and a fax filter. DigiCell i enables the use of Euro-ISDN features and allows the simultaneous use of two exchange lines.

DeTeWe introduced its Twinny DECT in March 1995. Twinny DECT has an optional loudspeaker and extended intercom function. Up to ten handsets and one exchange line are supported. Each handset weighs 220 g and provides a talk time of 10 h with a standby time of 100 h.

Philips markets its CP5000 and CP5002, which are being sold with one and two handsets, respectively; however, the CP5002 can support up to nine handsets. The CP5000 range is aimed at both the residential and the SOHO (small office, home office) markets and is available through retail outlets and via Deutsche Telekom.

Grundig launched its digital cordless offering, CP-830, in January 1995. The CP-830 handset weighs 200 g and offers a talk time of 6.5 h and a standby time of 40 h. Up to six handsets are supported by one basestation, and one handset can be simultaneously connected with up to four basestations. Like the Gigaset, the handsets also support automatic call time and charge evaluation.

The Siemens Gigaset and Grundig CP-830 are also marketed in the Netherlands. In Spain, Telefónica supplies digital cordless products from

Figure 10.1 Siemens Gigaset 1010 and 1030 DECT handset products.

Siemens. DECT-based products are also starting to appear on the Italian and Swedish markets.

In the UK, BT plays a key role in the market. Its cordless product range now includes the Gigaset from Siemens. However, this is currently premium priced.

The digital cordless telephone market in the rest of Europe remained fairly small in 1994 and 1995, but will now begin to grow as additional products become available and awareness increases. Siemens has been the only supplier of a DECT-based product for this sector until recently.

Business Use

DECT-based WPBXs can be found in the majority of European countries. Table 10.6 provides an overview of the products on offer in 1995.

The first supplier of a DECT-based WPBX in France was Ericsson with its Freeset system. In 1994, the company also started to distribute Freeset through its new joint venture company Eritelcom, created from an agreement with the French manufacturer SAT. The only other supplier of DECT-based WPBXs today is Alcatel. Alcatel launched its integrated DECT solution with the Alcatel 4220 and the larger Alcatel 4400 in 1994–95. The company has also developed adjunct DECT solutions, the Alcatel 4672 and 4674, which were also launched during 1994. Other vendors such as Matra are expected to launch DECT systems in 1996.

DECT-based cordless systems dominate the German market and several products are available. Philips supplies an Octo-Cellular solution (based on a DECT-like standard) to Deutsche Telekom. This solution works on the Octopus PBX range, which provides from 10 to 250 extensions. Manufacturers of true DECT products include Alcatel, Siemens, Philips, Ericsson and DeTeWe.

Siemens continues to market the Ericsson Freeset system as part of its Hicom Cordless 300 solution. It also markets the Gigaset range of DECT-based products, which are targeted at both the residential and business markets.

Table 10.6
DECT WPBX products available in 1995

Product	Maximum capacity	Configuration	Additional features/interfaces
Alcatel 4220	120 extensions 36 exchange lines	Integrated	Identical features to PBX, automatic handover, roaming
Alcatel 4400	800 extensions	Integrated	Identical features to PBX, automatic handover, roaming
Alcatel 4672/4674	800 extensions	Adjunct	Features of host PBX, automatic handover, roaming
Ericsson Freeset	5000 cordless extensions	Adjunct	Digital line to MD110 and AXE, analogue line to non-Ericsson systems; seamless handover, roaming, identical features to host PBX
Matra[a] MC 6270	180 handsets 24 basestations	Adjunct	PBX features, automatic handover, roaming
Matra[a] MC 6280	960 handsets 128 basestations	Adjunct	PBX features, automatic handover, roaming
Philips iS Mobile	960 cordless handsets 128 basestations	Integrated, Stand-alone	Automatic handover, roaming, SOPHO iS3000 feature set
Siemens Hicom Cordless 125	63 cordless users 8 basestations	Integrated	Automatic handover, roaming

[a]The Matra products were available as trial systems in 1995 with commercial launch in 1996.
Source: Dataquest (December 1995).

DeTeWe has developed the VariTel DECT cordless system, which can operate with up to ten intercommunicating handsets. It provides one internal speech circuit and one speech circuit to an exchange line or second basestation. Each handset weighs 220 g with a talk time of 10 h and a standby time of 100 h. Cordless working has also been developed for the Varix 200 ISDN PBX. The system should also function as a radio-based server for the Varix Content 840 or as an OEM for any other PBX connected via tie lines.

Other manufacturers, including Bosch Telecom, Matra, Hagenuk and Panasonic, have announced DECT products and these are expected to appear on the market in 1996.

DECT is the only cordless standard supported in Italy and Spain. Products are available in both markets from Alcatel, Ericsson and Philips. Both Ericsson and Alcatel have distribution agreements with the Spanish PTO, which is boosting sales of their cordless solutions.

In the Netherlands, DECT-based digital WPBXs are currently available from Ericsson, which like Nortel has an exclusive agreement with PTT Telecom, and Philips with its DECT iS Mobile solution for the SOPHO.

The leading WPBX manufacturer in Sweden is presently Ericsson, helped by its distribution agreement with Telia. The only other manufacturer currently offering a DECT WPBX is Alcatel.

In the UK, DECT-based systems are available from Ericsson, Alcatel and Philips. These suppliers have yet to develop a significant presence in the UK market, and will find competition difficult against the CT2 suppliers, who have a large installed base of systems into which to sell cordless functionality.

Of the smaller countries in Europe, the biggest market for DECT is Austria. Alcatel, Siemens, Philips and Ericsson all market their products in this area.

10.3   PHS

10.3.1   Origins

CT2 and DECT were well advanced by 1990, when standardisation of PHS began in Japan. PHS thus benefited from much previous work undertaken in Europe and Canada on cordless microcellular technologies. Standardisation was completed in 1993 and verification and market field trials were undertaken in Tokyo, Osaka, Sapporo and Takamatsu between October 1993 and March 1995.

To some degree the development and early local market success of PHS has been accelerated and enhanced by the traditionally high prices of telecommunications services in Japan, both wireline and cellular, compared, say, to Europe – interestingly, it was the opposite phenomenon, serious falls in the price of cellular service, that was a significant contributor to the commercial failure of public cordless services, telepoint, in the UK. Liberalisation of telecommunications service provision is occurring in Japan and could well impact upon the long-term development of PHS service; indeed cellular tariffs declined dramatically, by a factor of 3–4, between 1993 and 1995 [5]. It is unclear, from initial indications of market growth, whether PHS will be sufficiently established at an early enough stage to be able to weather the effects of such continuing price erosion.

10.3.2   PHS Equipment Availability

PHS equipment availability was initially focused solely on the Japanese market. However, PHS is seen as the first real opportunity for Japan to export a telecommunications standard, and to this end PHS has subsequently been actively promoted in the Asian arena – notably in Hong Kong, China and Australia – not only for public access, but also for residential and business applications. Other countries showing an interest include Singapore, Malaysia, Indonesia and Vietnam. Exports are being pushed by NTT and DDI, supported by the Japanese Ministry of Posts and Telecommunications.

Initially only trial systems were available outside of Japan, notably in Hong Kong and Australia, where trials have been undertaken (see below). Singapore was the first export market to create a mechanism for approval and operation of PHS products (for indoor use) outside of Japan, in October 1995. PHS has also been promoted as an unlicensed PCS technology in the USA, in the variant form of PACS-UA, described elsewhere in this book.

10.3.3   PHS Products

Personal Communications Services (PCS)

A key factor with PHS product availability has been the large number of Japanese manufacturers involved in the field. This has been stimulated by the attractiveness of the local market and the presence of three different public network operators. Today, there are around 19 manufacturers of PHS handset

Table 10.7
PHS handset manufacturers

Public network operator	Handset suppliers
NTT Personal	Sony
	Hitachi
	NEC
	Sharp
	Panasonic (Matsushita)
ASTEL	Tottori Sanyo
	Nippon Denso
	Nippon Musen
	NEC
	Sharp
	Panasonic (Matsushita)
DDI Pocket	Aiwa
	Casio
	Kyocera
	Kenwood
	Panasonic (Kyushu-Matsushita)
	Sanyo
	Toshiba
	Victor
	Mitsubishi

Source: from [5].

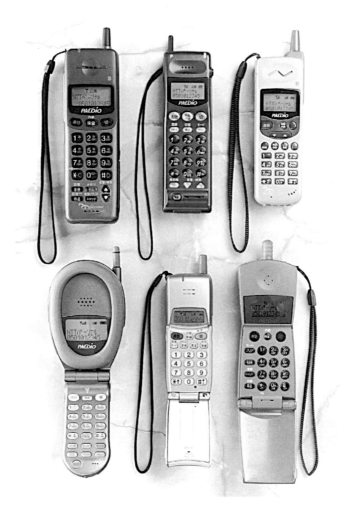

Figure 10.2   #PHS handset products. (Photograph courtesy of PHS International, Hong Kong.)

products, including Aiwa, Casio, Kyushu, Panasonic, Kyocera, Kenwood, Sanyo, Seiko, Toshiba, Victor, Mitsubishi, NEC, Sharp, Sony, Matsushita Telecom, Hitachi, Tottori Sanyo, Nippon Denso and Nippon Musen. Some example handset products are shown in Figure 10.2.

The first PHS networks were launched in July 1995 by NTT and DDI. They were joined by Astel in October 1995. Japan alone expects to have 40 million PHS customers by 2010. Different handset vendors are supplying these three operators, as shown in Table 10.7. Current products cost just less than US $400, but this could be subsidised in the future.

Business Use

PHS-based WPBXs are now appearing and are expected to steal share from the European products. PHS WPBX solutions are being manufactured by several vendors including NEC, Oki, Hitachi and Panasonic.

The "iox" digital cordless telephone system from Oki is a stand-alone system, but can be used as an adjunct to existing PBXs. It is available in three models, supporting from 128 to 480 ports. NEC provides the NEAX Wireless module, which supports a maximum of 500 handsets and 192 cell stations. The CX series of Hitachi PBXs offers digital cordless system functions by adding a cell station, handset, interface card and control software to any Hitachi PBX.

Wireless Local-Area Networks (WLANs)

Panasonic has developed a multimedia wireless system that supports LAN interconnection, wireless modem cards, PC conferencing and wireless internet connection. A maximum of 576 cordless telephones can be used on one PBX. Panasonic has also launched a PHS-based home cordless telephone that supports up to seven handsets on one basestation.

10.4 Global Competition

Telecommunications markets have become increasingly globalised over the past decade. The evolution from country-based to regional-based standards has been very rapid, and the 1990s are seeing this result in major competition, with new technologies such as mobile telephones and cordless phones leading the way. Since the launch of PHS in Japan in 1995, the technology has been promoted aggressively in the Far East, with the formation by Cable and Wireless of a new company, PHS International, based in Hong Kong to this end. The DECT manufacturing community has adopted a lower-key approach with an informal Global DECT Forum having been established to provide a mechanism for information exchange and mutual support in the promotion of DECT. Meanwhile, operators interested in adopting DECT technology for public access decided in 1994 to form a group to coordinate their activities, known as the DECT Operators' Group (DOG). As of October 1995, the DOG comprised some 20 operators, encompassing Europe and Asia.

10.4.1 Europe

Within the European cordless market, a number of factors point to the dominance of DECT. Sales of DECT products are helped by the fact that it is the only full ETS and has a permanent spectrum allocation under an EU Directive. In addition, most of the major PBX manufacturers have publicised their support of DECT and are steering the market with regard to technology type. However, all is not lost for CT2. It will remain a contender, but will probably be confined to niche markets; in telepoint countries, however, it will have far greater success.

Both DECT and CT2 product manufacturers cannot afford to be complacent.

They must and are continuing to adapt their products to meet the changing challenge set them by their customers. One way forward for cordless technologies is for them to coexist with and complement cellular systems, with several operators pushing in this direction. This will allow the provision of cost-effective personal communications for service providers and end-users. One such development favouring DECT is the announcement of dual-mode GSM/DECT telephones. Outside Europe, which technology will dominate is a question still to be answered.

10.4.2 America

In the USA, the wireless market is currently dominated by adjunct solutions. Growth in this area has been adversely affected by a lack of standards and the allocation of frequency spectrum.

At present, several cordless products are available on the market, including FCC Part 15 products such as AT&T TransTalk 9000, Ericsson Freeset, Nortel Companion, Siemens/ROLM ROLMphone 900 and Spectralink Pocket Communications System 200. Other products include the Southwestern Bell/Panasonic Communications and Systems Company FreedomLink solution. Several cellular companies have also jumped on the wireless systems bandwagon and are teaming with PBX manufacturers to offer wireless in-building solutions.

New frequency assignments began in mid-1994, including the allocation of 1910–1930 MHz to unlicensed PCS applications such as wireless PBX. Emerging standards in the USA have been described elsewhere in this book, including CT2-based (PCI), DECT-based (PWT) and PHS-based (PACS-UA) solutions. It is too soon to pick the winning technology in the US marketplace; we must wait a while longer for spectrum relocation and etiquette issues to be resolved.

10.4.3 Asia–Pacific

In the Far East, to date, the most popular digital cordless technology has been CT2, which has formed the basis of all telepoint services. CT2-based WPBXs, including those from Nortel, Peacock and Lucent, have, therefore, been the most popular. Ericsson has also been active in this market for a few years now, first with its CT3 offering and now with DECT in all markets where this standard is approved. Alcatel launched its DECT solution in the Far East in September 1995 and it is expected that others will follow in 1996.

The early success of PHS in Japan has been accompanied by strong promotion of the standard, particularly in the Asian countries – notably Hong Kong, where comparative trials with DECT were undertaken during 1995; a report on the outcome of these trials was expected by 1996, but at the time of writing this had yet to be published. The results of these trials could influence the outcome in the huge Chinese market. However, it is likely that China will explore both DECT and PHS, and will probably choose whichever technology offers the better opportunities for the development of its own domestic industry capability.

In India, DECT appears to be emerging as the preferred technology. As well as trials of equipment from European manufacturers, indigenous industry has also developed a DECT-based wireless local loop product, to address the enormous potential need for basic telephony services [6].

In Australia, two bands have for some time been available for cordless telephony, 861–865 MHz for CT2 and 857–861 MHz for other cordless technologies. In 1990 the potential for introducing DECT was recognised and an embargo established on new assignments for fixed links within the relevant frequency band. In 1993 AUSTEL recommended that spectrum be made available to allow introduction of DECT, and in 1995 the Spectrum Management Agency concluded that planning for the introduction of DECT could commence, with DECT operating on a non-interference basis alongside existing fixed links. Thus a proposed band plan for cordless telecommunications service in the 1880–1900 MHz band was published in December 1995 [7]. A technology-neutral approach has been adopted, based on spectrum sharing with fixed services, on a non-interference basis. Under the proposed band plan arrangements, any cordless equipment that can coexist with previously approved equipment standards may operate in the 1880–1900 MHz band. This means that PHS, US PCS technologies or any other emerging cordless system are candidates to use the band provided they can coexist with DECT; indeed, AUSTEL has decided to develop a standard for PHS.

10.5   Summary

This chapter has reviewed the origins and development of the digital cordless industry from small beginnings to its status in the mid-1990s as a significant element of the telecommunications industry. With forecasts of major market growth, the significance of this sector will increase dramatically over the next decade. Product availability to the various major standards has been described, for a range of applications, and availability in different geographical markets discussed.

The emergence and likely development of global competition has also been described. DECT is well established in Europe and has found ready acceptance in many other markets, with PHS emerging as a late but potentially powerful challenger; CT2 is likely in the future to assume diminishing overall importance. The outcome in terms of technology adoption in different geographical regions is likely in reality to be influenced more by political and commercial factors than by technology and performance capability.

Note   1 The chapter provides a snapshot of the market as at the time of writing.

References

[1]   "Cordless Telecommunications in Europe", WHW Tuttlebee (ed.), Springer-Verlag, 1990
[2]   CAI Schedule, CAI Founders Secretariat, December 1995
[3]   "Cordless Communications – No Strings Attached", Dataquest Europe, December 1995
[4]   Papers presented by operators at the DECT '96 Conference, IBC, London, January 1996
[5]   "Examining the Developmnts and Progress of the PHS Commercial Service in Japan", P White, Digital Cordless Communications Conference, IIR, London, January 1996
[6]   "DECT in India", informal presentation by M Harish, at the DECT '96 Conference, IBC, London, January 1996
[7]   "1.9 GHz Band Plan for Cordless Telecommunications Services in the 1880–1900 MHz Band", Spectrum Planning Document SPP 10/95, Spectrum Management Agency, December 1995

Part III

Technology

11 Audio Techniques

Dag Åkerberg and Julian Trinder

Traditional users of mobile radio systems in the past tolerated the poor speech quality often associated with an analogue radio channel in exchange for the benefit of mobility. This is not so, however, for today's cordless telephone user, who views the cordless telephone simply as another telephone and expects the same high speech quality associated with the fixed telephone network. This chapter on audio requirements and design addresses the requirements for and outlines the approaches adopted in the provision of good speech quality, in particular addressing the special requirements arising in a private automatic branch exchange (PABX) environment and/or over local, national and international public switched telephone networks (PSTNs), when one or both telephones use digital cordless rather than wired links.

Digital speech coding and transmission, compared to analogue, provides a speech level that does not vary with distance, has less noise and more delay, and requires higher transmission bandwidth. The distortion is partly of another kind, quantisation distortion. Thus traditional methods to evaluate the link quality are not completely satisfactory for digitally coded speech links, and new methods have been developed within standardisation bodies. The latter problem is emphasised when it comes to the evaluation of speech coding techniques suitable for radio transmission. For example, compliance to a frequency response mask requirement can have limited correlation to good results in a subjective perception test in a radio environment. The choice of speech coding method and evaluation of its performance, including transmission over a varying radio channel, is a complex task. In practice, a convergence towards 32 kb s^{-1} adaptive differential pulse-code modulation (ADPCM) has occurred for cordless telephony, being the standard adopted for CT2 CAI, DECT, PHS and other specifications.

Today's cordless handset is designed to be the only or prime telephone used. Thus the subjective quality under adequate radio conditions should ideally be on a par with that of wired apparatus. Other main requirements for speech coding for cordless systems with lightweight handsets are low cost, low current consumption, effective use of the limited radio spectrum and low delay. Generally a more frequency-efficient coding technique is more complex, and hence needs more processing power, has more delay and offers lower quality. Delay and/or low quality can be improved by higher complexity. The different schemes for digital speech coding that have been used or proposed for cordless telephony are reviewed in the first part of this chapter, section 11.1.

In providing telephony services, a voice transmission plan is used to specify acoustic and electrical levels and limits, and other network characteristics, in such a way that correct audio levels are received at the handset and that audio quality degradation due to distortion, noise, echoes and delay is kept within predetermined limits. The quality depends very much upon the transmission network. Analogue two-wire networks, especially older ones, have large variations in attenuation, delay and audio characteristics, while modern digital four-wire networks are much more tightly defined. The transmission plans for modern cordless telecommunication systems are based on well defined digital transmission. From this base, modifications can be made to fit analogue network specifications, which presently are not uniform across Europe. The voice transmission plan requirements for cordless systems are discussed in section 11.2.

Networks for ordinary telephony are recommended to contain mechanisms for echo control when the (one-way) delay exceeds a certain limit (25 ms); for example, international gateways are equipped with echo control devices when satellite links, which introduce long delays, are used. Digital cordless telephone systems introduce additional delays compared to an analogue, or digital, wired telephone. The radio multiple access method and/or the duplex operation use time division, which introduces delay; likewise, the digital speech coding method contributes further delay. The cordless system thus has to be specified, including possible internal echo control means if needed, so that this extra delay does not violate correct operation of the echo control at international gateways, or otherwise significantly reduce the subjective speech performance. These issues of delay and echo control are discussed in section 11.3.

Finally voice security and speech encryption, important issues from the user perspective, are discussed briefly in section 11.4.

The topics treated in this chapter are deeply interrelated and the requirements and solutions are dependent upon existing and evolving standards, technological possibilities and the constraints of the existing older and modern networks.

The aim of this chapter is to give a broad understanding of the basic principles and how to apply them to the design of cordless systems. For further details or more exact definitions, the reader is referred to the references.

11.1   Speech Coding

The evolution of the telecommunications industry has been shaped by a great number of market pressures, of which perhaps the most paradoxical is the pronounced affinity for voice communications, despite an extensive range of alternative telecommunication services. To meet the diversity of market demands, technology is increasingly moving towards digital transmission for all kinds of services. Digital format can offer rugged and reliable communications despite complex network routing without fear of signal degradation within or between the network nodes. In order to integrate improved telephony services within the evolving digital telecommunications industry, great emphasis has been given to the digital representation of the voice signal, which was previously largely handled in its analogue form.

Within the telephone unit the microphone and pre-amplifier circuits form a continuous linear electrical analogue of the incident sound pressure. The function of the voice coder is to convert the continuous analogue signal into a discrete digital representation suitable for digital transmission. At the remote location, a complementary voice decoder regenerates the continuous electrical analogue signal, which can then be amplified and converted back to a sound pressure wave by a linear acoustic transducer.

The complementary coder–decoder pair is often referred to as a "codec". Within each digital telephone unit the codec device usually handles two different signals – the coder handles the outgoing voice signal and the decoder handles the incoming voice signal. Depending on the type of codec and the signal characteristics, the two parts of the codec may at any one time be performing dissimilar and non-complementary signal processing functions. In a full-duplex telephone link the true codecs, which perform the complementary signal processing functions, are in fact divided between the two ends of the link.

Voice coding techniques for radio-based telephony (cordless and cellular) divide into two main groups – waveform coding and parametric coding. The principle of waveform coding is to reproduce as closely as possible the waveform of the continuous sound pressure analogue. It can be argued that since the human hearing mechanism is largely insensitive to waveform details and responds chiefly to short-term power spectra, waveform coding conveys a great deal of redundant information. Parametric coding, however, is designed to convey sufficient parameters of the voice signal such that at the remote location a waveform can be reconstructed whose short-term spectral characteristics closely match those of the incident signal. Significant savings in information rate can be made by using parametric coding rather than waveform coding, whilst still preserving excellent speech intelligibility and subjective quality. Examples of waveform coding techniques are pulse-code modulation (PCM) and delta modulation (DM). Examples of parametric coding are linear predictive coding (LPC) and its many variants.

The principles underlying some of the more popular voice coders will be outlined below. We start by describing PCM in some detail, since this is the staple digital voice coding scheme employed in the fixed PSTN; this also forms the basis for an explanation of adaptive differential PCM (ADPCM), the standard that has been widely adopted for cordless telephony. We then discuss delta modulation, a fundamentally simpler form of speech coding, a variant of which was used in early analogue cordless telephones in the USA and by Shaye Communications Ltd in one of the first proprietary CT2 digital cordless telephone products.

The DECT specification leaves an option for future addition of a second, more frequency-efficient, codec, when such becomes technically and commercially available. Parametric coding is likely to be used in this future codec. Some discussion of parametric coding, exemplified in the form of LPC and its extensions, is therefore also included within this section.

Finally the issues of speech quality assessment versus bit error rate (BER) are addressed. In general, channel coding of the speech data is not required.

The astute reader will have noticed that the tolerance of a voice coder (vocoder) to bit errors has remarkable diagnostic properties: unless the vocoder is uniformly susceptible to bit errors irrespective of function, the true information rate is less than the bit rate and there is therefore scope for improvement. It can also be argued that even if the bit error susceptibility has been carefully balanced, if the vocoder has a high tolerance to bit errors then there is also scope for improvement in the pursuit of ever lower bit rates.

To illustrate the latter point, one of the lowest bit-rate vocoders ever was a $1200 \, \text{bit s}^{-1}$ formant tracking vocoder algorithm developed by Dr John Holmes at the Joint Speech Research Unit at Eastcote during the 1970s. The vocoder produced high-quality speech by a combination of sophisticated techniques: analysis-by-synthesis formant tracking, pitch analysis including a mixed excitation voicing parameter, a parameter interpolation and plosive processing unit, and a parallel formant synthesiser based on an accurate glottal excitation model. This major step forwards in low bit-rate vocoder technology was achieved by recoding the synthesiser control parameters, which were already encoded into a low-redundancy format. The performance achieved by this method was close to optimum, any further reductions in bit rate at the parameter recoding level only resulting in high-quality inebriated speech! Unfortunately formant tracking vocoders became eclipsed by the emerging LPC-based techniques.

There is no clear-cut "best vocoder"; different applications require different vocoder techniques for a host of different reasons. The reasons include technology availability, time to market, market perceptions (whether valid or invalid), licensing restrictions, cost, current consumption, delay, voice quality and distortion, error tolerance, tandem capability, the ability to handle voice-band modem signals, DTMF signalling, music, background noise, a variety of languages, etc. Given the plethora of interdependent criteria, it comes as no surprise that the providers of cordless, cellular and mobile telephony have in the past chosen widely differing voice coding schemes. Diversity appears to be a trend set to continue well into the foreseeable future. The history of technology, however, has shown that diversity in the market is a two-edged sword, leading on the one hand to great strides of progress but on the other taking many casualties.

11.1.1   Pulse-Code Modulation

PCM coding is a method of waveform coding that can represent any band-limited analogue waveform to a specified precision in digital format [1]. The analogue waveform is usually band-limited (e.g. ITU Rec. G.712) then sampled at a uniform rate at least twice that of the highest frequency component in the waveform to be coded. Each sample is then quantised to the nearest level against a series of uniform steps. Each step is assigned a unique digital code, usually binary. The PCM digital format is a series of such codes, which define the sequence of levels selected by the quantiser at each sample in time. The decoder simply reconstructs each waveform sample based on the digital codes received and then band-limits the result to half the sampling rate in order to recover the analogue waveform.

A useful figure of merit for PCM coders is the signal-to-quantising-error ratio (SQER). This figure is the ratio of the root-mean-square (RMS) of the signal to the RMS of the quantising error. It differs from signal-to-noise ratio (SNR) in that the quantising error is a function of the signal itself, and is absent when the signal is zero.

If the range of the signal values to be represented is $+A$ to $-A$, optimal quantisation in uniform steps to α bits yields a resolution of

$$q = 2A/2^{\alpha}$$

If the quantiser operates by mapping each signal component to the nearest quantum level, the quantisation error function (ξ) will have a uniform distribution over the range

$$-q/2 \leq \xi \leq q/2$$

with probability density function

$$P(\xi) = 1/q$$

Accordingly the mean quantising-error power is

$$(\xi_{rms})^2 = \int_{-q/2}^{q/2} \xi^2 P(\xi)\, d\xi = q^2/12$$

so that

$$\xi_{rms} = q/(2\sqrt{3}) = A/(2^\alpha \sqrt{3})$$

Thus the signal-to-quantising-error ratio in dB is given by

$$SQER_{dB} = 20\log_{10}(\Sigma_{rms}/\xi_{rms})$$

where

$$(\Sigma_{rms})^2 = \frac{1}{T} \int_0^T S(t)^2 dt$$

For the voice signal it is normal to allow for $A = 4\Sigma_{rms}$ to avoid excessive clipping (probability $<10^{-4}$). Substitution in the above equations thus yields a value of $SQER_{dB} = (6.02\alpha - 7.27)$ dB. The quantisation error has uniform (non-Gaussian) distribution characteristics, is roughly white in spectrum and is correlated to the main signal. A PCM codec with range A, well matched to the optimum value of $A = 4\Sigma_{rms}$ and $\alpha = 11$, will thus have an SQER value of 59 dB. This is generally felt to be adequate precision for reasonable voice quality. For a 4 kHz bandwidth the sampling rate needs to be 8 kHz, which would imply a bit rate of 88 kb s^{-1}.

There are two main problems with linear PCM as described above. Firstly it requires high bit rates to achieve adequate precision, and secondly the peak signal level needs to be carefully matched to the quantiser input range to avoid excessive clipping or poor resolution. Both of these problems are overcome in a refinement of PCM, log PCM, which is described below.

Log PCM (ITU Rec. G.711) exploits the fact that the large peaks in the voice signal are generally infrequent and have short duration. Increasing the quantisation for large signals therefore does not have a significant effect on perceived quality. Smith [2] has shown that using non-uniform quantisation the performance of 11-bit linear PCM can be approached with only 7-bit resolution. Practical telephony systems usually standardise on 8-bit log PCM at a sampling rate of 8 kHz (64 kb s^{-1}), which is roughly equivalent to 12-bit linear PCM (96 kb s^{-1}).

Another variant of log PCM is A-law PCM, in which the central region of the logarithmic function is approximated by a linear section.

11.1.2 Adaptive Differential PCM (G.721, G.726)

In seeking to understand ADPCM (ITU Rec. G.721, G.726), we firstly discuss ordinary (non-adaptive) DPCM. These topics are reviewed in greater depth in [3].

In differential PCM the correlation between successive samples of the voice signal is exploited. The main effect of this positive correlation is a reduction in the variance $\langle \ldots \rangle$ of the first difference between samples. Since the variance is reduced, fewer bits are needed to code the first difference than would be required to code the raw sample, thus facilitating a lower bit rate. The quantisation of the first difference may be either linear PCM or log PCM.

If $X(r)$ represents the signal sample value and $D_1(r)$ the first difference, i.e. $X(r) - X(r-1)$, this reduction in variance may be easily derived, since

$$\langle D_1^2(r) \rangle = \langle [X(r) - X(r-1)]^2 \rangle = \langle X(r)^2 \rangle + \langle X(r-1)^2 \rangle - 2\langle X(r)X(r-1) \rangle$$

i.e.

$$\langle D_1^2(r) \rangle = 2(1 - C_1)\langle X(r)^2 \rangle$$

Hence the variance of the first difference between samples is less than the variance of the raw samples when the first correlation (C_1) is greater than 0.5. The variance would actually increase for first correlation (C_1) less than 0.5.

This restriction on first correlation (C_1) range can be lifted if a weighted difference is taken. Thus

$$D_1(r, a_1) = X(r) - a_1 X(r-1)$$

gives

$$\langle D_1^2(r, a_1) \rangle = \langle [X(r) - a_1 X(r-1)]^2 \rangle$$

i.e.

$$\langle D_1^2(r, a_1) \rangle = \langle X(r)^2 \rangle (1 + a_1^2 - 2a_1 C_1)$$

Differentiating the variance factor by a_1 yields a minimum for $2a_1 - 2C_1 = 0$:

$$\langle D_1^2(r) \rangle_{\min} = \langle X(r)^2 \rangle (1 - C_1^2)$$

The principle of weighted differences can be extended to exploit higher-order correlations:

$$D_n(r, a_1, a_2, \ldots, a_n) = X(r) - \sum_{i=1}^{n} a_i X(r-1)$$

It can be shown [3] that the variance $\langle D_n^2(r, a_1, a_2, \ldots, a_n) \rangle$ is minimised for

$$[\mathbf{A}] = [\mathbf{\Gamma}]^{-1}[\mathbf{\Sigma}]$$

where

$$[\mathbf{A}] = [a_1, a_2, \ldots, a_n]^{\mathrm{T}}$$

$$[\mathbf{\Sigma}] = [C_1, C_2, \ldots, C_n]^{\mathrm{T}}$$

and

$$[\mathbf{\Gamma}] = \begin{bmatrix} 1 & C_1 & C_2 & . & . & C_{n-3} & C_{n-2} & C_{n-1} \\ C_1 & 1 & C_1 & . & . & . & C_{n-3} & C_{n-2} \\ C_2 & C_1 & 1 & & & & . & C_{n-3} \\ . & . & & . & & & . & . \\ . & . & & & . & & . & . \\ C_{n-3} & . & & & & 1 & C_1 & C_2 \\ C_{n-2} & C_{n-3} & . & . & . & C_1 & 1 & C_1 \\ C_{n-1} & C_{n-2} & C_{n-3} & . & . & C_2 & C_1 & 1 \end{bmatrix}$$

Because the correlation matrix $[\mathbf{\Gamma}]$ is symmetric, it can be inverted by triangular decomposition, using various recursion algorithm techniques. Alternatively gradient algorithms can be used to obtain the weight vector $[\mathbf{A}]$ directly.

The optimum linear combination of previous samples for the weighted difference will vary with the correlation matrix of the voice signal. It is therefore important to update the matrix regularly in order to achieve minimum variance in the signal being presented to the quantiser. Such a scheme is called an adaptive predictor, since the weighted sum of previous samples is essentially trying to predict what the next sample will be.

DPCM that uses an adaptive predictor and an adaptive quantiser is called adaptive DPCM (ADPCM). The adaptive predictor keeps track of spectral characteristics (correlation between samples) and the adaptive quantiser keeps track of level changes by adjusting its step size or resolution to match the level of the difference signal, as discussed in the previous section on PCM.

Since its inception in the early 1970s [4–6] ADPCM has been adopted by the ITU as an international standard [7]. Because of its historically envisaged application in conjunction with $64\,\mathrm{kb\,s^{-1}}$ PCM in the PSTN, the G.721 coding algorithm is defined in terms of a transcoding function to be implemented in tandem with standard G.711 $64\,\mathrm{kb\,s^{-1}}$ log PCM codecs (μ-law or A-law) [8].

The processes of ADPCM encoding are presented diagrammatically in Figure 11.1. The first process within the transcoder is conversion to linear (or uniform) PCM. A difference signal is then obtained by subtracting a signal estimate that has been derived from earlier input samples. An adaptive 16-level quantiser then codes the difference signal to 4 bits every $125\,\mu\mathrm{s}$, thus halving the original data rate (8 bits every $125\,\mu\mathrm{s}$).

The inverse function of the adaptive quantiser then produces a quantised version of the difference signal, which is added to the signal estimate to give a reconstructed version of the input signal. Both these latter signals are used to update the signal estimate in the adaptive predictor. The adaptive predictor is based on a second-order recursive filter and a sixth-order non-recursive filter:

$$s_e(k) = \sum_{i=1}^{2} a_i(k-1)s_r(k-i) + \sum_{i=1}^{6} b_i(k-1)d_q(k-i)$$

where $s_e(k)$ is the new signal estimate at time k. $s_r(k-i)$ are the earlier samples of the reconstructed signal and $d_q(k-i)$ are the earlier samples of the quantised difference signal. Both sets of predictor coefficients, $a_i(k-1)$ and $b_i(k-1)$, are updated using gradient algorithms [7].

Many of the functions within the ADPCM decoder (Figure 11.2) are very similar to those in the encoder. The inverse adaptive quantiser reconstructs the

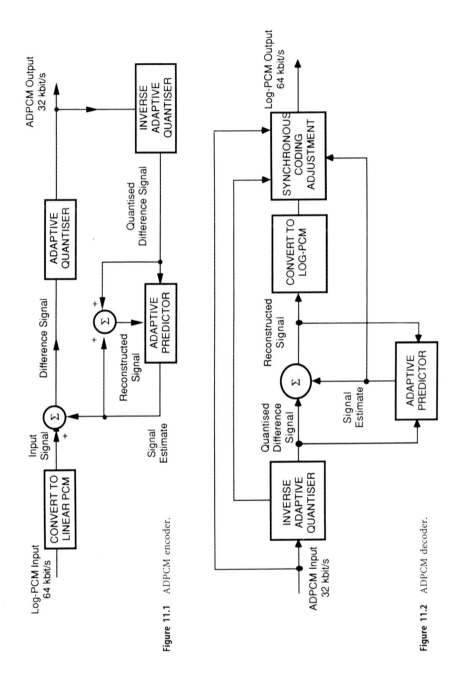

Figure 11.1 ADPCM encoder.

Figure 11.2 ADPCM decoder.

quantised difference signal, and the adaptive predictor forms a signal estimate based on the quantised difference signal and earlier samples of the reconstructed signal, which is itself the sum of the current estimate and the quantised difference signal. The synchronous coding adjustment function is an extra refinement that makes minor adjustments to the log PCM output values, so that quantising errors do not accumulate when ADPCM transcoders are concatenated.

As explained, ADPCM coding is fundamentally a reasonably faithful waveform coder and as such it retains a reasonably high quality of speech reproduction, comparable to that expected of the public switched telephone network (PSTN). The terms "toll quality" or "wireline quality" are often used to convey the quality requirement. Because of this quality and implicit compatibility with the digital PSTN, ADPCM has emerged as the primary speech coding standard for cordless telephones. The two ITU Recommendations commonly associated with ADPCM are ITU Rec. G.721 and G.726 (which incorporates both G.721 and G.723, i.e. both the transcoding algorithm and the timing requirements).

11.1.3   Delta Modulation

There are a great number of variants of delta modulation – linear delta modulation (LDM), adaptive delta modulation (ADM), continuously variable slope delta modulation (CVSDM), continuously variable slope delta modulation with syllabic companding (CVSDM-SC), digitally variable slope delta modulation (DVSDM), etc. [3, 9,10].

Delta modulators operate by sampling the analogue signal at very high rates, but with only one-bit (two-level) resolution. Their main virtue is simplicity, which reflects primarily in cost. At low bit rates (16 kb s^{-1}) voice quality begins to suffer, and the sound becomes rather rough in character, marginally acceptable for military communications but certainly not up to the public expectations of a toll-quality link.

We shall start by describing the basic principles of the linear delta modulator, illustrated in Figure 11.3. The band-limited input signal X_r is sampled at a rate f_0, which is much higher than $2W$, where W is the highest frequency component of the input signal. At each sample the analogue accumulator holds a previous estimate Y_{r-1} of the input signal, and the sign of the difference $X_r - Y_{r-1}$ is used to update the accumulator by an amount $\pm\Delta$.

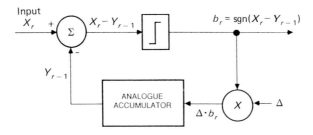

Figure 11.3   Linear delta modulation encoder.

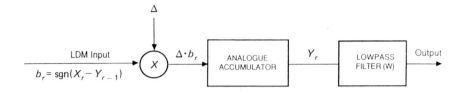

Figure 11.4   Linear delta modulation decoder.

With a careful selection of f_0 and Δ the accumulator can be made to track the input signal closely. The maximum signal gradient that can be tracked is $\pm\Delta f_0$, any gradient outside this range resulting in a slope overload. All features of the input signal will be lost during the slope overload condition. If Δ is too large then the smaller details of the signal will be masked. For a specific bandwidth W and sampling rate f_0 the quantisation Δ must be carefully selected according to the signal level.

The linear delta modulation decoder is particularly simple and is shown in Figure 11.4. The received data $(+1$ or $-1)$ is multiplied by an appropriate step value Δ to produce Δb_r, which is then accumulated to give a reconstruction Y_r of the original signal. This is then lowpass-filtered to its original bandwidth W, to remove components around the sampling frequency f_0.

Adaptive delta modulation (ADM) is a method of scaling the step size Δ or slope in order to accommodate variations in signal level. ADM techniques fall into a number of groups depending on whether the companding function (step-size adaptation) is instantaneous or gradual, and whether the function is driven by analogue signals within the encoder (CVSDM) or by the digitally encoded representation of the signal (DVSDM). The overall encoding function is equivalent to linear delta modulation of an input signal with compressed dynamic range. Within the ADM decoder a similar step-size algorithm ensures that the original dynamic range of the signal is restored.

A DVSDM scheme was adopted by Shaye Communications (subsequently acquired by AT&T, now Lucent) for their early proprietary CT2 cordless telephone product, the Forum Personal Phone. Their first type approved product, based on this approach, was brought to market in 1989. The reasons for the choice of DVSDM were primarily cost, power consumption and size – a DVSDM codec was at that time much simpler and cheaper to produce than an ADPCM codec and had a significantly lower power consumption. Since that time, with the widespread adoption of ADPCM as a speech codec standard, low-cost, low-power ADPCM chips have of course become available.

11.1.4   Linear Predictive Coding

Linear predictive coding is a parametric representation of the voice signal – only the principal features of the signal, such as short-term power spectrum and excitation function, are coded. This type of coding offers low data rate – for military communications $2.4\,\mathrm{kb\,s^{-1}}$ (LPC-10) is common, whereas for commercial applications $8–16\,\mathrm{kb\,s^{-1}}$ is typical.

To achieve low data rates, a very simple model of the voice signal is employed. The voice signal $S(z)$ is modelled as the output of a time-varying autoregressive filter $H(z)$ driven by an idealised impulse-like source $E(z)$:

$$S(z) = H(z)E(z)$$

$S(z)$ is the z-transform of the signal time series:

$$S(z) = \sum_{n=0}^{\infty} s(n)z^{-n}$$

$H(z)$ is the z-transform of the impulse response:

$$H(z) = \sum_{n=0}^{\infty} h(n)z^{-n}$$

$E(z)$ is the z-transform of the excitation time series:

$$E(z) = \sum_{n=0}^{\infty} e(n)z^{-n}$$

We can now expand $H(z)$ in terms of a pth-order non-recursive predictor $P(z)$:

$$P(z) = \sum_{k=1}^{p} a_k z^{-k}$$

and

$$H(z) = [1 - P(z)]^{-1}$$

so that

$$S(z) = H(z)E(z) = E(z)/[1 - P(z)]$$

i.e.

$$S(z) = P(z)S(z) + E(z) \tag{11.1}$$

The signal $S(z)$ is thus represented by a prediction term, $S'(z) = P(z)S(z)$, plus an error term, $E(z)$, which is the excitation function. Equation (11.1) may be regarded as the LPC voice synthesis equation.

The predictor coefficients a_k best fit the speech production model when the expectation value of the mean-square error $\langle E(n)^2 \rangle$ is minimised, tending to give a more idealised impulse source.

From the LPC synthesis equation:

$$\langle E(n)^2 \rangle = \sum_{n=1}^{\infty} \left(s(n) - \sum_{k=1}^{p} a_k s(n-k) \right)^2$$

this function can be minimised by setting the partial derivatives to zero:

$$\partial \langle E(n)^2 \rangle / \partial a_j = 0, \qquad j = 1, 2, \ldots, p$$

from which

$$r_{j0} - \sum_{k=1}^{p} a_k r_{jk} = 0, \qquad j = 1, 2, \ldots, p$$

where

$$r_{jk} = \sum_{n=1}^{\infty} s(n - j)s(n - k)$$

or in matrix form

$$r - [R]a = 0$$

i.e.

$$a = [R]^{-1}r$$

where a is the vector of predictor coefficients a_k, r is the vector of autocorrelations r_{j0} and $[R]$ is the matrix of autocorrelations r_{jk}.

Fortunately the matrix is symmetric, positive definite and approximately Toeplitz for quasi-stationary signals, so that a number of efficient techniques can be used to invert the autocorrelation matrix $[R]$ and find the best vector for the linear predictor a [11].

In a practical LPC system the predictor coefficients are often recoded into alternative more robust forms. One such scheme remodels the synthesis filter as an acoustic tube having "p" coupled segments of equal length, but with different cross-sectional area for each segment. The filter coefficients are then recoded as the reflection coefficients between segments. An alternative recoding scheme uses log-area ratios to describe the acoustic-tube model and hence the synthesis filter.

In addition to specifying the recursive filter, an LPC coder must also convey sufficient parameters relating to the excitation function $E(z)$ for an adequate resynthesis of $E(z)$ at the LPC decoder. There are a great variety of ways to code the excitation function. Simple LPC codecs use only amplitude, fundamental frequency and voicing (a one-bit parameter to distinguish between periodic and non-periodic excitation). Some of the more complex LPC schemes use PCM-type coding of the residual $E(z)$. Some LPC schemes reduce the residual coding still further by adding a long-term predictor to reconstruct the excitation function, rather like ADPCM. The synthesiser excitation function is then the sum of the long-term predictor output and a lesser signal, the innovation sequence, which is derived from a part of the LPC code format. Examples of codecs using LPC are the GSM codec and possibly also the emerging half-rate GSM codec.

The reader is referred to [11] for a fuller treatise on LPC techniques. As noted earlier, DECT will have an option for future introduction of a second coder. The advantage of using a parametric coder would be its relatively low data rate, less than that of 32 kb s^{-1} ADPCM; this could potentially permit an effective doubling of capacity over the radio interface once the technology allows the speech quality, cost and power consumption requirements to be met.

11.1.5   Algebraic Code-Excited Linear Prediction (G.729)

In 1990 the CCITT (now ITU-T) began an endeavour to formulate a new high-quality voice coder at $8\,\mathrm{kb\,s}^{-1}$ for personal mobile communications. The new $8\,\mathrm{kb\,s}^{-1}$ vocoder was to perform at least as well as $32\,\mathrm{kb\,s}^{-1}$ ADPCM (ITU-T Rec. G.726) despite its using only one-quarter of the data rate.

Two code-excited linear prediction (CELP) algorithms were presented in 1992, one by NTT and the other by France Télécom and the University of Sherbrooke in Quebec. Since both vocoders used a similar structure, it was possible to take the best features from each to form the new $8\,\mathrm{kb\,s}^{-1}$ standard (ITU-T Rec. G.729) [12]. This work was undertaken by an optimisation group composed of NTT, France Télécom, University of Sherbrooke and AT&T. The results of their work were presented for standardisation in 1995.

Because of the rigorous duplication of functions in the encoder and decoder (analysis by synthesis), the G.729 is also known by the snappy title "conjugate structure algebraic code-excited linear prediction vocoder" (CS-ACELP). The main features of G.729 are:

- An $8\,\mathrm{kb\,s}^{-1}$ data rate
- A 10 ms frame length and 5 ms look-ahead
- Approximately 15 ms delay (one way)
- Voice quality equal to $32\,\mathrm{kb\,s}^{-1}$ ADPCM (G.726) in the absence of bit errors
- Voice quality superior to $32\,\mathrm{kb\,s}^{-1}$ G.726 with channel errors
- Design criteria are met – performance under background noise, tandeming, etc.

The G.729 algorithm is rather complex and a full description is beyond the scope of this chapter. In broad outline G.729 is based on a 10th-order LPC analysis conducted once per 10 ms frame, the short-term filter information being encoded in line spectrum pair (LSP) format. LPC analysis is based on conventional autocorrelation methods.

The innovation sequence is composed of a fixed codebook part based on four encoded impulses and a pitch-based adaptive codebook. The codebook indices and gains are updated twice per 10 ms frame. The codebook search and adaptation are based on an analysis-by-synthesis method (hence the "conjugate structure") using an adaptive perceptual weighting filter. The adaptive weighting filter was introduced to improve the analysis of flat spectrum frames. The bit allocation of the code is shown in Table 11.1.

The partitioning of the encoded speech is thus:

short-term spectrum (LSP) = 22.5%

pitch information (delay) = 17.5%

innovation sequence = 60.0%

Encoder and decoder block diagrams are shown in Figures 11.5 and 11.6.

Table 11.1
Bit allocation of code

Parameter	Subframe 1	Subframe 2	Total
LSP coefficients			18
Adaptive codebook delay	8	5	13
Delay parity bit	1		1
Fixed codebook index	13	13	26
Fixed codebook signs	4	4	8
Codebook gains (stage 1)	3	3	6
Codebook gains (stage 2)	4	4	8
Total number of bits			80

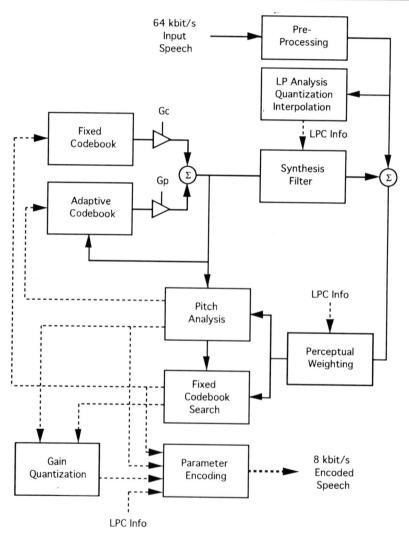

Figure 11.5 CS-ACELP encoder.

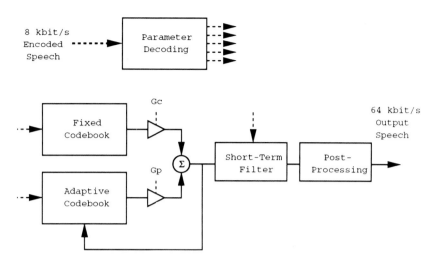

Figure 11.6 CS-ACELP decoder.

11.1.6 Enhanced Full-Rate (EFR) Vocoder

In early 1995, in an attempt to accelerate the development of PCS1900 products for the US PCS market, a group known as the North-American PCS Action Group (NPAG) composed of operators, service providers and manufacturers, met to select an improved vocoder for the GSM-based PCS candidate scheme. The general feeling across the industry was that the quality of the European GSM vocoder was not up to the standard required by the American market.

Candidate vocoder algorithms were offered by Motorola, Northern Telecom, Ericsson and Nokia. It was decided that the European full-rate GSM vocoder would be the default, the Nokia vocoder being selected as the enhanced full-rate (EFR) option.

The Nokia candidate had been developed jointly by Nokia and the University of Sherbrooke and was naturally based on the highly successful CS-ACELP (G.729) work to which the University of Sherbrooke had been a major contributor. The original concept from Nokia was an ACELP vocoder at $11 \, \text{kb s}^{-1}$ with a view to half-rate channel coding in the PCS1900 transmission format. This concept was subsequently dropped in favour of a $13 \, \text{kb s}^{-1}$ version of Nokia's ACELP algorithm, in order to eliminate compatibility problems at the "Abis" interface.

This somewhat surprising historical development from $8 \, \text{kb s}^{-1}$ G.729, through $11 \, \text{kb s}^{-1}$ EFR to the final $13 \, \text{kb s}^{-1}$ EFR further highlights the complexity of the market forces at play in the modern communications arena.

11.1.7 Low-Delay Code-Excited Linear Prediction (G.728)

In 1992 the ITU published a recommendation for low-delay speech encoding at $16 \, \text{kb s}^{-1}$ (Rec. G.728) [13]. The algorithm is based on an analysis-by-synthesis code-excited linear prediction structure in which only codebook indices are transmitted, as described below.

The log PCM encoded (G.711) speech signal at $64\,\mathrm{kb\,s^{-1}}$ is converted to uniform PCM and buffered into vectors of five consecutive samples (0.625 ms). These vectors form the basis for analysis-by-synthesis code-excited linear prediction using a minimum mean-square error (MSE) criterion evaluated via a perceptual weighting filter.

Each of 1024 codebook vectors is compared to the input vector and the codebook vector with the minimum MSE is selected. The selected excitation vector is passed through the encoder structure to update the backward gain adaptation unit (once per vector) and the backward predictor adaptation unit (once every four vectors).

The codebook indices are transmitted to the decoder at a rate of one 10-bit index every vector (0.625 ms), giving a net rate of $16\,\mathrm{kb\,s^{-1}}$.

The decoder uses an identical structure of codebook excitation, backward gain adaptation unit (updated once per vector) and backward predictor adaptation unit (updated once every four vectors) so that the output from the decoder's synthesis filter is a minimum MSE representation of the original unquantised speech vector, within the scope of the excitation codebook. After post-filtering, the reconstructed speech signal is converted back to log PCM (G.711) at $64\,\mathrm{kb\,s^{-1}}$.

The total delay of the encoder and decoder is less than 2 ms. LD-CELP (G.728) encoder and decoder block diagrams are shown in Figures 11.7 and 11.8.

11.1.8   Speech Quality

The measurement of speech quality is a non-trivial issue. One useful metric of quality is the quantising distortion unit (qdu), based on a reference point of 1 qdu for a $64\,\mathrm{kb\,s^{-1}}$ log PCM codec, with higher qdu ratings corresponding to progressively poorer speech quality. In a system containing a number of codecs, the overall qdu rating is calculated by adding the qdu ratings of the individual codecs. Qdu ratings for a range of speech coding techniques, collated from the published literature, are presented in Table 11.2.

For low-bit-rate speech codecs, measuring speech quality is much more difficult, owing to the nonlinear distortions of the speech signal, and has to be carried out using subjective tests. Two types of testing are often performed,

Table 11.2
Qdu ratings for various speech coding techniques

Coding scheme	Rating (qdu)
$64\,\mathrm{kb\,s^{-1}}$ log PCM codec (8-bit sample)	1
$56\,\mathrm{kb\,s^{-1}}$ log PCM codec (7-bit sample)	3
$64\,\mathrm{kb\,s^{-1}}$ log PCM codec (+$32\,\mathrm{kb\,s^{-1}}$ ADPCM transcodec)	3.5
$16\,\mathrm{kb\,s^{-1}}$ LD-CELP	3.6
$32\,\mathrm{kb\,s^{-1}}$ CVSDM	6
$13\,\mathrm{kb\,s^{-1}}$ GSM RPE-LTP	7–8
$16\,\mathrm{kb\,s^{-1}}$ CVSDM	10

Source: from [14, 15].

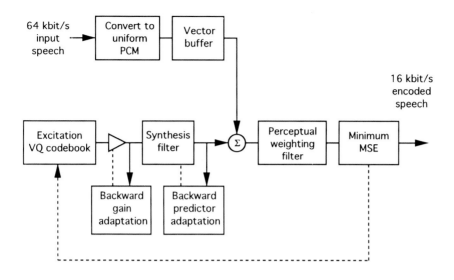

Figure 11.7   LD-CELP encoder.

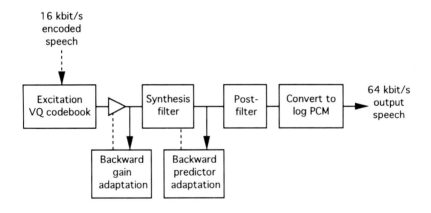

Figure 11.8   LD-CELP decoder.

measuring quality and intelligibility. The most widely used intelligibility test is the Diagnostic Rhyme Test (DRT), which uses pairs of rhyming words. The listener is asked to decide which of the two words was spoken. Not only does this test give an overall figure of intelligibility, but it can also highlight problems with particular types of sound.

The Mean Opinion Score (MOS) is commonly used for assessment of quality. Again, this is a subjective test and must be performed with a large number of listeners and speakers. Listeners are asked to assess quality on a scale of 0 to 4 (or 1 to 5), where the grades correspond to

- 4 (or 5) = excellent
- 3 (or 4) = good
- 2 (or 3) = fair
- 1 (or 2) = poor
- 0 (or 1) = bad

It is usual when carrying out these tests to include a reference codec, such as log PCM, whose performance is widely known. This can help when trying to compare results produced in different tests.

11.1.9 Codec Suitability

At present ADPCM has been adopted for use in all the mainstream digital cordless standards. The rate of progress in speech codecs, however, is rapid, as is plain from the discussions above. Speech quality is a very important factor in considering which speech codecs might find acceptance in future cordless systems, but not the only one. Power consumption is very important, as are other factors. A number of these factors are listed in Table 11.3, along with a comparison of ADPCM, G.728 and the EFR codecs – these two codecs have already been under consideration within ETSI RES 03 R within the context of DECT; other lower-rate codecs have not as yet been considered, for reasons of quality and complexity.

At present the power consumption of the newer codecs is too high for them to find acceptance; however, this may be expected to fall. Also at present no clear advantages accrue from their use, in terms of improved system traffic capacity or speech quality, and some disadvantages exist, e.g. speech longer delays.

Table 11.3
Comparison of ADPCM, LD-CELP and EFR codecs for cordless applications

Parameter	ADPCM (G.726)	LD-CELP (G.728)	Enhanced full rate (EFR GSM)
Bit rate	$32\,\text{kb s}^{-1}$	$16\,\text{kb s}^{-1}$	$13(-22)\,\text{kb s}^{-1}$
Delay	0.4 ms	1.9 ms	20(−60) ms
Speech quality	Good	As good as or better than ADPCM	Difficult to compare directly
Quantisation distortion	2.5 qdu	3.6 qdu	—
Fax via modem	$4.8\,\text{kb s}^{-1}$	$2.4\,\text{kb s}^{-1}$	Not transparent
Data via modem	$9.6\,\text{kb s}^{-1}$	$2.4\,\text{kb s}^{-1}$	Not transparent
DTMF transparency	Yes	Yes	Not fully for short tones
Processing power	Low	~30 Mips	~15–22 Mips
Power consumption	Typically 8 mW, or 0.5 mW if a dedicated chip	100–200 mW	70–150 mW
Availability	Wide choice	Available as code	Available as code
Cost	Low	Higher (less mature)	Higher (less mature)

11.1.10   Channel Coding

An issue closely related to speech coding is the possible need for radio channel coding. Channel coding means that the digitally coded speech bits are divided into blocks and additional bits added to each block, according to some algorithm that uses the speech bits themselves to derive these additional bits, to provide a capability of detecting and correcting any errors in the received speech bits. In the more complex speech coding schemes such as LPC, the bits have different importance and the more sensitive bits may need to be accorded better protection. In digital mobile radio systems the bits of the coded blocks are also normally interleaved. This has the effect that if the signal received by a rapidly moving mobile user experiences a fade, the fading dip will cause a few correctable errors in several blocks, rather than concentrated uncorrectable errors in one block. The channel coding of the pan-European GSM cellular radio system is a good example of this approach.

For cordless telephone systems at present, interleaving such as just described offers little gain, since the system cannot afford the inherent large delay that would be needed to make it effective against the slow fading resulting from the very slow movement that is typical for a cordless handset. Instead antenna diversity is employed as a cheaper and more efficient counter-measure against slow fading. Under these conditions, error correction, which increases the required transmission bandwidth, also is not considered useful provided the codec chosen does give more or less even weight to the coded speech bits.

Channel coding is therefore generally not employed for cordless telephones for the speech channel; it is however essential for the signalling and control data, and will become important for future multimedia applications. The application of channel coding for error control is discussed further in Chapter 12.

11.2   Voice Transmission Plan

The main difference between transmission plans for wired telephones and cordless telecommunication systems is the impact of specific speech codecs suitable for radio transmission and the extra delay introduced. This section introduces and explains the parameters in a transmission plan that are essential for the discussions on delay and echo control. For remaining parameters, often common with wired telephony specifications, as well as measuring methods, the reader is referred to the specifications for the various systems [16–19].

11.2.1   Definitions

Firstly, a number of basic definitions are needed. The main reference is the ITU *Blue Book* Volume III containing recommendations G.100–G.181 [20]. Another very useful reference is [21]. Loudness ratings and similar terms referred to below are expressed in dB attenuations. To assist in the understanding of the various definitions, a number of figures are provided. The definitions in this section apply generally to a telephone system; extension to the cordless terminal is considered in the following section.

Sending Loudness Rating (SLR) and Receiving Loudness Rating (RLR)

Figure 11.9 illustrates a terminal, e.g. an ordinary telephone, with a two-wire or four-wire interface to a PABX or PSTN extension. (The necessary filters and amplifiers are not shown in this general simplified diagram.) *SLR* and *RLR* express relations between sound pressure at the acoustic interface at the microphone and output transducer (M and T) of a terminal and the electrical signals in the network, e.g. at the extension interface (interface H in Figure 11.9) to a private or public switch. A two-wire interface is physically implemented by means of a hybrid, which permits signals to pass in both directions simultaneously. *B* is the echo balance return loss referred to the two-wire side of this hybrid (see below).

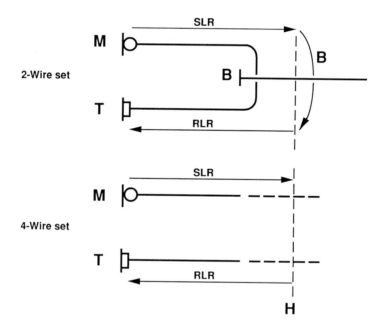

Figure 11.9 Simple telephone terminal: definitions of *SLR*, *RLR* and *B*.

Echo Balance Return Loss (B)

The echo balance return loss of the hybrid, *B*, refers to signals transformed to the two-wire side of the hybrid. For calculation of echoes, *B* is defined as 11 ± 3 dB, the mean value of 11 dB generally being used (see [20], Annex B/G.122 and Supplement No. 2 section 2).

Connection Loudness Rating (CLR)

Figure 11.10 illustrates end-to-end transmission in a simplified diagram. The figure is generalised for analogue and digital two- and four-wire transmission and mixtures thereof. The connection loudness rating (*CLR*) is the loudness loss between two electrical interfaces, as in this example between the extension line reference points close to the first switches. The *CLR* path can contain anything from a simple through-connection to a whole chain of PABX, national and international switches, satellite links, digital and analogue conversions, etc. If hybrids are contained in the transmission, echoes will occur, as indicated by the echo paths 1 and 2 shown in the diagram. Reference [20] uses the older expression, junction loudness rating (*JLR*), instead of *CLR*.

Overall Loudness Rating (OLR)

The overall loudness rating (*OLR*) defines the total transmission, sound pressure to sound pressure, from M to T, and is related to the previously defined terms by the equation:

$$OLR = SLR + CLR + RLR \qquad (11.2)$$

A value of *OLR* of 9 dB gives approximately the same sound pressure at T as at M. *OLR* is suggested to be 10 dB for digital networks. *OLR* for old analogue networks can vary significantly, e.g. 4–30 dB, and is mostly larger than 10 dB.

Sidetone Masking Rating (STMR)

The sidetone is an electrical path from the microphone in a handset to the speaker device. Its purpose is to substitute for the sound that normally travels in the air from the mouth to the ear, but is blocked by the handset pressed to the ear. *STMR* expresses sound pressure to sound pressure, but is defined with a different frequency weighting function than that used for loudness rating (*LR*). For digital systems *STMR* is specified to 13±5 dB. Taking the different weighting functions into account, an *STMR* of 13 dB approximately corresponds to the sound pressure obtained at the distant-end speaker with an *OLR* of 10 dB. For analogue two-wire sets, the *STMR* path is via the echo balance loss of the hybrid, path 1, and thus varies more than for (digital) four-wire sets. The sidetone is important, not only for compensating for blocking of the natural air acoustic path from the mouth to the ear, but also because it has a subjective masking effect on echoes, as discussed later.

Terminal Coupling Loss (TCL)

Echoes may also be generated by another path, path 3 shown in Figure 11.11, via the combined effect of the electrical and acoustic loss over the handset (*TCL*). *TCL* is defined at the same extension line reference point as *SLR* and *RLR*. An

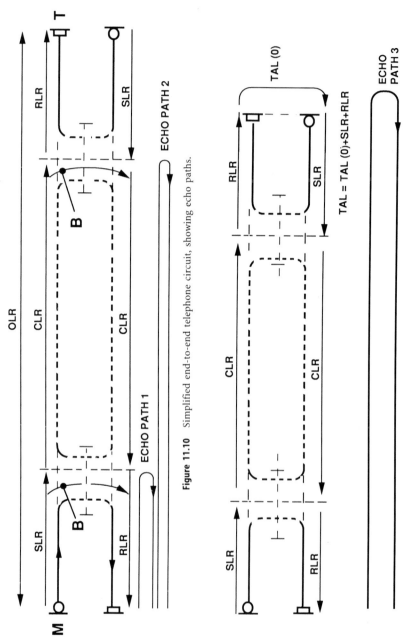

Figure 11.10   Simplified end-to-end telephone circuit, showing echo paths.

Figure 11.11   Acoustic distant-end echo path.

earlier definition, which occurs in some references and specifications, here denoted $TCL(0)$, is defined for the case $RLR + SLR = 0$. Thus

$$TCL = TCL(0) + SLR + RLR$$

$TCL(0)$ is that part of the telephone acoustic loss that relates to the specific acoustic–mechanical design of a handset.

- For digital transmission, CLR is $0\,dB$. Thus, since $OLR = 10\,dB$ (above), equation (11.2) implies that $SLR + RLR = 10\,dB$. Hence

$$TCL = TCL(0) + 10\,dB \tag{11.3}$$

- For analogue connections where $SLR + RLR$ is $-1\,dB$, such as in Sweden or the UK, then

$$TCL = TCL(0) - 1\,dB \tag{11.3a}$$

Overall Echo Loudness Rating (OELR)

The $OELR$ relates the sound pressure at the microphone M to the sound pressure of a returning echo at the speaker T of the same telephone. (Instead of $OELR$, the expression $TELR$ (talker echo loudness rating) is also sometimes used. There is in fact a small difference, $\sim 1\,dB$, between the two terms.)

- For echo path 1 in Figure 11.10 (short extension lines), $OELR$ is given by

$$OELR = SLR + RLR + B \tag{11.4}$$

- For echo path 2,

$$OELR = SLR + RLR + 2(CLR) + B \tag{11.5}$$

- For echo path 3 from Figure 11.11,

$$OELR = SLR + RLR + 2(CLR) + TCL \tag{11.6}$$

or

$$OELR = 2(OLR) + TCL(0) \tag{11.7}$$

Note that equation (11.4) has to be used with care for analogue transmission. If the extension line between the handset and the switch is long ($>2\,km$), there will be attenuation on the line. Thus, $SLR + RLR$ defined close the switch will differ from $SLR' + RLR'$ measured close to the terminal. In fact, for analogue transmission, since $SLR' + RLR'$ for path 1 will differ from $SLR + RLR$, the value for $OELR$ is given by

$$OELR = SLR' + RLR' + B \tag{11.4'}$$

According to national specifications for analogue lines, the value of $SLR' + RLR'$ (attenuation) can nominally be as low as $10\,dB$ below the value of $SLR + RLR$ for long extension lines. Thus the echo from path 1 can be larger at long extension lines than at short extension lines. Path 1 is the path that provides the sidetone in analogue telephone sets.

11.2.2   Transmission Plan for a Digital Cordless Telephone System

The transmission diagrams discussed above can be extended to incorporate a digital cordless telephone, by replacing the fixed telephone part (Figure 11.9) by the cordless equipment shown schematically in Figure 11.12. The cordless link is shown separated into its two distinct parts, that comprising the cordless portable part, CPP (the handset), and that comprising the fixed part connected to the network infrastructure. (Different terms are used by different authors in discussing these different components of a cordless system – see the "Glossary" for other terms used.)

A number of specific interfaces shown in Figure 11.12 are defined below (the notation adopted, A–H, does not follow any particular convention but is purely used for example). Interfaces A–E relate to the handset and E–H to the fixed part. The fixed part includes basestations and central control, and can range from a simple residential basestation to a large cordless PABX connected to the public network. Thus the network interface H can represent not only extension lines but also trunk lines of a PABX, ISPABX, PSTN or ISDN. This transmission plan has been exemplified with the well known ITU G.721 $32\,\text{kb s}^{-1}$ ADPCM codec for the radio transmission. Since the measurements are made at $64\,\text{kb s}^{-1}$ reference, any codec fits into the transmission plan, provided requirements on quality and delay are met. The interfaces A–H are defined as follows:

- *Interface A* is the acoustic interface at M and T of the handset.
- *Interface B* is the analogue electrical interface.
- *Interfaces C and G* are the linear 0 dB $64\,\text{kb s}^{-1}$ PCM digital reference used in specifying digital telephones. Interface C never needs to exist physically; however, interface G must physically exist if digital $64\,\text{kb s}^{-1}$ PCM connection (e.g. ISDN) is to be supported. It is practicable to include interfaces C and G in the diagrams since digital measurement methods refer to a $64\,\text{kb s}^{-1}$ PCM interface. There is a one-to-one relation between the digital signals of the $64\,\text{kb s}^{-1}$ PCM and the $32\,\text{kb s}^{-1}$ ADPCM. Thus under error-free radio transmission conditions the signals at C and G are equal.
- *Interfaces D and F* are $32\,\text{kb s}^{-1}$ ADPCM G.721 interfaces. The diagram can easily be modified for other choice of codec.
- *Interface E* is the radio interface carrying the $32\,\text{kb s}^{-1}$ ADPCM information.
- *Interface H* is the interface to the trunk or extension line of a private or local switch. H_1 represents the case of digital transmission, $64\,\text{kb s}^{-1}$ μ-law PCM. For analogue transmission, analogue-to-digital and digital-to-analogue converters, filters and amplifiers have to be included in the cordless radio exchange or central part, and the attenuation has to be adjusted according to national requirements on *SLR* and *RLR*. H_2 represents analogue four-wire connection and H_3 analogue two-wire connection.

STMR indicates the sidetone path of the handset. This does not necessarily have to be an analogue path, but the path must not contain substantial extra delay ($\lesssim 1\,\text{ms}$).

Compared to a standard wired telephone, delay will be introduced, especially at interface E, but also in codecs and other digital processing. This is not indicated explicitly in Figure 11.12, but is treated later in this chapter.

A cordless handset may be used as a loudspeaking set by placing it onto a table stand with built-in loudspeaker, microphone and voice-operated switch. This set

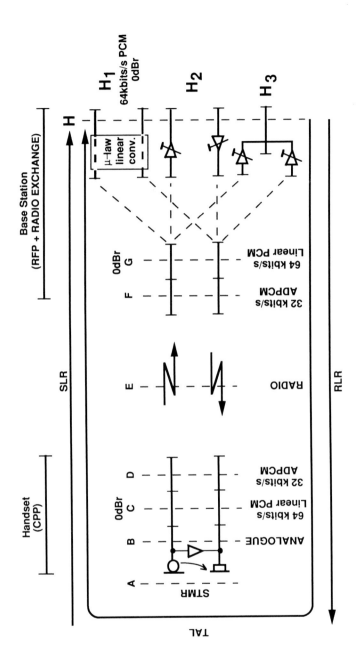

Figure 11.12 Transmission diagram for a cordless telephone system.

has to have the voice-operated switch at the B or C interface, as for ordinary wired loudspeaking telephones.

11.2.3 Transmission Plans for Specific Systems

Having established a frame of reference, Table 11.4 lists the requirements that arise in respect of *SLR*, *RLR*, *STMR* and *TCL* for specific systems, taking as examples the CT2 CAI and the DECT cordless systems. All the tabulated values are presented in dB. For digital interfaces, H_1, *SLR* and *RLR* are defined by the NET33 standard, and this is followed for DECT. For longer lines (>2 km) with analogue interfaces, H_2/H_3, the amplification has to be increased to compensate for extension line losses (see relevant specifications [16–19]). An example of the application of a voice transmission plan using the H_3 interface, the CT2 CAI high-level transmission plan, is included in the later chapter describing this standard.

Table 11.4
Example *SLR*, *RLR*, *STMR* and *TCL* requirements for specific systems

	SLR / RLR		*STMR*	*TCL*
	H_1	H_3, short extension lines		(*SLR* + *RLR* = 10 dB)
CT2 CAI	7±3 / 3±3	3.3±4.2 / −4.5±3.5	13±5	>34
DECT and PWT	7±3 / 3±3	National specification	13±5	>34 or >46[a]

[a]Two options are specified, with >46 dB recommended.

11.3 Delay and Echo Control

It is important to distinguish between two separate effects of delay – delay itself and delayed echoes.

One-way delays of up to about 100 ms result in no subjective interference to a telephone conversation. However, larger delays, e.g. a 260 ms one-way delay arising from a satellite link, are noticeable, and inexperienced users may start to talk simultaneously or think that the other person is slow in responding. The effect of pure delay itself, however, is not very large if an additional small delay, e.g. 10–15 ms introduced by a cordless telephone link, is added to already existing delays.

Delayed echoes are much more degrading to the perceived speech than the delay itself. Such degradation begins to occur for delays of a few milliseconds in the presence of very strong echoes. For long delays, weak echoes may also contribute a degradation to the link quality. Thus, imperfect means for control of such echoes can be more noticeable than direct effects of the delay itself.

When specifying allowable extra delay from a cordless telephone system, the limitation is not the total delay itself. For example, adding 25 ms (10%) to the delay of an satellite link will not be noticeable. The reasonable limits for extra delay depend on its influence on echoes and upon existing network echo control means and, if needed, the cost and complication of echo control means within the cordless system itself. Therefore the different possible echo paths have to be analysed and, in the case of satellite links, the requirements imposed by existing echo control means.

Three echo paths relevant for the cordless handset user have been indicated in Figures 11.10 and 11.11. These are shown in Figure 11.13 together with two more echo paths, relevant to the distant-end user, paths 4 and 5. In Figure 11.13, the left-hand terminal is a cordless telecommunication system as in Figure 11.12. T indicates the total extra one-way delay that is introduced compared to a wired telephone. EC indicates an echo control means that, for some cases, is needed on trunk or extension level at the linear interface G of Figure 11.12. The right-hand terminal is normally a wired telephone, but could also be a cordless device.

11.3.1 Criteria for Acceptable Echoes

Echoes do not influence the subjective perception of speech if they are sufficiently attenuated. Such attenuation comes from line losses and the hybrid return loss or the telephone acoustic path loss, as exemplified in equations (11.4)–(11.7).

The commonly used reference for minimum echo return loss as a function of the one-way delays from 10 to 300 ms is G.131 [20] (see Figure 11.14). In [20], the echo return loss is expressed in OLR. As previously mentioned, $TELR$ and $OELR$ are closely related, with less than 1 dB difference between them. It should, however, be noted that new subjective tests have been carried out on $TELR$ or $OELR$ requirements under conditions relevant to modern telecommunications, and these results may result in another revision of G.131 [20].

The curves shown indicate the 1% and 10% probabilities for subjectively encountering an objectionable echo; the 1% curves are used as design criteria (see [20], 2.3.1.1 Ideal rule A/G.131). The dashed line is for digital transmission and the continuous lines refer to transmission over a chain of 1–9 analogue four-wire circuits. The $OELR$ has to be larger for larger numbers of analogue circuits in the chain, owing to the increasing statistical spread of the line attenuations. For examples in this chapter, we assume digital or a single four-wire analogue transmission link.

As an example of applying Figure 11.14, we make an estimation of the maximum delay that can be accepted for a long-distance call between two analogue wired telephones. In this case it is the path 2 (or 5) that causes the echo. The $OELR$ for path 2 (see Figure 11.10) is given by

$$OELR = SLR + RLR + 2(CLR) + B$$

or

$$OELR = 2(OLR) + B - (SLR + RLR) \qquad (11.8)$$

In order to evaluate this expression, we note from earlier the value of 11 dB for B. Also, from Table 11.4 we find that nominally $SLR + RLR = -1$ dB in the UK or Sweden for analogue trunks (H_3). If we assume OLR is a typical value of 10 dB, this gives an $OELR$ of $2(10) + 11 - (-1)$ dB, i.e. 32 dB. Using Figure 11.14, following the curve for a single analogue circuit, we see that this corresponds to a 20 ms one-way delay as the limit for 1% probability of encountering unacceptable

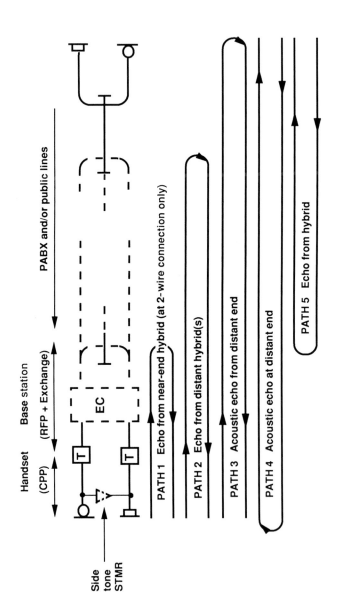

Figure 11.13 Echo paths relevant for a cordless telephone system.

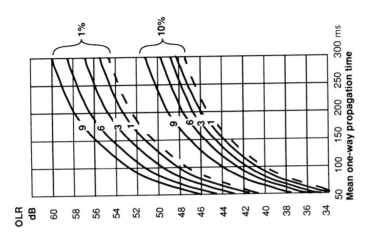

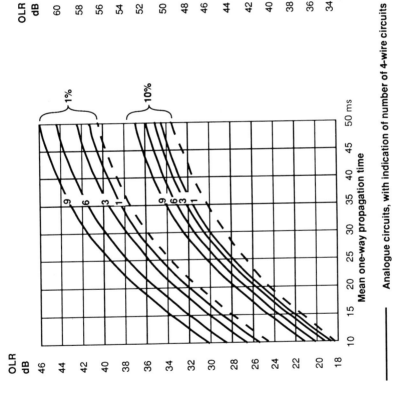

Figure 11.14 Minimum allowable overall loudness rating of the echo path as a function of delay. (*Source:* from [20], courtesy of ITU.)

echoes. Thus there is an ITU rule (see [20], 2.3.3.2.3 Rule M/G.131) that echo control devices are required in the network if the one-way delay exceeds 25 ms, otherwise an unacceptably high line attenuation has to be used. This situation arises most commonly when a connection includes a satellite link.

Figure 11.14 is useful for studying the echo paths 2–5, but not the echo path 1. Path 1 is not considered for a wired telephone, since it has no significant delay. This may not be the case, however, for a cordless device. For path 1 we refer to subjective tests in a report to ITU [22], which is a basic reference for impairment models being developed in ETSI. Figure 11.15 shows results in terms of MOS (Mean Opinion Scores). The tests were performed for 0, 2, 8 and 16 ms one-way delays and an $OELR$ ranging from 0 to 30 dB, with a direct sidetone masking $STMR$ of 15 dB. (Note that an MOS of only 2 is obtained for the case of zero delay and an $OELR$ of 0 dB. The reason for this is that the indirect sidetone is heard as a very loud direct sidetone. Such a situation occurs for an analogue telephone if $SLR' + RLR' = -11$ dB (long lines) and $B = 11$ dB.

An interpretation of Figure 11.15 gives the approximate limits shown in Table 11.5 for path 1, if one half-point of MOS degradation is allowed compared to 0 ms delay.

Table 11.5
$OELR$ values as a function of delay, for one half-point MOS degradation

Delay (one-way)	$OELR$ (dB)	
(ms)	Half-point degradation	No degradation
1	0	20
2	15	22
8	21	25
16	25	30

A cross-check between the 8 ms one-way delay results from Table 11.5 and the 10 ms one-way delay result in Figure 11.14 shows good correlation. Correction from 8 to 10 ms adds 2 dB to the 21 dB tabulated value, giving 23 dB; adding a few dB for the sidetone masking effect gives values close to the 25–26 dB of Figure 11.15.

11.3.2   Echo Control at International Gateways

A single satellite link adds 260 ms one-way delay and about 0 dB transmission loss. The allowed national delay is a maximum of 12 ms; if, for example, 4 ms are needed for digital processing in the transmission chain, then the remaining 8 ms correspond to 2000 km transmission. As seen from Figure 11.14, a total $OELR$ of 55–60 dB is needed to suppress 300 ms delayed echoes. The $OELR$ for the echo from the distant-end hybrid is only 20–30 dB without extra echo control means. It is for this reason that, when satellites are used, echo control is always used at the international gateways (see Figure 11.16 and [20], G.131, G.164 and G.165).

Echo control can be performed either by an echo suppressor, which is a speech-controlled directional switch, or by an echo canceller, which by means of adaptive linear filters makes an estimate of the transfer function of the echo path and uses that information to subtract the echo in the return path.

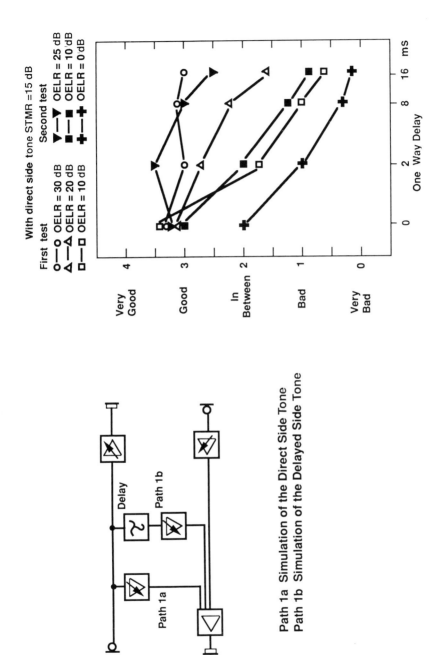

Figure 11.15   Mean Opinion Score test of delayed sidetone with direct sidetone.

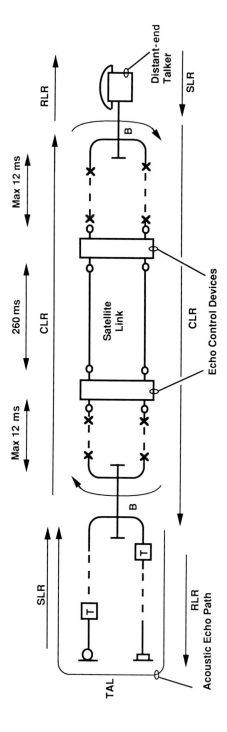

Figure 11.16   Schematic diagram of satellite-linked connection.

An echo suppressor operates by connecting different losses in the receive and send paths. During one-way talk from the distant end the receive loss is 0 dB and the suppressor loss typically 50 dB; thus no echoes are allowed through independent of their delay. No delayed acoustic echoes are therefore heard at the distant end. The only requirement is that the further delayed acoustic echoes from a cordless telephone are attenuated sufficiently so as not to be interpreted as a break in speech signals, which could switch the echo suppressor state. If $TCL(0) > 24$ dB, as mentioned above, the acoustic echo will be low enough to meet this requirement.

During double talk, that is when speech signals are simultaneously present at both ends of the link, a fixed 6 dB attenuation is employed in each direction. This means that echoes are heard, but these are partly masked by the local speech. This state of operation represents a weakness in the behaviour of an echo suppressor. It should be noted, however, that the echoes from the normally present hybrids (which have nothing to do with the extra delay of a cordless system) dominate over and mask the weaker further-delayed acoustic echoes by some 13 dB for the case of $B = 11$ dB and $TCL(0) > 24$ dB.

Echo suppressors often degrade speech due to the needed time constants in the switching. Therefore, more complex echo cancellers are instead being used in all modern networks.

The echo canceller employs adaptive processing to construct an estimate of the anticipated echo, which it then subtracts from the returning signal. The normal echo canceller has a control range of 30 ms, i.e. it can cancel echoes over this spread of delay. The canceller comprises a linear canceller component, capable of echo subtraction down to 25–40 dB, and a nonlinear processor that adds additional attenuation to a total of more than 65 dB. The latter acts like a centre clipper and cancels everything below its suppression threshold. In the idle condition the nonlinear processor is switched off (i.e. transparent).

One main advantage of an echo canceller, compared with an echo suppressor, is that full echo control is possible during both one-way and double-talk states. It is only the nonlinear processors at the two sides of the link that are switched on and off between the two one-way talk states (one state for each direction). During double talk, some echo cancellers have the nonlinear processors active, some not.

During one-way talk from the distant end, all echoes are suppressed that are within the 30 ms control range of the canceller; since the nonlinear processor is active, all echoes that exceed the control range, but are below the suppression threshold of the nonlinear processor, are totally suppressed. If $TCL(0) > 24$ dB, as mentioned above, the acoustic echo will be low enough, including an adequate safety margin, to be totally suppressed by the nonlinear processor. The acoustic echo will also be well below the limit to be detected as a break in signal for going to the double-talk state.

During double talk, if the nonlinear processor is not active, acoustic echoes outside the control range of the canceller will not be attenuated. However, the level of these echoes will be more than 30 dB below the level of the interfering double talk, and will be masked by this.

Analysis of the situation of international gateway links [23] shows that for analogue near-end extensions the distant-end user is protected from echoes over the delayed acoustic path provided $TCL(0) > 24$ dB for a cordless system. TCL must be more than 23 dB for an analogue two-wire interface ($SLR + RLR = -1$ dB). The digital case is more easily analysed, since the speech signal can never exceed

+3 dBm, and the threshold for break-in signals is −31 dBm. This gives the requirement that $TCL > 34$ dB for a digital interface ($SLR + RLR = 10$ dB). Note that the TCL is limited by the physical properties of the cordless handset. For the same handset, TCL will automatically change 11 dB between analogue and digital interface application, following the amplifier settings for SLR and RLR in this example.

(Note that the implementation of the nonlinear processor, as described above, is only given as an example in G.165 [20]. Thus to make, for example, DECT connections fully independent of future implementations of echo cancellers in the public network, it might be desirable to increase the TCL requirement to 55–60 dB. This higher TCL could be achieved by adding a nonlinear processor as described in G.165 in the microphone line of the cordless system. An advantage of this approach is that the requirement of acoustic attenuation in the handset and the setting of the suppression threshold of the nonlinear processor can be traded off.)

11.3.3   Echo Control for Non-International Circuits

Both the distant- and near-end user can suffer echoes through the paths shown in Figure 11.13. Having examined the requirements imposed upon a cordless system by the echo control at international gateways, we now examine the constraints that arise when a cordless system is used within a simple PABX or local, national, scenario where no echo control means is normally provided by the transmission chain. Examination of these constraints leads to a limit upon the maximum delay that can be tolerated from a cordless telephone system. It is also found that for cordless telephone systems, which indeed introduce significant one-way delay (e.g. $\gtrsim 3$ ms), a means of echo control must be incorporated within the cordless equipment. Thus echo cancellation is commonly used for example with DECT, but not CT2. To establish these conclusions, we firstly consider protection of the distant-end user and then the local user.

The innocent distant-end user can get echoes through the paths 4 and 5 of Figure 11.13. Path 5 concerns echoes from hybrids equivalent to those present in ordinary analogue telephones and is thus not considered further. For the echo path 4 via the delayed acoustic path of the cordless system, we refer to Figure 11.11. Applying equation (11.7), assuming an ideal OLR for analogue and digital connections of 10 dB and a value of $TCL(0) > 24$ dB, we can derive an $OELR > 44$ dB. Thus, from Figure 11.14, we can conclude that the largest permissible one-way delay is ~70 ms. Thus if the wired network gives a maximum of 25 ms (the maximum for a network without echo control), then 45 ms remains for the cordless system. Allowing for the case of a cordless link at each end of the line places a maximum limit on the cordless system one-way delay, T, of 22 ms. (This maximum delay figure may be increased by incorporating a soft echo suppressor in the cordless handset to increase the $OELR$.)

In considering the near-end user, three cases must be considered, the case of four-wire connections end to end, and the cases of hybrids at the far end and at the near end.

The first of these, path 3 only (Figure 11.13), is usual for analogue or digital trunk interfaces for modern PABX and modern public networks with ISDN terminals. The general analysis is the same as above for echo path 4 with cordless handsets at both ends.

For four-wire connection at the near end, with a hybrid at the distant end, the hybrid echo of path 2 will always dominate over the acoustic echo of path 3. In this case the delay T of the cordless system will be added to the delay (<25 ms) of the network. The slope of Figure 11.14, required OELR as a function of delay, over the interval of 10–25 ms is ~0.5 dB ms^{-1}. Thus, to maintain the subjective quality, additional attenuation of this order must be added. (For T below 5 ms the difference is so small that no compensation is needed.) This extra attenuation can be introduced in the cordless system as a soft suppressor in the speaker line. In this way the distant-end talker or outgoing levels are not affected at all. This simple suppressor unit can be in the handset or on trunk level in the central part and is activated by near-end speech. Thus, for connections that contain distant-end hybrids (which most do), a soft suppressor with ~0.5T dB attenuation (where T is in milliseconds) has to be employed in the speaker connection of the cordless system. The term "soft suppressor" is used to indicate that the speech in the suppressed direction is not completely suppressed – only some 6–18 dB of suppression occurs, compared to the 50 dB associated with a normal suppressor. This imposes less stringent requirements on, for example, switching times.

For the case of connection with a hybrid at the near end, i.e. analogue two-wire connections, the echo corresponds to path 1 of Figure 11.13. As explained earlier, for long extension lines $SLR' + RLR'$ can be as low as -11 dB and thus, if $B = 11$ dB, equation (11.4a) implies that the OELR can be as low as 0 dB. In this case echo control within the cordless system needs to provide attenuation of as many decibels as indicated in Table 11.5 for one-point degradation. For short extension lines, $SLR' + RLR'$ may be around -1 dB, resulting in the extra echo control attenuation being ~10 dB less than the tabulated values. If the echo control is implemented with an echo canceller, only ~4 ms control range is needed, leading to a full custom chip with less than 10 000 transistors.

The control range needs to be 10 ms if the connection is made via a digital PABX connected via two-wire interface to the PSTN and if the internal PABX one-way delay is the maximum allowed 5 ms [21] (2 ms is more typical). This extra control range leads to a requirement for a more complex, 20 000-transistor, integrated circuit.

The general requirement is thus for an echo canceller with 25 dB attenuation and 10 ms control range at trunk level in the cordless central part. If a soft suppressor also is used with e.g. 10 dB suppression, then the attenuation of the echo canceller can be decreased to 15 dB.

In many installations to PABXs an echo canceller may not in fact be needed, partly because the extension line will be short (e.g. $SLR + RLR \sim -3$ dB) and partly because the balance of the hybrid is very well defined (e.g. $B \sim 18$ dB). Thus, for a digital PABX, or other PABXs with well defined line impedances and four-wire connections to the PSTN, a simple soft suppressor (~10 dB) can provide the remaining attenuation needed.

For cases like a domestic cordless telephone, the line impedance is not well controlled. However, in most cases B is in the order of 15–18 dB or more. Thus, either the general case (with 4 ms control range) can be applied using an echo canceller and a soft suppressor, or alternatively an adaptive soft suppressor can be used. The soft suppressor can be adapted in steps of 6, 12, 18 and 24 dB. In most cases, settings of 6 or 12 dB will be used. Since this soft suppressor does not affect the distant-end user at all, but only the user of the cordless handset, the adaptation algorithm might be left as a manufacturer's proprietary feature.

11.3.4   Requirements and Solutions for Specific Systems

Having considered the principles relating to echoes and delay, we now briefly outline the ways in which these issues have been specified in specific cordless telephone systems in recent years.

The Common Air Interface (CAI)

The CAI system has succeeded in avoiding the complexities of echo control by minimising system delays. CAI embodies time-division duplex transmission, but not time-division multiple access, with the result that the one-way delay is sufficiently short that no soft suppressor is needed to compensate in echo path 2.

For echo path 4, when satellites are used, no specific requirements are set. Since the extra delay is so short, the statistical risk is low that the national delay budget of 12 ms will be exceeded.

The requirement on TCL is only 34 dB (with $SLR + RLR = 10$ dB) and is not intended to help for cases when the delay budget is exceeded.

The hybrid echo at two-wire connection does not require any echo control means in spite of the risk for quality degradation. However, the value $B = 11$ dB may be conservative. If instead 18 dB is used, degradation will only occur at long extension lines.

Digital Enhanced Cordless Telecommunications (DECT and PWT)

The full specifications on delay and echo control are found in [19] (Part 8). This reference also contains several informative annexes with examples of implementation and justifications of the requirements.

DECT is a TDMA system and the maximum one-way delay is specified to be 14 ms. There are two options allowed for the requirements on TCL.

- $TCL > 46$ dB at nominal setting of the volume control and $TCL > 35$ dB at maximum setting. This is the recommended option.
- $TCL > 34$ dB at every setting of the volume control.

Table 11.6
Echo loss requirements for DECT

	Echo path (two-way)	Echo loss
Requirement 1	0 to 4 ms	$TELR \geq 24$ dB
Requirement 2	0 to 70 ms	Extra echo loss ≥ 9 dB

There are two requirements for the network echo control. The echo from the network shall be controlled by inserting into the receiving speech path of the fixed path an echo loss meeting the requirements of Table 11.6. Requirement 1 only applies for fixed parts with an analogue two-wireline interface and is normally implemented as a simple echo canceller (see section 11.3.3). Requirement 2 applies to all interfaces and is normally implemented as a soft suppressor.

PWT has the same requirements as DECT.

11.4   Voice Security and Speech Encryption

Normal wired telephones do not use speech encryption, neither do analogue mobile telephone nor mobile radio systems, although the latter systems can rather easily be listened to with a radio receiver. In the case of a cordless telephone, used, for example, as the prime office telephone, it is vital to provide speech security for the radio link. Since the transmissions are digital, they cannot be accessed by normal radios. However, special radios designed to receive the specific digital radio transmission of a cordless telephone could easily find their way onto the market. Digital transmission, however, does make it rather easy to implement speech encryption. Such encryption may be performed simply by scrambling of the digitally encoded data according to some predefined encryption algorithm. This may be performed either in software or in an integrated circuit, possibly some additional gates on an existing speech coding chip. The key issue is in fact not the encryption process as such, but the distribution and security of the (individual) keys that tailor the encryption algorithm to the individual user or equipment.

Key distribution and security for a residential set or a closed PABX user group is rather uncomplicated. Both the user of the residential set and the communications manager at the office can, by simple instructions, program into the fixed system and the portable his or her own keys, which no-one else needs to know, and which do not need to be transferred.

It is not so easy to provide speech encryption for public telepoint systems or to roamers in private PABX systems (in this context, the term "roamer" is used to mean a user registered on a cordless PABX who is visiting another site whose cordless PABX is also linked in to the same private PABX network). In these cases the key has to be transferred via wire from the home location register (i.e. the control system at the "home" PABX) to the actual basestation or system covering the area visited. This is both more complicated and less secure than for the closed user group with limited mobility. Another possibility would be for the basestation to assign a temporary key to roamers. Transmitting the key over radio, however, would also appear to be insecure. A visitor at an office could perhaps have his or her temporary key transferred via a wired table stand in which the visitor's handset is inserted.

A reasonable level of security might be to provide encryption for the frequent use at the home location (office and/or residence) and to provide the roamer with less security or offer high security at a premium, if available. For actual specific cordless systems we find that CAI has no explicitly specified speech encryption, although encryption could be provided by the mechanism of defining a new codec type [16].

DECT has well specified authentication and ciphering algorithms and processes closely related to those of GSM (see [19], Part 7). The cipher key is securely transferred to the portable handsets during the authentication process, the latter being essential for secure billing.

11.5 Summary

In this chapter we have addressed the key audio processing requirements of cordless telephone systems. The relevant speech coding schemes were described – ADPCM, delta modulation, linear prediction and their derivatives – together with the issue of measurement of speech quality. The parameters associated with the voice transmission plan for a cordless system were defined and typical values presented. Working from these same definitions the significance of delay and echo control were explored, in order to explain how the limits on system parameters are derived. Echo control requirements for international and non-international links, involving cordless terminals, were established. This permitted us to illustrate the way in which the complexity requirements for echo control devices for cordless systems are derived. Requirements for specific systems were given as examples. Finally, the issues of speech privacy and encryption were presented.

Acknowledgements

The prior authorisation of the ITU, as copyright holder, to reproduce Figure 11.14 is gratefully acknowledged. The choice of excerpts from G.131 is, however, the authors' own and for which the ITU has no responsibility. The full text of the various ITU recommendations may be obtained from the ITU Sales Section, Place des Nations, Geneva, Switzerland.

References

[1]   "The Philosophy of PCM", BM Oliver, JR Pierce and CE Shannon, Proceedings of the IRE, Vol. 36, pp. 1324–31, October 1948
[2]   "Instantaneous Companding of Quantised Signals", B Smith, Bell System Technical Journal, pp. 653–709, 1957
[3]   "Digital Coding of Speech Waveforms: PCM, DPCM and DM Quantisers", NS Jayant, Proceedings of the IEE, Vol. 62, pp. 611–32, May 1974
[4]   "Optimum and Adaptive Differential PCM", RW Stroh, PhD Dissertation, Polytechnic Institute of Brooklyn, Farmingdale, NY, 1970
[5]   "Adaptive Differential PCM for Speech Processing", P Cummiskey, PhD Dissertation, Newark College of Engineering, Newark, NJ, 1973
[6]   "Adaptive Quantisation in Differential PCM Coding of Speech", P Cummiskey, NS Jayant and JL Flanagan, Bell System Technical Journal, pp. 1105–18, September 1973
[7]   "32 kbits/s Adaptive Differential Pulse Code Modulation (ADPCM)", ITU Blue Book, Vol. 3, Fasc. III.3, Rec. G.721, ITU, 1984
[8]   "Pulse Code Modulation (PCM) for Voice Frequencies", ITU Blue Book, Vol. 3, Fasc. III.3, Rec. G.711, ITU, 1984
[9]   "Companded Delta Modulation for Telephone Transmission", A Tomozawa and H Kaneko, IEEE Transactions on Communication Technology, Vol. COM-16, pp. 149–57, February 1968
[10] "Linear and Adaptive Delta Modulation", JE Abate, Proceedings of the IEE, Vol. 55, pp. 298–308, March 1967

[11]   "Linear Prediction of Speech", JD Markel and AH Gray, Springer-Verlag, Berlin, 1976
[12]   "G.729 – Coding of Speech at 8 kbit/s Using Conjugate-Structure Algebraic Code-Excited Linear Prediction (CS-ACELP)", ITU-T \ COM15 \ C \ 152E1.WW2 – 14.08.95 (Draft ITU-T Rec. G.729)
[13]   "Coding of Speech at 16 kbit/s Using Low-Delay Code Excited Linear Prediction", ITU-T Rec. G.728, ITU, Geneva, 1992
[14]   "Provisional Code of Practice for the Design of Private Telecommunication Branch Networks", OFTEL, The Office of Telecommunications, London, December 1986
[15]   GSM Recommendation 06.10, Annex 1, Table 2, ETSI, 1989
[16]   "Common Air Interface Specification to be Used for the Interworking Between Cordless Telephone Apparatus Including Public Access Services", UK Department of Trade and Industry, MPT 1375, London, May 1989, Amended November 1989
[17]   "Swedish Telecom Regulations on Radio Technical Requirements on Digital Cordless Telephones in the Frequency Band 862 to 864 MHz", Swedish Telecom Code of Statutes TVTFS 1989:103, Swedish Telecom, 1989
[18]   "Technical Telephone Requirements for a Digital Cordless Telephone", Swedish Telecom Specification, 8211-A:130, Swedish Telecom, 1989
[19]   "The DECT Specification", ETS 300 175, Parts 1–8, 2nd Edn, ETSI, European Telecommunications Standards Institute, Valbonne, France
[20]   "General Characteristics of International Telephone Connections and Circuits", ITU Blue Book, Vol. III, Fasc. III.1, Rec. G.101–G.181, ITU, 1984–85
[21]   "Overall Transmission Plan Aspects of a Private Branch Network for Voice Connections with Access to the Public Network", ETS T/TE 10-05, 6th Edn, ETSI, European Telecommunications Standards Institute, Valbonne, France
[22]   "The Effect of Delayed Side Tone on the Overall Sound Quality of the Telephone Connection", ITU COM XII–226, November 1987, ITU, Geneva
[23]   "On Echo Control When a BCT With T ms Inherent One-Way Delay is Connected via a Satellite Link", D Åkerberg, Report TY87:2017, Ericsson Radio Systems, Stockholm

12 The Radio Channel

Peter Hulbert and Bob Swain

Cordless telecommunications has today become a major consumer application field for radio communications. Its development is rooted in the practical demonstration of mobility by Marconi, but its commercial success has only been achieved by taking the results of the early pioneers, and the many who followed, to create products that the present-day market can accept in terms of applicability, user friendliness, quality, reliability, size, weight, appearance, flexibility, robustness, adaptability, supply, regulatory regime, cost and price. This list shows that the product purchasers and/or users are not interested in the radio channel technology. They will not marvel at the cleverness or originality of radio design, but they will expect it to do whatever they wish when they switch on, without having to undergo a training course, and without fail – and be cheap. Even the name "cordless telecommunications" eschews radio, and the basis of product performance comparison will be the wired telephone or terminal, with little, if any, licence given to the radio connection. This comparison is inevitable when cordless telecommunications is equated with communications in and around buildings, and not necessarily mobile in character.

In this context the design engineer's objective must be to make the radio part of the design responsibility invisible, i.e. he or she must aim for:

- Radio coverage of the nominated communication area.
- Communication reliability, in its widest sense, akin to that expected of a wired telephone or terminal.
- Cordless apparatus that is as easy to use as the wired equivalent.
- Physical characteristics that are suited for mobility, i.e. lightweight and pocketable.
- A radio system that encourages the widest application.
- A standardised radio interface that allows interworking between different manufacturers' products but does not stifle proprietary innovation.
- A radio design not constrained in its use by intellectual property rights (patents).
- A basic end-cost that will cause the various products to sell in numbers that will reap the full benefits of scale.

In this chapter, the radio issues are not addressed in the depth appropriate to detailed design. Instead, focus is given to those issues in which cordless telecommunications presently brings a somewhat different need, viewpoint or character. To the expert radio engineer, the omissions may seem awesome; to such we apologise in advance and recommend reference to their favourite collection of textbooks, or to references [1, 2]. To those others looking for a broader understanding and introduction to the radio aspects of unlicensed cordless communications, "welcome aboard".

12.1   Choice of Operating Frequency

Classically, the electromagnetic spectrum extends from direct current to light and beyond, but internationally the radio spectrum is considered to range from 9 kHz to 275 GHz. This radio resource may seem enormous in extent, yet only the bottom 60 GHz or so is commercially usable at present. Indeed, to achieve radio equipment costs appropriate to the mass consumer market, the exploitable spectrum presently is limited to below 12 GHz. The consequence is a multi-dimensional conflict of interests over the use of the prime radio spectrum allocations at national and international levels.

12.1.1   Spectrum Allocation

The 9 kHz to 275 GHz range has been allocated to services worldwide under the auspices of the International Telecommunications Union (ITU) with international agreement being achieved through regular World Radio Conferences (WRCs) at which changes and additions to the table of allocations can be made. This table is then administered by the Radiocommunications Bureau of the ITU.

Under these international agreements and recommendations, each national administration allocates spectrum to appropriately licensed and controlled users, service providers, researchers, etc. Some countries operate a policy whereby some low-radiated-power devices operating in certain bands do not require a licence. These devices include radio-controlled toys, alarms and cordless apparatus; such products benefit from not requiring a user licence procedure, an essential feature to facilitate a mass market.

In the USA this process is administered by the Federal Communications Commission, whilst in Europe the Conference of European Posts and Telecommunications Administrations (CEPT), together with the European Radiocommunication Committee (ERC), and its administrative organisation the European Radiocommunications Office (ERO), seek to coordinate European-wide policy towards the identification, allocation and timely provision of spectrum for apparatus expected to be standardised by ETSI for wide use and sale throughout Europe.

The range of services recognised by this regulatory process encompasses terrestrial and satellite point-to-point and point-to-multipoint (broadcast) systems, radar, navigation, astronomy, emergency services, radio and TV broadcasting and, of course, mobile services. This latter includes communications for vehicle fleet operators, paging, cellular radiophone as well as cordless communications. The distribution of these services with respect to frequency does vary, and furthermore in most nominated bands of frequencies more than one service is identified. Unacceptable mutual interference between sharing services is often possible. In this case one of the services is given primary status.

This in practice means that the secondary services cannot expect protection from interference from the primary service, nor must they cause unacceptable interference to the primary service. It can be readily appreciated, therefore, that, if the mobile service is accorded secondary status in a particular frequency band already used by the primary service in a particular area, then the use of the band for mobile services is virtually excluded if the above interference criteria are to be met. Such conflicts occur, for example, in those bands allocated on a primary basis to TV broadcasting (470 MHz to 862 MHz) or to terrestrial point-to-point and point-to-multipoint microwave links (e.g. 1.7 GHz to 2.3 GHz). In the face of the current burgeoning demand for radio spectrum for new terrestrial and satellite-based mobile and broadcasting services, the radio frequency regulatory authorities, at national and international level, have the very difficult task of judging the relative merits of one service with respect to another within the constraints of political and socio-economic policies and the perceived needs and preferences of the public at large.

To exemplify the problem, we consider the question: "Should 50 MHz of prime broadcast TV spectrum be vacated and transferred to serve the needs of mobility?" Mobility, after all, is serving business both large and small and making significant inroads into the consumer market. Indeed, projections for the year 2005 suggest that 50% of all telephone calls will involve a mobile link. Thus mobility will provide employment as well as facilities important to a country's well-being. However, broadcast TV offers hours of entertainment and edification per person per day at a minimal price to the viewer. A government that proposed such a swap policy would scarcely be popular. The classical response to this problem is to recommend the transfer of broadcast services onto multichannel, multi-choice, coaxial or optical-fibre networks feeding each home. Yet in a densely populated country the cost of doing so can amount to billions of ecus, which may far outweigh the economic benefits of an increase in mobile services and opportunities – further, market forces have in fact encouraged satellite TV broadcasting in competition with cable.

The question of who pays for the freeing-up of spectrum also arises – the taxpayer or a levy on the new service users? In recent years this question has been tackled in an innovative manner in the USA, with a new mechanism established whereby an element of the cost of new PCS products will be allocated to fund the migration of existing microwave fixed-link users into new spectrum to allow the establishment of the new PCS services, the process being managed by a newly established body, UTAM Inc. (see Chapter 8).

In principle, a new service could avoid these problems by identifying and requesting to occupy unused spectrum but, in the interests of regional standardisation, such spectrum should be widely available throughout many countries. This is a difficult task to achieve below about 30 GHz and almost certainly impossible below 10 GHz without major changes in present spectrum allocations. With these issues in mind, we consider the choice of frequency spectrum for unlicensed cordless telecommunications.

12.1.2 Constraints on Spectrum Choice

Cordless products are very sensitive to component cost. The relatively slow initial take-up of the analogue CEPT/CT1 cordless telephone products can be identified with their high cost differential compared to the more freely available and "lower-tech" analogue solutions, of which UK/CT0 and the illegal imports in other countries are prime examples. For cordless products to succeed in the mass consumer and small business markets, fundamental component technology must be mature and basic component production costs must have the potential to be reduced to levels competitive to existing products. The ability to exploit existing high-volume component technology developed for other mobile services in the region of 1 and 2 GHz has assisted cordless products in this regard. Other factors point to about 1 to 2 GHz being a favourable choice of frequency.

Man-made noise is one such factor. The unintentionally radiated radio-frequency noise produced by electrical equipment raises the effective receiver noise level well above that expected from consideration of thermal noise alone. Sources of such noise include electric motors, car ignition, radio-frequency heaters and domestic appliances. From this abbreviated list it would be expected that the level of noise will be a function of urban population density, and many researchers have confirmed this (see [1] for a comprehensive list). Thus no precise statistical definition can be given to the quantity but, as Figure 12.1 indicates, the importance of man-made noise significantly diminishes for radio equipment operating above 1 GHz. However, this tendency for man-made noise to push the desirable choice of frequency upwards is offset by the fact that free-space transmission loss increases in proportion to the square of the frequency, reducing potential range for a given transmission power. Thus, a frequency increase from 200 MHz to 2 GHz produces a 100-fold increase (20 dB) in transmission path loss for given range. In the range 1 GHz to 2 GHz these two factors tend to balance each other.

Thus, component cost, man-made noise and the fundamental physics of transmission loss all point to an optimum choice of cordless telecommunications

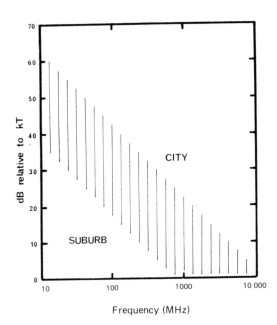

Figure 12.1 Frequency dependence of man-made noise. k = Boltzmann's constant, $T = 290$ K.

band between 1 GHz and 2 GHz. This has been reflected in national and regional spectrum allocations in recent years for a range of personal communications services, both cordless and cellular, in all regions of the world. Looking to the future, spectrum just above 2 GHz has been identified for future global third-generation services UMTS and FPLMTS, although already some of this spectrum is being used for earlier services, such as PCS in North America.

12.2 Spectrum Requirements

Cordless telecommunications supports much more than simple basic cordless telephony. It encompasses a wide diversity of applications and uses, as described in Part I of this book, including high-capacity, high-density cordless business communication systems designed for speech, data and multimedia wireless communications. The corresponding spectrum requirement is equally wide ranging, since it is a function of service and usage, the traffic density.

12.2.1 Traffic Density

Traffic density is generally expressed in terms of Erlangs per square kilometre ($E km^{-2}$) for low-density housing or, ideally, Erlangs per cubic kilometre ($E km^{-3}$), or alternatively Erlangs per square kilometre per floor, for high-rise office blocks. Typical traffic densities are given below [3] for the busiest period of the day (the "busy hour"):

- Residential suburban, 150 $E km^{-2}$
- Office (localised peak), 10 000 $E km^{-2}$ per floor

A cordless penetration factor for residential of 30% and for the office of 100% has been assumed in these figures. Also, of course, a basic speech telephony service has been assumed; the advent of multimedia services will increase these requirements.

Thus the office requirement is the determining factor in the spectrum requirement. Further, since a single WPBX system could service the requirement, it is clear that there is no scope for allocating radio spectrum based on growth of mean wide-area traffic densities. The full allocated spectrum must be available from the day of product launch, even if there is only one WPBX system in operation.

12.2.2 Spectrum Re-Use Principles

Spectrum was classically allocated on the basis of one radio carrier per user and an assumption that carrier frequency re-use range was greater than that determined from transmitter power and receiver noise sensitivity. Extrapolating this approach to the office traffic density referred to above would, for, say, 100 kHz effective bandwidth per two-way channel, require 1 GHz of spectrum allocation per floor, a requirement impossible to serve in the band below 2 GHz! It is for such reasons that cordless technology has adopted cellular frequency re-use techniques similar to those used in mobile telephone systems. Importing

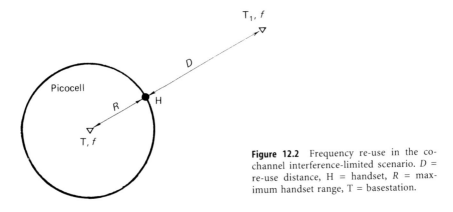

Figure 12.2 Frequency re-use in the co-channel interference-limited scenario. D = re-use distance, H = handset, R = maximum handset range, T = basestation.

these techniques into a building results in many small-radius cells per office floor, cells perhaps as small as 10 m radius. Naturally, a multi-storey building extends the cellular concept into a three-dimensional system with cells and corresponding transmitters located on every floor and forming part of a unified building WPBX system. To distinguish these cells from the relatively very large mobile telephone cells of 1 km or more radius, they have been termed "picocells". Such cellular techniques allow frequency re-use because care is taken to ensure that the distance between basestation transmitters using the same carrier frequency is not less than that required to achieve a given co-channel carrier-to-interference ratio. Consequently in a cellular system, the operational range of a radio link is limited by co-channel interference rather than by receiver sensitivity as in the earlier spectrum-greedy example. The situation of co-channel interference-limited operation is illustrated in Figure 12.2.

For a handset H to communicate successfully with basestation T on frequency f, the wanted signal power (C) from T and the interference power (I) from T_1 must be such that their ratio (C/I) exceeds a minimum value appropriate to the system technology used and the quality of transmission required. For 32 kb s^{-1} ADPCM speech encoding to the ITU G.726 standard, the threshold probability of bit error (P_e) allowable is 10^{-3}. Assuming the use of Gaussian filtered minimum shift keying modulation (GMSK) with a bandwidth symbol period product (BT) of 0.5 and the use of coherent detection, such a $P_e = 10^{-3}$ threshold requires C/I = 11 dB, after taking into account a reasonable engineering implementation margin. Unfortunately, the wanted signal also suffers from Rayleigh fading (assuming use of a narrow-band system), a consequence of the multiple transmission path environment encountered in mobile communication, multi-path propagation. Consequently, the threshold will be breached for about 65% of the time unless a fade margin is allowed for. Figure 12.3 shows that, for Rayleigh fading, a 20 dB fade margin without diversity results in the received signal level staying above threshold for 99% of the time. Naturally the interference signal also suffers from Rayleigh fading, but the acceptably small divergence between the mean and median probabilities of Rayleigh fading allows the statistics of interference to be equated to Gaussian amplitude distributed noise. Allowing for such a fade margin implies the need for a carrier-to-interference ratio, C/I, of 31 dB. In practice, antenna diversity is today widely used in cordless systems to

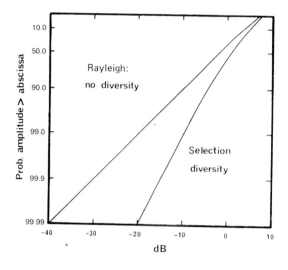

Figure 12.3 Effects of Rayleigh fading and the benefits of diversity.

avoid what would otherwise be an unacceptable range/performance degradation – see Figure 12.3.

The interference levels within a given cordless system will be a function of the physical distributions of basestations and active handsets as well as the choice of channel for every call. In an ideal conventional cellular mobile radio approach, basestations are considered to be deployed on a hexagonal grid; the spectrum is divided amongst these basestations to provide non-interfering channel assignments according to a pre-planned spatial re-use pattern. For reasons discussed in section 12.5, this approach has not been adopted for any business cordless telephone systems. The approach is rather to employ dynamic channel assignment (DCA) algorithms that seek to select the optimum channel from all (or a large subset) of those available (this approach is described in section 12.6.3). Nevertheless the pre-planned cellular grid approximation is, to a first order, a relatively simple representation to analyse and the results of such an analysis can provide a useful yardstick. Thus we now examine the WPBX scenario according to such an approximation in order to obtain a first estimate of spectrum requirements. In section 12.5 we discuss a computer simulation approach that more fully models the effects of DCA and irregular basestation deployment. These analyses and models are presented to indicate the types of approach that are employed in analysing or developing specific systems.

12.2.3 An Analytical Estimate of Spectrum Needs

For current purposes we assume that a handset experiences interference arising from, on average, six first-order interferers disposed in a hexagonal pattern, as shown in Figure 12.4.

In this scenario the distance between co-channel basestations and the reference handset is the re-use distance (D). The area within which all channels can be used, but not re-used, is given by:

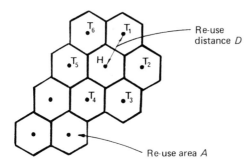

Figure 12.4 Fixed cellular scenario used for analysis. H = handset, T_1-T_6 = basestations all using same frequency.

$$A = \frac{1}{2}(\sqrt{3})D^2 \tag{12.1}$$

Assuming the wanted and interfering signals decay as the nth power of distance, then the carrier-to-interference ratio becomes:

$$C/I = 10\log_{10}[(1/6)(D/R)^n] \tag{12.2}$$

With a net busy-hour traffic density of T E km^{-2} per floor, the traffic offered in area A is:

$$E_a = TA \tag{12.3}$$

Combining all three equations generates:

$$E_a = \frac{1}{2}(\sqrt{3})6^{2/n}TR^2 \times 10^{2(C/I)/10n} \tag{12.4}$$

which is the traffic per re-use area per floor. The three-dimensional characteristic of the cells is introduced by a vertical re-use factor F, usually defined as the number of floors an interfering signal must penetrate before the resulting signal strength becomes of the order of or less than the horizontal components of interference. F depends on the floor penetration loss, which is a wide-ranging variable dependent on such factors as building construction, the opportunity for signals to propagate along stairways, etc. Excess loss values for floor penetration ranging from 6 to 15 dB have been quoted [4, 5], suggesting that the vertical re-use distance, in terms of number of floors (F), should range between $F = 5$ and $F = 2$ respectively to cover the extreme cases. Applying this factor to equation (12.4) generates the traffic per re-use volume:

$$E_r = \frac{1}{2}(\sqrt{3})6^{2/n}FTR^2 \times 10^{2(C/I)/10n} \tag{12.5}$$

Using Erlang-B tables, E_r can be converted to the number of channels, N, that a system has to offer to satisfy the demand for a required probability of successful call set-up. Typically this probability is 99%. The product of N and the specific bandwidth per channel (B_s) then produces the required estimate of spectrum requirement, B.

From the foregoing, and with a reasonable degree of accuracy, it can be shown that:

$$B \propto FTR^2 \times 10^{(C/I)/20} \tag{12.6}$$

In deriving this equation we have assumed a value for the propagation law of $n = 4$. This expression enables us to gauge the sensitivity of spectrum requirement to changes in parameters. For example:

1. Doubling the vertical re-use factor doubles spectrum requirement.
2. Doubling offered traffic density doubles spectrum requirement.
3. Raising operational cell radius range from 10 m to 32 m increases spectrum requirement 10-fold.
4. Reducing C/I threshold value from 31 dB to 21 dB reduces spectrum requirement by 3.2 times.

These examples show that the required spectrum allocation demanded of spectrum regulators is very susceptible to changes in the quality of communication and extent of service coverage per basestation.

Conversely, given a fixed spectrum allocation, B, and hence a fixed number of available channels, N, the proportionality of equation (12.6) shows the implicit dangers for system installers. When deploying a cordless system, a trade-off is made between offered traffic, T, and cell operating radius, R, i.e. the basestation layout. In making this trade-off, a margin must be allowed in the assumed traffic density, T, to accommodate the customer's likely communication growth demands, otherwise, for a fixed layout, growth will erode the C/I margin above the system performance threshold with a consequent reduction in quality of communication. For example, if growth in T disturbs the equality of equation (12.5) by a factor of 2, then the compensatory change in C/I is 3 dB, equivalent to a doubling in probability of exceeding threshold performance or indeed a doubling in outage probability. Pragmatically, however, most cordless business systems installed to date have tended only to support a low penetration of users – perhaps the customer's key mobile staff – and as penetration grows, additional basestations can of course be installed.

The spectrum requirement to service a given set of assumptions can be determined by application of equation (12.5); one example is shown in Figure

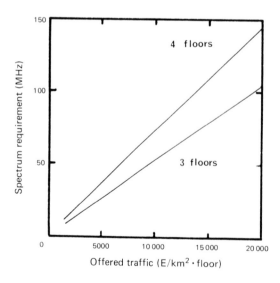

Figure 12.5 Example of spectrum requirement versus offered traffic. Cell radius = 15 m, propagation law $n = 4$, channel bandwidth = 100 kHz, no diversity, grade of service $GoS = 1\%$.

12.5. This result shows that of the order of 70 MHz is required to service an office offering $10\,000\,\text{E}\,\text{km}^{-2}$ per floor with four-floor re-use. The requirement can be reduced by decreasing operating range in classical cellular fashion, but there is also scope for improvement through the introduction of diversity. Figure 12.3 shows that switched diversity reduces the C/I requirement by 10 dB for a 1% outage probability. Item 4 above indicates that this improvement reduces spectrum demand by a factor of 3.2 times. Whether this significant reduction can be achieved in practice, however, depends on the extent to which frequency-selective fading is a significant factor resulting from the multipath radio environment typical of cordless telecommunications.

12.3   In-Building Radio Coverage

Before investing in a wireless PBX, a telecommunications manager needs to be assured that such a system will provide reliable communications throughout the nominated coverage area. If he or she is contemplating replacing, say, more than 30% of wired PBX telephones with cordless equivalents, then the manager could rightly demand that the usable radio signal strength must occur for more than 99% of the nominated area. This is a reasonable objective, for users of normal, wired, terminals have come to expect probabilities of call blockage (grade of service) of less than 0.5%. As far as the handset user is concerned, call blockage, whether arising from traffic congestion in the switch or from poor radio signal coverage, is unacceptable and the user is not likely to give much free licence in this respect in return for cordless mobility.

From the installer's point of view, the objective is to provide the necessary standard of radio signal coverage first time, every time, and, to be competitive with the wired alternative, the installer cannot use highly skilled, highly paid radio engineers versed in the art of radio propagation to achieve this. Thus simple-to-use tools have been developed to predict accurately transmission path loss, taking into account visible structural details of the building in question, preparatory to determining the positions of the fixed basestation transceivers. Such planning tools involve the use of a personal computer operated not by engineers but by salespeople.

There are two techniques available to provide good-quality radio coverage. The first uses small antennas distributed around the customer's premises, each antenna effectively providing a three-dimensional coverage volume, a picocell. The alternative is to radiate the signal from relatively poorly screened coaxial cables, "leaky feeders", laid along corridors or around corners, which enables the coverage to be tailored to meet specific, perhaps otherwise impossible, requirements. In both cases the requirement is for an expression that relates transmission path loss to distance between basestation and mobile terminal. Such a relationship for mobile systems can be expressed as [5]:

$$P_r = P_t - \bar{L} - 10n\log_{10} d \qquad (12.7)$$

where P_r = mean received power (dBm), P_t = transmitted power (dBm), d = direct distance between transmitter and receiver, $\bar{L}$ = path loss at $d = 1$ m (dB), and n = signal decay exponent.

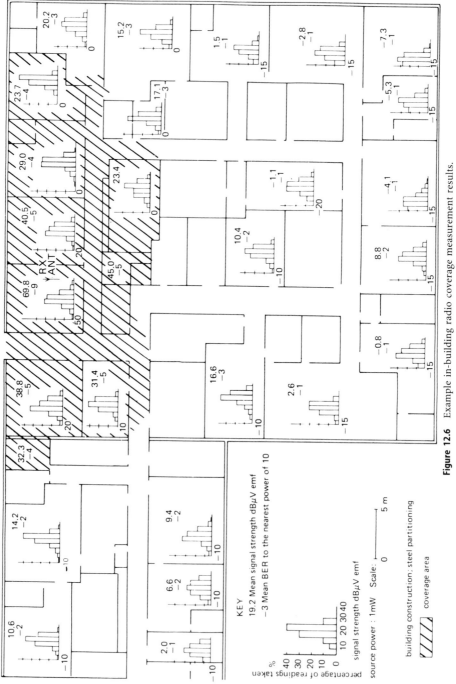

Figure 12.6 Example in-building radio coverage measurement results.

Power P_r may be subject to Rayleigh distributed amplitude fading due to the multipath propagation. The transmission path loss, L_p (dB), is the difference between P_t and P_r and hence

$$L_p = \bar{L} + 10n \log_{10} d \qquad (12.8)$$

Much propagation measurement effort has been expended worldwide to obtain parameters for equations of the form above that accommodate a range of building layouts and construction techniques. In the remainder of this section we discuss reported results that attempt to express in-building propagation, using antennas and leaky feeder techniques, in such a form.

12.3.1   In-Building Coverage Using Basestation Antennas

An example of propagation measurement is shown in Figure 12.6 for a reinforced concrete building with internal metal partitions. The hatched area represents the extent of usable radio signal coverage from the central receiver. Allowing for the fact that the experiment receiver is near the outside wall of the building, it clearly shows that coverage tends naturally to be cellular in a building of reasonably consistent structure. The received signal figures refer to the room-average value and the histogram to the variability around that value.

Plotting the room-average signal strengths against distance produces Figure 12.7, to which has been added the best-fit line computed to minimise the mean-square error of the data. Relating this result to equation (12.8), the gradient of the line identifies the signal decay exponent (n), where in this example $n = 5.7$. However, this is not a generally applicable result: reference [6] reports signal decay exponents for 14 buildings, using concrete or brick or plaster-board

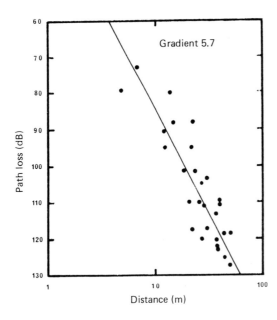

Figure 12.7 Path loss versus distance for the measurement results of Figure 12.6.

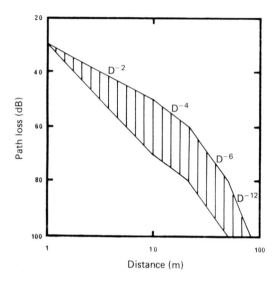

Figure 12.8 Classical ray theory mixed model for in-building propagation. Propagation law as indicated. (*Source*: adapted from [8].)

construction, that range between $n = 1.2$ and $n = 6.5$. The highest values tend to be associated with buildings using metal-faced partitions that impede transmission, whereas the lowest values occurred in a hanger-like workshop that appears to have acted like a waveguide [6]. Nevertheless there is no sufficiently consistent picture from which widely applicable conclusions can be drawn.

It has been postulated that for distances greater than, say, 40 m in a building the radio signal that has penetrated the internal clutter may often be exceeded by a signal that has taken a more indirect reflection route, perhaps by reflection from an adjacent building. The converse has also been suggested. In the latter case it is argued that the short-range results are dominated by localised reflections and the longer ranges are subject to relatively high clutter loss. Such a model based on classical two-path ray theory [7] would predict $n = 2$ at short range increasing to $n \geq 4$ at longer ranges. Certainly, mixed models of this type have been used to explain mobile transmission phenomena with some success, for example in Figure 12.8 and reference [8]. Either possibility is very dependent on the detailed layout and construction of the building, and in many cases the true picture could be a combination of both.

The work reported in reference [5] by Motley and Keenan sets out to investigate whether or not in-building data for ranges less than 40 m can be treated to produce a more generally applicable solution for path loss. The distance limitation was not considered overly restricting since the picocell radius in most high-capacity WPBX is expected to lay well within 40 m. This classic paper by Motley and Keenan has been extensively used as a basis for subsequent research, is intuitively comprehensible and yields good predictions compared to measurements; the work and its background is thus described here in some detail.

The approach taken was to apply equation (12.8) to the situation modelled in Figure 12.9 from which the need to penetrate (or bypass) walls and floors must be taken into account in predicting the path loss at 1 m, noting that it will vary widely from one receiver position to another, for the value of L reflects the mean clutter loss.

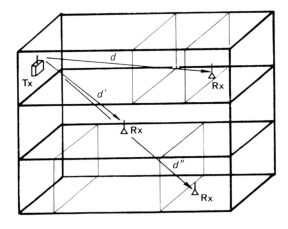

Figure 12.9 In-building propagation – effects of walls and floors.

To obtain a practical expression for path loss, therefore, the value of $\bar{L}$ must be related to structural features of the building. Even this degree of relative sophistication will still leave some variability in L due to furniture and people moving around. In reference [5] it has been shown that the variability can be described by a log-normal distribution with variance v and mean $\bar{L}$. Hence,

$$L_p = \bar{L} + 10n \log_{10} d + L(v) \tag{12.9}$$

In Figure 12.10 a typical set of results is shown for path loss related to distance for two frequencies (864 MHz appropriate to CT2 CAI, and 1728 MHz, which is close to the DECT frequency). These results relate to a multi-storey office block with reinforced concrete shell and plaster-board internal office walls; each floor

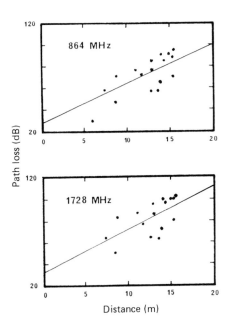

Figure 12.10 In-building path loss variation at CT2 and DECT frequencies.

was 50 m long and 15 m wide with a central corridor. Each point on the graph refers to the room-average path loss obtained by automatically recording a multitude of measurements through the office.

In reference [5] the results were processed to obtain the best-fit line and, by comparison with equation (12.9), values for $\bar{L}$ and n were obtained; these are recorded in Table 12.1.

Table 12.1
Best-fit propagation parameters

Frequency (MHz)	$\bar{L}$ (dB)	n	RMS error (dB)
864	27.5	3.6	12
1728	32.0	3.9	13.6

The high value for RMS error gives small confidence in the broad application of the results, but the signal decay exponent at least indicates that an assumption of $n = 4$ is reasonable for both frequencies. Indeed, the slight difference in value of n accounts for the difference in $\bar{L}$ being less than the 6 dB that theory would predict for a doubling in frequency. Closer analysis of the results was reported to show that similar but parallel regression lines could be obtained for each set of points associated with each floor. Consequently the path loss expression was modified to

$$L_\mathrm{p} = \bar{L} + 40 \log_{10} d + kf + L(v) \qquad (12.10)$$

where k = number of floors traversed, f = floor attenuation factor (dB) and $n = 4$.

To obtain the best-fit regression line, f was varied to minimise the RMS error of the data; the results are shown in Table 12.2.

Table 12.2
Revised best-fit propagation parameters

Frequency (MHz)	$\bar{L}$ (dB)	n	RMS error (dB)
864	10	4	3.9
1728	16	4	4.7

These results present a significant improvement in terms of RMS error, reduction in clutter loss and difference in $\bar{L} = 6$ dB, as would be expected from theory. Accordingly, these results indicate that it should be possible to characterise a range of building types in terms of a mean residual clutter loss ($\bar{L}$) and a floor attenuation factor (f) and apply these to an equation of the form of equation (12.10).

By applying equation (12.10) to each floor-corrected data point, the range residual clutter-loss values (L) were obtained. Their distribution showed a log-normal characteristic, from which the standard deviation ($v^{1/2}$) was determined. For the office block, standard deviations of 3 dB and 4.2 dB were obtained for 864 and 1728 MHz respectively.

In practice, the transmission path loss between dipoles spaced 1 m apart is 32 and 38 dB for 864 and 1728 MHz respectively, with an initial decay exponent of

$n = 2$ because the line-of sight path is dominant at small values of d. Accordingly the transmission path loss can be written as

$$L_p = 38 + 20 \log_{10} d$$

or

$$L_p = \bar{L} + 40 \log_{10} d + kf + L(v) \qquad (12.11)$$

whichever solution is the greater for 1728 MHz. This displays itself as a decay exponent of 2 up to around 10 m, then an exponent of 4 at greater ranges.

Clearly, to apply these expressions to a range of buildings requires a knowledge of $\bar{L}$ and f as functions of building type and construction. Typical results suggest $10\,\text{dB} < \bar{L} < 25\,\text{dB}$ and $5.5\,\text{dB} < f < 15\,\text{dB}$ for a range of buildings at 864 MHz. At 1728 MHz, $\bar{L}$ is 6 dB greater than for the corresponding 864 MHz result; likewise the floor attenuation factor at 1728 MHz is 3 dB greater.

The original work of Motley and Keenan has been taken further in the early 1990s by various workers. Specifically, work under various RACE projects has determined loss factors for a range of different wall and floor materials [9, 10]. Further, the basic model itself has been reconsidered and refined as part of the work of the COST 231 action[1]. This refinement comprises the introduction of a nonlinear function of the number of penetrated floors, to better reflect experimental observations [11].

Various other alternative models have also been investigated in COST 231. The one-slope model makes an assumption that the excess path loss, relative to the reference loss at a given distance from the basestation antenna, is proportional to the logarithm of the basestation–receiver distance. The linear-slope model makes an assumption that the excess path loss, with respect to the free-space loss, is proportional to this same distance. In these models, no attempt is made to give a physical interpretation to the parameters; rather the parameters are derived by fitting the models to in-building measurements, obtained at 900 MHz and 1800 MHz [12]. As computing power has increased in recent years, so also deterministic, ray-tracing, approaches to path loss prediction have been developed [13, 14]. Within COST 231 the performance of such different models have been compared [11]. Whilst offering some refinement in detail, the fundamentals described above have remained essentially valid.

12.3.2 In-Building Coverage Using Leaky Feeders

Radio coverage from a radiating coaxial cable is shown in Figure 12.11 and can be expressed by an equation similar in form to that for antennas, as follows:

$$P_r = P_t - \bar{L}_c - L(v) - 10n \log_{10} d - kf - XA \qquad (12.12)$$

where A is the attenuation per metre of the cable, X is the distance from the nearest point of the cable to the basestation termination and $\bar{L}_c$ is the cable coupling loss referenced to a 1 m radial distance from the cable.

To evaluate the various parameters, reference [5] studied a 900 MHz coaxial cable laid in the corridor of the building used for the antenna experiments. The measured signal levels along the cable are shown in Figure 12.12, displaying good agreement between the best-fit line and the specified cable attenuation factor.

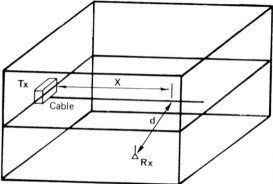

Figure 12.11 In-building coverage using a radiating coaxial cable (leaky feeder) antenna.

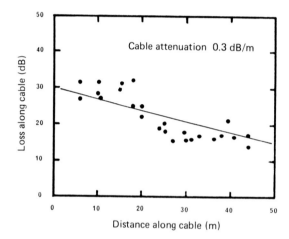

Figure 12.12 Longitudinal signal level variations for a leaky feeder antenna at 900 MHz.

The undulation in the results is a standing-wave effect caused by interaction between the internal and surface cable propagation modes.

Plotting the signal decay in a direction radial to the cable produced Figure 12.13, in which cable attenuation and floor factor f have been corrected for. As indicated, the signal decay exponent $n = 3$ and the mean coupling loss $\bar{L}_c$ at 1 m is 60 dB. Using a technique similar to that for antenna calculations, the variation in $\bar{L}_c$ due to building proximity effects was found to be log-normal with a standard deviation of 3 dB. Accordingly, for the selected cable at 900 MHz, the path loss relationship is given by:

$$P_r(\text{Rayleigh}) = P_t - 60 - L(\text{log-normal}, v) - 30 \log_{10} d - kf - 0.3X \quad (12.13)$$

where both k and f have the same meaning as adopted for the antenna results.

In practice, cordless systems in most environments to date have in fact used discrete basestation antennas, often with diversity, rather than distributed radiating cable antennas. However, the latter continue to be considered for some applications and for certain systems where they are perceived to offer certain potential advantages.

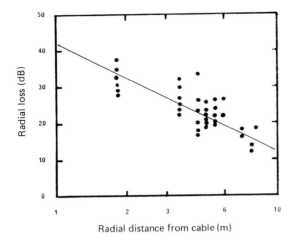

Figure 12.13 Radial signal level variations for a leaky feeder antenna at 900 MHz.

12.4   The Dispersive Channel

In the previous section the underlying concern was that the received signal strength should be sufficient to maintain acceptable communication quality for more than 99% of the time in the nominated coverage area in the presence of noise and co-channel interference. The implicit assumption, however, was that the modulated carrier signal could be reasonably equated to an unmodulated carrier signal. Whilst this assumption was reasonable for the early analogue modulated cordless telephone equipments, it is questionable for today's digital cordless systems, which have relatively large modulated carrier bandwidths. For these systems the dispersive nature of multipath radio transmission can be significant, especially in an outdoor environment.

12.4.1   Time and Frequency Dispersion

Mobile radio transmission channels are dispersive in both time and frequency. Time dispersion is the result of the received signal having been propagated over many different paths of different lengths and hence suffering different transmission delays and attenuation values. The end-result is that the received signal is a summation of many copies of the original transmitted signal with a spread of delays and amplitudes. Frequency dispersion, however, is caused by movement of the receiver relative to the transmitter either because one or other is in motion or because the signal-reflecting surfaces are themselves moving with respect to the receiver or transmitter even though the latter may themselves be stationary. Frequency dispersion appears as a variable Doppler shift of the transmitted signal components, which can cause rapid changes in their instantaneous phase, thereby causing problems for coherent signal-detection techniques and post-detection noise in non-coherent detection, e.g. discriminators. In the case of cordless telecommunications, most apparent motion is somewhat pedestrian(!). For example, the Doppler shift of a 1 GHz carrier frequency is less than ±6 Hz for a healthy walking speed of 6 km h^{-1}, and consequently the issue of frequency dispersion generally can be ignored. Exceptions to the rule occur when cordless equipment is operated in high-speed vehicles, for example trains, with reflections from passing buildings, or near to fast-moving traffic.

Let us return to time dispersion. The signal-distorting effects can be represented in the three dimensions of time, frequency and space. Remembering that the received signal is a summation of many independent copies of the transmitted signal, each relatively delayed in time and suffering differing attenuation, then at a given point, or position, for a particular frequency component of the modulated signal the copies will add vectorially. Any movement of the receiver (or transmitter or reflecting surfaces) will disturb this vector summation. The resulting signal level variations are shown in Figure 12.14a. This figure indicates the way in which antenna (space) diversity can be effective. For conditions of no movement, any change in frequency will equally disturb received signal vector conditions because of phase changes, giving rise to conditions exemplified in Figure 12.14b. This shows the channel response to a swept frequency source and the frequency-selective nature of multipath propagation with respect to an assumed system bandwidth. Figure 12.14c is representative of the same phenomenon in the time domain, many relatively delayed copies of the signal causing a spread of signals. If there is a direct line-of-

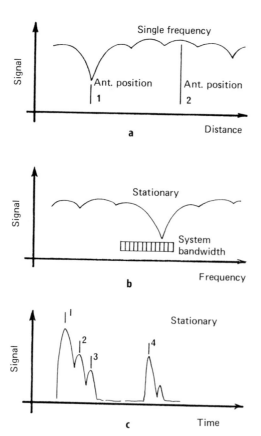

Figure 12.14 The effect of time dispersion, shown in **a** space, **b** frequency and **c** time domains.

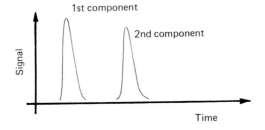

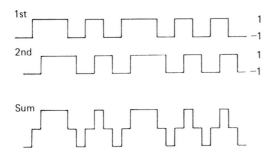

Figure 12.15 Intersymbol interference caused by delay spread.

sight component to the received signal, then this shows as the largest component near the origin.

Figure 12.14c is perhaps most useful in describing the effect of time dispersion on the received signal. If the delay spread is of the order of 30% of a transmitted symbol period, then significant signal degradation can occur, as Figure 12.15 shows diagrammatically.

12.4.2 The Impact of Delay Spread

It is clear from this simple demonstration that the signal can be corrupted considerably by the delayed symbols interfering with each other, the phenomenon of intersymbol interference. Moreover, the result shown is for a static transmission condition; in practice, the amplitude and phase of each of many delayed components of the received signal will undergo independent random variation. Such effects result in the existence of an irreducible mean bit-error ratio (BER). The significance of this effect upon signal reception has been found to depend on the delay spread of the received components (i.e. the second central moment of the power delay profile, s), the period of the transmitted symbol (T) and the sampling time. It has been found that the resulting irreducible mean bit-error ratios are broadly proportional to the square of the normalised RMS delay spread (normalised to the symbol period) [15–17]. It is of course possible for widely varying power-delay profiles to produce similar RMS delay values, so variability of receive system performance about the mean must be expected, particularly when one multipath component may have significant amplitude and relatively long delay with respect to all others. (Strictly speaking, BER proportionality to $(s/T)^2$ is only valid if the maximum excess delay is much smaller than one bit period.)

Clearly the RMS delay spread experienced is a function of the operational environment [18, 19]. Delay spreads in the range 50 ns (corresponding to measurements in the inner offices of large buildings) up to 300 ns (inside-to-outside office and residential buildings and telepoint situations) have been recorded [20]. The significance of such RMS delay spreads to system bit-error ratio performance becomes apparent when they are normalised to the symbol period. Early work [15] indicated that for GMSK modulation the mean irreducible bit-error ratio could exceed 10^{-3} for a normalised delay spread s/T = 0.05, which corresponds to 50 ns delay spread for a 1 Mb s^{-1} symbol rate. However, at this normalised delay the application of antenna space diversity (antenna spacing in the region of a wavelength for the enclosed environments of, say, an office block) would be expected to reduce the bit-error ratio to much lower values, allowing a normalised delay on the order of s/T = 0.2. Even with diversity, however, as the normalised delay spread s/T increases further, to 0.3 say, so the irreducible BER begins to become significant.

Such work gave rise to early concern about possible performance limitations of the DECT system in outdoor environments, given its high data rates and hence short symbol period T. Early thoughts were that operation without equalisation would be limited to distances where the RMS delay spread was less than 10–20% of the symbol length, which, in the case of DECT operating at 1152 kb s^{-1}, would correspond to a path length differential spread of only 50 m.

The theoretical controversy over outdoor performance was fuelled by a range of investigations presented in the COST 231 forum throughout the early 1990s, with a variety of theoretical papers published on the likely practical implications of delay spread and with different explanations offered as to the causes and significance of modelled results. Crohn et al. [16] from the Technical University of Vienna presented an analysis based on MSK modulation and fixed sampling time that indicated a very high residual BER; follow-on work gave explanations and an exact equation supporting these results [21, 22]. Other work, however, by Lopes, modelling a DECT system with adaptive sampling, in contrast indicated a much lower residual error floor [23], an encouraging result, subsequently confirmed empirically by the Vienna team itself using DECT testbed equipment. Recent collaborative work between these two groups has now resolved the apparent discrepancies, with a new understanding of the mechanisms at work, the development of more representative theoretical analysis and new efficient modelling techniques [17]. The latter have yielded results in harmony with empirical measurements from real DECT hardware. The conclusion of this work is that with a realistic model for the DECT system the error floor is some three orders of magnitude lower than for MSK with fixed sampling. For small delay spreads the BER can be approximated as $c(s/T)^2$, where c is a constant dependent upon the Gaussian filtering, the receiver filtering and the sampling time.

The effects of time dispersion on DECT in various practical scenarios were studied by COST 231 and reported in [20]. This work indicated that, whilst delay spreads above 100 ns could cause difficulties using basic equipment with no antenna diversity, RSSI-driven antenna diversity would allow delay spreads of up to 250 ns to be accommodated, compared to 400 ns with ideal BER-driven diversity. The first DECT products demonstrated very satisfactory performance, with operational ranges comparable to CT2 products – some 50 m in cluttered indoor environments and over 200 m outdoors, using omnidirectional antennas. However, lingering uncertainty remained over whether multipath dispersion would cause problems over long-range deployments of a few kilometres, such as could be required in telepoint or wireless local loop applications. To some degree it was to resolve these uncertainties in a pragmatic manner, and to prove in practice the capability of DECT to support such applications, that the telepoint and wireless local loop trials in Hungary, described in Chapter 5, were instigated. Several other long-range WLL trials have since been undertaken in different countries. In practice, such trials have demonstrated the capability of standard DECT equipment to support long-range operation as required for wireless local loop. For such operation, directional antennas are employed, as well as antenna diversity, the combination of these factors serving to constrain the observed delay spread in most deployment scenarios. These operational results are encouraging with respect not only to DECT but also for the application of other cordless technologies in such outdoor applications.

To date, manufacturers have not implemented equalisation in their cordless telephone or cordless technology local loop products, although this remains an option for DECT applications in severe multipath scenarios. A number of measures have been proposed to improve the multipath tolerance without modifying the DECT standard [20], of which adding a simple equaliser would probably be the most cost-effective.

12.5   System Capacity and Grade of Service

An approach to determining the requirements for providing in-building coverage was described in the previous section, a requirement to allow installation planning. When installing large indoor cordless communication systems, another issue arises, namely that of planning the sizes of the microcells in order to ensure adequate handling for the traffic densities predicted for that particular installation. An empirical approach to this for the pre-planned re-use approximation was presented in section 12.2 in order to understand system dependences on various parameters. However, in today's cordless systems, re-use is not pre-planned but automated. Once such automatic channel assignment is incorporated, its operation becomes too complex for simple empirical evaluation and computer-based models come into their own. The rationale for the adoption of such strategies is initially outlined below, followed by a description of a common approach to the problem of modelling system capacity and grade of service, *GoS*.

12.5.1   Automatic Channel Assignment Strategies

As discussed in section 12.2, the provision of cordless communications in buildings requires picocellular re-use strategies in order to satisfy the call traffic demands without excessive spectrum requirements. Moreover, multi-storey buildings imply three-dimensional re-use patterns to ensure acceptable interference levels from all directions. The number of degrees of freedom that this affords is high, yielding great potential for efficiency through optimisation but at the expense of considerable effort. This is best illustrated with reference to consideration of interference between floors. As mentioned in section 12.2, the re-use factor for floors, F, is estimated to range between 2 and 5. Interference between floors is a complex phenomenon. Optimum re-use is obtained through staggering the locations of the co-channel users on different floors. Depending on the areas of the picocells and the two-dimensional re-use pattern employed, it is feasible that basestation location staggering could permit some co-channel re-use on adjacent floors. Potential planning difficulties exist – e.g. where an installer needs to deploy a new WPBX in a shared office block that already has some systems operational. Clearly it is impracticable to re-plan the existing systems to accommodate the new one.

It is in part considerations such as these that have led to the adoption of fully self-organising anarchic approaches. In these approaches, all basestations may choose from all channels. Optimal (or sub-optimal) channels are chosen on the basis of carrier-to-interference ratio, i.e. dynamic channel assignment (DCA) – this theme is expanded in section 12.6 on usage of the radio channel.

The benefits of such an approach are considerable. Firstly, it is unnecessary to perform detailed measurements to assess the propagation characteristics of a building to plan re-use. Secondly, an existing system will automatically and immediately adapt to the incorporation of new basestations or of a new system in adjacent proximity. This is true whether one considers the deployment of a new additional WPBX or of cordless local loops in an area where existing cordless systems are already deployed.

The problem of installation amounts to one of economics, i.e. deploying the minimum density of basestations necessary to support the predicted traffic levels (with an appropriate allowance for growth). For a given basestation density the supportable traffic will depend upon the propagation characteristics of the building. For the somewhat amorphous channel re-use that arises with fully adaptive systems there are no break points (i.e. the cellular re-use patterns do not change in discrete numbers), so there is, on average, a smooth monotonic relationship between the attenuation law and the supportable traffic density.

Modelling studies [24] have shown that the number of basestations required exhibits a very strong dependence on the propagation model assumed, in particular the standard deviation of the path loss at any given range. Indoor propagation can be very different from outdoor in that partitions can introduce sharp propagation thresholds. In particular, the standard deviation of interfering signal levels can be quite small due to guaranteed screening effects. The implication of this is that in practice fewer basestations may be required than might otherwise be expected.

12.5.2 Approaches to System Capacity Modelling

In this section, we outline a commonly adopted computer modelling approach. Firstly, the area/volume to be modelled is determined. Towards the centre of an office floor the cordless links may be subject to interference from any direction. Near the external walls, however, interference will generally only come from directions over an approximate 180° span. Thus a simple model will be subject to noticeable edge effects – i.e. blocking will be less significant near to the edges. Because the requirements for a cordless system will generally be determined by the worst-case conditions, the results relating to equipments operating near to the edge may be ignored. Edge effects tend to decay only gradually with increasing distance from the boundary; thus quite large floors need to be modelled in order to obtain useful results from a significant proportion of their area.

In one modelling approach [25] this problem has been tackled by the use of an abstraction that eliminates edges altogether! The concept is to wrap each boundary of the model around to its opposite boundary. This is achieved simply by defining distance parameters modulo a deployment width distance parameter. This approach is appropriate for hexagonal as well as rectangular deployments. (The effect is akin to that observed on some video games where an object may disappear off one edge of the screen and reappear on the opposite side.) The approach may also be used in three-dimensional office blocks with each floor wrapping independently. This method allows accurate modelling to be performed with relatively modest floor areas; results can be taken from the entire area.

Once the modelling area/volume has been determined, it is necessary to decide the appropriate propagation model to be used. Early research into cordless systems attempted to create models of office buildings which included partitions of specific materials in specific locations, with propagation modelled using ray tracing, scattering and diffraction techniques. Whilst such an approach is becoming of renewed interest, as the computational power of PCs has increased in recent years, the approach still yields results specific to a given location and office structure. Also, the complexity of the approach prevents usefully large models (in terms of numbers of links and operating time) to be considered. Thus, as in other fields, a degree of abstraction is adopted for large system modelling, in this case using a propagation model based upon the averaged results of a large number of measurements, as described in section 12.3. Given such a propagation characteristic, the attenuation over paths between basestations and handsets may be readily modelled.

The usual approach is to generate, in numeric form, a representation of a cordless system with basestations positioned in some defined manner and handsets distributed across the deployment. This representation is then set to operate for a period of time with calls generated at random times and from random sites. At the end of the model run, various statistics can be derived such as the proportion of call attempts that succeeded, the proportion that were blocked and the proportion that were prematurely curtailed. Other parameters of interest include the average carrier-to-interference ratio and the average utilisation of basestation channels.

A useful method of modelling for this type of system is the discrete event simulation. In this approach, model time is stepped forward in non-uniform increments determined by the time at which each event should take place. Future events are stored in time order in a list, along with the event times, thus forming an event queue. The model operates by successively removing, reading and executing the event at the head of the queue.

New calls are generated according to a Poisson process (i.e. they start at random time) and have durations that may be fixed at a median value or are random according to some appropriate probability distribution. A common approach is to start every new call by placing a new handset at a randomly determined new location. It is then possible to calculate its distance from all other active basestations to determine the interference levels on all channels. The handset then selects a channel according to the dynamic channel assignment algorithm and attempts to set up a call. Once the call has been set up, the effect of interference from this new call on all other links can be evaluated.

A significant amount of computational effort is required for this type of model. The statistical nature of the answers required often dictates the need for a large set of results in order to obtain reliable values. This is particularly true of grade of service estimates (blocking probabilities) where figures of the order of 0.5% are commonly of interest. Obtaining a reliable figure could require some 20 000 calls to be modelled.

The benefits of modelling can be seen by reference to an illustration. In one model an examination of blocking was done, comparing interference in a one-storey building with that in the centre floor of a three-storey building for a 16-channel system. The excess floor attenuation was taken as 10 dB and 21 dB was assumed for the minimum required C/I (i.e. two branch diversity was assumed). The ratio of traffic capacity between the two deployments was found to be ≈ 1.6, rather than the figure of ≈ 2 that section 12.3 would predict. This illustrates both the power of DCA in assigning frequencies at separated locations on adjacent floors and also the benefit of modelling to take account of the specific case where only three floors need be considered (a relatively common requirement).

A new interest is emerging at the time of writing in such modelling techniques as cordless technologies have matured and are beginning to be deployed for wireless local loop applications. Radio regulatory authorities are beginning to grapple with the concept of licensing radio-based local loops, which must support a certain guaranteed grade of service, but which may use the same spectrum as private systems. Such modelling is being used in order to assist the debate as to whether indeed shared spectrum is viable or whether separate public and private allocations should be used. Preliminary modelling work at Roke Manor Research has demonstrated that only in a very small and unlikely proportion of deployments will any mutual interference be seen between public and private systems.

12.6   Usage of the Radio Channel

In mobile radio communications, and particularly cordless applications, there is a basic need to set up many independent mobile handset to fixed basestation communication links within a confined volume – confined in the sense that all these independent links, in attempting to use the same spectrum allocation, could interfere with, and potentially destroy, each other unless appropriate measures are taken to avoid it. The need is to ensure that multiple communication links can gain access to the limited spectrum resources in a orderly manner that avoids the worst effects of mutual interference and yet in a manner that achieves the optimum spectrum use efficiency. The process of doing this is known as "multiple access" and, taken in its widest sense, it also includes duplex (two-way) transmission.

12.6.1   Duplex Transmission

Conversation between people is (generally) two-way across a common, literal, air interface from mouth to ear. The imposition of a radio link between them then raises the problem of providing a concurrent two-way link in a manner that ensures conversation is uninterrupted by artificial restraints – for example, one person only being allowed to talk at any one time (simplex).

Conventionally in many radio transmission applications two separate links are created on separate carrier frequencies for each direction of transmission. This is known as frequency-division duplex. The CEPT/CT1 analogue system is an example that uses the two frequency bands of 914–915 MHz and 959–960 MHz in this way to support 40 duplex channels. One of these bands is used for transmission from the handset and one for transmission from the basestation, the spacing between the bands being necessary in order to permit adequate filtering within the equipments to allow simultaneous transmission and reception in the two bands.

Discussions with the UK frequency regulatory authorities during 1981 for the nascent digital CT2 CAI standard indicated that a two-frequency allocation in the 800–900 MHz region would not be possible, although a single contiguous band was potentially available. Consideration of how to exploit this opportunity resulted in the adoption of time-division duplex (TDD) transmission, a technique subsequently widely adopted in other cordless systems because of its inherent benefits.

In TDD a single carrier frequency is used to provide two-way communication (handset to base and base to handset) by dividing the transmission time into two equal parts, each supporting unidirectional transmission of a digitised sample of speech (say 2 ms) that is compressed by bit-rate doubling into half the sample period (1 ms) before transmission. The intervening gap is filled by the return transmission using the same technique. After reception, the compressed sample is expanded back to its normal period and then decoded. Of course, this technique can also be used in multi-burst time-division multiple access systems (see below).

The use of TDD has a number of benefits:

- The transmission path characteristics are identical in each direction because the same carrier frequency is used. Thus the received communication quality is closely similar to that transmitted, thus enabling in-built transmission quality monitoring.
- Antenna diversity at the basestation will optimise both directions of transmission.
- Handset and base units have identical RF circuit and component requirements.
- Cordless telecommunications systems that use TDD do not compete with conventional cellular mobile radio systems for scarce paired-band spectrum resources.

For these reasons, TDD has been widely adopted in cordless systems, alongside a range of different multiple access techniques.

12.6.2 Multiple Access Techniques

Three basic forms of multiple access technique are today in common usage:

- Frequency-division multiple access (FDMA)
- Time-division multiple access (TDMA)

● Code-division multiple access (CDMA)

These are reviewed below from the point of view of cordless telecommunications.

Frequency-Division Multiple Access (FDMA)

FDMA is the access technique in which the available spectrum resource of P MHz is subdivided into n discrete channels each of bandwidth P/n MHz (see Figure 12.16).

When FDMA is employed, the associated radio transmission and receiving equipment must ensure that the signal occupying, say, channel $(n - 2)$ does not unduly spread into the adjacent channels. Similarly, channel $(n - 2)$ expects equivalent protection from channels $(n - 3)$ and $(n - 1)$. This mutual protection is provided by a guard band between channels, by appropriate modulation techniques and by the use of selective RF filters that attenuate signals that otherwise would fall outside the allocated channel. Nevertheless, because of spectrum economy, filter characteristics and technical prudence, this isolation is not perfect and some energy leaks from one channel to the next. A need to compromise between the various factors is of course indicated, but the guard band in practice constitutes a spectrum overhead and in the interests of spectrum efficiency it is desirable to minimise it.

The transmission capacity $(b_r \text{ bit s}^{-1})$ of each channel determines the necessary modulation bandwidth (B) but the need for a guard band results in $B < P/n$ and hence capacity falls below the maximum attainable, assuming a given modulation technique.

Consequently the effective spectrum efficiency becomes n/P traffic channels per MHz. Typical examples for European FDMA cordless systems are 20 channels/MHz (CEPT/CT1, analogue transmission, frequency-division duplex operation) and 10 channels/MHz for digital CT2 CAI. This apparent discrepancy in spectrum efficiency[2] is more than compensated for by digital transmission's greater robustness to co-channel interference.

From an implementation viewpoint, FDMA has the disadvantage that each radio-frequency carrier only conveys one Erlang of traffic, with the result that

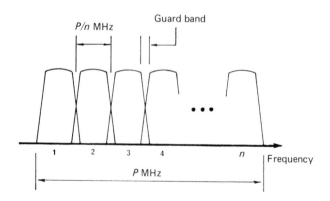

Figure 12.16   Frequency-division multiple access.

basestations needing to offer E Erlangs of capacity require E radio transceivers per basestation. The radio-frequency combining of such channels is also a requirement, although for cordless equipment with its low radiated powers this is not a major technical or cost problem. Indeed, combining can be achieved using multi-element antenna designs. On the other hand, the radio channels so created can be considered to be narrow-band, and hence insensitive to time dispersive propagation appropriate to cordless use, have a peak to mean radiated power ratio of 2:1 (time-division duplex assumed) and introduce negligible processing time delay that might otherwise raise implications for overall speech transmission with regard to echo delay (see Chapter 11).

Time-Division Multiple Access (TDMA)

In contrast to FDMA, which offers the user access to a fraction of the available spectrum for 100% of the time, TDMA allows each user to access a wider spectrum allocation for a fraction of the time. TDMA is a technique in which the available spectrum resource of P MHz is occupied by one modulated carrier and each of n channels (of traffic signal bit rate b_r bit s^{-1}) gains access to that carrier for $1/n$ of the time in an ordered sequence. The process is shown diagrammatically in Figure 12.17 where n channels are serially and synchronously combined onto a single carrier, after compressing each T seconds of traffic signal (speech or data) into a period approximately $1/n$ of the time. In a TDMA system serving multiple independent users a guard time (G) is often left between individual time slots, thereby easing the requirements for perfect instantaneous synchronisation and allowing a buffer for a modicum of timing drift between independent terminal equipments. In addition the guard space will cater for differential absolute transmission path delay for different terminal-to-basestation ranges. The guard space is generally identified as being equivalent to a number of bit periods.

Figure 12.17, however, shows the inherent need in TDMA systems to have a header sequence. The header sequence ensures faithful recovery of the traffic signal by the receiver through the provision of carrier recovery (CR) and bit-timing recovery (BTR) sequences to aid synchronisation of the receiver's carrier and bit-timing clocks. These are followed by unique identification of, and synchronisation to, the traffic signal by means of a unique word (UW) and channel identity (CI) sequences. The example shown is classical in showing the header functions as discrete entities; however, in practice, some of them are merged into one sequence (e.g. CR and BTR), sometimes omitted (no CI) or perhaps distributed over several frames in the case of distributed frame synchronisation words.

As a result of these transmission overhead bits, the transmission bit rate B_r is greater than that demanded by the traffic signal (nb_r) and, in order to maintain acceptable spectrum efficiency, care is exercised in deciding the size of the guard space and the need for, form and integrity of the header functions. In this context integrity refers to the ability of the unique word and channel identifier to resist being inadvertently mimicked by the traffic signal sequence or not being recognised because of transmission errors within them. Robustness against these effects is generally determined by sequence length – the longer the better.

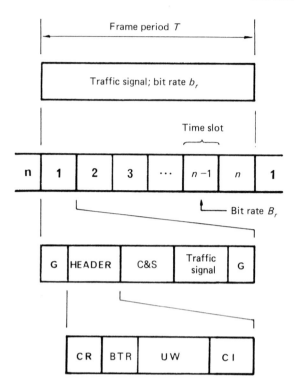

Figure 12.17 Time-division multiple access. CR = carrier recovery, BTR = bit-timing recovery, UW = unique word, CI = channel identity, C&S = control and signalling, G = guard space.

An alternative to reducing, and perhaps weakening, the header sequence is to increase the frame period T, thus associating more traffic signal bits with each header sequence, thereby raising the traffic-to-header ratio. This is an effective measure to restore spectrum efficiency (in terms of channels per MHz) but the penalty is an increased one-way processing delay of T seconds. As discussed in Chapter 11, excessive processing delays have implications on the quality of speech transmission in respect of speech echo performance, with the possible need for echo control devices in both terminal and basestation equipment. Hence a trade-off between complexity and performance against TDMA frame period results.

The modulation bit rate for a TDMA system is much greater than the inherent traffic signal bit rate per channel and typical cordless TDMA systems have transmission symbol rates in the range $300-1500\,\mathrm{kb\,s^{-1}}$ (time-division duplex assumed) with corresponding wide transmission bandwidths. Consequently under certain operating conditions, cordless TDMA systems may suffer from the effects of transmission path time dispersion common to all so-called wide-band systems (see section 12.4). The wide transmission bandwidth also requires a peak-to-mean radiated power ratio of $2n$:1 to maintain a given performance under thermal-noise-limited conditions.

The prime advantage of a TDMA system clearly resides in its ability to transmit n channels, or offer n Erlangs of traffic capacity, from a single basestation transceiver. The implication for basestation equipment cost in a high-density business communication system is significant and the full potential will be retained through careful matching of traffic density requirements and basestation location within a business entity. TDMA also offers a means of providing dynamic capacity allocation in which more than one time slot can be concatenated. The transmission capacity offered to a single terminal can, therefore, range from one to n time slots to form higher-capacity links for specific applications, of which data transmission is an important case.

Code-Division Multiple Access (CDMA)

Code-division multiple access (CDMA) is often cited as an appropriate multiple access technique for communication systems subject to high interference levels and multipath fading. It is natural, therefore, for CDMA to be considered for cordless applications.

In this technique, borne out of anti-jamming military requirements, the traffic signal of bit rate b_r bit s^{-1} is multiplied by a much faster pseudo-random (spreading) sequence such that the final modulated symbol rate is mb_r bit s^{-1}. The resulting wanted signal, post-modulation, is spread across a bandwidth m times larger than the inherent requirement of the traffic signal. This bandwidth is also much wider than the likely coherence bandwidth of any multipath fading, hence averaging out the frequency-selective and signal strength variations in the fading signal. The multiple access feature of CDMA arises through the use of a different spreading sequence for each traffic signal; hence n channels require a set of n spreading sequences with suitable cross-correlation properties.

At the receiver the wanted signal is de-spread by mixing the incoming sequence with a synchronised, locally generated, version of the original spreading sequence, with the result that the receiver outputs the wanted traffic signal. This de-spreading process, however, re-spreads any non-synchronous, different sequence, interference (including very strong unmodulated carrier wave signals). The end-result is that the C/I power ratio is effectively enhanced by a factor of m times. This enhancement is referred to as "processing gain". If $m = 100$, the enhancement is 20 dB for a single co-channel co-system interferer. If there were 10 interferers, however, the enhancement would fall to 10 dB. A vital element in CDMA systems is the use of forward error correction (FEC) coding as discussed later in section 12.8. It has been shown [26] that the coding gain afforded by FEC is obtained without loss of processing gain. Thus the required energy per bit/noise spectral density E_b/N_0 is reduced by the coding gain, making the processing gain more useful. Let γ be the required E_b/N_0 and m the processing gain. In simplistic terms we see that the number of users that can be supported in a single cell, N, is given by $N = m/\gamma$, since this number can each receive just enough E_b/N_0 to operate in the presence of interference from all of the others. A typical figure for γ with convolutional coding in the indoor environment is 4.5 dB. We then have $N/m \approx$ 1/3. Thus the number of simultaneous users that can operate is one-third of the bandwidth expansion factor. Comparing TDMA and CDMA this means that the efficiency within a single cell is only 33%.

On the face of it, this would rule out CDMA as a more complex system than FDMA and TDMA providing poorer spectral efficiency. However, where CDMA wins out is in the frequency re-use environment. Unlike traditional FDMA and

TDMA, it is possible to use CDMA in a single-cell re-use pattern. When this is done, the overall increase in interference levels due to the surrounding cells has been found to be between 50 and 66%. This corresponds to a re-use efficiency, F_r, of about 60%. Thus we now state $N = F_r m/\gamma$ for the multi-cell case. The re-use efficiency of FDMA and TDMA systems is the reciprocal of the re-use cluster size, which is typically 4 or 7. Once this factor is taken into account, the capacity of CDMA becomes comparable to that of FDMA and TDMA. The use of discontinuous speech transmission with CDMA can yield further improvements in capacity (up to a factor of 2).

One difficulty with CDMA is the so-called "near–far" problem. This arises when nearby interfering sources swamp a wanted signal from greater range – i.e. where a nearby interferer exceeds the level of the wanted signal by a value exceeding the processing gain. This problem can be effectively solved through the use of automatic control of transmitter power (particularly needed for the cordless terminals) to equalise the received power levels at the receiving station.

The very high re-use efficiency available from CDMA arises only when every cordless terminal communicates via the basestation to which it has the smallest RF path attenuation (or to which it can transmit the smallest power). This is practicable in the case of cellular mobile radio where a single operator has their own allocation of spectrum and can regulate the handoff regime. In the unregulated cordless environment, however, considerable problems could arise. Consider two CDMA-based WPBX systems (for different businesses) operating in close proximity and using the same RF frequency. It is possible that a cordless terminal operating near the boundary between the two systems could be at limit range to the nearest basestation within its own system and therefore be transmitting near to maximum power. On the other hand, a basestation of the other WPBX could be at very close range on the other side of the partition, and receive devastating interference from that terminal. For further information on CDMA principles, the reader is referred to reference [27].

The choice between multiple access schemes is a complex function of equipment complexity, regulatory situation, fixed network requirements, system robustness, interference susceptibility, spectrum need and cost of implementation. Since no one solution meets all the criteria, it is perhaps of little surprise that both single and hybrid solutions have been adopted. Some solutions, within a "free-market" environment, have been developed reflecting the technology bias of the originators, whilst others have represented the compromise implicit in a "democratic" standardisation process. Thus, for example, the multiple access process adopted for the Digital Enhanced Cordless Telecommunication equipment seeks an optimum solution to needs and requirements by using time-division duplex transmission with a combination of 12 time slots per carrier TDMA and 10 carriers per 20 MHz of spectrum FDMA – i.e. a TDD/TDMA/FDMA system.

12.6.3 Dynamic Channel Assignment

To achieve high orders of traffic capacity with good spectrum efficiency, it is a requirement of cordless equipment that in establishing a communication link it must be able to seize any available channel. For TDMA this means any free time slot and for FDMA any free radio-frequency channel. With this requirement, cordless telecommunications realises the full advantage of trunking, the process familiar to fixed telecommunication networks. Its use ensures that all available channels are assignable to any user, thereby maximising the use of spectrum and maximising the probability of a call being established. Cordless systems often operate under co-channel interference-limited conditions and in this regard dynamic channel assignment (DCA), through the use of appropriate channel selection methods, can be used to select the channel with the least, or acceptable level of, interference [28]. Thus again channel usage and call success probability are maximised.

Since there is no central control of channel selection in cordless equipment, the DCA process is essentially autonomous to the terminal equipment in residential, business and telepoint applications (although information about channel availability as perceived at the basestation may be communicated to the terminal). Clearly some standardisation of the DCA procedure is necessary to avoid mutually destructive interactions between independent equipments, but at the same time to maintain the advantages of DCA.

Bearing in mind that the destructive effects of co-channel interference depend upon the C/I power ratio, rather than the absolute interference power, then channel selection procedures tend to select the channel with the least interference coupled with some measure of channel quality indicative of C/I power ratio, for example error detection measurement of a regular (known) link sequence. It is also wise to adopt random channel search techniques, although this requirement may be modified if paging techniques are used to identify calling channels. From its initial application to cordless telecommunications, the development of DCA techniques and algorithms has developed into an active area of research during the early 1990s, with the recognition of its wider potential application to cellular radio and public personal communications systems.

In public access cordless or mobile cellular systems, the interference environment is generally rapidly varying. Hence, a channel chosen simply on the basis of current interference level may not remain useable for very long. Thus, such simple DCA algorithms suffer limitations. Longer-term interference statistics are generally less variable. This can be exploited through the use of priority tables, which rank channels based on the past performance. The priority may then be used to decide which channel should be used for a new link, rather than basing the decision solely upon the prevailing short-term conditions. This approach has been successfully developed for UMTS in the RACE ATDMA project [29].

12.7 Modulation Schemes

Traditionally, modulation was analogue, and commonly FM, as for the first-generation analogue cordless telephones. Digital modulation has been widely adopted for voice communications in cellular radio and soon (as DAB) in audio broadcasting; the development of digital techniques for cordless telecommunications during the 1980s contributed to this trend. The reasons for the adoption of digital techniques are many, but the following are the more important ones:

- The development of low-rate (i.e. less than 64 kb s^{-1}) toll-quality voice coders. This improves the spectral efficiency of the transmission.

- The emergence of spectrally efficient digital modulation formats. These new formats permit the necessary transmission rates to be accommodated within modest bandwidth channels.
- The enhanced tolerance of digital format transmissions to co-channel interference.

Together, the above can result in a spectral efficiency for digital transmission which exceeds that for analogue in spite of the generally wider minimum channel spacing for digital systems. In addition the following advantages arise:

- The flexibility of digital modulation formats in permitting time-division multiplexing and/or duplexing.
- The ease of switching channels in digital format.
- The ease of multiplexing signalling data onto the transmission.
- The inherent voice privacy and potential for security.
- The flexibility and efficiency for adding data services (potential for ISDN compatibility).

Finally, the availability of the technology in the form of high-performance silicon and of the relevant tools to develop digital systems has made them technically feasible and economically viable. The rapid and continued advances in silicon technology have served to accelerate the widespread adoption of such techniques. In reviewing modulation for cordless operation, therefore, we will consider only the digital formats.

12.7.1 Digital Modulation Requirements

The desirable elements for digital modulation may be summarised as follows:

- Spectral compactness (to maximise the number of channels per MHz for a given adjacent channel rejection performance)
- Good error performance in noise (for range capability)
- Good error performance against co-channel interference (for channel re-use)
- Low power consumption
- Ease of implementation

Digital modulation formats can be divided into those which have a constant transmitted envelope and those which do not. Constant envelope modulation permits transmitter amplifiers to operate in a non-linear Class C mode, permitting high efficiencies and simple implementation. Those modulation types which require some variation in the transmitted envelope need a linear transmitter amplifier with its attendant cost in terms of power and complexity. Those modulations requiring high transmitter linearity can sometimes yield particularly good spectral efficiency. This may justify the additional cost/complexity for some systems, but for a cordless system where cost is vital this is not yet the case. With the emergence of new techniques for power amplifier linearisation, Cartesian and polar loop schemes [30], this may change in the coming years, but is still some way off.

We focus, therefore, on the constant envelope class of modulation formats. For this class, of necessity, it is the phase angle of the transmitted signal that is varied. In the same way as for the case of analogue modulation, we can modulate either the phase or the frequency of the output. We initially consider phase modulation.

12.7.2 Phase Shift Keying

The simplest and most common form of digital modulation is phase shift keying (PSK). A PSK modulator is very straightforward and consists of a simple mixer in the transmitted path of the carrier, as illustrated in Figure 12.18.

Demodulation is achieved in the receiver by performing the same process. In order to do this, it is necessary to produce a replica of the transmitted carrier in the demodulator. This may be done by phase locking a local oscillator to phase information derived from the received signal. The need for carrier estimation implies that PSK must be demodulated coherently. In general, coherent modulation schemes provide better performance in additive white Gaussian noise (AWGN) than non-coherent. They are, however, usually more complex.

The amplitude spectrum of unfiltered PSK is rather wide and decays very slowly with offset from the centre frequency, having a sinc function shape. Simple filtering of the modulated signal can greatly improve its spectral containment without compromising power efficiency but results in a non-constant amplitude signal waveform.

In CDMA systems, the need for power control tends to imply the need for a linear amplifier. Thus filtered PSK is a key contender for CDMA modulation. However, for reasons of linearity requirement, PSK has not been adopted for digital cordless systems based on FDMA or TDMA.

12.7.3 Frequency Shift Keying

Frequency shift keying (FSK) was an obvious candidate for early digital cordless systems. With FSK the frequency is varied in sympathy with the baseband waveform as illustrated in Figure 12.19.

All frequency modulation systems are characterised by their modulation index, this being defined as the ratio of the peak frequency deviation to the highest frequency component in the modulating signal. For digital systems, the highest modulating frequency is one-half of the bit rate. In order to shape the transmitted spectrum, a filter is commonly placed between the modulation sequence source and the voltage-controlled oscillator (VCO) input. This greatly improves the shape of the spectrum and, in this case, does not result in any

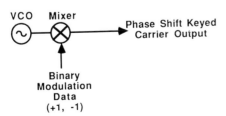

Figure 12.18 PSK modulation.

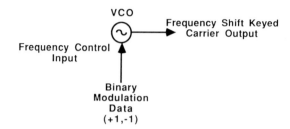

Figure 12.19   FSK modulation.

envelope modulation. Such filtered FSK modulation is employed in CT2 CAI, with a modulation index being specified as about (but not necessarily exactly) 0.5.

It is possible to demodulate FSK with a simple limiter discriminator as used for conventional analogue FM signals. This is not always used, however. There is a class of FSK-type modulation schemes that can benefit from coherent demodulation. This class is generically known as continuous phase FSK (CPFSK). The best known in this family is Gaussian-filtered minimum shift keying (GMSK). (MSK modulation may also be viewed as a special case of the summation of two filtered PSK signals, see [31].) This modulation was adopted for the GSM cellular radiotelephone system. It is therefore appropriate to give some attention to this type of modulation. First we will examine the simpler, unfiltered, modulation, MSK.

12.7.4   Minimum Shift Keying

Minimum shift keying (MSK) may be viewed as FSK with a modulation index of exactly 0.5. Thus, for example, a $32\,\mathrm{kb\,s^{-1}}$ MSK modulator would have a peak deviation of $\pm 8\,\mathrm{kHz}$. This choice of modulation index for MSK is significant because it determines that the accumulated phase change over a single bit period is $90°$. It follows that for every bit period the phase either precesses or regresses by this amount. This leads to the result that the carrier phase on alternate bit periods is either varying between $0°$ and $180°$ or between $90°$ and $270°$. Thus it is possible to demodulate the data by examining the sense of the in-phase and quadrature components of the carrier. A carrier recovery circuit is required to generate a phase reference for demodulation in this way. Of course, the MSK waveform can be demodulated using a simple limiter discriminator, as for other types of FSK. Coherent demodulation yields a theoretical 3 dB performance advantage over simple limiter discriminator detection in AWGN, however.

An important feature of the above properties of MSK is the fact that the phase and quadrature components of the carrier are only varying at one-half of the bit rate. It is therefore possible to apply narrow-bandwidth filtering to the modulating signal without loss of performance. Moreover, it follows that the transmitted spectrum may also be filtered in this way.

12.7.5   Gaussian-Filtered Minimum Shift Keying

In Gaussian-filtered MSK, the baseband modulating signal is filtered to restrict its transmitted spectrum. For relatively modest filtering the MSK demodulation technique can be applied in unmodified form. However, heavy filtering (bandwidth × time, *BT*, products[3] less than 0.5) results in intersymbol interference. The effects of this can actually be exploited to provide improved performance in noise, but only at the expense of increased complexity in the demodulator. If the demodulator is not adapted to operate with the heavy filtering, a significant degradation of bit error rate (BER) performance in noise will result.

In general, the standards for cordless telephony have been generated with a view to placing the minimum of unnecessary constraints on manufacturers. A choice of heavy filtering (e.g. *BT* < 0.4) would necessitate the use of coherent demodulation for acceptable noise performance. For this reason, a figure of 0.5 is more common. This permits the use of non-coherent demodulation, albeit with slightly poorer performance than coherent.

The coherent demodulation of GMSK relies upon the modulation index being essentially exactly equal to 0.5. For the simple FSK modulator shown in Figure 12.19 this will not generally be the case, since the circuit will be subject to design tolerances leading to errors in modulation index of anything up to ±10%. (To implement GMSK modulation requires a more complex circuit, for example a vector modulator, as discussed in Chapter 15.) On the other hand, limiter discriminator demodulation places no such constraint upon the modulator, even poorer tolerances being acceptable. This potentially creates something of a difficulty in attempting to generate common standards for cordless telephony. One manufacturer could decide to use a fully coherent system requiring a complex modulator circuit whereas another might choose to use simple limiter discriminator detection and so be able to use the simple modulator configuration of Figure 12.19. Both could be compliant with the specification and coexistence would guaranteed. However, the transmitters of the second manufacturer's scheme could not be received by receivers of the first manufacturer's approach. Such a problem is amenable to solution, but should not be overlooked.

The above factor, taken together with the very limited improvement in range offered by coherent modulation (because of the fourth-power range law for the in-building environment) and considerations of implementation complexity, perhaps explains why some standards makers have appeared to date to favour simple filtered FSK modulation for cordless telephone applications.

12.8   Error Correction and Detection

Cordless communications are moving from the use of analogue to digital transmission; this allows the support of digitised voice and data services. Analogue voice transmissions are subject to perturbations from noise from the receiver and/or from the radio propagation environment. For digital voice transmission, the effect of noise is manifested in data errors. These, in turn, cause degradation to the quality of the voice signal when converted back to analogue form. For a data service, errors may be even more serious, with the potential to corrupt totally the sense of a message. More fundamental is the fact that errors in the digital signalling channel of a cordless telephone would cause the basic speech link to collapse. Thus error control is a basic prerequisite in digital cordless systems.

Error control is a generic term for all the procedures that may be employed to ensure that the data errors or the effect of data errors experienced by the user/ system are kept to an acceptable level for any error rates likely to be produced by the raw communications link under normal operating conditions. Moreover, error control serves to prevent unpredictable system operation whenever error rates exceed those which can be handled as part of an active communications process.

There are two basic strategies for error control, namely error detection and correction: the former ensures that the occurrence of data errors is recognised in order that some other appropriate action may be taken; the latter actually corrects data errors (up to a certain limit, determined by the capabilities of the error-correcting code employed). Both techniques involve sending error control data bits in addition to those data bits required for the transmission of the raw information.

Error detection results in a relatively modest coding overhead, in terms of error control bits. In this case information is transmitted with a small number of error detection bits appended (commonly referred to as a checksum). When used in conjunction with a communications feedback protocol (such as automatic repeat request (ARQ)), error detection can result in a very effective means of error control. If detected without errors, the communication has been successful and the recipient can confirm this by transmitting an acknowledgement message. If not, no acknowledgement will be sent and the transmission source will re-send the transmission until an acknowledgement is received. Such an approach is attractive in that it adapts automatically to the prevailing channel conditions. It is, however, appropriate only for the transmission of non-voice data (system signalling and data communication services), since it is subject to variable user-to-user transmission delays. Typically, a checksum approach is used for signalling in a cordless system. CT2 CAI, for example, uses block transmissions with 48 data bits and a 16-bit checkfield (15-bit cyclic redundancy checksum plus 1-bit parity). In a signalling scheme the acknowledgement generally takes the form of a transmission embodying a response to the command in the original message. The principle, however, is fundamentally the same as ARQ. Approaches to error control for specific cordless systems are described in the relevant chapters later in this book.

For voice transmission, error control, if employed, would need to be implemented using forward error correction. In this case the coding overhead is significant (commonly 100%, since, typically, half-rate coding schemes are employed). Such an approach is used within the GSM cellular radio system, although in that case such protection is only afforded to the more sensitive of the (parametrically coded) speech bits. The benefit of such an error correction approach is that a significant proportion of transmitted errors (either in the information or in the redundancy bits) can be corrected. For the normal digital cordless telephone application employing waveform coding such as ADPCM and where rapid radio channel fluctuations are reasonably limited, error coding is not normally employed. However, error control techniques are clearly essential for the new emerging multimedia applications.

Further information on error control strategies and techniques can be found in [32].

12.9   Protocol Layers

"Terms of treaty agreed to in conference; formal statement of transaction" – so does *The Concise Oxford Dictionary* define "protocol". In human terms a protocol is a means of formalising and standardising communication, and its meaning, across what otherwise would be a barrier to the full and free flow of information and understanding.

Whenever information must be passed across an interface between otherwise independent, perhaps different in kind, communicating entities, clear unambiguous understanding demands the use of protocols. This is true for modern communications networks of all forms including cordless telecommunication networks, particularly when they act as multi-service channels between users, e.g. a computer to its terminal, an ISDN to a telephone handset. The creation of a protocol hierarchy (or architecture) is the essential element to the logical partitioning of processing and communication resources between networks and/ or users in the establishment of a communication service.

In endeavouring to formalise such matters the Open System Interconnection (OSI) reference model was established, which embodies a layered structure approach to protocol specification. The use of such a structure has several benefits:

- The overall problem of logical communication definition is thereby divided into smaller, more manageable, units.
- Each layer has a homogeneity of purpose and procedure, permitting layer-to-layer logical communication and independent modification.
- Each layer, therefore, has independence of function except for the need to interact with adjacent layers.

The standard OSI reference model has seven layers as tabulated and defined in Table 12.3. Relating this to cordless telecommunications, the lowest three levels can be defined as in Table 12.4. Protocol layers 1, 2 and 3 serve to support the creation of a functional data link through the cordless network, whereas the higher layers are concerned with supporting communication between the end-users/networks. Thus the cordless link provides them with transparent non-interventionist layer-to-layer communication.

Essentially the OSI structure is a user–network interface architecture that does not preclude modified structures being used within the cordless network. An

Table 12.3
Standard OSI reference model

Layer	Description
7	Application layer
6	Presentation layer
5	Session layer
4	Transport layer
3	Network layer
2	Data link layer
1	Physical layer

Table 12.4
Lowest three layers of OSI model related to cordless telecommunications

Network layer	Furnishes the means to create, maintain and release the network path/connections between the terminating users/networks. Terminal-to-terminal signalling and control, data flow control and priority control belong to this layer. It offers these facilities to layer 4
Data link layer	Possesses the means to create, maintain and release the logical (data) link. It provides framing and synchronisation (hence data transparency), sequencing of bits and sequences, error detection and correction, and offers these facilities to layer 3
Physical layer	Identifies the radio methods, radio channel allocation, performance monitoring, fault detection, channel shaping and coding, and other functions and procedures necessary to create, maintain and release the radio channel. To layer 2 it provides a service of transparent bit transfer

example of this is shown structurally in Figure 12.20 for the DECT standard, where layers 1, 2 and 3 have been rationalised as:

- The physical layer (PHL) and the medium access control layer (MAC), a mix of OSI layers 1 and 2
- The data link control layer (DLC) and network layer (NWL), formed from OSI equivalents 2 and 3

These layers are described further in Chapters 19 and 20, where the DECT specifications are outlined, and in the next chapter, where the network aspects of cordless telecommunications are considered.

In this chapter we have focused on the ways in which a transparent communication can be effected at the lowest protocol layer, the physical layer, to support the transmission of a wide range of services, requiring these other higher protocol layers.

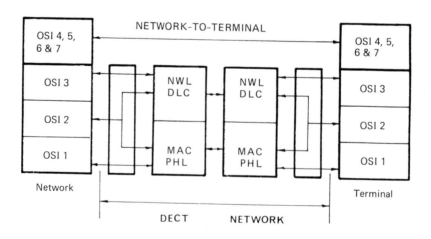

Figure 12.20 DECT protocol structure mapped onto the ISO model.

12.10   Summary

The early part of this chapter has described the issues relating to spectrum requirements for cordless telephone systems, the characteristics of in-building propagation and the dispersive nature of the radio channel. After examining the fundamental constraints upon spectrum choice, an approximate theory was presented to derive some "rules of thumb" in terms of interrelationships between spectrum requirement and, for a given grade of service, a range of parameters – traffic density, cell size, range propagation law, floor attenuation (for multistorey buildings) and required C/I ratio. Detailed propagation studies were then described, which have led in recent years to new propagation models for the in-building environment. The ways in which the performance of high-density cordless installations are quantitatively predicted by the use of computer simulations was then described.

In the latter parts of the chapter the issues relating to the choice of parameters for the radio system needed to match the radio channel were described. This included a review of the usage of the radio channel, introducing the concepts of duplex transmission, multiple access schemes and dynamic channel assignment. This was followed by a look at various digital modulation schemes – PSK, FSK, MSK and GMSK. The issue of error control was then briefly discussed. Finally, the concepts of communication protocols were introduced and the application of the OSI ISO model to cordless telecommunications described.

Overall, this chapter has attempted to survey the very broad issues of the characteristics and usage of the radio channel, drawing out in particular those aspects peculiar to the cordless telecommunications application. The basics of radio communication technique have also been presented for those readers less familiar with these and a range of references provided for the interested reader to pursue.

References

[1]   "Microwave Mobile Communications", WC Jakes (ed.), Wiley, 1974
[2]   "Advanced Digital Communications, Systems and Signal Processing Techniques", K Feher (ed.), Prentice-Hall, 1987
[3]   "DECT Reference Document", ETR 015, ETSI, Sophia Antipolis, Nice, 1991
[4]   "Radio Coverage in Buildings", AJ Motley and AJ Martin, Proceedings of the National Communications Forum 1988, Chicago, IL, pp.1722–30, October 1988
[5]   "Radio Coverage in Buildings", JM Keenan and AJ Motley, British Telecom Technology Journal, Vol. 8, No. 1, January 1990
[6]   "Characteristics Building for Propagation at 900 MHz", SE Alexander, Electronics Letters, Vol. 19, No. 20, p. 860, September 1983
[7]   "Radio Propagation Studies in a Small City for Universal Portable Communications", DMJ Devasirvatham, conference record of the IEEE Vehicular Technology Conference 1988, pp. 100–4, Philadelphia, PA, June 1988
[8]   "Additional Simulations of the Traffic in the TCS/BCT Test Installation", D Åkerberg, ECTEL/ ESPA Working Group Document No. BCT 18/87 (Ericsson Radio Document No. TY 87:2062), 1987
[9]   "Radiowave Propagation Model", A Falcon (ed.), RACE Mobile Telecommunications Project, December 1991

[10]   "Channel Models", Issue 2, R Gollreiter (ed.), R2084/ESG/CC3/DS/P/029/b1, RACE Advanced
        TDMA Mobile Access (ATDMA), May 1994
[11]   "COST 231: Evolution of Land Mobile Radio (Including Personal) Communications", E
        Damosso (ed.), Kluwer, 1996
[12]   "Radiowave Propagation in Office Buildings and Underground Halls", J Lähteenmaki,
        Proceedings of the 22nd European Microwave Conference, pp. 377–82, Espoo, Finland, 1992
[13]   "Ray Optical Indoor Modelling in Multi-Floored Buildings: Simulation and Measurements", D
        Cichon et al., Proceedings of the International Symposium on Antennas and Propagation,
        pp. 522–5, Newport Beach, USA, 1995
[14]   "Strahlenoptische Modelierung der Wellenausbreitung in urbanen Picofunkzellen", D Cichon,
        PhD Thesis, Forschungsberichte aus dem Institute für Höchfrequenztechnik und Electronik
        der Universität Karlsruhe, 1994
[15]   "The Effects of Time Delay Spread on Portable Radio Communications Channels With Digital
        Modulation", JCI Chuang, IEEE Journal on Selected Areas in Communications, Vol. SAC-5, pp.
        879–89, June 1987
[16]   "Irreducible Error Performance of a Digital Portable Communication System in a Controlled
        Time-Dispersion Indoor Channel", G Crohn, G Schultes, R Gahleitner and E Bonek, IEEE
        Journal on Selected Areas in Communications, Vol. SAC-11, pp. 1024–33, 1993
[17]   "On the Error Floor in DECT-like Systems", L Lopes, A Molisch, M Paier and J Fuhl, presented
        at the First European Personal and Mobile Communications Conference, Bologna, Italy,
        November 1995
[18]   "Multipath Time Delay Spread in the Digital Portable Radio Environment", DMJ
        Devasirvatham, IEEE Communications Magazine, Vol. 25, pp. 13–21, June 1987
[19]   "Delay Spread and Signal Level Measurements of 850 MHz Radio Waves in Building
        Environments", DMJ Devasirvatham, IEEE Transactions on Antennas and Propagation, Vol.
        AP-34, pp. 1300–5, November 1986
[20]   "Digital Mobile Radio", E Damosso, Springer-Verlag, 1996
[21]   "Phasor Explanation of Error Mechanisms for MSK Transmission Over a Slightly Dispersive
        Radio Channel", G Schultes and J Fuhl, Proceedings of the Melecon Conference, Antalya,
        Turkey, pp. 38–42, 1994
[22]   "Error Floor of MSK Modulation in a Mobile Radio Channel With Two Independently Fading
        Paths", A Molisch, J Fuhl and P Proksch, IEEE Transactions on Vehicular Technology, 1996.
        See also: "Bit Error Probability of MSK Modulation With Switched Diversity in a Mobile Radio
        Channel With Two Independently Fading Paths", A Molisch, J Fuhl and P Proksch,
        Proceedings of the Personal Indoor Mobile Radio Communications Conference, PIMRC, pp.
        1223–7, Toronto, Canada, 1995
[23]   "The Performance of DECT in the Outdoor 1.8 GHz Radio Channel", L Lopes and M Heath,
        Proceedings of the 6th IEE International Conference on Mobile and Personal Communications,
        pp. 300–7, 1991
[24]   "The Effect of Pathloss Models on the Simulated Performance of Portable Radio Systems", RC
        Bernhardt, IEEE Global Communications Conference, Dallas, TX, 1989
[25]   "Digital European Cordless Telecommunications System Simulation Performance Results", AP
        Croft, S McCann and WHW Tuttlebee, IEE Mobile Radio and Personal Communications
        Conference, Warwick, December 1989
[26]   "Spread Spectrum Communications – Myths and Realities", AJ Viterbi, IEEE Communications
        Magazine, pp. 11–18, May 1979
[27]   "Spread Spectrum Systems With Commercial Applications", RC Dixon, Wiley-Interscience,
        1994
[28]   "Simulation of Cordless Communications Systems", F Al-Salihi and LP Straus, Electronics
        Letters, Vol. 24, No. 12, pp. 742–3, June 1988
[29]   "Dynamic Channel Allocation for ATDMA", M Frullone and P Grazioso, RACE Mobile
        Telecommunications Summit, Cascais, Portugal, November 1995
[30]   "Linearisation of Class C Amplifiers Using Cartesian Feedback", A Bateman and RJ Wilkinson,
        IEEE Workshop on Mobile and Cordless Telephone Communications, London, September
        1989
[31]   "Minimum Shift Keying: A Spectrally Efficient Modulation", S Pasupathy, IEEE Communica-
        tions Magazine, pp. 14–22, July 1979
[32]   "Error Control Coding: Fundamentals and Applications", S Lin and DJ Costello, Prentice-Hall,
        1983

Notes

1   COST 231 is a pan-European collaborative research forum that has been coordinating European work on radio channel propagation in the 2 GHz region for future personal communications applications. It was the follow-on action from COST 207 that determined propagation models for GSM.

2   To take into account such factors, spectrum efficiency is commonly specified in terms not of channels/MHz but of bits/Hz per spatial volume – this incorporates the effects of modulation scheme, multiple access technique, spectrum re-use strategies, etc.

3   The bandwidth of a bit shaping filter is usually defined in terms of the bandwidth $\times$ time product, BT. The bandwidth is that of the filter and the time is a symbol period. Thus for $BT = 0.5$, the 3 dB filter bandwidth is equal to the frequency corresponding to the highest bit rate.

13 Cordless Access Networks

Andrew Bud

Cordless technology began to crash the personal mobile communications party in the late 1980s. It brought with it a new way of designing wireless networks, one that derived from its varied roots in the wireless PABX, telepoint and domestic business, which contrasted with the stately, well integrated traditions of the cellular telephony business. One of its innovations was its affirmation of the cordless access network as a system in its own right, distinguishable and distinct from the fixed network, to whose mobility and call-switching services it could offer access.

The value of cordless technology as a flexible network access method had begun to be recognised in the mid-1980s. Its advantages in terms of achievable traffic density, lack of requirement for spectrum planning, potential unlicensed operation and very low cost were very attractive to a number of telecommunications players. These included PABX manufacturers, aspiring competitive mobile communications network operators and fixed network operators looking for new markets. Each had their own agenda, priorities and objectives, which were notably divergent. All recognised that collaboration in the standardisation of the air interface would be essential in order to gain access to spectrum and markets. But all recognised also that any corresponding attempt to standardise the fixed network would be counter-productive, either because it conflicted with their business interests (as in the PABX business), or because they were too weak with respect to the other forces at play (such as GSM), or simply because it was felt to be premature, in the light of forthcoming regulatory change.

In this way, the cordless access network came to be designed and specified as an independent system. It comprises an air interface with the power to transfer certain types of information in standardised ways. It implements procedures that are easily linkable to those of common fixed networks. It is based throughout on the concept of *interworking* with the fixed network, whether for call control, call switching or mobility management. This reliance on interworking means that a single cordless access network technology can be applied to many applications and many fixed networks.

In Europe, both CT2 and DECT were implicitly defined this way. The US PCS standards have followed this model too, although PHS in Japan has a slightly different approach, owing to its different history. This chapter will not dwell on the differences between these technologies, but where specific examples are appropriate the DECT standard, with which the author is most intimately familiar, will be used.

This chapter therefore begins by reviewing the different kinds of cordless networks, as they are shaped by the different applications. The general functional requirements of cordless networks are then described.

The technology of cordless access networks is relatively new and has been spawned almost entirely in the development laboratories of private industry; thus there are at the time of writing very few detailed technical references in the public domain, other than those given at recent conferences [1-3].

13.1   Types of Cordless Networks

To structure a review of cordless systems, it is essential to classify the different types of network. Networks can be classified by many parameters - number of lines, area covered and technical performance are examples.

For present purposes, an effective way to distinguish between different types of system is to ask: "Who owns the fixed part?" The answer implies a conveniently clear definition of the services and the technical implementation. Although telecommunications deregulation and new business models will rapidly and dramatically blur these boundaries, this approach provides an effective and valid model. Using this approach leads to the study of cordless networks in three main groups:

- *Domestic systems* - owned by individuals or families
- *Business systems* - owned by productive or service organisations such as companies or governments
- *Public access systems* - owned by public telecommunications operators

Each of these systems is now briefly described, highlighting their key features.

13.1.1   Domestic Systems

Domestic cordless telephones satisfy the huge market for simple mono-cell mobility, bringing calls into the garden or bathroom. They are also real cordless access networks, implementing a large proportion of the functions described in section 13.2. When standard access profiles, such as the DECT GAP (see Chapter 20) become ubiquitous, the distinction between a domestic handset and those used for much more powerful services will simply disappear. Furthermore, features such as intercom between two portables and, in future, integration into wide-area mobility networks will transform them into sophisticated telecommunications terminals.

13.1.2   Business Systems

In general, business networks have three principal characteristics:

- Several lines of attachment to the public network
- Considerable traffic between internal extensions

- Usage by people who are very often mobile

Conventional fixed business networks are either key systems, for up to 20 users, or PABXs supporting anything between 20 and 5000 extensions. The main functions of these systems are the switching of calls between extensions, and the management of access by users to the limited number of outside lines. Cordless business systems, such as described in Chapter 2, must support these functions and, in addition, the management of mobility, and the requirements of security and identification. These aspects are developed in more detail in the second part of this chapter.

A cordless business network may offer various levels of integration with the company's and with the public operator's fixed network. It is convenient to discuss this integration under three broad categories of system: the adjunct, the stand-alone wireless PABX, and the network integrated system.

Adjunct Systems

An adjunct system is a cordless mobility front-end that can be attached to an existing legacy PABX. Its role is to provide mobile access to a certain number of the PABX's existing extension lines, and so the adjunct is in principle responsible only for those functions directly related to mobility, such as location and authentication. It needs to be effectively transparent to all call-related and supplementary service signalling.

In theory, the adjunct is an attractive type of network for the customer who wishes to add mobility to his network, without touching the investment (usually substantial) already sunk into a standard PABX. There are a number of technical aspects that complicate this simple scenario.

Firstly, it is sometimes difficult to offer an adequate level of feature support to the mobile user at a reasonable cost. The reason is that the adjunct must be connected by a signalling link to the existing PABX. If the adjunct is supplied by the manufacturer of the PABX, then this link can support the manufacturer's proprietary inter-switch or remote concentrator protocols, and features can be adequately offered by the adjunct to the mobile user, too. If, however, the adjunct is supplied by a third party, only open standard protocols are available for interconnection. For small numbers of lines, two-wire analogue connections can be used to mimic exactly a subscriber line, on which features may be supported by an in-band keypad protocol. For larger systems, this becomes technically and economically impracticable and a trunked interface is necessary. However, the feature support available over digital trunk lines, using protocols such as DPNSS or QSIG, is quite modest. Solutions to this problem exist, but are complex and intricate.

The second complication with an adjunct manifests itself in larger systems, where the number or geographical dispersion of the coverage cells is such as to require more than one adjunct in a given corporate network. In this case, either inter-adjunct roaming is simply impossible, or else some means of inter-adjunct mobility networking is necessary. This inter-adjunct networking, based on proprietary signalling via switched or permanent circuits between adjuncts, creates an overlay mobility network. Owing to the low level of integration between the adjuncts and the host PABXs, roaming traffic must often be routed on this overlay network back to the home adjunct, which can give rise to problems such as high traffic levels and extra delay.

For these reasons, adjuncts are often engineered with substantial processing intelligence. They always contain substantial switching capability as well, in order to support inter-cell handover, and need to have a high level of reliability and product quality. Many manufacturers therefore base their adjunct products on PABX chassis, but this often leads to a high cost per line, since customers effectively have two PABXs to serve their mobile users.

Stand-Alone Wireless PABX

A wireless PABX is a private switch into which the access functions of an adjunct have been fully integrated. The radio basestation control cards are integrated as line cards in the PABX rack, and the cordless air interface signalling protocols run as software either on the line cards or on the PABX processors. The mobility management application software runs on the PABX, and network roaming is supported by the proprietary inter-switch networking protocols. Network management of the cordless infrastructure and terminals is fully integrated into the switch's OA&M system.

The wireless PABX offers the customer a very attractive technical solution, providing that his or her PABX supplier offers the cordless option as an upgrade, and providing that he or she has the necessary release of switch software to permit upgrade. Unfortunately, these conditions can easily not be met. Furthermore, a wireless PABX has a high development cost, owing to the need to graft the mobility functionality onto an existing software load, and the potential market is much less broad than that for the adjunct and therefore, even when available, the cost of a wireless PABX can be high.

One solution to this trap is for a manufacturer to specialise and integrate a generic adjunct product with the specific PABX, so as to limit the switch-specific hardware and software investments without losing the benefits of integration described above. In many markets, this is likely to become the dominant approach.

Network Integrated Wireless PABX

A wireless PABX network is capable of offering mobility and roaming only between nodes of that network. As standards and technology for personal communications consolidate, wireless PABXs will be integrated into geographical personal communications networks. These networks will be based upon broad acceptance and open standards, and will be driven from the public service market. These networks will therefore be of two types: mobile networks, such as GSM, and AIN-based fixed service networks, as envisaged by the Cordless Terminal Mobility (CTM) initiative (see Chapter 6).

The integration of cordless PABXs with the GSM network has been the subject of study and standardisation since 1990. The objective is to provide the customer with all the services of his or her GSM subscription, especially incoming call delivery on a mobile number, when he or she is using a cordless telephone attached to the office PABX.

This means that the PABX must attach to a GSM MSC with MAP-like signalling, typically using the A interface. Unfortunately the A interface is based on CCITT Signalling System No.7, which is not ideal for PABX use. Therefore work is in course in Europe to develop an interface for PABXs with A-interface properties, but based on the simpler ISDN DSS1 signalling stack. It also implies that the terminal must support the GSM security mechanisms, which are always end-to-end, and therefore the development of the DECT GSM interworking profile (GIP) (Chapter 20) is a key step in this direction.

13.1.3 Public Access Networks

Public access networks based on cordless technology have developed in a number of different directions, depending upon the maturity of the technology and on local regulatory conditions.

In Europe and Asia, the first type of service was telepoint, a simple personal communications service for geographically localised access, often limited to outgoing calls only.

More recently, in Europe and Japan, development has focused upon geographically more comprehensive personal communications services, rivalling those of cellular mobile networks in urban areas, based upon CTM-like approaches.

With the maturation of the technology, new markets are opening throughout the world for wireless local loop networks based on cordless technology.

Each of these types of network will be briefly reviewed.

Telepoint

Telepoint was the world's first public service based on cordless access. It permitted customers to make (and in some cases receive) calls in certain, limited areas of some cities, using handsets selling at around US $200. The background to telepoint in Europe and Asia is described in more detail in Chapters 3 and 4 of this book.

The service developed first in the UK, driven by pressure from companies who had invested in the creation of CT2 technology. The service offered was initially limited by their licences to outgoing calls only (that is, originated by the cordless handset owner). The service they received from the PSTN was limited to the rental of business subscriber lines on normal terms and their initial investment in network support, other than the basestations, consisted only of billing and administration facilities. The simple structure of these early networks is shown in Figure 13.1.

The base terminals were highly autonomous and were responsible for the local authentication of handsets and for the collection of billing and maintenance information. Such information was polled by the network management centre at regular intervals. The operation of the network strongly resembled that of a credit or debit card system using smart cards for low-value transactions.

For political and regulatory reasons, these UK companies were forced to build overlay networks, operating separately from the existing operators of wired and mobile cellular networks, and therefore they did not have favoured access to

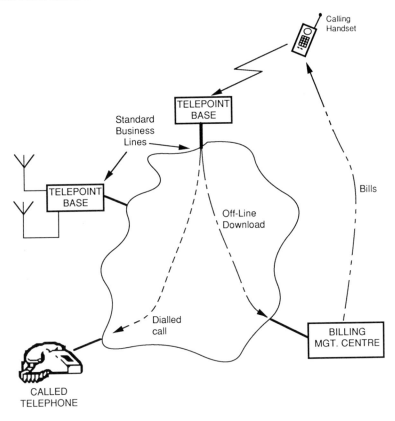

Figure 13.1 Early telepoint network structure.

fixed network resources. The services were also developed as high-risk ventures, with a cautious, low-investment approach to the development of the new networks. Such considerations defined the character and topology of the UK telepoint networks. They also sealed the commercial fate of the service, which failed to compete effectively in a market populated with vigorous and highly developed cellular telephony operations. By 1994 all the UK telepoint networks had closed. The heritage of this service is still visible in the CT2 and DECT standards, which support a number of functions necessary for the off-line management of subscriptions, credit and authentication.

With the signature of a Memorandum of Understanding in March 1990, many European PTOs also committed themselves to building telepoint networks. In most cases small experimental networks were tried, but the results were not sufficient to lead to commercial service. Only in France has the Bi-Bop network operated by France Télécom attracted nearly 100 000 customers with its near-total coverage of several major cities and its two-way calling service, which resembles PCS (see next section), in an environment in which competition from cellular telephony was initially weak. The other success story was Hong Kong, which peaked in 1995 with over 170 000 customers in a society characterised by a small densely populated cityscape and high pedestrian mobility. However, this number had dropped to 100 000 a year later, reflecting aggressive competition from cellular networks, which led to subsequent closure. Telepoint has also made an impact in other urbanised but not motorised economies in Asia, such as Vietnam and China, as described in Chapter 4.

Cordless Personal Communications Services

With the development of powerful cordless air interfaces such as DECT and PHS, the concept of personal communications services (PCS) based on such interfaces has matured. The most spectacular example of a network of this kind is in Japan, where a concerted commitment to PHS has created the world's largest operational cordless PCS systems. In Europe, both traditional and new entrant operators are experimenting with PCS services based on DECT access, and it is likely that some licensees of the US PCS bands will use cordless technologies for their public services (see Chapter 8).

The services offered by these networks are broadly similar: the objective is to provide ubiquitous outdoor coverage in urban areas, full two-way calling, and future continuity of service between public areas and private zones. Hence for this sort of application, cordless technology is positioned as a direct alternative to digital cellular technologies. The relative merits and balance of advantage between cordless and cellular access technologies are not the subject of this chapter, but there are a number of important issues in the implementation of cordless PCS networks that merit analysis.

Backhaul

Backhaul is the means for connecting the cordless radio basestations to a central concentrator. In public PCS systems based on short-range cordless technology, the requirement for near-ubiquitous coverage means that the number of basestations is very high, while the revenue-earning traffic on each backhaul is often quite low. For this reason the backhaul technology is perhaps the most critical issue affecting the design and economics of a public cordless network. Operators have attempted to solve this problem in different ways, depending upon their technical and regulatory degrees of freedom.

For operators who already own fixed public networks, or to whom regulators have granted equal access to such networks, local loop copper twisted pairs are the optimum solution. They are ubiquitous and have low marginal investment. It is true that this approach imposes some limitation on the traffic capacity of one basestation, since in practice local loop copper pairs support only ISDN U-interface speeds (192 kb s^{-1}) unless costly techniques such as HDSL are used, but this is a significant limitation only in the context of advanced data services.

Cable television operators also have a readily usable infrastructure, providing that the percentage of homes passed in their concession area is sufficient to support PCS coverage levels. The installation of dedicated wire or fibre in existing cable trunking or the use of multiplexed transmission capacity on existing cable have a low enough marginal cost to make PCS networks feasible. CGE Cofira in France has led the experimentation of this type of approach in Europe.

It is technically feasible to use wireless links as backhaul; cellular networks have used microwave point-to-point links for many years. However, the cost of microwave link technology is today still too high for the cordless PCS environment, and the limitations imposed by the need for line-of-sight links too restrictive to be feasible.

A particularly attractive solution that has been developed in the DECT environment is the use of cordless links as backhaul. In the wireless basestation (WBS), the user–network DECT links are connected by separate DECT backhaul circuits to a centralised fixed basestation. The economic advantages of this approach are substantial; however the WBS can add end-to-end delay, and the only way to avoid this is to make less efficient use of spectrum resources.

Mobility Service Control

Cordless networks for PCS must implement sophisticated centralised functions to support handover, roaming, security and service management. Unlike cellular networks, however, there is as yet no established, proven reference network architecture for cordless PCS networks, so that each operator still has to invent its own or swallow a complete proprietary solution from a manufacturer. This will certainly change rapidly as networks and standards emerge, but at the time of writing the discussion of this aspect is inevitably a balance between the theoretical and the anecdotal.

There are fundamentally three architectures worthy of review: the overlay network, the IN approach, and the cellular network graft.

The overlay network is the approach taken by those who are using cordless access to provide personal mobility services via a fixed network that cannot support personal mobility. The switches, signalling and databases of the fixed network either do not have the necessary functionality, or else the fixed network operator does not make it available to the newcomer.

In this case, the operator has a range of choices. In the case of France Télécom's early Bi-Bop service, an overlay signalling network based on X.25 was constructed, to enable the basestations to communicate in real time with centralised database servers via the existing fixed network, but without its intervention. Early experiments in Germany and Italy used a similar approach, based on extensions to existing payphone overlay networks.

In the case of Nortel's CT2Plus service, a complete network of cordless mobile switching centres was created, strongly resembling a cellular network. Many initial experiments by fixed network operators have followed this path, pragmatically using proprietary cordless PABX technology to create small-scale overlay networks. In future, this approach will be adopted by operators possessing only a licence for cordless access PCS.

The IN approach is an alternative available to operators who already possess or are building a fixed network with intelligent network capabilities. The mobility functions are treated as an extension to the IN capability set, to be implemented in the network at modest extra investment. The result is that the cordless access network becomes just another means to purchase the fixed network's services, which include mobility. Cordless users have available to them all the features and services already implemented on the fixed network, together with its interconnect and numbering plan.

This is an exceptionally attractive approach for both operators and customers, since the investment is limited to the cordless access network, rather than a complete overlay network, and the customer can access a service set that is already familiar. There are a number of serious regulatory and political difficulties with this approach, owing to the dominant positions of existing IN operators in many markets, but such a discussion is beyond the scope of this chapter.

At the time of writing the leader in this field is Telecom Italia, whose declared intention is to offer DECT-based cordless access to its existing IN network in major cities throughout the country, with an aggressive rollout plan starting in 1996. It anticipates 3 million users by the year 2000. It is in a particularly strong position to do so, since its local exchanges (SGU) are all equipped with service switching (SSP) capability, and it intends to integrate DECT protocol and mobility functionality into these SGUs.

An IN approach also lends itself to the implementation of Cordless Terminal Mobility (see Chapter 6), and the two will normally go hand-in-hand.

Faced with the threat from fixed network operators entering the PCS business using cordless access, the mobile operators have a response. They are already operators of large, mobility-oriented intelligent networks, and the introduction of a new means of wireless access is for them a logical technical step. This is one of the dynamics underlying the third approach, the cordless graft onto a cellular network.

In this case, the approach taken is to attach the cordless access network at a convenient reference point in the fixed network, usually the A interface. In the GSM architecture, this is the point of interconnection between the basestation controller (BSC) and the mobile switch (MSC). To the switch, the cordless users can be made to appear exactly as though they were normal cellular users, so that the customers receive their full cellular service profile. Alternatively, if the operator so chooses, they can be established as a separate class of users, with a dedicated service set and tariffing structure, but the technology does not force this approach.

This approach today raises a number of technical issues; since it is based upon the propagation of existing cellular-based architectures and standards back into the cordless domain, it introduces some constraints upon the cordless network. Two examples of this, drawn from the DECT/GSM interworking arena, are useful. There is a requirement to support the GSM smart card (SIM) in the cordless terminal in order to support GSM user authentication, which cannot be spoofed by an interworking unit. There is also the need for some rather strange air interface protocol athletics to permit interworking between DECT's cordless model of portable-controlled forward handover with GSM's network-controlled backward handover procedures. More detail on the DECT/GSM interworking profile is given in Chapter 20.

Many GSM operators are experimenting with this sort of integration at the time of writing (see Chapter 16). There is a certain feeling that the key to this market will be the availability of dual-mode (GSM/DECT) handsets, permitting seamless passage from one access network to another. There are also a number of regulatory and political issues, since it requires a rethink of the scope and limits of the licences awarded to cellular operators. These licences have historically been technology based, and in many cases exclude CTM-like services. Such barriers are unlikely to survive the impact of cordless technology.

Wireless Local Loop Networks

An important new type of cordless access network is the wireless local loop, discussed in Chapter 5. This is aimed at providing transparent access to conventional switching fabrics, by-passing the need for local loop copper. It is of great importance both in developing countries, where the priority is to increase the absolute penetration of telephones as fast and as cheaply as possible, and in developed countries, where it represents a key entry strategy for a competitive access operator seeking to establish a position against a dominant established operator who owns local loop copper.

There are many point-to-multipoint technologies suitable for wireless local loop (WLL) applications, and cordless technology is just one possible solution. In this chapter, no attempt will be made to discuss the relative merits of these solutions, but attention will focus on the particular characteristics of wireless local loop cordless networks.

The main difference between this service and all the others discussed in this chapter is the secondary role of mobility. Wireless local loop hinges on cost and quality. Of course, no WLL operator is blind to the opportunities offered by the potential to add mobility to his service offering, and a competitive access provider needs all the service differentiation he can get, but he cannot afford to compromise his core service for it. So the key network issues for a wireless local loop are those related to investment and cost of ownership. These include:

- Customer premises equipment and distribution, perhaps the single costliest part of the network.
- Operations and maintenance, to guarantee service quality, minimise outage and limit the cost of maintenance and repair interventions.
- Network architecture, to minimise the number of costly sites and the quantity of repair-prone technology distributed in the field.
- Trunking and concentration architecture, to obtain an acceptable grade of service with the minimum investment in transmission.
- Integration to the fixed network, to ensure service transparency and low interconnection charges.

Each manufacturer's system is different, and each operator puts his own stamp on the solution, but the service priorities listed above lead to a number of general considerations, described below.

The customer unit is a difficult element, since it is under enormous cost pressure, yet must provide a telephone socket indistinguishable from that offered by the dominant PTO. For this reason, many manufacturers put central office subscriber line cards into the CPE, guaranteeing thereby full compatibility for a range of supplementary services (metering pulses, calling-line identification, etc.) as well as extensive OA&M functions. The cordless air interface must therefore support the information flows needed to control these line cards, which implies a functionality scarcely dreamed of when the cordless access standards were designed. Where an interface has a high degree of flexibility, this enhancement is nevertheless relatively easy, as the development of the radio local loop access profile (RAP) for DECT illustrates.

A minimum number of high-capacity sites is most desirable, since this minimises the outlay on site preparation, concentrates technology into easily maintainable units, and maximises the trunking efficiencies on the backhaul transmission. Unfortunately, this approach implies large coverage radii, which are difficult to achieve with cordless technology. High-gain antennas are required at both the base sites and at the customer premises, with near line-of-sight between them, and this complicates the distribution of the service at the customer premises.

A high degree of centralisation of functions at the interface to the fixed network is also desirable. The speech transcoding required on the air interface is best carried out there, in order to maximise transmission efficiency on the backhaul, and other functions such as security and authentication, billing and service interworking are best concentrated there in order to ease the maintenance process.

The interface to the fixed network is an area where the unique character of the WLL cordless network is particularly clear. Personal communications networks are equipped with interfaces and protocols compatible with CTM or with cellular networks, whilst typical WLL systems are designed specifically to operate as access networks for public switches. As such, they offer the appropriate interfaces – either proprietary to the public switch manufacturer, or one of the emerging standards for these interfaces such as the European V5.1 and V5.2 standards. These access standards, whilst rich in OA&M capability, unfortunately do not support any kind of mobility, and this will create architectural difficulties for operators intending to extend their WLL service into the PCS area.

One area of particular difficulty in the wireless local loop application is the issue of traffic interworking (section 13.2.4) and service transparency. In order to compete with copper, the WLL operator would like to offer a channel that is transparent to the service. Unfortunately, using cordless technology, this is very expensive in spectrum and network resources. The alternative, to provide the user with a service-based interface to the network, will be frequently more cost-effective, but requires both customer and operator to accept a different business and service model. This will be one of the challenges facing operators of cordless WLL networks.

13.2   Network Functionality

Having discussed the character of specific kinds of cordless networks, we now go on to examine the functional capabilities of a cordless access network itself. The network must offer four elements of service:

- Call control interworking
- Mobility management
- Supplementary service support
- Traffic interworking

These will now be briefly reviewed.

13.2.1   Call Control

In a cordless access network, the call control function has two roles. Firstly, it has to establish the right services at the time of call set-up. Over a cordless link, services are not supported transparently, and a speech call, a modem call or a fax may be treated in very different ways by the digital cordless access network, even though they then converge onto the same kind of PSTN circuit. The call control processes must therefore inform the other end of the cordless link of the service that has been requested, and if necessary negotiate a mutually acceptable compromise. Only when both ends of the cordless link are sure that the right low-layer radio protocols are available, and that the necessary interworking unit exists and is available, can that interworking unit be commanded to begin call set-up over the fixed network.

At this point the call control function has to give the user access to the circuit routing and switching machine that lies behind the access network. It is conceptually simplest to view the cordless network itself as operating without call switching functions, and therefore all such functions must be offered by the attached network, be it a PABX, PSTN or GSM network. These networks have specific user-to-network interfaces (UNI) for call control that cannot, in general, be transported transparently over the air interface: How could a digital cordless telephone directly execute pulse-disconnect dialling, for example? Thus the air interface protocol serves to control the interworking unit that operates the fixed network UNI.

These two roles are complementary and separate, but are linked by the concept of a call control "instance". In the first phase, the instance's characteristics are negotiated and agreed, and then in the second phase the instance's state evolves to reflect the state of the fixed network.

This shows that the call control function must have two important characteristics. Firstly it must include a negotiation mechanism that permits a discussion about services to converge rapidly. This is not easy, since the range and variety of services that may be offered in modern telecommunications networks is very wide, once we take into account that a cordless access network can be used to access LANs, the Internet and at least three different sorts of voice circuit network (PSTN, ISDN, GSM).

Secondly, its internal states must be capable of accurately reflecting the call states on a variety of different networks. The way in which a telephone call evolves on the PSTN is different from the ISDN, which is different again from the GSM. It is important that the cordless access network does not obscure this evolution of events from the user, and this means that the cordless network call states must be a superset of those imposed by the attached network.

This second problem, at least, is not entirely unique to cordless access networks, and is shared to some degree by the ISDN, and therefore many cordless networks use signalling derived from the ISDN standards.

13.2.2 Mobility Management

Managing the mobility of user terminals is a task characteristic of a cordless access network. Often, such networks provide access to a fixed network that is incapable of dealing with terminal mobility, and therefore the whole task falls on the cordless network. In other cases, the fixed network is mobility-aware, in which case the cordless network need only interwork the relevant information into a suitable format.

A mobility management machine consists of a database, an application process and a protocol engine. The database contains details of the user, and his or her service profile and location. The application process defines how this information should be used to control access to service, and what measures should be taken to maintain the database. The protocol engine is responsible for executing mobility-related transactions with the user, to receive and send mobility-related information.

Typical database entries include the current location of the user, the services to which he or she is subscribed, the characteristics of his or her terminal, the secret parameters used to confirm identity and the status of bill payments.

A typical application process might be responsible for the periodic confirmation of the user's location, the choice of where to page a user, the management of a temporary identity assigned to the user and the decision to authenticate the user on certain sorts of call set-up. More generally the application process is responsible for the script that governs the mobility exchanges when a caller requests a service, before the call control machine is invoked.

The protocol engine is the heart of the cordless network's mobility capabilities, and it may implement a varied set of functions. As an illustration, those implemented in the DECT specification are briefly described below:

- *Paging* This is used to deliver an incoming call to a user. Paging may occur throughout the entire network, or only in location areas around where the user is attached. In systems intended for unlicensed operation, such as DECT, there is no dedicated paging channel, but instead a connectionless broadcast channel associated with every active circuit, to which a quiescent handset can listen on a low duty cycle.

- *Handover* One of the most important services offered by a cordless access network is continuity of service throughout the coverage area. In order to support this, even when a user is moving, it is necessary to provide handover between cells. In the best networks, such handover is "seamless", meaning that it is inaudible to the user. In DECT, there are three different types of inter-cell handover. One, called bearer handover, occurs at the very lowest level in the protocol hierarchy, and requires the handset to activate a second channel to run in parallel with the existing one until a precisely timed switch is executed. The second type, called connection handover, operates at the data link layer and is consequently more flexible, permitting a range of performance/quality balances. The third type is called external handover, and is designed to permit handover between two separate cordless access networks, both of which are connected to a fixed network capable of managing mobility. In addition, procedures are supplied (see below) that support interworking to fixed networks such as GSM, which have different models of handover.

- *Location registration* This process informs the fixed part that a portable part has moved into a particular area of the network, and provides the fixed part with certain useful information about the portable part. It has two objectives: it is intended to enable a multi-cluster cordless network to page a handset only in the appropriate cluster, to minimise redundant paging traffic; it also enables a fixed network to acquire from its databases relevant data about a portable before the user makes a service request.

- *Location update* This procedure is used by the fixed part to force the portable to formally register its location, in case the portable tries a call set-up without having done so.

- *Attach and detach*   To inform the network of the status of the handset, to accelerate the handling of incoming calls.
- *Temporary identity assignment*   By substituting a short temporary number from a local numbering plan for a long globally unique identity, paging messages are kept short and unambiguous, but these temporary numbers are a resource that must be managed, by means of this protocol.
- *Portable identity interrogation*   This enables the fixed part to request the various aliases by which the portable is known, so as to understand what sort of user is requesting service.
- *On-air subscription and desubscription*   To simplify the process of exchanging the several multi-digit numbers involved when a terminal is subscribed to a network, by using on-air processes instead of the manual keypad operations otherwise necessary.
- *Portable authentication*   Enables the fixed part to confirm the portable's identity, to avoid fraud and theft of call time.
- *Fixed part authentication*   Enables the portable part to confirm the fixed part's identity, to avoid spoofing.
- *Key allocation*   Procedures for the maintenance of authentication and cipher keys.
- *Ciphering control*   Controls the application of the ciphering function.
- *Network parameter management*   This enables the fixed network to load the portable with certain parameters, and to request them later on. This is used to assist the process of handover in certain fixed network architectures such as GSM where, for example, the portable must pass information from the old basestation to the new one.
- *Roaming support*   This permits a portable to determine if it can obtain service from an in-range fixed network, and it permits a fixed network to determine if there is a reasonable likelihood that a portable is authorised to obtain service. This support is offered by means of a sophisticated identity architecture: each network identity includes a code indicating the issuing authority, together with the specific subscriber number. The cordless network fixed stations broadcast a list of issuing authorities with whom roaming agreements are in place, and the portable can contain multiple numbers, issued by a number of different authorities. A first screening can rapidly be done on both sides by comparing these issuing authority codes. A session key procedure is also recommended, to permit authentication when roaming without compromising the security of the system.

13.2.3   Supplementary Services

Supplementary services may be inherent in the functionality of the cordless access network, or they may be offered by the fixed network. In the latter case the cordless access network must provide rapid, efficient and transparent access to them.

As cordless networking develops, new cordless-specific supplementary services will be devised. For illustrative purposes, it is useful to look at those defined in DECT:

- *Queue management* Provides a mechanism for queuing service requests to an access network that is temporarily congested, and for keeping the customer informed of the queue status.
- *Indication of subscriber number* Informs the user of the number on the fixed network allocated to his or her terminal by the cordless access network.
- *Control of echo control functions* Enables the portable to control the echo cancellers in the cordless network fixed part, on the basis of some local knowledge about the destination number and its routing.
- *Cost information* Enables the user to request comprehensive tariffing information before selecting a cordless access operator or setting up a call, and to monitor call cost information during and after the call, to keep track of expenditure.

Accessing the fixed network supplementary services can most simply be done using a keypad protocol. This only requires the cordless access network to transfer key presses during a call, which the fixed network will then interpret. The cordless network is transparent.

However, some networks do not support keypad protocols – GSM is an example. In this case, the fixed network uses a signalling protocol with states, and this must be implemented in the cordless network interworking unit. If the handset is only capable of supporting keypad protocols, then a network-specific translator is required in the interworking unit, and supplementary service status is lost when a terminal moves between different cordless access networks attached to the same fixed network. Alternatively, the handset may implement these states, providing that the air interface protocol supports, as DECT does, the signalling necessary to transfer these states. In this case the supplementary services are supported end-to-end, and the terminal MMI for these services is constant whatever the attached network, but the introduction of a new service requires, at the very least, a modification to the standard and a software upgrade in the terminal, making this a very rigid approach.

13.2.4   Traffic Interworking

A cordless network is constrained by radio spectrum and performance limitations to transmit information in special ways. These may not necessarily have anything to do with the way the information was originated, nor with the way it is transferred to the fixed network. Therefore, a key function of a cordless network is traffic interworking.

A few examples will suffice. Voice is transmitted on fixed digital networks using PCM at $64\,\mathrm{kb\,s^{-1}}$, but this is very redundant, and if transferred to the ether would represent an unacceptable waste of spectrum. Therefore all digital cordless systems transcode the speech information to a lower rate, typically using G.726 ADPCM at $32\,\mathrm{kb\,s^{-1}}$.

This channel is however unsuitable for modem data, which is degraded above $4.8\,\mathrm{kb\,s^{-1}}$. Therefore a different traffic interworking is required for this service. Two approaches are used. One method constructs a $64\,\mathrm{kb\,s^{-1}}$ bearer from two $32\,\mathrm{kb\,s^{-1}}$ bearers, occupying twice as much spectrum as a telephone call, and transmits raw PCM. The other approach extracts the data, assembles it into packets, and sends it over special packet-mode bearers, protected with error correction. In this way the data arrives error-free, at much greater spectral efficiency.

These approaches are however unsuitable for ISDN videophone traffic, which requires an isochronous $64\,\text{kb s}^{-1}$ bearer with very low residual bit error rate. In this case, an approach is to construct a special bearer protected by FEC and ARQ, using the equivalent of two or three telephone channels.

These examples illustrate how a cordless network must be capable of distinguishing the service requested and installing the appropriate traffic interworking functions, at both ends of the link.

13.3   Summary

This chapter has offered an introduction to some of the principal network issues raised in the design and application of digital cordless access networks. It has described the continuum that exists between the simplest cordless telephone set and personal communications systems. It has shown that the cordless access network is a new and powerful tool in the development of personal mobility services and has explored the functions implemented in such networks.

References

[1]   Papers presented at the Digital Cordless Communications Conference, organised by IBC, London, September 1995
[2]   Papers presented at the Wireless LANs Conference, organised by IBC, London, September 1995
[3]   Papers presented at the DECT '96 Conference, organised by IBC, London, January 1996

14 Cordless Data and Multimedia

Frank Owen and Andrew Bud

Cordless data communications in the local-area environment are now a business reality. Since 1990 several commercial networks have been launched onto the marketplace. Applications range from mainstream office through to those niche environments where simple data transfer to mobile terminals dramatically increases productivity.

The underlying requirement for integrated communication systems has been a driving force in the commercial environment since the late 1980s. Cost, compatibility and increased economies of scale promote the integrated solution. Today's cordless data products begin to incorporate this integration, particularly those based on DECT. The next generation of cordless data standards under development, both in Europe (HIPERLAN) and in the USA (IEEE 802.11), will lead to a seamless combination of high-information-rate services on both corded and cordless mediums. This chapter provides a general tutorial on cordless data, presenting the services and application areas after discussing the benefits from a commercial viewpoint. The network architectures and specifications of existing cordless products are reviewed, with reference to the available communications media. The new standards are outlined and the chapter concludes with a description of the opportunities for cordless multimedia products, including early cordless video systems.

14.1 Benefits of Cordless Data

Cordless voice started the mobility revolution and digital cellular telephony is now an everyday experience for many mobile professionals and, increasingly, consumers. The market is therefore not only aware of the benefits of personal mobility, but has become accustomed to wireless communications, and users expect further services including data and wider application areas.

In addition to the key benefit of personal mobility, cordless data communications also offer other specific benefits:

- *Simple installation* This is particularly important if the network is of a temporary nature or if a specific location is hard to wire, or represents a hostile environment.

- *Rapid installation* Cordless LANs available today are virtually an 'out-of-the-box' solution enabling peer-to-peer networks to be set up for temporary projects and meetings.
- *Network expansion* Cordless data communications may be flexibly employed on construction sites, in warehouses and in other large buildings. They are also suitable for connecting networks in adjacent buildings, for example in campus environments.

Concerning cordless services, there is a basic migration path from voice (typically enabled by a PBX with DECT cordless handsets), towards data and the integration of these two for various applications, followed by a higher level of integration involving still pictures and video.

While the future for cordless communications is attractive, it should not be inferred that corded systems are no longer relevant. Wired Ethernet LANs and others will continue to play a major role in corporate data communications, and not merely for historical reasons – a wireless interface will initially find it hard to compete on a combined speed/price front for many applications. Instead, it is likely that cordless LANs will complement existing wired infrastructures with the additional benefits of cordless data communications outlined above.

14.2   Data Services and Attributes

All existing commercial or domestic data equipment that requires a communication link could, in principle, implement this as a cordless connection. However, data services vary in their connection requirements, and certain services are more suited to cordless transmission than others

Table 14.1 summarises briefly a number of key characteristics of various traditional and generic data services together with the communication link attributes necessary to support them and which affect the design of any cordless network connection. These requirements are typical values and illustrate the differences between services rather than providing precise figures.

Table 14.1
Requirements of different cordless data applications

Service	Information rate (bit s^{-1})	Packet size (bits)	Delay limit (s)	Integrity (BER)
Voice	8–64 k	continuous	0.01	10^{-2}
Digital facsimile	64 k	10–50 k	10–1000	10^{-2} [a]
Keyboard	1–10	10	0.5–2	10^{-4}
VDU display	10–20 k	100–1000	1–10	10^{-6}
Videotelephony	32–256 k	continuous	0.1	10^{-2} [a]
File transfer	64 k–10 M	0.1–2 M	1–1000	10^{-11}
CAD graphics	64 k–10 M	1 k–2 M	1–1000	10^{-11}

[a]Before error control. Video compression techniques continue to advance rapidly and sub-32 kb s^{-1} systems are under development.

14.2.1   Voice

Digitised voice can be considered a data service in an integrated services system, requiring a constant information transfer rate and constant, low delay. With high-bit-rate digital speech encoders, as currently used for cordless telephony systems, a large bit error rate (BER) over a voice link can be tolerated owing to the high redundancy in speech. Voice is likely to be one of the most mobile of the cordless services provided and is expected to make up a high proportion of all cordless services.

14.2.2   Facsimile

A number of standards exist today for facsimile transmission and it is likely that there will be continued growth of this service. The latest facsimile standard defined by the ITU for digital networks is known as Group IV. Facsimile has an almost constant data rate but, in common with a large number of other data services, has a greater information flow in one direction of the link than the other. Delay is not as critical as with other services and a high link error rate can be tolerated for short periods. Retransmissions providing error correction add a small amount to the delay.

14.2.3   VDU and Keyboard Terminal

The conventional computer terminal has significantly different communications requirements in each link direction. Keyboard entry produces a low data rate but requires minimum delay (tens of milliseconds with distant-end echo). VDU bit rate is very low or zero until a screen refresh is required or moving images are involved, when a large communications bandwidth is required. Surveillance systems and other site control systems have similar data characteristics to those described above.

14.2.4   Wide-Band Services

Examples of wide-band services usually requiring data rates exceeding $64\,\mathrm{kb\,s^{-1}}$ include the following.

Video Transmission

This is desirable where the visual information in a conversation plays an important part. Video transmission may be point-to-point (e.g. video telephony), point-to-multipoint (e.g. video conference or broadcast), or simply unidirectional (e.g. video surveillance).

High-Speed Computer File Transfer

In a large networked computer system there is often a requirement for large quantities of computer data or programs to be moved around the network – for instance, from a central storage device. This service is typically provided today by a wired LAN such as Ethernet, operating at up to 10 Mb s^{-1}.

CAD/CAM Graphics

In the modern factory design environment there is a significant requirement for high-speed graphical presentation of designs. Often a dedicated link or very high-speed LAN (hundreds of Mb s^{-1}) is used to minimise VDU screen refresh time.

14.3 Applications

For mainly historical reasons, cordless data services have tended to be associated with standard office environments with relatively short distances between terminal and network infrastructure. In practice, however, the early market need has been the provision of access over longer distances, for example into factories and warehouses or throughout a hospital complex – in many cases data communications are required over a relatively large site or campus.

Specific examples of existing cordless data applications include:

- E-mail, fax services and Internet access for notebook users
- Direct portable access to patient databases and medical records in hospitals
- Portable electronic point-of-sale (EPOS) terminals within retail shopping outlets
- Forklift truck roaming in warehouses with on-line access to a central database
- Portable 3270 terminal emulators for host queries and transactions
- Education projects, particularly for campus-wide network access

The integration of cordless LANs with specialised hand-held terminals such as bar-code readers is already enabling cost-effective solutions to be developed for inventory control in distribution and retail markets. Other applications under development in the network connections field include on-line traffic management in large public areas such as railway stations, airports and harbours.

Cordless data communications are clearly demanded by the new generation of palmtop computing devices, which are as portable as a cordless telephone. Similarly, dedicated cordless terminals are also being developed for particular market sectors. An example are portable PCs with reduced keysets to simplify the user interface for use in industrial environments. These may be mounted inside vehicles or used in handheld form for production-line test, control and display.

Most of these specific applications have one common factor, namely that the files being transferred are relatively small, typically around 1–4 kbyte. There seems at present to be less requirement for mobile CAD/CAM or other resource-intensive applications.

14.4 Architectures

The cordless data systems further examined in this chapter and common to the above applications and services possess a network architecture that supports either or both of the following:

- Access to a backbone local-area network (LAN), i.e. cordless clients communicating with corded servers
- Cordless clients working in a peer-to-peer mode architecture

Most organisations requiring a LAN will have already implemented a wired backbone, either Ethernet or Token Ring. However, the expanding market need to access more information, and to employ higher-performance and more cost-effective communications, dictates that the backbone grows branches to reach further into the organisation. Thus there is a need for the capability to extend networks quickly and flexibly, and perhaps only for short durations.

If a building has a structured cabling system, expansion may not be a major problem. However, if cabling cannot be extended, or even implemented if walls are thick or if the building is protected or on a short lease, then cordless access provides a fast and economic alternative. Indeed, the short time it takes to commission a cordless network often makes this the preferred solution in environments even where the requisite wiring is in place. For example, a peer-to-peer mode network can be set up in a matter of hours for small workgroups and then expanded or shrunk depending on demand. Indeed, an impromptu peer-to-peer cordless LAN can be set up in a meeting using notebook PCs and a popular software program such as LANtastic.

Depending on the application, the most cost-effective topologies are likely to be along the lines shown in Figures 14.1 and 14.2. Figure 14.1 shows LAN adapters and access points enabling cordless communications between client PCs and a backbone wired Ethernet or Token Ring network. In Figure 14.2 a peer-to-peer mode architecture is illustrated with a wired single gateway to further networks.

A cordless microcellular network based, for example, on DECT may be represented in a similar way. In this case the client PCs use the cordless access protocol to communicate with radio basestations that comprise a microcellular network. These stations connect via a hub to a backbone system, or an Ethernet segment. A microcellular system additionally allows users to roam extensively throughout a site.

The architecture of Figure 14.2 illustrates the way that existing single-link spread-spectrum and DECT products are typically deployed.

14.5   Transmission Media

There are several different cordless media that can be considered for short-range data communications, the principal ones being:

- Infrared optical 850–950 nm (line-of-sight and diffuse)
- Millimetric radio >30 GHz
- Microwave radio >1 GHz
- VHF/UHF radio <1 GHz

Each of these media has specific characteristics that makes it more suited for use in some applications and environments than others. UHF radio and the low microwave spectrum provide good area coverage at reasonable transmitter power levels and at an acceptable cost to the mass market. As radio frequencies increase, the radio component technology cost rises significantly, but additional

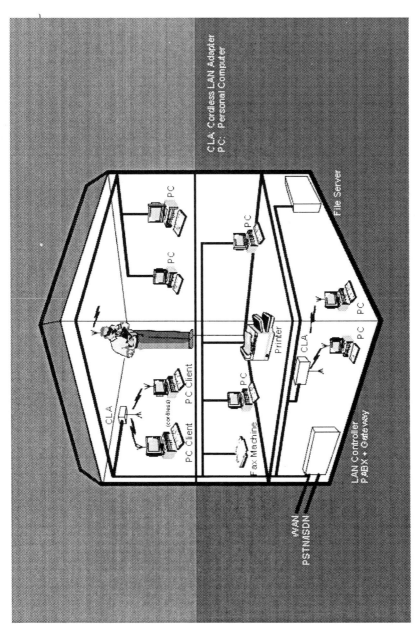

Figure 14.1   Topology of cordless client access to a wired backbone server.

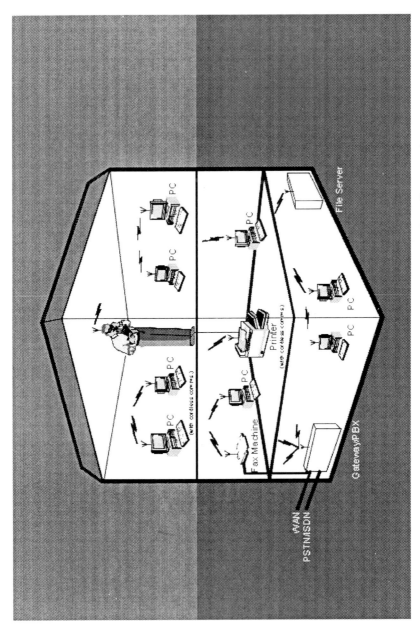

Figure 14.2 Topology of a local-area peer-to-peer mode cordless network.

system advantages appear. These include more available bandwidth and therefore faster transmission, better definition of coverage area through use of directive antennas, and a potentially higher degree of frequency re-use resulting in greater capacity. Trends in recent years have seen the frequency capability of radio technology rapidly increasing for a given cost, thus making higher-frequency systems more feasible. Diffuse infrared is also a suitable medium for cordless communications. Bandwidth is generally limited but the technology is not expensive and a great deal of remote-control equipment makes use of this simple but effective technology.

14.5.1   Infrared Systems

Infrared links have been developed as line-of-sight links between PCs, as well as between PCs and peripherals, using the same technology principles as used in television remote-control units. Infrared systems have the potential for very high-bandwidth transmission, particularly in directional line-of-sight systems (up to $10\,Mb\,s^{-1}$ versus a few hundred $kb\,s^{-1}$ for diffuse infrared). A feature of this technology is the fact that it is not subject to regulatory controls. However, the need for a line-of-sight or near-line-of-sight link is a clear limitation, and complete coverage solutions are not fully effective. Other disadvantages are the inability of infrared to penetrate walls, floors, or ceilings and the fact that systems may suffer from interference sources, such as sunlight.

A simple but effective IR data system that has been well publicised is the Olivetti active badge system. The system works by equipping all users of a building with a discrete, infrared badge that is in continuous, bi-directional communications with sensors placed in the monitored environment. These sensors are, in turn, connected to the server of the system (typically a PC), and the whole system is able to localise each badge, dynamically updating this information as the wearer moves through the environment. By accessing the central database, the system provides any client connected to the server with information of the movements of each badge wearer and allows people to be found very quickly and conveniently. As users have to do nothing apart from wear their badge, the system provides a simple but highly effective method of establishing a link between users and complex environments. A more detailed description of another early infrared system developed by Philips may be found in [1].

14.5.2   Radio Systems

Most commercially available cordless LANs are based on radio transmission media. Radio spectrum is legally regulated by governmental authorities, e.g. the FCC in North America and PTTs or other governmental bodies in Europe. These organisations have traditionally granted licences for "approved" equipment to use parts of the spectrum. These approvals are based on standards, and customers are therefore legally compelled to choose equipment that conforms to those standards. This means that the transmission and communication protocol standards for radio LANs are fundamentally different from those of wired LANs, precisely because they are legal instruments rather than market options. The way in which this regulation and standardisation process operates in different world regions thus has heavily influenced the emergence of preferred systems for cordless data as it has for voice.

Radio Technology in the USA

In the USA, spread spectrum has emerged as the dominant technology for cordless data systems. This has arisen from an FCC ruling some years ago that has allowed spread-spectrum transmissions at up to 1 W output on a minimal-regulation basis. This ruling has encouraged both innovation and competition, with many manufacturers offering different proprietary systems. Spread-spectrum technology[1] was originally developed for military applications and modulates the data signal such that it occupies an increased radio bandwidth, hence the term "spread". It is this spreading of the data over a much wider bandwidth that firstly allows multiple users and further protects it from eavesdropping and interference.

Spread-spectrum wireless LANs may employ two related techniques – direct-sequence spread spectrum (DSSS) or frequency-hopping spread spectrum (FHSS). DSSS spreads the transmission by multiplying the user data by a higher-rate pseudo-random codestream, which allows multiple transmitters to operate in the same area without interference. The receiver deciphers the data since it knows the spreading code employed by the transmitter. In practice, for DSSS systems, vendors limit the spread and use a fixed code to give transmission rates of the order of $2\,\mathrm{Mb\,s^{-1}}$. In FHSS the transmitter jumps or hops from one frequency to another at a specific rate and sequence, which appear to be random but which are known to the receiver. The fact that spread-spectrum systems may have various unregulated characteristics has prevented the establishment of a true standard for cordless LAN and other applications in North America – the technology has been implemented in different ways by different manufacturers. Nonetheless, a considerable reservoir of experience in this field has been built up in recent years in the USA.

Spread-spectrum systems are licensed in the USA to operate in the ISM (industrial, scientific and medical) frequency bands, allocated for unlicensed wireless applications, as described in Chapter 8; these bands are 902–928 MHz and 2.40–2.484 GHz. In approving a personal communication service (PCS) in the 1.8–2.2 GHz range, the FCC has ruled that users may employ some form of spread-spectrum technology – the OCDMA and Omnipoint technologies described elsewhere in this book are such examples.

Radio Technology in Europe

In Europe the 900 MHz band is not suitable for use for spread-spectrum WLANs, since these frequencies are reserved for GSM. However, CEPT have recommended use of low-power (100 mW) applications in the 2.45 GHz band and defined the technical parameters for such applications in recommendation T/R 10-01.

This recommendation specifies the use of spread-spectrum modulation and it gives total power and power density limits for both FHSS and DSSS modulation, together with a minimum aggregate bit rate of $250\,\mathrm{kb\,s^{-1}}$. It does not address details of the modulation techniques and therefore does not cover the design or operation of the equipment being tested; instead it describes a common set of measurements to be applied to the various types of equipment, those employing both FHSS and DSSS modulation. It also defines limits of spurious emissions. The T/R 10-01 requirements have been met by various US vendors, such as Lucent, thus enabling these systems to be sold in Europe with some success.

In a market ruled by open international standards, Europe has, however, created a powerful open standard for cordless LANs – DECT. The technology behind this standard is fundamentally different to that outlined above, since DECT is a radio interface standard specifically designed [2, 3] to provide integrated cordless voice and data systems. It has also been specified to allow interworking between cordless terminals of different manufacturers, so removing the inherent disadvantage of incompatibility experienced by spread-spectrum systems.

DECT-based systems provide an elegant and cost-effective way for mobile terminals and their users to work with the resources of wired LANs. Section 14.7 describes the operation of DECT cordless local-area networks in more detail.

14.6 Cordless LAN Standards

The computer industry is one in which compliance with standards can only have a beneficial effect for end-users, because this is the only way of ensuring that hardware and software from different vendors can be effectively integrated to create a single operational system. This desirability of a common standard for the computing communication industry was acknowledged in the USA by the creation of the IEEE 802 committee, which developed the Ethernet and Token Ring corded LAN standards. Following such successful earlier work, a new committee, IEEE 802.11, was established in 1990 to develop a common cordless standard.

The methodology by which standards are set in the USA and Europe, however, could hardly be more different, as has already been described in earlier chapters. In the USA several vendors usually launch products onto the market and in time a *de facto* standard may emerge. Innovative technology may come quicker to the market than in Europe, but optimal economies of scale cannot be achieved until there is a clear leader and a *de facto* standard. In Europe the standardisation and regulatory process is well ordered, but takes time. When standards like GSM and DECT are the result, then the wait is well worth while, but when it takes too long, in a flexibly regulated market, many manufacturers take a pragmatic approach and implement proprietary solutions.

In Europe it was recognised in the early 1990s that dedicated cordless LAN standards were needed. Thus ETSI in 1991 established an Ad Hoc Wireless LAN committee to study the requirement and propose a way forward. The report issued by this group recommended a three-phase approach. Firstly it acknowledged the contribution that early spread-spectrum products from North America could make, and recommended that a means to allow these onto the European market be facilitated; it was this action that led to the development of the CEPT recommendation. Secondly, it recognised that the existing DECT standard had been developed with a wide range of medium-data-rate applications in mind, and thus reconfirmed that DECT-based solutions be allowed for this segment of the market. Thirdly, it recognised the requirement for a high-performance radio LAN solution, exceeding the capabilities of DECT or of the spread-spectrum systems. ETSI RES accepted these recommendations, with the result that a new ETSI Sub Technical Committee, RES10, was established to develop the new "high-performance radio LAN", or HIPERLAN, standard.

14.6.1 The IEEE 802.11 Standard

The IEEE 802.11 committee began its initial work on a standard for cordless communications in 1990, but at the time of writing the most optimistic forecast for publication of the standard was some time in 1996. Having said this, integrated circuits to the standard are already available, with the remaining specification changes expected mainly to impact only the system software. Meanwhile the various vendors continue to track progress and modify their systems accordingly. The goal is to define the wireless communications protocols that reside at the lower levels of the OSI seven-layer model – more specifically, the definition of a set of unified medium-access-control (MAC) protocols. These protocols must work independently of the underlying physical layer, i.e. with the 802.3 Ethernet standard. At the physical layer, the IEEE has specified three options: DSSS radio, frequency-hopping radio or infrared transmission may be employed. The maximum RF transmission power is set at 1 W, but most portable devices will operate at a much lower level, typically 1–200 mW.

One of the key issues concerns collision avoidance. Use of exactly the same techniques as employed on corded LANs would not be efficient owing to the lower cordless transmission rates and the so-called "hidden terminal" problem. A centralised system can prevent collisions by taking rigid control of all nodes and only allowing one client to communicate at one moment in time. Alternatively, a distributed control scheme can be employed that enables flexible peer-to-peer operation and the kind of impromptu workgroups and mini-networks referred to earlier.

The IEEE 802.11 committee has selected the Distributed Foundation Wireless Media Access Control (DFWMAC) [4] protocol, the lowest level of which supports asynchronous communications between multiple stations, i.e. it supports the basic medium access that allows for automatic medium sharing between similar and dissimilar systems. Contention between multiple stations wishing to access the same medium is resolved through the CSMA/CA mechanism, the function of which in DFWMAC is similar to that in Ethernet. A CTS/RTS protocol can be added that exchanges short messages to reduce the loss from collision. When a central control point (access point) is used, this can operate access on a polled basis as well, which allows collision-free access and guaranteed capacity to support synchronous services.

At the same time, it is worth noting that interoperability only applies in a peer-to-peer relationship. Products that comply with IEEE 802.11 may also communicate with each other over the backbone part of the network, i.e. from one cell or workgroup to another.

14.6.2 The DECT Data Standards

The work on DECT standards specifically for data applications began in August 1993, and it yielded the world's first complete standard for wireless LANs in December 1995. The 1992 DECT base standards had already included the technical features necessary to support advanced data communications, and so this new standardisation work focused on formalising a series of air interface profiles, together with the interworking conventions to fixed networks, so as to complement the general-purpose base standard with application-specific precision. These data profiles are members of a much broader family of application-oriented "access profiles" for DECT, as described in Chapter 20.

The standards have two objectives. The first is to guarantee interoperability between portable units and fixed DECT infrastructure, even when these two parts are supplied by different vendors. This is accomplished by unambiguously defining the air interface, making some DECT options mandatory, whilst prohibiting others. The second objective is to define conventions for network interworking at the fixed part and for service interfaces at the portable part, so that a consistent, standardised end-to-end transport service is available to applications authors and network providers. In this way, the market can clearly distinguish between suppliers of portable terminal, installers or operators of infrastructure (including public operators) and providers of service together with their related terminal application software. These relationships are illustrated in Figure 14.3. The PC market is the model for this sort of disintegrated structure, which maximises investment and market growth.

The RES03/Data committee, charged with designing these profiles, took a structured, phased approach to the task. Rather than attempting to create one massive, all-purpose standard, inevitably of great complexity and long gestation, the committee chose to build a family of profiles; each successive profile used its predecessor as support, and then added new higher-layer capabilities in order to offer more sophisticated services. This approach is illustrated in Figure 14.4. The result is a series of standards that are targeted at different data services, segmenting the data market to permit optimised performance with minimum investment.

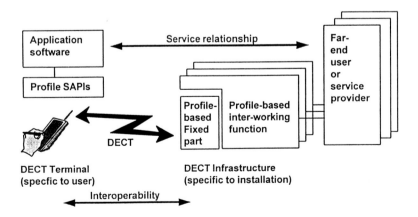

Figure 14.3 DECT data transmission as an extension of existing IT infrastructure. (*Source*: from [5].)

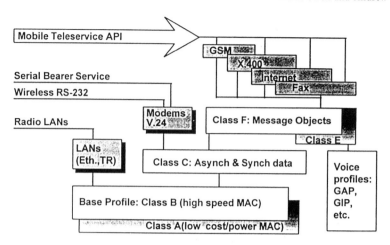

Figure 14.4 DECT data profile architecture. (*Source*: from [5].)

The standards are divided into two mobility classes, 1 and 2. Class 1 is designed for smaller, simpler applications in which there is very tight administrative integration between the users and the network administrator, such as might occur in a small or medium-sized private network. In this case, the terminal identity is sufficient to enable the network to infer all aspects of the user identity, requested service, requested communication end-point, interworking requirements and so on. This greatly simplifies the terminal, but becomes unmanageably rigid in larger networks, in which hundreds of roaming users equipped with multifunctional terminals try to connect to a variety of networks and end-points. For these users, class 2 mobility provides the full power of the DECT network layer.

The basic standard is the A/B profile (ETS 300 435). This standard specifies a packet-oriented, error-corrected data transmission protocol with interworking to Ethernet and Token Ring LANs. Type A operation is for single-slot, simple and low-speed operation (up to $24.6\,\mathrm{kb\,s}^{-1}$), whilst type B supports higher performance through the use of multiple slots, asymmetric connections and fast establishment of network-initiated links. Type B is capable of offering throughput well in excess of $500\,\mathrm{kb\,s}^{-1}$, qualifying it as a medium-speed cordless LAN; it is for this reason that ETS 300 435, published in December 1995, is justifiably described as the world's first interoperability standard for cordless LANs.

The C standards build upon the A/B standards but, by adding an HDLC-like link access protocol and a standard PAD function, they are capable of supporting serial data with high reliability. The C2 standard (ETS 300 651) then defines two interworking conventions. One is designed simply to replace a piece of RS-232 cable. It transfers data and control line information transparently, and is primarily applicable to the private domain where the user knows *a priori* what is attached to the other end of the link. The second convention is for cordless data communication equipment (DCE) and creates the concept of a DECT modem. This part of the standard strongly resembles a combination of the GSM non-transparent data service and the generalised packet radio service (GPRS), only with much higher performance: a C2 DECT modem based on the B-type lower services can offer throughputs in excess of $200\,\text{kb s}^{-1}$, faster even than an ISDN BRA V.120 link, which it also strongly resembles. This will probably be the basis for future public access data services, in both mobile and wireless local loop applications. In future, new interworking conventions will be designed – offering, for example, direct support for Winsock 2 without the use of TCP/IP.

The D standards represent an alternative approach to serial data transmission, for use in cases where isochronicity is paramount. It is comparable in function to the GSM transparent services or the ISDN V.110 protocol. At the time of writing, development is not complete, but it is anticipated that it will permit trade-offs between erasure rate and latency and permit differential protection for coded bit-streams. One of its principal applications is likely to be the secure telephone (STU) used by the US Federal Government.

The F2 standard (ETS 300 *xxx*) builds upon the C2 standard to create a service for the transport of message-oriented information such as fax, e-mail and paging. It includes a protocol, called the Multimedia Messaging Service (MMS), for interworked communications with message servers in the fixed network. The MMS is unusual because it is tightly integrated with the DECT network layer – indeed, so tightly that it has no states of its own, and is therefore ideally suited for implementation in low-cost, low-power, limited-memory portable units. Its interworking annexes are headed up by one for normal real-time Group III fax, in which the far-end fax machine is treated as the message server. Annexes for store-and-forward and other value-added services fax can use practically the same interface. Annexes for interworking to the GSM SMS service, Internet e-mail protocols and http:// for Web access are planned, using essentially the same protocol. It will thus be possible to build very versatile personal terminals capable of accessing a wide range of services using the same portable-optimised protocol core.

The E2 standard (ETS 300 *xxx*) uses the same MMS protocol as the F2 standard, but its underlying transport is completely different. Instead of using the A/B/C data profile, it uses spare capacity in the normal DECT voice-call signalling channel, to enable telephone and in-call messaging applications. The interworking conventions and service presentation are identical to the F2 profile, but the speed is up to 100 times slower. Thus it is well suited for text messaging, and permits seamless upgrade to the F2 profile if circumstances require.

14.6.3 The HIPERLAN Standard

Work began on the ETSI HIPERLAN standards in 1992 and the first standard was issued for Public Enquiry (part of the ETSI standardisation procedure) at the beginning of 1995 as a draft prETS 300 652. Some 200 comments were submitted to ETSI from the Public Enquiry phase, most of which were resolved at the RES10 Sub Technical Committee meetings in December 1995 and February 1996. At the time of writing, the standard is expected to go to vote in late spring 1996, which could result in formal adoption and publication by the end of 1996. On this timescale the first HIPERLAN products could be expected in 1997.

This HIPERLAN standard has been standardised with several key target characteristics in mind (see Table 14.2), including the aim of providing a cordless bit rate of up to $20\,\text{Mb s}^{-1}$, which will enable access to high-rate multimedia

Table 14.2

Target characteristics of the initial HIPERLAN system

Parameter	Value
Operating range	50 m indoors
User bit rate (async.)	10–20 Mb s^{-1}
Services supported	Asynchronous and time-bounded
Network topology	*Ad hoc* or hub-based
System capacity	1 Gb s^{-1} per hectare per floor
Security	Protected against eavesdropping – comparable with wired LANs
Mean power consumption	Few hundred milliwatts
Size	PCMCIA type III card: 85 × 54 × 10.5 mm^3

Source: from [6].

services. This figure comes from the goal of having access rates that are at least as good as those of wired Ethernet, which has a nominal bit rate of 10 Mb s^{-1}.

Radio spectrum for HIPERLAN systems has been identified in Europe in the 5 GHz and 17 GHz bands. The first HIPERLAN system has been specified to operate in the frequency band 5.15–5.3 GHz, allocated by CEPT on a pan-European basis. Attempts are now in course, spearheaded by Apple, to convince the FCC to allocate similar spectrum in the USA.

Table 14.3 summarises a few other key technical parameters of this standard.

Table 14.3

Technical parameters of the initial HIPERLAN standard

Parameter	Value
Maximum transmit power	1 W eirpep (equivalent isotropically radiated peak envelope power)
Low-bit-rate (LBR) header rate (signalling)	1.4706 Mb s^{-1}
LBR modulation	FSK
LBR deviation	±367.6 kHz
High-bit-rate (HBR) transmission rate	23.5294 Mb s^{-1}
HBR modulation	GMSK
HBR $B \times T$ product	0.5
HBR bit period	42.5 ns
Coding	BCH (31, 26)
Interleaving depth	16
Maximum burst length	47 blocks of 496 (31 × 16) bits
Protocol/signalling overhead	Typically 2000 HBR bits (access cycle, LBR header, training sequence)
Mobility of nodes	1.4 m s^{-1}

HIPERLAN has been specified to offer a service equivalent to an IEEE 802 series wired LAN – i.e. an anisochronous MAC service. This will make it particularly well adapted for broad-band services developed to run over existing Ethernet and Token Ring LANs. It employs non-preemptive priority multiple access (NPMA) as a channel access mechanism. This permits prioritised contention for the channel, meaning that high-priority stations can prevail over lower-priority contenders, where the priority is set by rules, taking into account the service requested and the previous history of the station. A low-priority station can become a high-priority station if it is refused access long enough, whereas a user of time-bounded services, such as speech, is almost immediately a high-priority station. The priority is signalled by special fields at the beginning of the transmitted packet. The elegance of this solution is that priority and contention resolution is built into the channel access mechanism and does not require an independent arbiter to manage it. The system is ideally suited to non-hierarchical *ad hoc* networks, rather than simply to centrally managed systems.

HIPERLAN also contains a novel networking mechanism in its MAC that permits one HIPERLAN to use another as a transit network. Thus if a user is isolated from the wide world but is able to join a HIPERLAN in which another member is also connected to another HIPERLAN linked to a fixed network, then this networking mechanism provides routing support to enable the isolated user to transmit his or her information over the two HIPERLANs out onto the fixed network. This approach, which lies at the heart of the Internet, is entirely novel in the area of wireless networking and, again, is ideal for supporting unmanaged *ad hoc* networks.

Applications for HIPERLAN systems broadly divide into two categories: asynchronous and those where there is a time constraint. File transfer, for example, is an asynchronous application, since there is no fixed time relationship between the two ends of the link; data may be delayed because of the network load, but provided the delay does not exceed certain boundaries then it does not matter. Applications such as telephony and video are time bounded – there is no need for synchronous transmission, i.e. small delays are possible, but their duration must be short enough so as to be undetectable by the user.

It is anticipated that the second HIPERLAN standard will represent an extension of the initial capability to support high-rate wireless ATM to the workstation; a substantial group of companies are working to this end, including Lucent, Ascom, IBM, Nokia, Thomson, Bosch, Intracom and DASA (Deutsche Aerospace), and wireless ATM is now formally part of the RES10 list of work items. At the time of writing these HIPERLAN variants are simply referred to as Type 1 and Type 2, although specific names are likely to replace this rather terse terminology in the future.

14.7   DECT Cordless LAN Technology

As explained earlier, DECT technology is seen as inherently offering a medium-performance wireless LAN capability. The first DECT product actually to be launched on the market in Europe was in fact not a cordless telephone but a DECT cordless data product, the Olivetti NET[3] system, in March 1993 [5].

The DECT standard is a universal air interface standard, being specifically defined to access existing and future voice and data networks: LANs, PBXs and X.25 for private networking, as well as the PSTN, ISDN and GSM for public applications. It offers a wide range of services owing to the use of wide-band radio technology and can accommodate the high traffic densities encountered in private and public applications. For data applications, DECT is a scaleable technology; it enables raw transmission channels of $n \times 32\,\text{kb s}^{-1}$, using multiple time slots, and supports both symmetric and asymmetric data transfer.

14.7.1 The DECT Architecture

The DECT radio architecture is described in detail elsewhere in this book. To summarise for present purposes, DECT has been initially assigned 20 MHz of spectrum in the 1.88–1.90 GHz band, divided into 10 carriers, each of which has a 1152 kb s^{-1} aggregate bit rate. As illustrated in Figure 14.5, TDMA technology is used to make a further time-domain division into 24 time slots, which normally provide 12 duplex channels, i.e. 12 slots for sending and 12 for receiving. A transmission channel can therefore be formed by the combination of a time slot and a carrier frequency, supporting a total of 120 duplex channels. Individual channels can be used to carry 32 kb s^{-1} ADPCM speech, or by using dynamic slot aggregation DECT can support higher data rates of typically up to 250 kb s^{-1} full-duplex, with the normal symmetric channel arrangement, or up to 400 kb s^{-1} half-duplex, using asymmetric channels. (These are conservative figures based on available product, rather than theoretical maximum values.)

As illustrated in Figure 14.5, packets transmitted during these time slots contain 424 bits comprising bits for preamble and synchronisation, signalling, cyclic redundancy check of the signalling bits, and user data. The final 8 bits are used to detect partial interference at the trailing edge of the packet caused by non-synchronised TDMA interference. Chapter 19 provides a more detailed description of the time slot structure.

Because of the time-domain separation, the same frequency can be used on different time slots. For periods when time slots are not being used to send user speech or data, the terminal is free to perform other tasks such as channel selection, interference detection and intercell handover (when necessary). By monitoring check bits in the signalling part of each burst, both ends of the link determine whether reception quality may be about to degrade and initiate an intra-system frequency–time slot handover to maintain a clear channel.

A key feature of DECT is its capability to establish and close down connections on each or many of its time slots at great speed. A 10-slot connection, with a gross throughput of 320 kb s^{-1} full-duplex, can be set up, inverted or resized in around 30 ms, comparable to the head seek-time on a hard disc. This means that DECT can offer a true packet mode service, based upon a response time that is two orders of magnitude faster than most other connection-oriented networks. Clearly the technology is best exploited when transmitting large blocks of data rather than small low-level packets, and this is one reason for the trend to interworking at higher levels described in section 14.6.2.

In typical on-site radio LAN applications, the lower-layer functionality of the DECT transmission system is hidden from the user application; this is offered a packet-mode, high-throughput bearer, accessed through industry-standard APIs. (Microsoft/IBM's NDIS and Novell's ODI are the industry-standard APIs for access to LAN adapters of any technology.) Once abstracted in this way, the customer can have specific application software written by anyone who understands these APIs, rather than being restricted to radio communications software experts. This has proven a major advantage of DECT cordless LANs compared to, for example, proprietary radio modems.

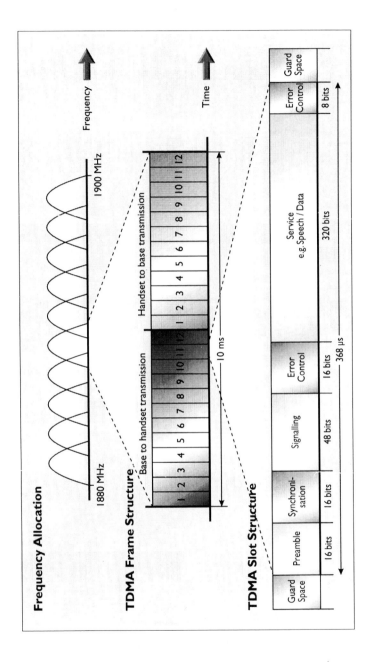

Figure 14.5   DECT timeframe and slot structure.

14.7.2 Topology

DECT has a simple topology, which is principally the same across the wide range of applications that this technology serves. There are four principal elements:

- *Mobile intelligent terminals* (voice and/or data), which communicate by radio with fixed radio basestations.
- *Radio basestations*, which form a network of microcells.
- *Intelligent switching facility* that has wired connections to the radio basestations. For voice applications this would typically be a PBX having a cordless capability. This part of the system is responsible for locating the position of the terminals and in making the connection to the relevant basestation.
- *Communications network* that is being accessed – a LAN, the PSTN, ISDN, or other appropriate network.

Figure 14.6 shows how this four-part topology is exploited in the case of Olivetti's NET^3 DECT cordless LAN system. In this case the terminal is a desktop PC or notebook that connects to a very small antenna. Cordless communications take place with basestations, which are connected using regular unshielded twisted pair cable to a PC-based hub. This provides bridging to Ethernet and Token Ring networks and runs remotely accessible management functions. Thus cordless clients do not involve additional LAN-based resources. As a system, channel set-up is established in some 30 ms and error rates of 10^{-9}, as needed for computer network applications, are provided.

One can also view the wireless part of the system as having a microcell star architecture. Each cell is centred on its radio basestation and can typically cover from 5000 to 10 000 square metres in buildings with concrete walls, whilst a coverage of more than 100 000 square metres per cell is normal in open-space installations. Cordless client terminals may be up to 1000 m distant for the basestation and typically 10–20 client terminals are connected per basestation. Typically five basestations may be connected to a low-end hub. The basestation-to-hub wiring distances are typically up to 400 m. In addition to the classic

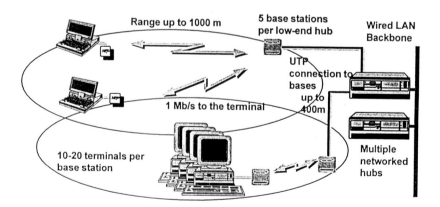

Figure 14.6 Topology of the NET^3 DECT wireless LAN. (*Source*: from [5].)

client–server model, NET^3 can also be implemented in a peer-to-peer architecture, which is often the optimal approach for workgroup situations involving relatively few users in a small area.

With over 1000 systems shipped by the start of 1996, NET^3 has demonstrated the maturity and suitability of DECT for multimedia and data communications. NET^3 is conformant to computer industry standards – ISA and PCMCIA versions are available, it supports Novell and Microsoft interfaces as well as Ethernet and Token Ring bridging, and is essentially a networking product [5].

14.7.3   Performance

Speed

Comparisons of the speed and performance of cordless LANs with conventional Ethernet or Token Ring must be done with care since the figures for raw bit rates are meaningless without reference to the different network architectures. For example, on Ethernet LANs it is widely accepted that in practice the quoted 10 Mb s^{-1} basic data rate is reduced to around 3–4 Mb s^{-1} with the use of carrier sense multiple access/collision detect (CSMA/CD) protocols, i.e. clients must wait before they are able to transmit. Similar delays are caused by token passing. Further, even this reduced rate cannot be supported simultaneously with multiple users.

Cordless products available today and described earlier are optimised for small file transfer and in general there is no significant difference in throughput performance between corded and cordless LANs for files of up to several tens of kilobytes. Many applications exist meeting this criterion.

With DECT, which establishes a connection-oriented channel between the server and any client workstation, the bit rate at the server may be as high as 11 Mb s^{-1} although the bit rate between any one cordless client and the server is significantly less. In practice, the effective bit rate at cordless terminals is slower than with corded, but it is more than adequate for many of the short file transfer applications described earlier. Experience with working installations suggests that the most meaningful figure to use is the throughput speed in terminal emulation mode, which is 25 kbyte s^{-1} or 200 kb s^{-1}, a rate that is around three times higher than the speed at which files can be transferred across an ISDN B channel (64 kb s^{-1}).

Security

Some concern has been expressed that unauthorised users could listen in to the signal of a cordless LAN and gain access to confidential information; this is certainly not the case with a DECT-based system. DECT employs a European certified encryption mechanism and the algorithms have had their security independently verified by a number of European PTT research laboratories. Since these algorithms form part of the standard, safety from eavesdroppers is maintained, even in multi-vendor systems. The same applies to the mechanisms used for terminal authentication, which prevents unauthorised access. It is generally agreed that it would be easier to read the screen of the monitor, or to pick up unencrypted emissions from the PC or wired cabling system, than to break the DECT encryption codes.

Another aspect of security is the cordless LAN's susceptibility to interference or jamming. Pan-European spectrum has been allocated exclusively to DECT; this, combined with the robustness of the DECT protocols and its design for high-capacity networks, means that DECT cordless LAN systems will suffer in this respect much less than systems operating in the unprotected ISM band. In practice, DECT cordless LANs have proved very robust.

14.8   Cordless Multimedia

DECT was conceived as a combined voice–data standard and DECT systems have already been applied to a very wide range of voice and data applications that take in mobile communications in the office and home, as well as PCS access and wireless local loop. Thus DECT may be viewed as perhaps the first multimedia wireless standard [7].

The multimedia applications of the future will require support for both asynchronous and time-bounded applications. DECT can perform very effectively in this area for two reasons:

- The technology is intrinsically time bounded.
- The appropriate bandwidth may be allocated to the service requirement.

If we consider a parallel with asynchronous transfer mode (ATM), the transmission technology that is immediately associated with multimedia services, then this point becomes clearer. Since ATM is intrinsically asynchronous, it needs the very high bandwidth in order to be able to handle time-bounded services like desktop video, and, because this bandwidth is relatively cheap, it can also be used for low-speed applications like telephony.

With radio the situation is the reverse. Connections are switched so transmission is synchronous and bandwidth is relatively expensive. It therefore makes no sense to use a lot of bandwidth for telephony simply because video is needed from time to time. What is needed is a bandwidth-on-demand solution so that transmission speeds are optimised for the different services.

This concept and functionality was incorporated from the start in the DECT cordless standard. As explained earlier, DECT has 12 duplex channels and these can be aggregated on demand. Thus duplex telephony can proceed at $32\,\mathrm{kb\,s^{-1}}$ and higher rates can be established (virtually on-the-fly) up to $12 \times 32\,\mathrm{kb\,s^{-1}}$ ($384\,\mathrm{kb\,s^{-1}}$), which is the H.320 standard [8] for desktop duplex video. The principal reason why this capability is not yet widely exploited is probably the fact that the market is not yet ready for these innovative cordless services. DECT is only just beginning to make its mark in single-service solutions (voice and data). In principal there are no technical reasons preventing further terminal combinations, and multimedia devices of this type will certainly emerge in the near to medium term.

In Japan, the PHS standard proponents have promoted strongly the capability of PHS for multimedia applications, perhaps reflecting different perceptions of the growth of multimedia markets in Japan and Europe. Given the underlying commonalities and similarities between DECT and PHS, it is to be expected that both cordless technologies will find increasing application to multimedia PC-based applications. Indeed, it is possible that an analogy with the increasingly common incorporation of modems into PCs may exist, which could drive the incorporation and acceptance of cordless data communications technology much faster than observers have to date anticipated.

14.8.1 Cordless Modems

A key enabler in accelerating the introduction of cordless networking is the introduction and growth of the PCMCIA (also now commonly known as PC card) standards. Originally conceived as an interface technology for PC memory cards, the interface has evolved into a standard capable of supporting both wired and wireless communications. In the digital cellular radio field it has been adopted as a standard for interconnection of PCs to mobile phones to enable wide-area wireless networking, already available on several European GSM networks. Indeed wireless Internet access using such technology is already a commercial service in the UK. As noted earlier, HIPERLAN also envisages products in PCMCIA format.

Cordless technology is of course intrinsically simpler and cheaper than either GSM or HIPERLAN; thus it is no surprise that PCMCIA modems for most of the cordless standards are either already available or under commercial development. The first PCMCIA modem to be introduced was based upon CT2 technology and introduced in the Netherlands to complement the Greenpoint public access telepoint service. Similarly, public cordless data access was the driver in France, with the French Bi-Bop telepoint network offering data access fairly early after service introduction, based on an Apple product, the so-called "PowerBop" variant of its PowerBook notebook computer. DECT PCMCIA products have also been developed, such as that shown in Figure 14.7, the PCMCIA version of the NET^3 product, which can be reconfigured as a basestation or terminal, or indeed as a repeater, simply by changing its software load. PHS PCMCIA modem products are also under development, with applications also envisaged running on handheld computers – personal data assistants (PDAs) – as well as notebooks and larger PCs.

14.8.2 Cordless Video

The growth of wireless data services is essentially driven not by the availability of a technology capable of transporting data, but rather by the availability of useful end-user applications – whether we consider the outdoor or indoor environment, this is so. It is for this reason that extensive work has been undertaken in the recent past aimed at supporting video applications over low-rate wireless data channels; envisaged applications range across security and surveillance, the mobile videophone and PC-based video conferencing in the wireless office.

Cordless Video With Today's Technology

Newly available, miniature, high-performance, low-cost, CCD camera devices are well suited to such applications, enabling the construction of both small and

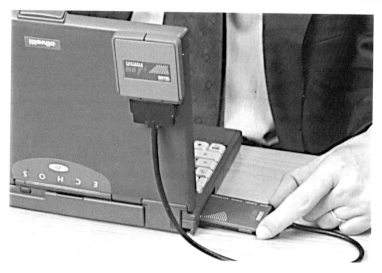

Figure 14.7   The NET³ PCMCIA DECT module. (*Source*: Olivetti.)

economic wireless camera devices using today's technology. Indeed, viable prototype systems have been constructed, transmitting the compressed video data over the 32 kb s^{-1} channel offered by the DECT standard [9, 10]. One such video link system, developed at Roke Manor Research in 1994 (Figure 14.8), allowed transfer of standard QCIF (Quarter Common Interface Format) images

Figure 14.8   Prototype DECT-based video system. (*Source*: Roke Manor Research [10].)

(176×144 pixels) at a rate of 2.5 frames per second using very low-cost, readily available, technology. The image compression algorithms exploited spatial redundancy in the image using two-dimensional discrete cosine transforms (DCT) and quantisation to obtain a lossy data reduction. In this equipment the camera was only slightly larger than a standard DECT telephone, accommodating additional digital image compression circuitry. Reflecting a philosophy of using available and cost-effective technology, the received image at the other end of the link was displayed on a standard PC, with image decompression performed in software; thus, at the monitoring end, the additional circuitry in the DECT equipment comprised simply a $32\,\mathrm{kb\,s}^{-1}$ to RS232 adapter.

Higher frame transmission rates could be achieved by also employing temporal compression (frame-to-frame differencing) – the latter is more suited to a videophone application, where there is much more frame-to-frame continuity (e.g. head and shoulders) than for security applications, in which single frame events may be significant. In contrast, it is clearly possible to transmit a much higher-quality single image, at lower frame rate, instead. Also, for security applications, several sequential high-quality images may be stored in memory and subsequently transmitted.

Cordless Video in the Future

In contrast to the above approach, research within the CEC-sponsored Mobile Audio Visual Terminal (MAVT) project under RACE focused upon the development of very high-performance, and hence, by definition, much more complex, video compression algorithms, again capable of supporting non- "head and shoulders" types of images but at higher frame rates – 25 Hz – than has been possible with today's technology. The aim has been eventual international standardisation as part of the ISO MPEG4 activity, aimed at rates of $p \times 8\,\mathrm{kb\,s}^{-1}$.

As with the Roke Manor work, standard Siemens Gigaset DECT telephones were employed in the demonstration equipment. The initial MAVT demonstrator was completed in 1994 and was a PC-based solution, employing multiple DSP engines – five TMS320C40s. During 1995 this was migrated to a portable PC device using the MVP processor – multiple DSPs on a single device – TMS320C80. Although the algorithms are very complex, it is easy to appreciate that international standardisation should result in large markets for various applications and that the size and cost of a standard video compression chip will fall to acceptable levels for the mass market.

The message of such developments is clear. Reasonable-performance cordless video, adequate for many applications, is already commercially viable. Whilst the cost and complexity of high-quality cordless video is at present prohibitive, the rate of progress in digital silicon technology is such that even this, too, will be feasible within a few years. Even in the short term, improvements in frame rate may be expected. The incorporation of inexpensive video cameras into multimedia PCs is already beginning. As cordless PABX infrastructures proliferate, simple cordless interfaces to these, complementing or substituting the wired LAN infrastructure, may be commercially developed – whether this happens will be influenced by the rate of the WPABX market growth and by the way competition and cooperation shape up between the telecommunications and IT industries.

14.9 Summary

This chapter has explored the requirements, applications, benefits and technologies available to support cordless data communications. The main technologies and standardisation activities discussed were:

- Spread-spectrum ISM band systems
- IEEE 802.11
- HIPERLAN
- DECT cordless LAN technology

Spread-spectrum systems have secured a strong early place in the market, despite the variety of proprietary standards. In the future it is to be expected that DECT-based systems will find increasing acceptance, given their potential for wide-area on-site coverage and interoperability. In the longer term, IEEE 802.11 systems, once standardised, may gain acceptance in North America and likewise HIPERLAN in Europe for "high-end" applications.

The data applications of today's existing and emerging cordless technologies, including new multimedia applications, have also been described. In the coming years we can expect to see a variety of new applications emerging, demonstrating that technologies such as DECT, PHS, etc., are suited to much more than simply voice applications.

Acknowledgements

Frank Owen wishes to thank the management of Philips Communication Systems for permission to publish this chapter.

References

[1]   "Cordless Telecommunications in Europe", WHW Tuttlebee (ed.), Springer-Verlag, 1990, Chapter 8
[2]   "Data Services in DECT", A Bud, IEE Mobile Radio and Personal Communications Conference, December 1989
[3]   "DECT – Integrated Services for Cordless Telecommunications", FC Owen and CD Pudney, IEE Mobile Radio and Personal Communications Conference, December 1989
[4]   "Universal Wireless LANs", C Links, W Diepstraten, et al., Byte, Vol. 24, No. 4, 1995
[5]   "DECT and Wireless Multimedia", A Bud, DECT '96 Conference, London, January 1996
[6]   "HIPERLAN: the High Performance Radio Local Area Network Standard", GA Halls, IEE Electronics and Communications Engineering Journal, pp. 289–96, December 1994
[7]   "Multimedia Over DECT", A Carmine, Philips Telecommunications Review, Vol. 52, No. 4, October 1995
[8]   "Narrow Band Visual Telephone Systems and Terminal Equipment", ITU-T Recommendation H.320, International Telecommunication Union, March 1993
[9]   "Wireless Office Data Communications Using CT2 and DECT", JJ Spicer, GA Halls and G Crisp, IEE Colloquium on Personal Communications, January 1993
[10]   "Videophone in Cordless Transmission Over DECT", S Parry, Electronics Weekly, 22 June 1994

Note

1 The principles of spread-spectrum technology are described in Chapter 13. Chapters 25 and 26 provide a detailed description of two commercial spread-spectrum systems proposed for unlicensed PCS applications.

15 Handset Architectures and Implementation

Yasuaki Mori and Javier Magaña

Most cordless telephones sold today are still based on analogue technology (e.g. CT0, CT1/1+). However, digital technology (e.g. CT2, DECT, PHS) has been introduced onto the market and is rapidly becoming increasingly cost competitive, with over 3 million units sold by 1996 and with huge projected markets. The primary difference between analogue and digital cordless telephones lies in the technology used for conveying the speech signal over the radio medium. The former uses analogue frequency modulation (FM) for conveying the speech signal, with transmission and reception typically occurring over two separate frequencies between the portable phone and its base. Privacy has always been a concern with analogue phones since the use of analogue FM means that conversations may be mistakenly picked up by neighbouring systems or even monitored by using standard tunable FM scanners. Another drawback of analogue systems is their susceptibility to interference, thereby reducing the link quality.

Digital cordless telephones, in contrast, use digital signal processing (DSP) technology for conveying speech in a digital format, with transmission and reception usually occurring over the same frequency channel between the portable phone and its base. Compared with analogue systems, digital systems can provide higher capacity via more efficient use of the available frequency spectrum, and because of their digital nature provide greater security against eavesdropping and interference. Offsetting the advantages of digital, however, has historically been the higher relative cost compared to analogue systems. The early 1990s, however, saw major semiconductor and radio component suppliers enter the wireless market with increasingly integrated off-the-shelf solutions for the various new standards, signalling the start of the "consumerisation" of digital cordless telephony.

This chapter outlines the various equipment architecture options, looks at the fundamental differences in electronic implementation between the various standards[1] and briefly examines future trends. It focuses primarily on the portable handset (or "cordless portable part", CPP), which is regarded as the main challenge for designers; also there is considerable commonality of circuitry between the CPP and the basestation (or "cordless fixed part", CFP). It should be noted that there is no obvious "right" architectural solution for a cordless telephone, and this is reflected in the fact that different manufacturers have implemented different approaches for existing products and no doubt will continue to do so in the future. This chapter is not intended to be an exhaustive guide on cordless telephone design but rather an introduction raising some of the important design issues.

15.1 System Architecture

At the simplest level, both analogue and digital phones have the same fundamental architecture (Figure 15.1): the man–machine interface, the audio processor, the controller, the radio receiver and transmitter, with associated frequency synthesiser, and the power source.

The man–machine interface and power sources are relatively standard across product offerings. Rechargeable batteries are the most common source used, though some offer the option of primary cells. The man–machine interface is essentially the numeric keypad, special function buttons and in many cases includes a liquid-crystal display (LCD).

Frequency bands are different from system to system and each band presents its own implementation challenges; nevertheless, there are several basic radio transceiver architectures available for implementation in the different standards, described later in the chapter.

The audio processing and protocol processor blocks are, however, very different between the various standards. These differences are especially visible between analogue and digital systems, hence some of the major cost differential (see Figures 15.2 and 15.3). Some of these issues are discussed in sections 15.6 and 15.7.

15.2 Baseband Design

Before delving further into discussion of the radio design, we briefly preview the elements that comprise the baseband side of a cordless telephone.

An example architecture for a cordless portable part (CPP) can be represented by the block schematic diagram shown in Figure 15.2. This block diagram depicts a very general embodiment of a CT2 two-stage superheterodyne-based design utilising a dual-channel synthesiser with I/Q (in-phase/quadrature) modulation and with the receiver and transmitter sections sharing a common aerial. In contrast, Figure 15.3 represents an example of an 46/49 MHz, US CT0 analogue-based design.

15.2.1 Baseband Controller

The audio and protocol processing functions in an analogue cordless, CT0 and CT1/1+, telephone are relatively straightforward and are handled by standard 4- and 8-bit microcontrollers. This simplicity is principally due to the fact that each voice link is handled by dedicated up and down links between the CPP and CFP, and apart from frequency and channel selection, most of the work is done by the radio block. There is also little audio processing apart from noise filtering and gain control, performed in analogue circuitry.

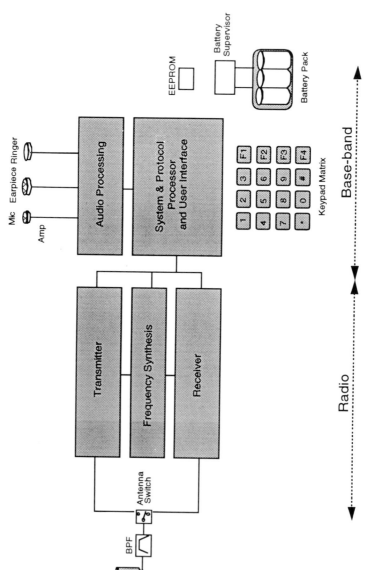

Figure 15.1 The basic cordless telephone block diagram.

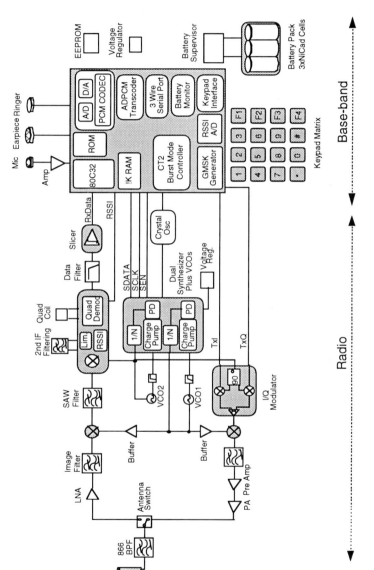

Figure 15.2 Digital cordless telephone block diagram (CT2).

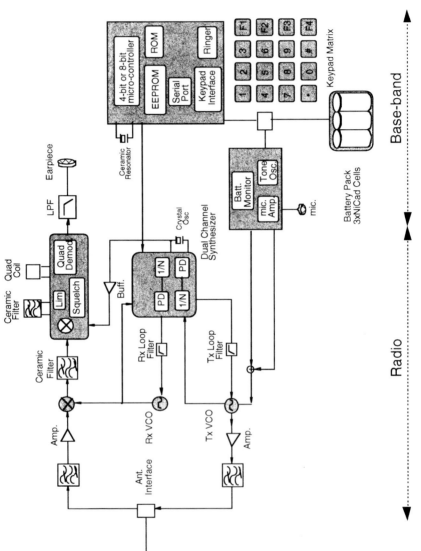

Figure 15.3 Analogue cordless telephone block diagram (CT0/US).

On the other hand, the baseband control functions embedded in digital cordless systems such as CT2, DECT and PHS are complex. Voice is digitally encoded and decoded and, unlike analogue cordless systems, digital systems use time-division duplexing (TDD) to put the up and down links onto a single frequency channel. Most systems go further, using time-division multiple access (TDMA) or spread spectrum to fit several voice channels onto one frequency as well as adding encryption capabilities. Fortunately for the digital cordless phone designer of today, several silicon manufacturers have introduced highly integrated chip offerings that already embody most of these complex functions. A more detailed discussion of digital cordless baseband controllers can be found in section 15.7.

15.2.2   System Software

The first-generation digital cordless phones did not have access to circuits specifically dedicated to CT2, DECT or PHS and had to make do with standard microcontrollers and general-purpose digital signal processors. This meant a relatively large software overhead just to have the devices behave in the desired manner before addressing the issues of user interface and features.

With today's dedicated circuits, much of the voice coding and protocol procedures are already embedded in hardware; some can even function without a microcontroller and ROM after initial set-up. With much of the lower-layer functionality already embedded in silicon, the designer can spend more time on the man–machine interface. With time-to-market pressures being what they are in the consumer market, many manufacturers are looking to consultants as well as silicon vendors to provide some level of software support to help expedite product development – such issues are considered further in section 15.7.5.

15.2.3   Program Memory

The size of program memory depends on the standard and the complexity of the man–machine interface; in digital systems this can easily be 64–128 kbyte of code. Some silicon vendors offer integrated chips with on-board ROM to enable manufacturers to reduce chip count and manufacturing cost further. However, it usually takes several preproduction runs before software code becomes stable enough to economically justify fixing the code onto the on-chip ROM. Manufacturers therefore use external EPROMs or one time-programmable (OTP) EPROMs to satisfy their initial production runs.

There have been discussions regarding integrating erasable ROM technology (FLASH or EPROM) onto the integrated controller as is done with microcontrollers. This would enable manufacturers just to replace the integrated chip with FLASH or EPROM by one with Mask-ROM when the code is stable. One of the problems with this approach, however, is the cost of such devices. It is not uncommon to find the EPROM version of popular microcontrollers priced several times higher than the one with Mask-ROM. The reason behind this is that FLASH or EPROM fabrication technology is more expensive than random logic technology.

There are cases where both on-chip and external ROM are used for applications requiring more sophisticated and complex user interfaces. One example is support of an LCD where, in order to reduce cost, manufacturers incorporate the character table directly in ROM, eliminating the additional cost of an external LCD controller.

15.2.4 Non-Volatile Memory

Common in almost all designs that require non-volatile storage of registration, system identification codes or electronic serial numbers is an EEPROM. For residential use, the EEPROM can be used for storing "one-button" dialling and support for telephone number memory. Programming the EEPROM is usually via a simple three-wire serial interface.

15.2.5 Microphone Interface

Almost all cordless handsets use an electret-type microphone owing to its small size, low cost and excellent sound quality. The frequency response of electret microphones is usually flat over the audio range of 20 Hz to 20 kHz; however, once the element is mounted in the plastic housing of a typical handset, the frequency response can be quite different. In addition, the volume level, or gain, of the microphone can vary significantly between various handset designs because of variations in the distance between the microphone element and the user's mouth. This is particularly true of small pocket handsets, which are targeted towards truly portable applications. For proper design of the external microphone pre-amp circuit, the characteristics of all of these elements must be known. Unfortunately, the acoustic frequency response and gain of the microphone must be measured in the actual telephone handset housing to be meaningful. This is not an easy measurement and requires expensive and complex equipment. It is generally done by independent laboratories that specialise in this type of acoustic testing.

15.2.6 Earpiece

There are two types of earpieces that are used in phones today – the dynamic type and piezo type. Both are small enough to satisfy the requirements of most pocket-sized phones, and the decision on which one to use is usually based on the circuitry needed to drive them. The dynamic-type earpiece presents a low impedance (around 120 ohm), while the piezo-type earpiece presents a higher impedance and therefore requires a larger voltage signal to drive it. In some cases the piezo device is preferred since it can also be used as the ringer element. Adjustment of the earpiece volume is usually programmable via software.

15.2.7 Alerting Element

Almost all pocket-sized telephones today use a piezo element as the ringer to alert the user of incoming calls. These elements provide sound levels sufficient to meet the requirements of pocket-sized telephones, but because they present a high impedance they require a large drive signal. User adjustment of the ringer level is programmable via software.

15.2.8 Keypad

The size of the keypad matrix used in a cordless telephone application varies, but in most cases a 4×5 keypad matrix allowing up to 20 keys is provided. The keypad contact area on the printed circuit board (PCB) is usually implemented using standard carbon ink silk screening, allowing for a low-cost implementation.

15.2.9 Supply Regulation

The immunity of the oscillators in the cordless telephone to noise on the power supply lines and to other noise sources is important, particularly with regards to VCO phase noise. Thus most designs have a voltage regulator dedicated to the VCO. It is not uncommon to find at least two voltage regulators in a design, one for the baseband circuitry and one dedicated to the VCO. By contrast, the power amplifier is frequently driven directly from the battery, maximising power efficiency.

15.2.10 Battery Monitoring

Provision is made in most handset designs for monitoring the state of the battery voltage during operation. Such circuitry is primarily used to enable the remaining battery life to be approximated so that the user can be informed, usually in the form of an audible beep or lighting of a LED, that the battery charge is "low" and that it needs to be recharged, thus freeing the user from having to keep track of talk time.

15.2.11 Low-Battery Supervision

Low-battery supervisory devices are used to monitor the battery voltage continuously and to force a shutdown before the battery voltage goes below the minimum operating voltage of the baseband processor. The device used is normally powered "on" all the time even during shutdown to guarantee that the phone never accidentally transmits an illegal signal when the battery is too low to guarantee proper operation. Thus, whereas battery monitoring is normally something that is handled via software, low-battery supervision is normally a hardware function, which places the baseband microcontroller in reset mode when a low battery is detected and is only exited when the battery voltage goes above a set threshold.

15.2.12 Shutdown Timer

In order to reduce power consumption of the phone while waiting for an incoming call, it is not uncommon to find a simple one-second timer that is used for turning off the main oscillator during the idle time between channel scans. This reduces power consumption of the phone while waiting for an incoming call and is therefore useful for extending battery life. The shutdown timer is something that is easy to incorporate on-chip and is something that is becoming increasingly available.

15.2.13   Batteries

The largest single component in a portable cordless telephone is its power source. The size of the battery is determined by its voltage and capacity. An important consideration for a CPP is its operational duration before battery recharging is required. In general the average power consumption and the chosen battery capacity determine the talk and standby times. It is for this reason that component manufacturers looking to extend talk time per charge as well as battery life incorporate battery-management functions such as power-down and slow-down modes. A talk time of more than 24 hours and a standby time of 100 hours can be achieved even with standard NiCd batteries in today's DECT telephones [1].

Battery cells are categorised as either primary (i.e. disposable) or secondary (i.e. rechargeable). Although rechargeable batteries are the most common power source used, some designs have offered primary cell options. Because of the extended usage of cordless telephones, the inconvenience as well as the replacement cost of primary cells is something the consumer finds unacceptable; therefore, a majority of cordless telephones on the market today are powered by secondary cells, of which the nickel–cadmium (NiCd) battery is the most common. NiCd batteries, unfortunately, suffer from memory effect that reduces their lifespan. Nickel metal hydride (NiMH) and lithium ion (Li^+) are newer technologies that overcome this problem and are now starting to appear in cellular phones and portable PC products and have made their debut in digital cordless phones with the advent of PHS products using them.

In general, most portable products – including cordless telephones – are powered by battery packs that involve more than one cell connected in series. The nominal voltage of a cell is governed by its electrochemistry – this is why alkaline batteries produce 1.5 V per cell whereas NiCd batteries produce 1.2 V per cell. Since the battery contributes to most of the system weight, some handset designs use just one cell, in conjunction with a DC/DC voltage converter to provide the voltage necessary for operation.

15.2.14   Mechanical Considerations

The importance of the mechanical and acoustic design must be recognised. Like any portable product, cordless telephones are subjected to a large amount of physical abuse, which they will be expected to withstand both cosmetically and functionally over their lifetime of over five years. The ability to withstand drop tests from heights of around 2 m onto a concrete floor poses real problems for the designer when attempting to produce a small, lightweight, attractive product. In recent years, advances in material technology have helped considerably in this task, with polycarbonate becoming more popular than some of the older plastics such as ABS.

15.2.15   Acoustic Considerations

It is dangerously easy to overlook the interaction of the mechanical and acoustic designs in the belief that the electronic circuitry can always be tailored to the specification requirements. Acoustic transducers, like many other components, require impedance matching, not just electrical but to their mounting cavities and the outside air path, putting clear demands and constraints on the mechanical design and styling. Delay is the other important parameter, being expressed as a loop time from the mouth reference point to the microphone, over the radio link to the line interface and back again to the receiver earpiece and the ear reference point. The distinction between the reference points and the transducers becomes more significant as designers work towards smaller pocketphones, since a typical maximum allowable delay is 5 ms and the speed of sound is approximately $30 \, \text{cm} \, \text{ms}^{-1}$, the nodal distance between the mouth and ear being outside the designer's freedoms!

15.3   Radio Receiver Architecture and Design

In this section, the different basic options available for the implementation of the receiver are discussed. Keep in mind that first-generation analogue cordless systems need two frequencies (up and down links) to establish a voice link while digital systems require only one.

15.3.1   Superheterodyne Receiver

The classic implementation for a radio receiver is the superheterodyne architecture illustrated in Figure 15.4. The principle of operation of the superheterodyne concept is that the local oscillator (LO) frequency is chosen to mix with the desired RF signal, thereby translating the wanted signal to the intermediate frequency (IF). Tuning is accomplished by varying the LO to mix

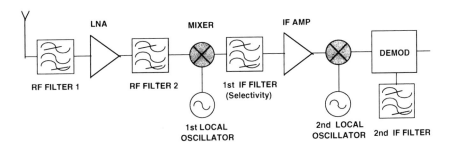

Figure 15.4   Superheterodyne receiver.

different input channel frequencies into the IF passband. The mixing process generates both the sum and difference frequency products. Thus, there is an input frequency other than the desired, which can also mix to produce an output at the IF. This is known as the "image" and is one of the classic problems with the superheterodyne architecture.

The image response can be reduced by the use of RF filter 2 prior to the mixer input. The required selectivity for this filter and the choice of frequency for the IF represent a very important trade-off. The RF filter circuitry is simplified by the fact that the fractional bandwidth for any of the bands for narrow-band cordless telephony is very small; this means that the RF filter 1 is chosen on the bandwidth requirements.

It is, in principle, possible for the IF frequency of a superheterodyne receiver to be higher than the RF tuned frequency. The choice of a high IF increases the separation of the image from the wanted frequency, thereby simplifying the task of rejecting it in the RF filter. On the other hand, it is usually more difficult to obtain the necessary channel selectivity, gain and demodulation processing within an acceptable power budget if the IF frequency becomes very high. Given the high operating frequencies applicable to today's digital cordless telephones, this would generally appear inappropriate.

The choice of frequency for the IF is also influenced by the technology available in relation to the modulation and signal bandwidth of the applicable transmission standard. For example, the DECT standard involves a very wide-band spectrum (>1 MHz) for which SAW filters are now readily available and are typically used. This would permit operation with a much higher IF than would be possible using a conventional crystal filter. Moreover, the wider bandwidth for DECT implies lower specified sensitivity (i.e. –83 dBm as opposed to –98 dBm for CT2 for the specified bit error rate) and thus a requirement for less gain, again resulting in an easing in the IF circuitry requirements. The bandwidth for CT2 on the other hand is ~80 kHz, for which SAWs might be impracticable from a cost viewpoint but, as will be explained later, have been implemented in first-generation CT2-based products. However, the operating RF frequency for CT2 is about half that for DECT, so a lower IF could be countenanced.

One way of overcoming the limitations of the superheterodyne is to introduce a second frequency conversion as depicted in Figure 15.4. In this case gain is divided between the first and second IFs, easing the problems of stability. The selectivity of the first IF filter needs to be good if the first IF amplifier is to have a significant amount of gain, otherwise there exists a risk of high-level signals, falling within the band of the first IF but outside the second IF, causing limiting prior to the second-stage filtering. If there is need to use only restricted selectivity in the first IF, this must be at least good enough to provide adequate rejection of the image frequency of the second conversion. In this case, the gain in the first IF must be kept to the minimum necessary to ensure acceptable overall noise performance.

15.3.2 Filter Technology

Very significant advances have been made in IF filter technology over recent years. The old concept of the IF filtering requirements being provided by inter-stage coupled tuned circuits disappeared in the 1960s with the arrival of filter modules, but these were just a miniature prepacked realisation of the same technology. Crystal filters are, of course, well suited to the narrow-band IF function but their cost is a major obstacle to their use in a cordless telephone application, and they are really confined to the high-cost professional communications market. This is where ceramic filter technology comes into its own.

Reduction of labour costs and component count are becoming mandatory requirements of cordless telephones based on digital technology. While higher integrated ICs offer one solution, components that need no alignment are also attractive. The availability of low-cost ceramic filters makes it possible to design a minimum-alignment radio front-end. From the first offerings of these piezoelectric devices, which were hungrily snapped up for home entertainment transistor radios, a whole technology has grown up and at present theirs is the dominant position in cheap IF filtering applications. In quantity, most ceramic filters cost less than US $1. Another reason for their popularity is that these easy-to-install three-terminal filters deliver good selectivity while lowering parts count, eliminating tuning and providing a convenient inter-stage matching impedance. Indeed, the available performance possible from a ceramic filter was a prime input considered by those committees charged with the task of producing type approval specifications such as BS 6833.

One drawback associated with ceramic filters, however, is bandwidth. High-frequency (i.e. >10.7 MHz) ceramic filters with bandwidths greater than 10 kHz have proved difficult to produce and are therefore not currently available. This led early CT2 telephone designers to use SAW filters instead to establish the first IF passband – an expensive alternative; in most CT2 designs the SAW filter is the most expensive component on the RF front-end. Of course, one way to overcome this drawback is to move the total burden of selectivity to the second-stage IF filter and use a ceramic filter with a wider bandwidth to establish the first IF passband. Unfortunately, particularly when using discriminator-based detection of FM signals, the noise power at the output of the demodulator increases with a wider first IF bandwidth and results in a loss in signal-to-noise ratio (SNR), which in turn degrades the receiver's adjacent channel selectivity in strong signal areas.

The reader will observe in Figure 15.4 that a second RF filter is shown interposed between the LNA and the first receive mixer. The purpose of this filter is to prevent noise generated in the LNA at the image frequency from being mixed into the IF bandwidth and desensitising the receiver. This second filter is not required, nor indeed is any filtering for the sake of image rejection, in a direct conversion (zero-IF) architecture, which we consider next.

15.3.3   Direct Conversion Receivers

A direct conversion receiver as depicted in Figure 15.5 operates on a similar principle to the superheterodyne except that now the antenna signal is directly mixed with an orthogonal LO in-band (I) and quadrature (Q) pair in order to generate a complex baseband signal at zero frequency – hence the term "zero IF" – which is then low-pass filtered and digitised with a high-precision analogue-to-digital converter (ADC); further filtering and demodulation may then be done in a digital signal processor.

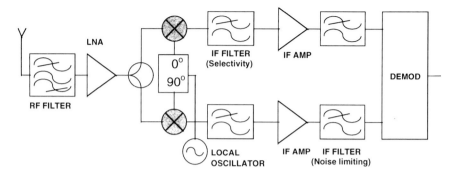

Figure 15.5 Direct conversion (zero-IF) receiver architecture.

The effect of mixing down to zero frequency is effectively to fold the spectrum about zero frequency, which results in the signal occupying only one-half the bandwidth. The process of mixing the received signal with the I and Q components of the LO may be understood simplistically as a way of resolving the ambiguity in the instantaneous frequency of the received signal: if the quadrature component leads the in-phase signal, the frequency is positive; if it lags, the frequency is negative. A more rigorous explanation is to say that the in-phase and quadrature components represent the real and imaginary parts of the complex envelope of the signal [2]. Another helpful way of looking at the problem is to view the two IF components as providing the X and Y coordinates of the tip of the vector in the signal phasor diagram.

The zero-IF architecture possesses several advantages over the normal superheterodyne approach. Firstly, because the IF is at zero frequency, the image response frequency is coincident with the wanted signal frequency. This results in the selectivity requirement for the RF filter being greatly eased. Secondly, the choice of zero frequency means that the bandwidth for the IF paths is only half the wanted signal bandwidth. Thirdly, the channel selectivity filtering can be performed using simply a pair of low-bandwidth low-pass filters.

Although the superheterodyne and the zero-IF approaches have been presented as separate, it is possible to combine them as shown in Figure 15.6

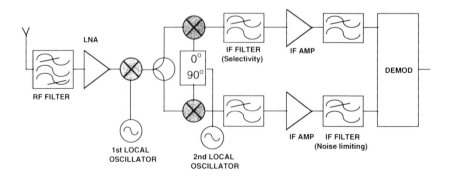

Figure 15.6 Double conversion with zero-IF receiver architecture.

in a double conversion receiver where the first frequency conversion is to a high-frequency IF and the second to zero IF. This approach can be attractive under some circumstances.

The potential benefits of using digital signal processing (DSP) in radio receiver design were recognised some years ago [3], but it is only with products such as digital cordless and cellular telephones that such concepts have been successfully commercialised for mass-market consumer products. It is the lure of migrating cost and complexity from the analogue/RF part to the digital signal processing section that has been a prime driving force for some manufacturers to develop zero-IF or direct conversion designs.

15.3.4 Demodulation

Until now the demodulator has been shown as an undefined functional block for both architectures. The modulation formats employed or proposed for current and emerging cordless systems are constant-envelope schemes, namely FM, FSK and GMSK. Given this fact it might be supposed that the IF chain of a receiver could be allowed to hard limit since there is no amplitude information in the transmission which would be lost. For non-coherent demodulation this is indeed the case and the superheterodyne architecture of Figure 15.4 may be used with a limiting IF amplifier; no automatic gain control (AGC) is needed. Non-coherent demodulation is in fact commonly used in both CT2 and DECT systems in the form of a quadrature-type discriminator, the popularity of this method being that it helps minimise the cost of the end-product by providing more than adequate performance using very simple and relatively inexpensive circuits. A common problem, however, in almost all applications that use this type of demodulation is DC offset at the demodulator output due to frequency pulling and/or VCO drift. This offset directly leads to a loss of SNR and as a result has prompted RF as well as baseband manufacturers to integrate the slicer or DC recovery functions on-chip where component tolerances can be better controlled and DC compensation techniques can be employed for maintaining a consistent level of performance. Products now on the market that have integrated this function on-chip typically provide a bypass capability, allowing the designer the freedom to use external recovery circuits if required.

In a coherent demodulator all information in the phase and amplitude of the signal plus noise is exploited; this means that the amplitude modulation must be preserved. Thus, in a superheterodyne receiver designed for coherent demodulation, it would be necessary to apply AGC to the IF. To some extent this argument, at present, is somewhat academic, in that the sensitivity improvement offered by coherent reception is at most a few dB which, because of the near fourth-order propagation law over the radio path, translates into a very small improvement in operating range. Such a small improvement does not justify the increased power consumption currently needed to implement a coherent receiver. As technology develops it is possible however that this situation may change and that coherent demodulation may begin to be utilised particularly in high-performance applications.

Coherent detection of FM digital signals invariably requires I/Q demodulation, so a superheterodyne implementation of a coherent receiver would look like the double conversion receiver with a final conversion to zero IF described earlier. It is a moot point whether such an architecture is a double conversion receiver or a single conversion receiver followed by a demodulator that happens to contain a frequency conversion. Generally, if most of the gain and all of the AGC circuitry and selectivity resided in the first IF, then the latter view would apply. On the other hand, if a significant proportion of the gain, selectivity and AGC circuitry occupied the second IF, then the receiver would be viewed as a double conversion circuit.

For the twin IF chains of a direct conversion receiver, AGC is always required. This need arises because, although, when combined, the IFs compose the complex envelope of a constant-envelope signal, individually the amplitudes of the I and Q IF components vary between zero and the envelope peak at rates much lower than the highest signal bandwidth frequency. The additional complexity of this AGC is a minus point for zero IF with non-coherent demodulation when it is compared against a superheterodyne, which requires no AGC.

An additional requirement of the newer systems is the requirement for a received signal strength indication (RSSI) signal to permit the baseband controller to measure the signal or interference level on any given channel, to perform dynamic channel allocation. In an architecture embodying AGC, an RSSI signal is essentially implicit in the AGC signal, although the linearity requirements of AGC are in fact usually less stringent than those for RSSI. Compensation for such nonlinearity may under some circumstances be undertaken in software however, since the RSSI signal is used by the controller within the transceiver to choose a quiet channel to set up a new call. In a superheterodyne receiver, where AGC is not a requirement of the fundamental architecture, separate RSSI circuitry must be provided. Such circuitry is indeed incorporated in some superheterodyne receiver chips now on the market.

The component industry has for many years provided a wide range of catalogue integrated circuits for the superheterodyne architecture, and it is hard to justify a costly and lengthy full custom design for these functions. However, the zero-IF architecture is slightly different. The first commercially successful application of a zero-IF design in consumer products was that in the STC VHF wide-area radio pager launched in 1979. This was a fully custom-designed integrated circuit using a high-frequency bipolar process and triggered immediate interest in the architecture. From this start the semiconductor industry now produces catalogue parts for zero-IF receivers. However, most of these products have been targeted at the pager market and therefore have only supported data rates below $1200\,\mathrm{b\,s^{-1}}$. Because a zero-IF architecture requires a quadrature mixer, which uses two RF mixers and a broad-band phase shifter, the high-performance circuits required – not to mention their power consumption at high frequencies – have restricted their use to low-frequency applications. However, in light of recent zero-IF DECT product announcements, this situation may change.

15.4   Radio Transmitter Architecture and Design

The task of generating an RF signal is generally much simpler than that of receiving one. A transmitter fundamentally consists of three main components –

a local oscillator, a modulator and a power amplifier (PA). These components may sometimes be combined into common circuits, for example a frequency synthesiser with an in-built modulator.

15.4.1 Local Oscillator

The problem of generating a carrier at a high frequency is largely one of frequency control, since devices offering useful gain at all frequencies currently considered for cordless communications are readily available. The main approaches for providing an accurately defined output frequency are direct generation from a crystal reference or frequency synthesis, conventionally employing a phase-locked loop (PLL).

The crystal oscillator approach has some fundamental limitations. Firstly, it is in essence a single, fixed, frequency technique and can support a maximum output frequency of not higher than around 300 MHz, limited by the availability of suitable crystals. The fact that the frequency is not easily changed is particularly unfortunate for superheterodyne receivers since a separate oscillator will generally be required for receive and transmit, except for FDD systems where an IF could be chosen at the difference frequency between receive and transmit. The upper limit on frequency is traditionally overcome by the use of harmonic frequency multipliers. These circuits produce all the harmonics up to and beyond the wanted frequency, with the wanted frequency being selected by a band-pass filter. A fairly exacting specification has to be placed upon the filter in order to restrict the levels of the unwanted components.

Whilst acceptable for some of the early analogue cordless telephones, for second-generation digital systems single-frequency operation would represent an unacceptable restriction. Frequency synthesis is therefore employed in these systems. Figure 15.7 shows a typical phase-locked synthesiser loop. In most designs, the frequency synthesiser generates both the main LOs for all channels supported. One LO frequency for each channel is required in a single conversion transceiver, while a second offset frequency is usually required in a dual conversion receiver. The transmit frequency can be obtained by mixing an offset frequency with the main LO or by having the LO operate at the transmit frequency. By far the most common radio architecture used in today's cordless telephones is based on a dual conversion receiver with the transmit frequency obtained by mixing with an offset frequency as shown in Figure 15.2.

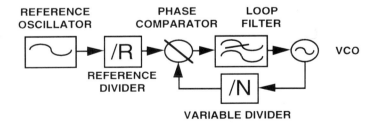

Figure 15.7 Typical phase-locked synthesiser loop.

The output from the voltage-controlled oscillator (VCO) running at the final frequency is divided by a counter circuit to a frequency suitable for comparison with a reference source. These two signals are fed to a phase comparator, which produces an output proportional to their phase difference. This output is passed through a low-pass loop filter to provide a DC control signal for the VCO. This feedback system operates to minimise the loop phase error, thereby phase locking the output VCO frequency at the desired multiple of the comparison frequency. The comparison frequency is often selected to be equal to the channel spacing, with the effect that adjacent values of N, the division ratio, specify adjacent channels at the output frequency. By switching the division ratio, the output frequency can thus be made to switch between channels. Because the comparison frequency is low, it is possible to choose the reference oscillator frequency as appropriate to provide good frequency stability.

The requirements of CT2, for free channel selection, and of DECT, for frequency switching between time slots at a basestation, impose a requirement for rapid switching between channels, more so for DECT than for CT2. The switching time of a PLL synthesiser is inversely proportional to the bandwidth of the loop, and thus one might expect a wide loop bandwidth to be employed. A frequency synthesiser will however generate not only the desired frequency, but also unwanted spuriae, notably reference-frequency sidebands. These arise from modulation of the VCO by any reference-frequency components from the phase comparator that remain after the loop filter. A wide loop bandwidth results in poor attenuation of these components and hence a higher level of unwanted spuriae. Thus the issue of synthesiser design is complicated for the newer systems, and certain refinements and subtleties must be employed to achieve the requirements of rapid switching and spurious-free signal generation. Further discussion of the basics of frequency synthesis may be found in [4, 5].

Until the late-1980s, the maximum input frequency of the variable divider limited the output frequency of this type of synthesiser to a few hundred megahertz; however, advances in prescalar technology have today made final frequency synthesis practicable beyond 10 GHz (see e.g. [6]).

15.4.2   Modulation

As mentioned earlier, CT2 and DECT both employ constant-envelope frequency modulation schemes. If only non-coherent demodulation is to be supported, then the modulation can be achieved by directly modulating the synthesiser VCO as depicted in Figure 15.8. This figure is the same as shown in Figure 15.2 except that, instead of using I/Q modulation, direct modulation of the VCO is shown.

15.4.3   Direct Modulation

The modulation applied at the VCO input for changing the output frequency proportionally to the filtered NRZ (non return to zero) transmit data is sensed by the phase detector, which in turn produces a voltage equal to the modulation (albeit divided by N), but opposite in phase. This error voltage must be filtered by the low-pass filter to keep the modulation from being cancelled. In any case, because the transfer function from modulation input to VCO output represents a

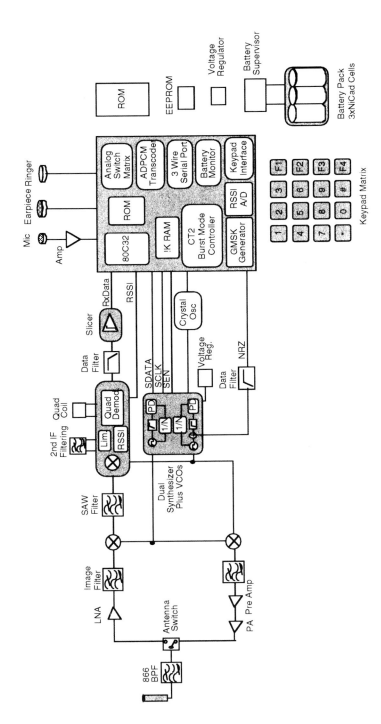

Figure 15.8 Direct modulation of synthesiser VCO.

high-pass characteristic, the lower-frequency components of the modulation will be attenuated by the operation of the phase-locked loop but, provided the loop bandwidth is not too high, this will represent only a small proportion of the signal energy. This effect can be ameliorated by phase or frequency modulating the synthesiser reference by an amount scaled by the division ratio, a technique known as two-point modulation [7].

15.4.4   I/Q Modulation

With increasing demands for personal communications service, bandwidth-efficient modulation techniques are being increasingly used. With such techniques the phase path of the modulation signal is restricted to 90° increments, with the result that the modulation will produce a narrower power spectrum, which in turn provides both better bandwidth efficiency and reduced adjacent channel interference. This is difficult to achieve in a cost-effective manner for the direct modulation approach due to component variations in the external NRZ filter. A more common approach is to use I/Q vector modulation, either at the final frequency or at a lower frequency that is then up-converted to the final RF as shown in Figure 15.2.

The case where the modulation phase path is restricted to 90° phase increments is referred to as minimum shift keying (MSK) modulation. In order to produce a narrower power spectrum than that possible with MSK, Gaussian-type filtering is applied and the modulation is then referred to as GMSK. Gaussian filtering is used because of its low overshoot impulse response and its impulse response symmetry. The parameter used to define the amount of filtering applied to the baseband data is $B \times T$, where B represents the –3 dB bandwidth of the equivalent Gaussian low-pass filter used to smooth the modulating data and T is the bit period of the data. Thus a BT of infinity (i.e. no filtering) gives MSK.

The generation of the GMSK filtered in-phase (I) and quadrature (Q) components is suitable for integration since the I and Q paths can be generated using two ROMs – one for the I component and one for the Q component – containing the possible phase trajectories within a given interval and two DACs; each produces a baseband signal representing the in-phase and quadrature components of the desired signal.

After low-pass filtering to remove DAC-related clocking spuriae, these signals are up-converted with in-phase and quadrature signals from an RF oscillator. These up-converted signals when combined generate the desired constant-envelope modulated signal. The modulation process is much easier with the transmitter I and Q baseband signals being digitally constructed before filtering and mixing with the LO followed by RF amplification to produce the required output. An important benefit of this approach is that the modulation index is very accurately controlled, allowing the use of a higher BT value than would be used in the direct modulation approach. For example, with direct modulation a BT of 0.5 is recommended whereas for an I/Q generator a BT of 0.8 could be used instead.

15.4.5   Transmit Power Amplifier

The power amplifier stage of a cordless terminal is relatively straightforward: the required power is modest and the tuned band is small. The output stage would be expected to operate in Class C (non-linear switching mode) to provide high efficiency. The SNR of a received signal in a TDD or TDMA environment can be affected by transients generated when the high-frequency carrier is bursted on and off. Hence, a particular requirement for cordless systems is the ability to achieve smooth gating of the transmitted carrier output to restrict the adjacent channel interference. This can be achieved by the use of control of the DC bias of the output stage(s).

15.5   Antenna Diversity

Antenna diversity has been shown to be an effective technique for minimising the effects of multipath fading in mobile radio communications, thus providing better communication quality without increasing power or bandwidth. This is an area where a CT2 TDD system has been shown [8] to provide better performance than an offset frequency system such as that used in CT1 phones. The possible performance improvements did not go unnoticed by one CT2 manufacturer, which incorporated diversity into its first-generation CT2 product with two aerials at the basestation as well as in the handset; early DECT products have also incorporated antenna diversity, which has proven very effective against effects of time dispersion/delay spread.

The basic principle of antenna diversity is based on comparing the received signal strength between two aerials and choosing the one with the strongest signal for transmission as well as reception. In testing done between CT1 and CT2 telephones [8], where both were configured with antenna diversity, testing showed that a TDD system could support over five times as many cordless telephones as compared with the frequency offset operation of a CT1 phone. Incorporating diversity, however, increases the cost, and thus simple, cost-effective, diversity implementations have in general been adopted.

Having considered the basic cordless transceiver architecture and design, the next sections attempt to show the interrelationship of the product design with specification requirements and technology developments for a number of phases of product evolution.

15.6   Analogue Cordless Implementation

Broadly speaking there are two major branches of analogue cordless technology. CT0 operates in frequencies below 49 MHz and is found in countries such as the USA, Spain, France and the UK. CT1/CT1+ operates in the 900 MHz band and is found in countries such as Germany, Austria and Switzerland. The USA also has its own blend of 900 MHz cordless telephony operating in the ISM (instrumental, scientific and medical) band, as described in Chapter 8.

This section will discuss the UK CT0 and CT1/1+ as representative examples of the low-band and high-band analogue cordless technology implementation.

15.6.1   UK CT0 Technology

The first successful cordless telephones were analogue low-band designs, which in the UK were produced to the requirements of British Telecom Technical Guide

47. A large part of this specification was devoted to the detailed requirements for the telephony performance to maintain the operational quality of the wired network. It also reflected the need for the radio parts to be realisable in a simple cost-effective manner. Thus, in line with the technology and radio spectrum availability at the time, the portable transceiver had a VHF low-band, single-channel transmitter in the 47 MHz range and an MF receiver in the 1.7 MHz range. This meant that products could use the techniques of contemporary radio systems, such as radio control for model aircraft, to achieve a low-cost, low-power-consumption, robust design that did not stretch the basic technology or require costly pioneering development work.

There are many good examples of the typical design of UK CT0. In most cases, the circuit architecture of the 47 MHz FM transmitter is conventional in using a crystal oscillator, which is frequency modulated, multiplied and filtered and amplified, while the 1.7 MHz FM receiver is of the traditional superheterodyne type. Realisation of both transmitter and receiver is typically in standard miniature discrete components with one integrated circuit and a ceramic IF filter. Numerous variable resistors, capacitors and inductors are used typical of Far-East-sourced radios. The basic assembly is on a printed circuit board produced to good but not limit standards, in keeping with the objective of a cost-effective robust product that does not demand advanced techniques and materials.

Simplicity was the key theme of the UK CT0 products. The specification of a fixed factory-set channel (one of eight) with analogue FM in the 1.7 MHz and 47 MHz bands achieved its purpose of enabling low-cost products to become quickly available, but this very success was also the root cause of the commercial limitations of the product range. Call blocking, interference, lack of privacy and short operating range all result from the specification and can be only marginally improved by the use of advanced implementation technology.

It was against this background that the specification for the CEPT CT1/CT1+ was produced. This has achieved considerable success in parts of Europe, notably Germany and Switzerland, even though the demands on the implementation technology are much more severe than those of the low-band designs and therefore more expensive.

15.6.2   CEPT CT1/CT1+ Technology

As often happens in life, the response to a condition is frequently an over-reaction, with the pendulum swinging in the opposite direction; this was arguably the case for the CEPT CT1/CT1+ where the prime objective – to make the maximum possible use of the limited frequency spectrum availability – resulted in a specification that could initially only be met by complex designs and high cost. The CT1 specification defined operation on 40 paired duplex channels in the 900 MHz band using analogue FM; CT1+ was an upgrading of CT1 to 80 channels and a shift of frequency to avoid contention with GSM operation. Channel bandwidth is 25 kHz with a 45 MHz duplex separation between transmit and receive channels. This frequency allocation is combined with tight requirements on receiver sensitivity, spurious response rejection and inter-modulation rejection to permit operation on adjacent channels with transceivers in close proximity, to achieve the required high-user-density figures. The

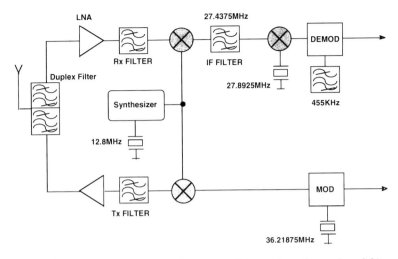

Figure 15.9   RF block schematic of an Ascom CT1+ product. (*Source*: from [9].)

complexity required to meet these stringent specifications is ably illustrated by the block schematic of an Ascom product, illustrated in Figure 15.9.

Referring to this block schematic, immediately apparent is the need for three crystal oscillators. The high-stability frequency synthesiser has to be very agile to scan all 80 channels in a reasonable time period and still has to achieve a stability of ±5 kHz in set-up mode and ±2.5 kHz (2.5 ppm) in operating mode. Full analogue duplex operation means that a duplex filter is obviously required, but additional filters are also used in the separate transmit and receive channels to achieve the required specification performance. The transmitter design is required to achieve very low adjacent channel power, a maximum of 10 nW (−50 dBm) under normal conditions, which impacts on the design of the modulator and mixer as well as placing strict phase noise requirements on the frequency synthesiser. Receiver performance specification is also extremely tight, with the very high adjacent channel performance requirement of 53 dB placing exacting demands upon the IF filter performance.

Generally, the design is much closer to that of professional communications equipment than the first-generation cordless telephones and it is a very creditable reflection on both the design development and the manufacturing that commercially successful CT1/CT1+ products have been achieved. Emerging improvements in technology have resulted in design improvements, allowing cost reductions – however, the cost base of CEPT CT1 telephones has always remained significantly above that of CT0 products. It is the regulatory specification that dictated the basic product cost range and it is this situation that the next generation of digital designs has begun to be able to exploit.

15.7 Digital Cordless Implementation

Despite early uncertainties as to when or how digital cordless would ultimately replace the initially cheaper analogue cordless telephones, especially the low-band CT0-based systems, many semiconductor, radio component and telephone manufacturers have recognised that this is a key market and have invested very heavily. Whereas the pioneers of cordless had to rely on expensive nascent technologies, such as general-purpose digital signal processors, today's designers are able to purchase standard products or ASIC solutions specifically optimised and tailored to the various standards. The combination of technological advancement and aggressive competition from suppliers has made the cost of digital cordless systems accessible to the mass market. The fact that the US market, the largest as well as the most cost sensitive of markets, has seen a phenomenal growth of the more expensive 900 MHz ISM market seems to indicate that there are perceived needs for the advantages offered by digital technology.

15.7.1   ADPCM

Digital speech encoding is the most fundamental difference between digital cordless telephones and earlier analogue generations. CT2, DECT and PHS have all adopted the standard 32 kb s^{-1} ITU ADPCM algorithm derived from 64 kb s^{-1} PCM algorithm defined for network interfaces. Some of the reasons behind this choice were its availability as an international standard, the use of PCM coding in digital switch back-planes, its good speech quality and its inherent resistance to interference. Before the arrival of dedicated and optimised ADPCM coders, designers used voice encoding based on delta sigma modulations (e.g. CVSD) or, depending on voice quality requirements, had to resort to more expensive general-purpose digital signal processors for implementation of ADPCM.

15.7.2   FDMA Versus TDMA

The content of a product specification is determined by its objective; in this respect, "cordless telephone" is too general an objective for specification purposes. For CT2, the emphasis was placed on the domestic and small business applications, whereas for DECT the large business cordless PABX application was the primary driving force; this difference is widely accepted as the main reason for the CT2 specification being FDMA and the DECT being TDMA. Both systems, however, use time-division duplex (TDD) to achieve a pseudo-duplex speech path over a single RF channel. TDD gives the operational advantage of a symmetrical transmit and receive propagation path and hence the ability to get the full benefit of aerial diversity. A major benefit in the implementation is that there can be a simplification in the RF frequency synthesiser, particularly when zero-IF radios become reality, where no transmit/receive switching is required.

There is a further simplification and relaxation of specification possible in the frequency synthesiser for a TDMA design since there are fewer but wider RF channels than in an FDMA design. The wider bandwidth of the TDMA channels clearly places different demands on the filter technology but generally this presents no real problem and the front-end aerial filter is no different than for FDMA. Relaxation of the RF channel frequency tolerance in a TDMA design is offset by the need for a very accurate control of the channel time slots in the basestations, necessary in order to avoid fast slot slip in relation to unsynchronised neighbours. Although the clocks controlling these time slots operate at frequencies where technology is well understood and characterised, the accuracy demanded must not be underestimated and great care must be taken in the implementation of this function. For an FDMA design with more and narrower-bandwidth RF channels, the RF frequency stability requirements are demanding. Dynamic channel allocation (DCA) and call set-up timings do require the capability of fast switching between channels from the frequency synthesiser, which certainly means designing at the technology limits. For wireless PABX applications using TDMA, where a basestation may be required to support conversations on adjacent time slots, very limited switching time is available and either a dual-synthesiser architecture or rapid synthesiser switching performance is demanded.

This latter point illustrates a significant difference in basestation architecture between FDMA and TDMA designs. In multichannel TDMA basestations, only one radio transmitter and receiver is required compared to one per channel in the FDMA approach. It is this feature which makes TDMA more attractive for multichannel applications, such as wireless PABX or wireless local loop. However, a simple domestic single-channel basestation has to carry the complication of the synchronised time slot capability. This is more a basic architecture issue than implementation technology, although there is a strong link between architecture complication and the affordability and justifiability of the implementation technology.

Having considered the basic differences between FDMA and TDMA of modern cordless specifications, there is a large area of commonality in the digital signalling and digital speech coding in respect of the implementation technology. The requirements of the digital signalling functions are well served by the semiconductor processes in CMOS integrated circuits and the ever-increasing processing power of microprocessors and ASICs. Even in a business application, the demands on the signalling channel are a small part of the total communication channel, which is dominated by the speech path. There are two exceptions to this, however – firstly during the radio link establishment and call set-up phases, which are short periods of intense signalling, and secondly in a cordless data terminal application. This latter case will not be pursued in this section since it is really only a different use of the digital speech channel for carrying standard data formats with or without additional error control and management.

In considering a cordless telephone system design, it is important to get the distribution of intelligence correct between basestation and the handset. Even though technology enables very low-power digital processing, it is dangerously easy to impose a heavy burden of system management onto the portable unit, resulting in a significant power consumption problem. This can become more user apparent in domestic-type applications where the ratio of active to stand-by use is very low and the background processing power consumption dominates.

15.7.3 CT2 Technology

Design partitioning between hardware and software realisation of the signalling functions is an important issue involving the generally conflicting aspects of unit cost, development cost, development time and product size. Most initial products, however, had a similar compromise in this area, employing a low-power CMOS microcontroller and a mixture of custom and catalogue logic parts. This compromise has evolved as the level of integration has increased in products now available on the market, reflecting the factors below.

Firstly, in recent years semiconductor manufacturers have begun to target microcontrollers towards specific market segments. Thus, in addition to devices containing the standard assortment of RAM, ROM and general-purpose I/O ports, products became available with application-specific peripherals (e.g. a DTMF generator for the telecommunications market). Many manufacturers also now offer an ASIC facility whereby users may configure their own microcontroller from a core processor plus a selection of peripherals.

Secondly, in a parallel and complementary development, as the complexity of ASICs has increased, many semiconductor manufacturers have introduced popular microcontrollers to their cell libraries. Unlike the original CT2 pioneers, today's manufacturer can design his own semi-custom ICs containing a microcontroller, RAM, ROM and custom high-speed signalling logic.

Early manufacturers were constrained to the use of general-purpose DSPs with embedded software for speech coding. These devices have migrated to custom architecture devices [10], optimised for function and offering reduced power consumption; indeed, such cells are now available for integration in more complex ASICs. Further, highly integrated digital baseband/controller devices have also now become available, incorporating most of the digital/baseband and control functions on a single chip.

An example of such offerings is the CT2 PhoX component family from AMD, which incorporates all the low-frequency audio and digital circuitry necessary to meet the requirements of the CT2 CAI specification. Functions integrated into this baseband controller include an audio interface to the microphone or public switched telephone network (PSTN), a ringer and earpiece driver, a G.721 ADPCM codec for converting the analogue speech signals into digital signals, an industry-standard 8-bit microcontroller with on-chip ROM and RAM used for performing all control and man–machine interface functions, a TDD burst mode controller (BMC), which manages the radio link according to the CT2-CAI physical layer specifications, and other miscellaneous blocks, which provide control signals to the radio front-end and provide system power management.

15.7.4 DECT Technology

Just as late entrants to the CT2 market have benefited from the extended technology gestation period, so also DECT has clearly benefited from the slightly later product entry compared to CT2 by being able to take advantage of the later technology intercept with its improved performance semiconductor processes. This has been particularly important for DECT, which operates at higher frequencies and supports fairly complex medium access control (MAC) layer functions with regards to handoff, RSSI table and slot management requirements, and concatenation of time slots, for example. Lessons learnt in CT2 development have transferred well to DECT, allowing a faster climb up the learning curve for both equipment and semiconductor designers. By the mid-1990s a wide range of components, modules and software were available supporting DECT product development, well summarised in [1].

RF Front-End, PA and IF Technology

The scarcity of radio spectrum was a key feature in the definition of the DECT standard and, from the outset, it was clear that operation above 1.8 GHz would be required. For low-power, low-cost portable transceivers, RF frequency stability can be a major problem and operation much above 900 MHz was approaching the limit of practicable narrow-band systems in the late 1980s. This was a lesson well learnt from the CEPT CT1+ experience; CT2 benefited from the wider 100 kHz channel, but further widening for 1.8 GHz operation could not really be justified on the user density versus spectrum allocation issue.

The wider channel spacing of 1728 kHz in the DECT system has allowed the use of (and indeed encouraged the development of) cheap SAW filters in the receiver and a general easing of the IF design compared to narrow-band superheterodyne designs. For zero-IF designs, the IF filtering and bandwidth are less of an issue but a single conversion from 1.8 GHz to baseband is questionable. It remains to be seen if zero-IF product announcements by some component vendors will truly result in manufacturable zero-IF-based products. Another alternative, previously described in section 15.3.3, is the single superheterodyne or single-IF approach where the RF input is down-converted to a first IF, after which it is then directly converted to baseband. An example of this architecture is depicted in Figure 15.10. This approach reduces the cost of the transceiver by eliminating the need for a dual synthesiser. However, it is important to keep in mind that this can only be made possible by providing the capability to reprogram the synthesiser automatically between transmit and receive bursts. To date, most first-generation DECT products have been based on the two-stage superhet approach, but will more than likely progress to single-IF architectures in the future.

Advances in design approaches and semiconductor performance have been driven by the development of new higher-frequency systems – DECT and DCS1800 – in recent years. In terms of the RF circuitry, early DECT products have tended to adopt relatively conservative, but robust, superhet design approaches. Discrete technology has been used in many early designs, based on silicon bipolar or GaAs technologies; an F_t of more than 10 GHz is typically needed to meet the low-noise-figure requirement at low current levels for the LNA function. Again, discrete transistors or integrated bipolar/GaAs solutions have been employed for the mixers, with PIN diodes used for RF switching. For the power amplifier, discrete transistors – bipolar/GaAs – have been used, as well as hybrid monolithic/MMIC GaAs solutions. Integrated LNAs, switches and PAs in GaAs technology are available with acceptable performance – at present however, as well as having a requirement for an additional negative supply, these have a relatively high cost. Integrated designs may be expected to proliferate in the coming years as more manufacturers introduce such devices onto the market and competition forces down prices, such that they offer a clear and sustainable price advantage over discrete approaches.

The variety of approaches to synthesiser designs reflects the wide choice of prescalars and synthesiser ICs available; integrated VCOs are also now available, although discrete implementations are still most common.

IF amplifier and demodulator ICs are available based on both FM and direct conversion approaches. Prices of the associated IF SAW filters are expected to

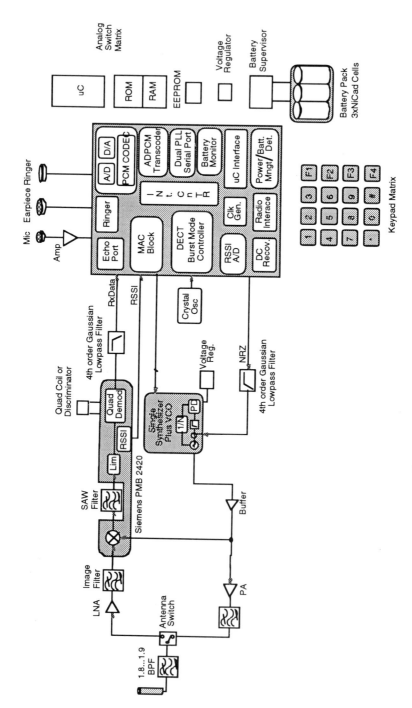

Figure 15.10 Single superhet DECT block diagram.

drop further with the more widespread adoption of low-cost plastic packaging; insertion losses of such packaged SAW filters are also expected to reduce.

Table 15.1 summarises the main suppliers of DECT RF/IF technology as at mid-1995. Given the growth of the DECT market, this table will almost certainly be incomplete at the time of publication, but serves to illustrate the wide availability of components from the major international semiconductor and filter manufacturers.

Table 15.1
Major RF/IF technology suppliers (non-exhaustive) as at mid-1995

	RF silicon	RF GaAs	RF/IF filters	Synthesiser/ PLL ICs	IF amp/ demod
Mitsubishi		•			•
Motorola		•		•	•
Murata			•		
National Semiconductor	•	•		•	•
NEC	•				•
Philips	•			•	•
Siemens	•	•		•	•
S&M (Siemens–Matsushita)			•		
Toshiba				•	•

Data from [1].

In addition to RF components, as described above, a number of complete DECT RF modules are available; initial suppliers have included Philips, Murata and Alps. Availability of such modules is allowing companies with no RF design capability to develop their own customised DECT products for niche markets.

Baseband and Controller Technology

The approach to DECT was certainly more ambitious than CT2 in respect of the implementation technology demands, reflecting to some degree its later market entry and to some degree the focus on the business market; an example of this is the echo control function at the basestation audio line interface. As explained in Chapter 11, echo control at the audio line interface is necessary when the two-way loop delay round the transmit–receive path in a cordless telephone basestation and portable configuration exceeds ~5 ms. The burst structure of the DECT system gives a delay of at least 10 ms, which results in the need for echo cancellation in the basestation, in contrast to shorter delay systems such as CT2 or PACS-UB. This resulted in added cost in first-generation solutions, as this function was initially implemented in general-purpose DSPs. However, as DSP technology has matured, this functionality has been integrated with other digital functions in the basestation, minimising its cost impact.

Similarly, whereas initial CT2 products had to rely on programmable DSP devices with embedded software for the ADPCM coding, DECT manufacturers were able to benefit from the increasing technology maturity in the form of off-the-shelf chips. Early development of such devices was encouraged by the almost universal adoption of ADPCM as the coding scheme of choice for cordless telephones.

Burst mode controllers for DECT have clearly required custom development, and are available today from several sources as, of course, is the basic microcontroller IC, used for overall control and MMI functions.

More recently, as digital technology has continued to advance and device geometries shrink, several IC manufacturers have developed high-complexity devices integrating several or all of these coding/baseband/control functions onto a single chip (labelled as multifunction ICs in Table 15.2). As well as single-channel multifunction ICs, i.e. a single coder/BMC/microcontroller, multichannel devices are also available, intended for wireless PABX or WLL basestation applications. Table 15.2 summarises the main suppliers of DECT back-end technology; this table will probably have dated even more rapidly than that for RF component suppliers, given the larger number of digital IC manufacturers likely to enter the market.

Table 15.2
Major back-end technology suppliers (non-exhaustive) as at mid-1995

	ADPCM IC	Burst mode IC	Controller IC	Multifunction ICs
Advanced Micro Devices				•
Mitsubishi			•	
Motorola	•	•	•	
NEC			•	
Philips	•	•		
Siemens				•
Sitel				•
Texas Instruments	•	•	•	
Toshiba	•		•	
VLSI Technology	•			•

Data from [1].

Embedded DECT software and algorithm support is today widely available, particularly from IC manufacturers who offer microcontroller or DSP cores for custom integration. In addition, several companies offer software packages supporting the basic DECT protocol functionality. Companies offering software modules include S3, Sitel, Siemens, Symbionics and Texas – again other suppliers are entering the market.

15.7.5 System Software Development

With today's dedicated circuits, much of the voice coding and burst formatting is increasingly embedded in hardware. Some such devices can function without a microcontroller and ROM after initial set-up. With much of the physical layer already done, the designer can spend more time on protocol development and on the man–machine interface.

For example, the DECT specification not only addresses the physical interface layer, such as the transmitter/receiver burst frame structures and channel bit rate, but also the higher layers of protocol needed for link establishment and secure user authentication procedures. In this regard, a DECT telephone can be viewed as an embedded system (i.e. one involving a microcontroller or processor running something resembling a small operating system) requiring software overhead not normally found in CT0/CT1 cordless telephones. Thus, in order to meet their time-to-market demands and accommodate the protocol require- ments inherent in DECT telephones, many manufacturers are looking to consultants as well as silicon vendors for providing some level of software and hardware support to help expedite their product development.

The availability of standards documents has enabled test companies such as Hewlett-Packard, Symbionics and others to provide test equipment for verifying conformance at the physical layer, but the lack of software protocol analysers specific to a standard can leave software conformance to the programmer's interpretation. The use of a layered protocol structure means that the software can involve hierarchical levels where bugs can feed on other bugs, and even when the hardware and software seem to work alone, integrating the two can create a set of problems all its own. Therefore, having the means for debugging at the system level is an absolute requirement.

In general, software for embedded systems is normally designed and implemented with cross-compilers on a PC and then downloaded for debugging onto a target system under control of an in-circuit emulator (ICE) that has been optimised for a particular family of microcontrollers. With the migration towards system-level integration, the microcontroller itself is now also being integrated along with peripherals and memory. This fact causes a great problem with regard to available emulators supporting the exact mix of custom peripherals and I/O, since the mix varies from application to application. One solution to this problem is to have an emulator company develop an emulator version supporting the peripheral mix specific to the application needs; however, this can be expensive since the mix will undoubtedly change for the next customer. To this end, the approach taken by some semiconductor manufac- turers is to provide a hardware platform that addresses the customer development needs using available emulator technology with little impact on the programmer. This section provides an overview of the development environment and addresses what second-generation telephone manufacturers are doing to address their development needs.

For any embedded design there are four ways in which the software development can be accomplished. These are use of:

- In-circuit emulator (ICE)
- ROM emulator
- EPROMs
- One time programmable ROMs (OTPs) and flash memory

A simulator is another possible option; however, complete simulation of a target system is not always possible or in many cases a realistic endeavour.

In-Circuit Emulator (ICE)

In most embedded applications a software kernel (or executive) is implemented to fulfil the real-time requirements of the application. Depending on the complexity of the task at hand, this can be something as simple as a software loop idling in a loop and handling asynchronous events only as they occur, or a small multitasking kernel with a corresponding task scheduling algorithm. As the complexity increases, so too does the need to have a tool with greater debug capability. This is where an ICE plays a critical role in reducing code development time.

An ICE provides the capability of debugging on the target system without placing restrictions on the target hardware or application program. This debug tool provides a convenient way of viewing program execution and program flow by providing a number of debug features, such as support of trace buffers and instruction and memory/data breakpoints, with no penalty in real-time execution. In an ICE environment, the target board's ROM and RAM spaces are overlaid with emulation memory and the target microcontroller is physically substituted by an emulator pod. Program downloading to either target or emulation memory is possible with execution occurring under PC control. Usually the ICE is connected to the target board during the entire debugging phase and hence requires that a number of ICEs be available to support adequately a number of users. With the push towards finer lead pitched SMT packages – such as found on thin quad flat packages (TQFP) – it is becoming difficult to couple the emulator pod reliably and inexpensively directly onto the target device. This, coupled with a small target board, is what is leading silicon manufacturers to provide development platforms.

ROM Emulator

An alternative to an ICE is a ROM emulator. Unlike an ICE, which replaces the target microcontroller with a pod, a ROM emulator physically replaces the ROM on the target board. Program downloading from a PC is provided with more expensive ROM emulators providing trace capability in conjunction with logic analysers and disassemblers specific to the target microcontroller. However, if plans are to use the ROM emulator in troubleshooting a finished, but malfunctioning, product, access to all control signals required for accessing target board RAM or peripherals needs to be accommodated in the final design – a requirement not easily accommodated on small PCBs.

EPROMs

EPROMs/OTPs provide the least expensive means for testing the application program in a target system. A drawback, however, is that modifications require a time-consuming process owing to the lack of in-circuit programming capability. When modifications are needed the EPROM is removed from the target system, reprogrammed and then inserted back into the target system, thus requiring frequent reprogramming. Another inconvenience with EPROMs as well as with less expensive ROM emulators is their low visibility of program execution. These shortcomings are usually overcome by embedding test code in the application program that provides a view of program flow. An example of this might be the toggling of port pins or lighting of LEDs as indicators of program status or flow.

Flash Memory

In the cellular phone market where service and cost are based on features supplied, flash memory is used where ease of in-circuit programmability is traded off for the cost of the memory. Flash memory differs from OTP or EPROMs in that they are capable of being programmed in-circuit without having to remove them from the board. Thus, provided the microcontroller or processor can accommodate a small system program on-chip, flash memories provide a way of debugging production PCBs and a convenient way of upgrading a product's software in the field. Although used in cellular telephones, the use of this kind of memory is currently too expensive for use in consumer-based cordless telephones and will more than likely not be used in the near term.

15.7.6 Electromagnetic Compatibility

The rapid spread of portable products supporting higher switching bandwidths is bringing to the forefront electromagnetic compatibility (EMC) concerns. For example, in Europe, the EEC EMC Directive 89/336/EEC states that:

> Apparatus . . . shall be so constructed that: (a) the EM disturbance it generates does not exceed a level allowing radio and telecommunications equipment and other apparatus to operate as intended; (b) the apparatus has an adequate level of intrinsic immunity of EM disturbance to enable it to operate as intended.

With this in mind it is becoming apparent to many that if the concept of PCS is to be developed to its full extent without penalising – or being penalised by – other services using the RF spectrum, consideration for EMC must be given up-front to the design of second-generation cordless telephones based on digital technology. All products sold in Europe after the end of 1995 have had to bear the "CE" marking, indicating compliance with EEC Directives. What this means to system manufacturers is that their products are required to stay within specified limits of conducted and radiated emissions.

For many CT0 manufacturers, meeting the minimum requirements has not been much of a problem, as evident in the PCB material used and in the amount of shielding used in a CT0 product. However, with the processing requirements for protocol handling and link quality management, second-generation digital telephones are becoming more like embedded systems. The nature and physical size of these products requires that digital circuitry and sensitive RF circuits be in close proximity. This increases the electromagnetic coupling between components and makes it more difficult to break the differential and common-mode current paths that are the primary sources of radiated emissions. This coexistence of mixed signal technology is further complicated in a CPP where space constraints limit device separation and shielding is restricted by end-product package requirements.

Aside from the EMC implications, the proximity of high-speed logic to sensitive RF circuitry can potentially desensitise the receiver, which, if not identified early in the design phase, can add cost and delays to the delivery of the end-product. In addition, current trends towards higher system-level silicon integration is bringing with it smaller SMT packages with higher pin counts and finer pin pitches, all of which are making circuit trace layout more critical, especially when analogue and digital functions are in such close proximity. The push towards digital communications technology is mandating higher processing requirements, which in turn means higher clock rates with faster rise and fall times. With the development of higher-performance systems pushing CMOS technology bandwidths well into the gigahertz range, design for EMC is something that second-generation telephone manufacturers have had to consider early in the design process.

One aspect of EMC is radio-frequency interference (RFI), which is primarily due to PCB traces acting like antennas to pick up radio-frequency signals, such as those from radio and TV stations or other electromagnetic radiators such as PC products. As a general rule, the higher the frequency, the more susceptible a circuit is to RFI.

For example, some RFI problems in CT2 units have been encountered due to direct rectification of the RF burst envelope. Because CT2 is a TDD-based system with an RF burst frequency of 500 Hz, some developers have encountered problems of 500 Hz noise getting into the microphone path. Simply stated, this means that the RF burst envelope has somehow been converted to an AM signal – the antenna feed line or microphone leads are the primary suspects since the feed line (i.e. coax line) can itself radiate and the microphone leads can form a quarter-wave receptor at the frequency of operation. In many cases, the microphone lead length is the main culprit since the resonant characteristic of the lead length can act as a high impedance. Careful layout of signal traces and placement of wiring thus can play an important role in meeting time-to-market demands.

RFI in the form of burst detection in a DECT system is less of a problem than with CT2 since the burst rate is 5 ms. This means that even if detection of the RF burst envelope does occur, the 200 Hz detected tone lies outside the audio band (i.e. 300 Hz to 3.4 kHz) and is easily filtered by the baseband filters.

15.8   Future Trends

Silicon RF technology will continue to be developed, stretching the power/ frequency performance and enabling cost-effective portable products above 2 GHz with new processes such as SiGe. Although gallium arsenide (GaAs) has found some application in cordless telephones, it has not as yet displaced silicon – looking to the future, no clear adoption of one technology in preference to the other is apparent.

Changes in product technology are more likely to be seen in design architectures where the move will continue towards more digital implementation. It is already possible to visualise a transceiver in which it is only the front-end mixer and LNA that are analogue designs and the remainder digital; however, the more extensive use of digital signal processing technology means more power consumption, which may be considered unacceptable for a portable product. Today, the processing done via DSP techniques can be considered a small percentage when compared with the power requirements of the PA. Since power conservation leads directly to longer talk and standby times, and reduced weight, baseband manufacturers are addressing the power consumption issue by supporting power-down modes of just about every functional block that resides on-chip and supporting sleep modes that allow the processing to occur at a lower operating clock frequency when minimal processing is required.

To realise cost-effective personal communication systems, it is important to reduce the cost and power consumption of portable handheld units. For these purposes CMOS has emerged as the process of choice. A good example is the large-scale digital integration of the $\pi/4$ QPSK modulator used in PHS and in the North American Digital AMPS cellular standard (IS-54). This has shown promise in realising compact, highly stable and maintenance-free portable units. Another example is the level of integration found in direct sequence spread-spectrum cordless telephones.

An industry trend in the portable market in general is the rapid migration towards 3 V power supplies – with technologies optimised for 3 V operation, power consumption can be halved, compared to 5 V, at the same performance. This, together with the application of new battery technology, is enhancing customer appeal by making more complex terminals truly pocketable rather than just portable. Yet another industry trend is in higher system-level integration. Since any IC device consumes less power when its external pins are not switching, bringing more RAM and ROM on-chip, as already done in a number of offerings from CT2 and DECT baseband manufacturers, will further improve perceived performance.

Possibly the most interesting trend, however, may have nothing to do with the technical issues discussed in this chapter. We are seeing that more and more functions are being integrated into fewer and fewer chips – the baseband chip count in second- and third-generation CT2 and DECT phones is now down to two or three components. Radio integration is somewhat further behind but is catching up. Systems expertise and experience in radio products should not be discounted in the process of product development, but clearly more and more silicon and component suppliers are offering essentially "shrink-wrapped" baseband and radio solutions to cordless telephone manufacturers. Questions of whether these solutions are manufacturable aside, phone manufacturers should not interpret this as a conspiracy by component vendors to take over their business – rather, these are actions born out of market necessity. Digital cordless telephones will be a mass market no matter what, competition will be intense and investment levels high. The winners of digital cordless telephony are likely to be those who can reach the market and serve customers' needs and not necessarily those who own the technology.

15.9 Summary

This chapter has sought to give the reader an insight into the product architectures and implementation technologies adopted in cordless telephone design, from the viewpoints of RF, baseband, controller and other technologies, particularly as they relate to early digital cordless products.

From the early analogue products operating at VHF frequencies with just a few channels, we have moved rapidly into an era when digital designs are offering improved performance at ever-lower prices. In reality, this continued trend is a basic requirement if the enormous market forecasts predicted for cordless telephony are to materialise.

In the next few years we can expect further steps in baseband integration, resulting in the single-chip baseband product. This will be accompanied with a wider choice and adoption of integrated solutions for the RF aspects, displacing discrete designs as such offerings become more cost-effective. The major architectural change on the horizon is the possible adoption of the zero-IF approach with increased digital processing, although the timing of this is hard to predict – as market growth increases the significance of the cordless market, so resources available to develop such innovations, which promise high returns, may increase.

Acknowledgements

The authors gratefully acknowledge the contribution of Brian Bidwell, whose chapter on this topic in the earlier book, "Cordless Telecommunications in Europe", formed a valuable basis for the current chapter.

References

[1]   "DECT Reference Document", ECTEL Open Forum on Components and Technology, Final Version, May 1995
[2]   "Radio Receivers", W Gosling (ed.), Peter Peregrinus (IEE), London, 1986
[3]   "Architectures for Digitally Implemented Radios", AP Cheer, IEE Colloquium on Digitally Implemented Radios, London, October 1985 (and other papers in this colloquium)
[4]   "Phaselock Techniques", FM Gardner, Wiley Interscience, New York, 1979
[5]   "Frequency Synthesisers, Theory and Design", V Manassewitsch, Wiley-Interscience, New York, 1980
[6]   "A 10.7 GHz Frequency Divider Using a Double Layer Silicon Bipolar Process Technology", MC Wilson et al., Electronics Letters, Vol. 24, pp. 920ff., 1988
[7]   "Synthesiser Review for Pan-European Digital Cellular Radio", RA Meyers and PH Waters, IEE Colloquium on VLSI Implementations for Second Generation Cordless and Mobile Communications Systems, London, March 1990
[8]   "Diversity Advantages for Cordless Telephones", KM Ibrahim and A Karim, Electronics Letters, Vol. 19, No. 45, July 1983
[9]   "Cordless Telecommunications in Europe", WHW Tuttlebee (ed.), Springer-Verlag, 1990
[10] "Algorithm Specific Speech Coder Architecture for Second Generation Cordless Telephones", P Dent, R Bharya, R Gunawardana and JM Baker, IEE Colloquium on Digitised Speech Communication via Mobile Radio, London, December 1988

Note

1   In this chapter the different standards are referred to as follows: CT0 = analogue cordless systems using carrier frequencies below ~49 MHz, e.g. the 1.7/47 MHz eight-channel UK standard or the 46/49 MHz 10-channel US standard. CT1 = the 914/959 MHz 40-channel CEPT CT1 standard used in most other European countries. CT1+ = the 885/932 MHz 80-channel upgrade specification designed to replace CT1 when its frequencies become occupied by GSM.

16 Future Evolution of Cordless Systems

Walter Tuttlebee

Digital cordless technology has evolved from a concept to an established market, of different maturity in different regions. At the start of 1996 around 600 000 units each of CT2 [1] and PHS [2] handsets had been sold and were in use, alongside some 2000 000 units of DECT handsets [3]. Digital cordless is no longer a future market, but a present reality – however, it is not a *static* reality.

In Europe, in particular, the success of DECT is driving the rapid development of new products and evolutionary applications – services and facilities that were originally envisaged when DECT was conceived, but were not expected to be its initial market focus [4]. The advent of cordless access service (CAS) licences in Hong Kong, PHS public access in Japan and PCS in the USA are driving similar evolution of digital cordless in Asia and America. In the next few years such public access applications in fixed networks and the integration of cordless technologies in cellular/PCS networks would seem set to be significant. Such market opportunities will be facilitated or constrained by regulatory and political factors, as well as by competition from other technologies and by the emerging degree of competition, collaboration and consolidation between the telecommunications operators, fixed, cellular and new operators.

The potential revenues from public access are such that these markets are likely to be the primary ones driving DECT evolution in the next few years, in the same way that terminal technology development has been driven by the cellular market, although as volumes of cordless terminals grow to be comparable to or exceed cellular terminals this could change. In this chapter we therefore primarily address future cordless evolution from this public access perspective. This is not to say that private systems usage and development will be insignificant; indeed, the business systems and domestic markets may be the ones initially demanding multimedia higher-rate access.

We primarily consider these issues from a technology perspective, although with the recognition that, whilst technology is an enabler, other market factors are likely to be of major import in determining the actual adoption of cordless public access. We begin by considering a range of drivers and enablers, against the backdrop of digital cellular radio – an important adjacent market. We then consider synergies of digital cordless and the cellular and fixed network, describing a number of experimental trials undertaken in the mid-1990s. We finally discuss possible near-term evolutions, notably activities relating to dual-mode cordless/cellular integration, from both the handset and network viewpoints.

16.1   Drivers and Enablers

From initial concept to mature and established products, digital cordless took about a decade in Europe – a similar gestation period to digital cellular radio. Recognition of the early pioneering work in Europe in personal communications stimulated competitive approaches in North America and Japan in the late 1980s and early 1990s, with the result that by mid-decade the pace of innovation in product and service development had accelerated significantly. Such developments reflect market and service development as drivers, and technology innovation as an enabler. In considering such evolution drivers and enablers, we consider not just digital cordless but also the adjacent areas in personal communications that are influencing its future development.

16.1.1   Digital Cellular Radio

Cellular radio has had a profound effect upon the development of the mobile communications market, in its widest sense, as outlined in Chapter 1. From nowhere in the early 1980s, by the mid-1990s in some countries the number of new cellular subscribers had already exceeded the number of new fixed-line subscribers. Digital technology has accelerated the effect of this impact and has extended service possibilities. Digitisation of cellular radio has changed the make-up of the cellular radio marketplace; prior to digitisation, which did not begin until 1992, some 75% of the world's cellular markets were based upon the US AMPS standard (or its TACS derivative), whereas today the European Global System for Mobile (GSM) standard has emerged as a *de facto* digital standard for new networks in many regions of the world. Other standards, notably IS-95 CDMA, are still fighting to compete as air interface options. From the infrastructure viewpoint, the GSM platform is recognised as being unparalleled in maturity and capability and has been adopted for emerging satellite personal communications services (SPCS) and as a basis for many licensed PCS networks in North America. The size, growth and investment in the cellular radio market mean that it will continue to be a major factor influencing the development of the digital cordless market.

16.1.2   Mobility as a Basic Service

The rapid growth of cellular radio since its introduction in the mid-1980s has affected every region of the globe, with systems already used in less-developed regions lacking basic wired infrastructures as a primary means of telecommunications, for those who can afford it. Worldwide familiarity with cellular radio and the mobility it affords is creating a new expectancy, with mobility likely to develop into a basic service prerequisite in many markets in the early years of the next century. The advent of satellite-based personal communications services over the next few years will reinforce such trends.

16.1.3 Telecommunications Liberalisation

The last decade of the twentieth century is seeing liberalisation of telecommunications service provision worldwide, not only in mobile but also in fixed services. New fixed operators are emerging, offering fixed service provision by radio, rather than by wireline, in competition with long-established monopoly PTTs. Likewise, mobile service operators see increasing liberalisation as a means of competing with established incumbent fixed service providers. Such change in the market environment inevitably creates opportunities for new entrants and threats to existing incumbents. New operators in particular are examining all available technologies to identify possibilities to address these opportunities, whilst existing incumbents cannot afford to ignore these possibilities themselves if they are to compete effectively in the future world. In Europe mobile telecommunications is already operating in a fully liberalised market, with fixed telecommunications liberalisation taking effect from 1998 at the latest, with some countries already implementing a policy of full and open competition. In the USA the PCS spectrum auctions during 1995 and 1996 have resulted in the advent of many new operators. In Asia, competition is seen as a way of accelerating service development and provision. Increasingly a technology-neutral, service-driven, approach is being adopted by governments and regulators worldwide.

16.1.4 Fixed and Mobile Service Integration

The liberalisation referred to above, in some countries, will soon begin to erode the remaining distinction between fixed and mobile service. In such markets, and indeed in other more conservative ones, telecommunication operators are aggressively seeking new means of increasing revenues or of creating a unique offering, by introducing new service concepts. Innovations such as a single number that routes calls to the user regardless of location, via fixed or mobile network, as appropriate, have already been introduced in some countries, using simple technology in advance of more advanced solutions that have been promised for some years. Operators who have both fixed and mobile networks have a capability to offer new services, providing differentiation from single network suppliers, e.g. by bundling services in different ways for private and corporate customers.

16.1.5 Increasing Market Segmentation

Increasing market segmentation in developed countries offers opportunities for new niche services that existing large operators may consider too small to be worth addressing, but which can offer profitable business for the focused new operator. Already in some countries we see mobile operators focusing on different customer segments, with appropriately targeted marketing and service tailoring.

16.1.6   Growth in Data Access

Requirements for data telecommunications access is being increasingly demanded in many countries, reflecting a recognition of the growing importance of the Internet and the desire for high-rate access at $28.8\,\text{kb s}^{-1}$. Digital cordless technologies are well suited here as a wireless technology, with their higher data rate transport capabilities, compared to cellular radio.

16.1.7   Wireless Access in Developing Countries

The market success of digital radio technology to support mobility in developed countries has led to rapid technology cost reduction, with the result that wireless can at last offer an economic alternative to the wired local loop connection, offering the prospect of an acceleration of telecommunications provision in the less-developed regions, for both voice and data, as well as advanced services in the developed world.

16.1.8   Global Competition

The development of regional standards and attempts to export technology are already creating a competitive environment that fosters innovation and encourages technology and service evolution. This is evidenced in the new impetus for wireless multimedia and local loop applications; new evolutionary possibilities can serve as a competitive force. Indeed, competitive service capability assessment has been overtly used in the USA to evaluate and accelerate the development of potential PCS technology contenders.

The enormous global markets forecast for radio-based telecommunications continue to encourage further rapid technology innovation. As well as advances in the base technologies, new architectural and product concepts have emerged in the early 1990s, such as wireless relay stations, which extend the capabilities of cordless systems, and multi-mode handsets, capable of supporting, for example, both public cellular and private cordless applications in a single terminal.

16.2   Digital Cordless in Public Networks

Is digital cordless a complement or a competitor to digital cellular? How real is the role and market for digital cordless outside of the private environment? Will it meet the requirement for wireless local loop?

The success of digital cellular radio has created a sense of strength and self-sufficiency amongst the cellular operators, in many ways well justified. Certainly the mobile operator has developed from being a supplier of niche services to having the potential in the coming decade to supplant the wired operator. The failure of some digital cordless telepoint services has served to create scepticism over the potential of cordless technology for public access. However, the emerging global interest in cordless access services requires that the opportunities for public access be reassessed. Examples given in this section suggest that digital cordless offers some tangible opportunities for both fixed and mobile network operators to use cordless access to leverage added value from their existing investments.

16.2.1 The Operator's View

The capability and low cost of digital cordless technology, compared to digital cellular or others, cannot be ignored in a rapidly liberalising market environment – data rates are higher and cost lower. Public operators licensed to offer services in a competitive environment have appreciated such potential benefits. Some of them decided in 1994 to form a group to coordinate their activities, forming the DECT Operators' Group (DOG). Membership of the DOG is open to operators who would like to offer access to an existing public network using DECT technology and to DECT service providers who could offer services using more than one network operator. In October 1995 the DOG had some 20 members from 12 countries (see Table 16.1), including both traditional fixed and mobile operators. The membership from Scandinavia, the UK, Hong Kong and Germany reflects the existing higher levels of liberalisation of the telecommunication markets in those countries.

Many of these operators have undertaken a range of DECT access trials based on their wide variety of network infrastructures during 1994–96, with many of these presented at the Telecom '95 (Geneva) and DECT '96 (London) conferences. Operators in the more liberalised markets clearly perceive potential for digital cordless to complement and enhance their existing fixed and/or mobile networks. To accommodate the anticipated growth in DECT public access, especially supporting ISDN and other high-rate multimedia services, additional spectrum between 1900 and 1920 or 1930 MHz has been proposed in Europe.

Table 16.1
Members of the DECT Operators' Group, October 1995[a]

Country	Operators
Denmark	Sonofon
	TeleDanmark
Finland	Telecom Finland
	Helsinki Telephone Company
France	SFR/CGRP
Germany	DeTeMobil
	Deutsche Telekom
	E-plus Mobilfunk
	Mannesmann Eurokom
	RWE Telliance
Hong Kong	Smartone Mobile Communications
Hungary	Westel Radiotelefon
Italy	Telecom Italia
Norway	Telenor
Spain	Telefonica
Sweden	Telia
Switzerland	Swiss Telecom PTT
United Kingdom	British Telecom
	Mercury Communications
	Orange PCS

[a] The DOG merged with the manufacturers' organisation, the DECT Forum, keeping the latter's name, in September 1996, to create a stronger worldwide body.

16.2.2   Cordless and Cellular

The growth of mobile telephony has been both extremely rapid and sustained. In urban city centres, with current trends, cellular spectrum will be insufficient for some operators in the coming years, if forecast growth rates are achieved. As cellular usage has grown, so the benefits of microcellular technology, pioneered for digital cordless, has been increasingly appreciated, in terms of the spectrum efficiency offered, as discussed in Chapter 12. Thus smaller and smaller cells have been introduced in cellular networks, particularly as new systems around 2 GHz have been introduced.

In Europe, the adoption of DECT systems in corporate and private environments has encouraged cellular operators to explore the potential for public DECT microcellular coverage to extend their cellular infrastructures to provide additional capacity in high-user-density urban centre hot-spots and inside buildings, shopping malls and other "semi-public" locations. The capacity advantages of digital cordless, over DCS1800 say, with similar size cells, arise essentially from the use of dynamic channel allocation. To offer such combined service requires the availability of dual-mode (cellular/cordless) handsets and offers the user the benefit of a single handset that can operate as a personal telephone in all environments. It also requires the ability to interface cordless basestations to the cellular infrastructure. Dual-mode PCS handsets are also a high priority in the US marketplace, as discussed in Chapter 8.

A technical alternative to the dual-mode cellular/cordless phone is the so-called cellular home basestation (HBS) concept. In this approach a low-power cellular basestation may be used effectively as a cordless basestation, connected to the user's PSTN socket or PBX, to provide the functionality of a cordless telephone in the home or workplace. When within range of this basestation, the user's telephone then simply accesses the wired network, at no additional cost to him or her over the ordinary telephone, rather than the user's normal public mobile network. This concept again enables a user to access the PSTN when in range of his or her HBS or the public cellular network when away from home using a single handset. The weakness in this approach, of course, is the higher complexity and cost of such an HBS compared with a conventional cordless basestation, as well as the fact that digital cellular systems are not designed for high-density deployments as may be encountered in the office scenario.

Such differences arise from the fact that digital cellular telephone systems and digital cordless systems have been designed with essentially different criteria in mind. Table 16.2 summarises some of the resulting characteristics of the two types of system.

An examination of Table 16.2 readily indicates a strong degree of complementarity between digital cellular and cordless technologies. The former is generally better suited to large-area coverage, but has associated with it a higher cost for infrastructure and handset elements. Cordless, by contrast, is better suited to high-user-density, small-area, coverage with corresponding low-cost infrastructure and terminal elements; it is also better suited to the provision of higher-rate services. Recognising this complementarity, several operators have undertaken trials of cordless access, to assess its potential as an extension to their existing cellular networks; some of these are described below.

Table 16.2
Comparison of digital cellular and cordless technologies

Parameter	Cellular	Cordless
Range/cell size	Designed for large cell radius	Optimised for small cells, in-house cells
Frequency planning	Fixed, inflexible	Dynamic, adaptive
Spectrum efficiency	Moderate	Very high
Traffic per subscriber	20 mE per subscriber[a]	200 mE per subscriber
Data capability	Low (e.g. 9.6 kb s^{-1})	High (e.g. 32 kb s^{-1})
Robustness to interference	Moderate	High
Handset technology complexity/cost	High	Low
Basestation complexity/cost	High	Low

[a]E = Erlangs.

Cordless as an Analogue Cellular Network Extension

One such example is that reported by Westel Radiotelefon, who have interconnected islands of DECT access into their NMT 450 analogue cellular network [5]. The first trial offering a neighbourhood access began in Vaszar, Hungary, in December 1995. Vaszar is a small village, with poor wireline telephone supply (the local exchange is manual and only manned for some 6 hours a day!), located some 8 km from the nearest NMT basestation. The small size of the village, some 1.5 square kilometres, means that it can be virtually covered with only five DECT basestations; 11 dBi gain omnidirectional antennas are used on three basestations, a 10 dBi directional antenna on a fourth and a 2 dBi omnnidirectional antenna on the fifth. The size of the village means that the limited mobility offered is useful to its inhabitants.

The architecture of the trial is shown in Figure 16.1. Local traffic is routed by the local DECT switch, which collects usage data and subsequently transfers it to the existing network billing system. In addition to basic POTS (plain old telephone service), the system also offers enhanced services normally supported by the NMT network such as call waiting, call forwarding, conference calls and "do not disturb", as well as value-added services such as "voice-mail" and "audiotex". The low additional infrastructure cost and opportunity to generate additional traffic over the existing NMT infrastructure are attractive features to the operator; in a market where the fixed infrastructure is poor, it offers the opportunity to access a new market segment with low-cost additional investment. The Vaszar pilot has been successful and commercial operation began in 1996.

Cordless as a Digital Cellular Network Extension

Interconnection of DECT with a GSM cellular network was trialled in Germany by Mannesmann, an operator with both fixed and mobile networks, during the latter half of 1994 [6]. In this trial (Figure 16.2), a DECT system was directly attached to the D2 GSM network via a protocol converter at the GSM A interface; DECT terminals were each assigned a D2 number. Standard DECT and GSM terminals were used, instead of the dual-mode terminals that would be employed in a mature system. The trial was successful in verifying both system interworking and user acceptance factors.

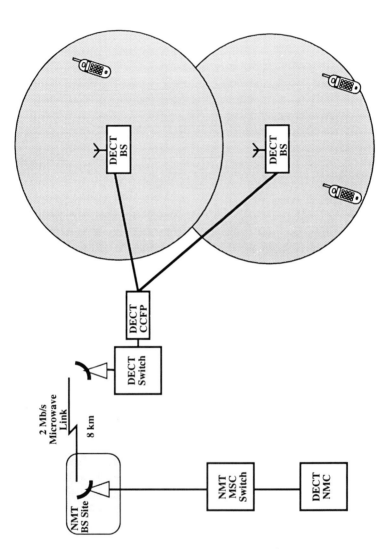

Figure 16.1 The DECT–NMT network architecture in Vaszar, Hungary. (*Source*: adapted from [5].)

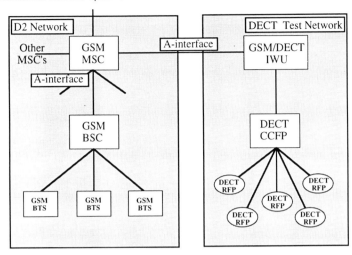

Figure 16.2 The DECT–GSM network architecture in the German Mannesmann trial. (*Source*: adapted from [6].)

Mannesman have also examined the applicability of DECT as a means for wireless local loop access, with the added value of local mobility, and concluded that DECT is indeed suited to such a dual application. To support the use of a portable handset in the house as well as in public areas, Mannesmann concluded that the wireless relay station was desirable, a concept specified by ETSI in the early 1990s [6]. In this application, rather than providing range extension, the wireless relay station supports the use of a portable handset within the customer's premises, avoiding the need for a conventional wired phone as envisaged in the conventional single-application wireless local loop.

16.2.3   Cordless and Fixed Networks

Wireless local loop, other than for rural applications, has only been economically viable since the beginning of the 1990s. It has been stimulated in Europe initially by the need to improve rapidly the telecommunication infrastructures of the former Eastern bloc countries and by the European regulatory liberalisation plans. It has been facilitated by the drop in radio technology costs associated with the mass markets brought about by cellular radio and digital cordless telephones. The rationale for wireless local loop has been more fully presented in Chapter 5.

Requirements in developed and developing countries could be expected to differ, in terms of the need for advanced services; however, certain developing countries are increasingly looking to adopt leading-edge solutions when investing in telecommunications infrastructure, as well as looking, where possible, to create local production capability.

Digital cordless technology is well suited to such requirements, in its ability to support the advanced services, wide bandwidths, security features, etc., required in a public network. In the case of DECT it is also a mature technology, amenable to technology transfer for local production, as has already been proven by product developments in India [7]. In Europe DECT has proved itself amenable to quick and cost-effective local loop deployment, with many operators having undertaken service trials as described in Chapter 5. However, there exist a number of potential competitor technologies to digital cordless, which cannot be ignored; these are discussed later in this section.

A further important factor, when considering evolution, is the potential requirement for service integration; adoption of a technology for wireless local loop that can also support mobility clearly offers advantages in some markets. This concept of "city mobility", as it is sometimes called in Europe, is similar in many respects to some interpretations of PCS in North America – it is essentially the evolution of an economic and common wireless infrastructure that can support both wireless local loop and public mobile services. Such a concept is particularly attractive to the fixed operator wishing to benefit from regulatory liberalisation, as a route to offering a limited mobile service without the high costs associated with a conventional cellular network infrastructure. It is of course also attractive to the mobile operator seeking to build a "fixed line" business, without wasting premium spectrum.

Some initial approaches to wireless local loop used analogue cellular technology, for example NMT 450 in some parts of former East Germany and Spain. Specific technologies have also been developed, such as the TDMA technology used by the new public network operator Ionica in the UK, the Motorola WLL technology [8] and spread-spectrum solutions from players such as DSC, Qualcomm and Interdigital Corporation. During the early 1990s wireless local loop suppliers and operators have increasingly focused on two main technology solutions for wireless local loop – digital cordless (in Europe DECT), and code-division multiple access (CDMA). These two technologies are summarised and compared at a top level in Table 16.3.

The shift in focus towards cordless and CDMA essentially represents a recognition of market requirements and the desire for future-proofing of investment in new network infrastructure. Whilst, for example, analogue cellular can support a basic speech service, the requirement for data services in the future is increasingly accepted, especially given the enormous and rapid growth of Internet access traffic in the past two years. This has led operators to wish to support $28.8 \, \text{kb s}^{-1}$ data access as a minimum. Recognising also that ISDN is finally here and growing, ISDN basic-rate access is also frequently cited as an important service need, although still at present mainly for business customers. The desire for a standardised technology is also strong in Europe, although less so in other parts of the world. These factors have encouraged manufacturers down twin paths of product development based on digital cordless and CDMA.

Several manufacturers already have local loop products based on CT2 and DECT. CT2-based products have been available for some time and have found limited success, mainly in short-range rural wireless local loop applications. DECT products have tended to focus more upon the mainstream second operator market currently emerging in Europe and upon developing regions. Initial products without equalisation have demonstrated good performance in some WLL trials, as described in an earlier chapter. A number of measures have been proposed to provide improved multipath tolerance without modifying the DECT standard – although it remains to be seen whether manufacturers will decide to go down this product evolution route. The great strength of DECT for WLL at the present time, compared with CDMA, is the maturity of its technology, with its

Table 16.3
A basic comparison of digital cordless and CDMA WLL technologies

Parameter	CDMA	Digital cordless
Point-to-point link range (using directional antennas)	Extended ranges readily available – 20 km	Typically up to 5 km (i.e. standard equipment without WRS[a])
Area coverage – cell radius (with sectored basestation antennas)	Several kilometres	Typically 1–2 km
Spectrum efficiency	Very good – efficient spectrum re-use, especially with broad-band CDMA	Good – DCA allows effective spectrum re-use
Robustness	Interference and multipath immunity inherently provided; FEC has no capacity penalty	Adequately robust for many environments and applications; FEC required for data services requiring low BER
Services	High-quality, very low-delay, voice; flexible service provision and bandwidth-on-demand	High-quality voice; POTS and basic ISDN BRA
Flexibility	Range/capacity trade-off without increasing power	WRS allow flexibility (at expense of spectrum efficiency) – see below
Mobility support	Possible, in principle	Possible[b]
Spectrum allocations	New spectrum needed	Spectrum allocated[c]
Standardisation situation	Proprietary systems	Standardised products – allow greater competition and choice of supplier
Technology maturity	Low	High
Technology complexity	High	Low
Economies of scale	Low	High
Cost	High	Low
Availability	Soon	Now (1996)

[a]Dedicated line-of-sight narrow beam links up to 15 km to a WRS have been demonstrated.
[b]With DECT systems, both GAP and RAP will be supported from a common infrastructure.
[c]Additional spectrum of 20 MHz, 1900–1920 MHz, may be made available for DECT public access in Europe.

associated economies of scale and hence cost benefits, together with the fact that a full standard has been defined, the radio local loop access profile (RAP) – see Chapter 20 – which includes full definition of ISDN WLL services, O&M features, etc.

CDMA technologies for civil applications were initially developed in North America, where a strong reservoir of technical capability in this field was released from the "peace dividend" at the end of the Cold War and where the regulatory environment encouraged the commercialisation of new technologies. Proprietary CDMA WLL products have thus been under development for some years and are still not widely available, although some systems are under evaluation by operators. Some European manufacturers have in fact developed a dual strategy for product development; Siemens, for example, are offering a DECT-based system, DECT*link*, launched in 1996, alongside a CDMA-based product, CDMA*link*, to be launched in 1997, within the framework of a common system architecture.

The capability of DECT for WLL applications has been enhanced by the development of the wireless relay station (WRS) [9, 10] – see Chapter 19. The WRS – also sometimes referred to as a wireless basestation or radio network repeater – has the function of a conventional basestation but requires no fixed, wired, connection. The equipment thus simply requires power and acts as a relay to provide range or coverage extension, without the need for the telecommunication cabling otherwise required if an additional conventional basestation were to be installed. It is thus ideally suited to the provision of cost-effective infrastructures for low-density traffic applications.

A wireless relay station equipment contains both fixed radio termination and portable radio termination elements. The fixed termination element acts towards a handset or fixed access unit just like an ordinary radio fixed part whilst the portable termination element acts like a handset or a fixed access unit towards the basestation and is locked to the closest basestation. The wireless relay station contains interworking between its fixed and portable termination elements, including transparent transfer of the higher-layer DECT protocols. The equipment must comply with the general fixed termination identities requirements for basestations – installing a wireless relay station to a DECT infrastructure must be done under the control of the system operator/installer/owner who provides the necesary system identities, access rights and authentication/encryption keys. Such relay links may be cascaded and, compared to a normal basestation, a relay station may introduce capacity restrictions to the services offered. The wireless relay station concept is not limited to DECT but is applicable to other cordless technologies, most obviously and easily to the North American PWT system, which is closely related to DECT. Thus wireless relay stations may well find application outside of Europe.

16.3 Near-Term Evolution

16.3.1 Technology Development

The technical performance and capabilities of digital cordless systems will continue to evolve in the short term, driven by the competitive pressure and potential of new markets and applications. For example, the development of PWT for the North American marketplace has helped accelerate several new features, which will be incorporated in the second edition of the DECT base standard. Enhanced receiver sensitivity has already been seen, with successive DECT products developing from a value of -86 dBm to -89 dBm; products with a sensitivity of -92 dBm are expected. Similarly with PHS, higher-sensitivity, higher-power, basestations have been developed providing extended range. The advent of an extended preamble sequence for DECT will allow single receiver selection diversity to be applied, offering enhanced performance at higher mobile speeds – necessary, for example, in the PWT-E application. The direct handset-to-handset communication mode supported by PHS is applicable to other technologies and forms part of the second-edition DECT specification. The possibility also exists of a handset operating in a repeater (WRS) mode. Many of these enhancements do not require great technological advances, merely changes to the technical specifications and regulatory scenarios.

The size and cost reduction of digital cellular handsets has in recent years been stimulated by, and contributed to, the growth of the cellular market; many of the technology advances driven by cellular have directly benefited cordless. We have thus seen major progress in cordless products during the early 1990s. To recognise this requires only to compare today's DECT or PHS handsets with the early CT2 products, described in [11]. Technology trends in handset design have been well described in the previous chapter.

Figures 16.3 and 16.4 complement this by showing pricing trends for the semiconductor content and the handset product, for the examples of CT2 and DECT, the earliest and the highest-volume digital cordless technologies, respectively [12]. The forecast dramatic year-on-year price reductions reflect anticipated growth in market volumes, and are commensurate with the historical data, i.e. the substantial product price reductions already seen. Such price reductions indicate one of the contributory factors to the thinking behind dual-mode handsets – the cost premium of adding a cordless capability into a cellular handset is rapidly shrinking.

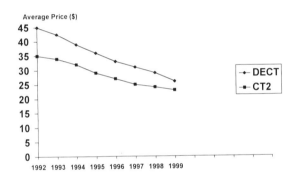

Figure 16.3   Semiconductor content of cordless telephones – pricing trends. (*Source*: from [12].)

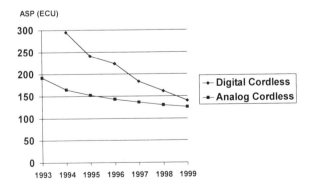

Figure 16.4   Cordless telephone handsets – pricing trends. (*Source*: from [12].)

16.3.2   Dual-Mode Handsets

In the remainder of this section, addressing short-term evolutions, we focus upon the dual-mode integration potential of cordless and cellular technologies[1]. We focus on DECT and GSM[2] below as an illustrative example, for three main reasons:

- These technologies are the most mature ones, in terms of air interfaces and network infrastructure, with standardisation having begun in the mid-1980s.
- The widespread global acceptance of GSM as an installed digital cellular standard means that there is worldwide interest in GSM dual-mode evolution.
- The first dual-mode digital cellular/digital cordless products to be developed have combined these two particular technologies.

Outside of Europe, the concept of public and private PCS supported by a single dual-mode handset has been advocated by the FCC in the USA – e.g. PACS-UA or PACS-UB with PACS, PWT with PWT-E, etc. Similarly the combination of PDC and PHS has been proposed in Japan. In principle, such combinations could arise – many of the issues discussed below are generic and could thus be relevant to these alternative options. Compared with current production volumes of GSM and DECT, however, it is likely that dual-mode handsets for the North American technologies are likely to lag in development timescales, simply because of the reduced economies of scale for some time at least. In Japan, where PHS appears positioned to compete with PDC at present, it is not clear that a market requirement exists to drive dual-mode handset development in the short term.

We firstly consider the requirement for dual-mode terminals in Europe.

The Requirement for Dual-Mode Cordless/Cellular

From the viewpoint of the operator, the combination in a single handset of cordless and cellular offers the provision of effective indoor coverage for cellular operators with low-cost cordless infrastructure, which was designed for this purpose. It also allows such operators to add capacity with DECT subnetworks, deploying cordless coverage selectively in densely populated areas, for which it is well suited. In summary, it gives operators an additional low-cost option with which to trade off capacity and service in a flexible manner.

This latter factor of traffic capacity is particularly important – if an operator is limited in spectrum and the only way in which he can secure more traffic is to install cordless access, then he may be strongly motivated to do so; if he has other alternatives, e.g. as a GSM operator he is given additional DCS1800 spectrum, he will be less inclined to take this route. Thus, regulatory policies on spectrum allocation, as well as on cordless/cellular network interconnection, have the potential to severely affect the development of the dual-mode market. Another important factor for an operator to consider when planning solutions for indoor coverage is that the telephone usage for an indoor user is typically higher than for a mobile user (see Table 16.2), a statistic that again favours the cordless technology, designed with this in mind. Having said these things, the dual-mode handset may be seen by an operator as an opportunity or as a threat, depending upon his current position and environment.

From the viewpoint of the user, a dual-mode terminal is attractive as it offers a single take-everywhere terminal, usable in all environments, public and private. This was the early concept of telepoint, but cordless technology on its own could not deliver this promise. From the user perspective, a cellular handset that can also be used with his or her home cordless basestation offers a benefit over one that does not. In the home/office environment the terminal would default to the cordless air interface, using unlicensed spectrum, with no airtime charges, allowing cost savings over cellular to such users, as well as a longer battery life and/or smaller, lighter handset. If regulators allow direct interconnection between corporate networks supporting cordless access and cellular networks, as already exists in the UK and Scandinavia, the support of corporate PABX features (e.g. private numbering plans) over the cellular network offers added user benefits.

From the viewpoint of the manufacturer, the continuing cost and size reductions of terminals has at last made them feasible. In Europe many of the major manufacturers have both GSM and DECT technologies available, and a merging of the two might be seen as a logical synergy. However, in the early days of the digital cellular market, most manufacturers preferred to focus scarce development resources onto standard high-volume terminal products. By the mid-1990s, with increasing demands emerging from operators for dual-mode products, most major manfacturers recognised the potential significance of this market. Figure 16.5 offers an indicative view of the conversion/penetration of the cellular handset market as a function of price premium that a dual-mode handset might command. If such a market view is correct, then a manufacturer adopting a low-margin, low-premium pricing strategy for the dual-mode handset could use it as a competitive weapon to steal market share, in the same way that low-priced digital cellular handsets were used to do exactly this when GSM was first launched in Germany.

Handset Design

Early approaches to dual-mode DECT/GSM handsets have adopted a very simple and basic approach, that of combining existing DECT and GSM module designs in a single handset (Figure 16.6). Whilst these RF and baseband circuits may be readily re-used, the major product engineering effort involved has related to the development of a single ergonomic user interface and of the necessary control software. The first such terminal, developed by Ericsson in support of trials by the Swedish operator Telia, was submitted for Type Approval in late 1995/early 1996. Several other GSM handset manufacturers are reported to be developing similar products.

Assuming the dual-mode market does become a reality, dedicated dual-mode chipsets and passive components will be developed, to allow both cost and power consumption to be minimised. Such integration will begin with integration of combined baseband functionality, followed some time later by integration of the RF – clearly this will be easier for some types of dual mode, operating in a common frequency band, than for others.

Recent years have seen increasing interest in the so-called "software radio", embodying common hardware and programmable to different modes or standards by software; such a concept has already in fact been implemented

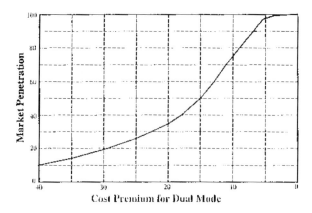

Figure 16.5   Cellular handset market penetration versus price premium. (*Source*>: from [13].)

Figure 16.6   Dual-mode DECT/GSM handset. (Photograph courtesy of Ericsson.)

for some PMR radio products [14]. Nonetheless, the complexity of today's cellular standards is such that it is likely to be still some years before this becomes fully feasible and cost effective for a combined cordless/cellular handset.

Interworking Issues of DECT/GSM

DECT and GSM are particularly well suited to dual-mode integration and interworking because of the close links through the standardisation process, with many common perspectives and approaches having been taken as the standards were developed. As well as the complementary differences noted in Table 16.2, DECT and GSM have many commonalities. Both technologies support two-way calling, authentication, encryption, handover between basestations and the necessary support of location registration that this implies. When considering their integration therefore, the issues that arise are which of the two options for any function to support. These issues relate to the interworking of the infrastructure, rather than to the handset, and are considered in the next section. However, important issues that do relate to the handset include the baseline profile functionality (generic access profile (GAP) and GSM interworking profile (GIP))[3], provision of robust control in switching between cordless and cellular modes and the choice between the GSM subscriber identification module (SIM) or the DECT authentication module (DAM).

Support of the GAP is necessary to support home, public and office access; likewise, support of the GIP is necessary to allow access to DECT systems directly connected to a GSM PLMN. As dual-mode terminals are not tied to a single application, such open standards are essential, if the terminal is to meet its potential functionality to the full.

Mode switching relates to terminal operation as it migrates between different geographical areas of cellular-only coverage, areas of both cellular and cordless coverage, and areas of cordless-only coverage. The handset should detect the presence of the cordless system and check its identity to determine whether the handset has access rights to the system, if so transferring access to the cordless system (assuming this is its preferred mode). This could be done by sending location registration messages to the GSM network, or to the new DECT network if GAP is supported; if GIP is supported, then location registration would be handled automatically by the GSM network. Support of GIP will allow interworking of voice, supplementary services and short message services. (Standardisation of external handover will form part of the second edition of ETS 300 370 – see Chapter 20.)

The use of the GSM SIM card seems to have found favour for the first dual-mode handsets; common players have in any case been involved in the development of the standards for the SIM and DAM cards, and a multi-application smart card is being defined.

16.3.3 Infrastructure Evolution

As implied above, two main approaches to interworking have been explored, the first making use of the GSM mobility management based on the GSM interworking profile (GIP) and the second based on interworking between a corporate network supporting cordless access and the cellular PLMN.

The GIP Approach

The GIP solution is essentially the architecture that was adopted in the Mannesmann trial described earlier. In this architecture the DECT basestation subsystem is connected via an interworking unit directly to a GSM mobile switching centre (MSC) via the GSM A interface. The DECT subsystem is viewed by the GSM network as if it were an ordinary GSM basestation subsystem. All services normally offered by the GSM network, including GSM phase 2 services, are supported in the DECT environment – authentication (GSM based, including SIM), GSM mobility management, short message service, supplementary services, etc. Consideration of the architecture and the different options for mapping GSM and DECT cell coverage, indicated in Figure 16.7, reveals that it can support a number of new types of applications – for example, PABX-type features can be supported by the GSM network such that a small business could find it more cost effective to use such an externally provided service, rather than needing to purchase its own PABX.

The Corporate Network–PLMN Solution

The alternative architecture that has been explored is the interworking of the corporate network and the GSM network (Figure 16.8); as noted above, such interconnection is not yet allowed by regulatory authorities in some countries. The direct interconnection between the corporate network and the PLMN allows the services offered in the two environments to be integrated, thus supporting for example short code dialling, private numbering plans, as well as a range of new services, tailored for the corporate user.

Combined Solutions – The Swiss Platform

The two approaches just described are not necessarily mutually exclusive and an interesting trial combining such interworking possibilities was demonstrated by the the Swiss PTT in Geneva at Telecom '95 [16]. The combined architecture of the Swiss platform is shown in Figure 16.9.

The rationale of this architecture is to use GSM for basic countrywide coverage, augmented by DCS1800 to increase capacity in urban areas, further augmented by DECT to provide high-capacity indoor coverage.

The combined GSM and DCS network is already a fully operational commercial network; the DECT extensions for Telecom '95 provided coverage of the different geographical areas indicated in the diagram – the Swiss pavilion (using the GIP base interworking), Telecom '95 and Geneva airport (using the corporate network–PLMN interworking). The user is reachable on a single number regardless of his or her location or air interface. The results of the trial were viewed favourably by the Swiss PTT and have been reported widely at several conferences. The Swiss PTT plan to provide direct PLMN interconnection to several large corporate networks during 1996 and also to implement the GIP interface to their GSM platform as soon as equipment to the final GIP standard becomes available.

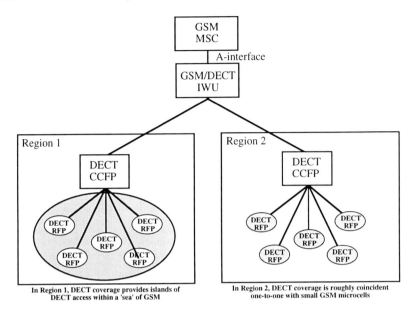

Figure 16.7   DECT/GSM interworking based on GIP. (*Source*: from [15].)

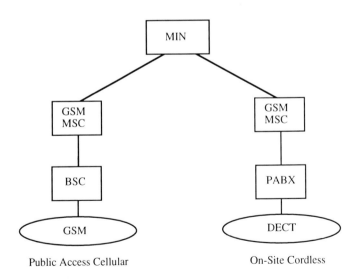

Figure 16.8   DECT/GSM interworking based on corporate network interworking. (*Source*: from [15].)

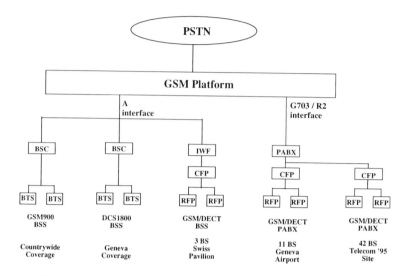

Figure 16.9   The Swiss GSM/DCS1800/DECT platform. (*Source*: from [16].)

Fixed Network Development

As described in Chapter 6, Cordless Terminal Mobility (CTM) aims to support full mobility across a range of cordless environments, evolving as it is from a fixed network perspective. However, as described above, infrastructure trials to support the development of dual-mode cordless/cellular systems have not awaited such developments. Similarly, wireless local loop systems supporting local mobility have also been deployed, such as already described in Chapter 5.

The purpose of such experiments has been twofold – firstly to prove the technical feasibility of new services and secondly to probe the market and explore the potential requirements, in advance of full liberalisation in Europe by 1998. It is envisaged that many mainstream network operators will be content to await the availability of standardised solutions based on CTM, whilst in principle the experimental architectures described above can migrate to adopt the interface standards emerging for CTM. Certainly in Europe the evidence of recent history, of GSM cellular and DECT domestic cordless for example, would suggest that standards can encourage the development of a mass market. Elsewhere, in America and Asia, it will be interesting to see whether the route of cordless evolution follows a similar or divergent path.

Given the cost of providing widespread radio coverage with cordless microcellular basestations and given the emerging proliferation of competing network operators, an approach that most effectively utilises all cordless basestations must be attractive, and of potential benefit to all. A minimum set of standards that will allow the interworking of a federation of networks to provide a consistent service to an extensive population of cordless (or dual-mode) terminals, such that a user can access service almost anywhere where there exists a basestation, would seem a practical and pragmatic way forward.

16.4 Summary

In this chapter we have reviewed current experiments and near-term developments in cordless telecommunications. Although from a primarily European perspective, many of the issues discussed have generic application to North America and Asia.

Changes in the environment arising from regulatory change, market expectations, technology capability, increased competition, enlarged and more segmented markets, new requirements and applications all suggest that the future evolution of cordless systems will continue to be both fast moving and to some degree unpredictable.

Combinations of cordless technology with cellular and fixed networks, with cordless wireless access and city mobility, promise new services and heightened competition. Cordless technology development will continue apace, with continuing and significant cost reductions. If the regulatory environment allows, the advent of dual-mode handsets could influence market shares not simply for cordless, but for the (currently) larger cellular handset market. Certainly the potential capabilities of dual-mode cordless/cellular systems have now been well demonstrated, in the context of European systems. The degree to which similar developments will occur with Japanese and North American standards is at present uncertain.

The evolution path of future digital cordless systems will be influenced not simply by technology, but also by a host of other environmental factors, such as the degree of perceived competition between fixed and mobile, and existing and new, operators and the rate with which dual-mode terminals and systems become a reality in the marketplace. Such issues themselves will be very dependent upon regulatory policies, spectrum availability, etc. This chapter has served to outline technical possibilities – it remains to be seen how market and environmental factors shape the channel down which the technology will flow.

References

[1]   CT2 volumes for non-telepoint applications are taken from D Trivett, Dataquest; CT2 telepoint volumes from sources given in Chapters 3 and 4
[2]   PHS volumes are taken from Financial Times Mobile Communications, No. 188, 8 February 1996
[3]   DECT volumes are taken from ECTEL figures, quoted as part of the presentation "The DECT Review", P Olanders, DECT '96 Conference, London 1996
[4]   "DECT Common Interface Services and Facilities Requirements Specification", ETR 043, ETSI
[5]   "Connection of DECT to NMT 450", J Veres, presented at the DECT '96 Conference, London, January 1996
[6]   "DECT Between Residential Access and Mobile Networks", B Kluth, presented at the DECT '96 Conference, London, January 1996
[7]   "DECT in India", informal presentation by M Harish at the DECT '96 Conference, London, January 1996
[8]   "Where There's a WILL", R Dettmer, IEE Review, pp. 145–8, July 1995
[9]   "DECT Wireless Relay Station (WRS)", ETS 300 700, ETSI
[10] "Application of DECT Wireless Relay Station (WRS)", ETR 246, ETSI
[11] "Cordless Telecommunications in Europe", WHW Tuttlebee (ed.), Springer-Verlag, 1990
[12] "Digital Cordless Communications for the Residential Market", D Trivett, presented at the IIR Digital Cordless Conference, London, January 1996

[13]   "DECT/GSM Dual Mode – Development and Customer Satisfaction Issues", Y Neuvo, presented at the DECT '96 Conference, London, January 1996
[14]   "A 5kHz Channelling VHF Linear Modulation Receiver for Volume Production", SM Whittle, BJ Whitmarsh and RA Hillum, 6th International Conference on Radio Receivers and Associated Systems, IEE, Bath, September 1995
[15]   "DECT/GSM Interworking", T Ryberg, presented at the DECT '96 Conference, London, January 1996
[16]   "The Integration of DECT into the Swiss GSM Platform", P Zbären, presented at the DECT '96 Conference, London, January 1996

Notes

1   Considerable interest also exists in developing dual-mode cellular/cellular terminals of different standards – for example, dual-mode AMPS/DAMPS terminals have already been developed for the North American market. Here we just consider cordless/cellular dual mode.
2   DECT and GSM are considered here essentially in their generic forms – i.e. for GSM read also DCS1800 and PCS1900, and for DECT read also PWT.
3   The generic access profile (GAP) and GSM interworking profile (GIP) are described in Chapter 20.

Part IV

Technical Standards

17 CT2 Common Air Interface

Richard Steedman

The CT2 Common Air Interface Specification [1] is designed to ensure interoperability between second-generation (digital) cordless handsets and basestations from different manufacturers. Originally published as a UK MPT specification [2], it became adopted as an Interim European Telecommunication Standard (I-ETS) in 1990, with a second edition published in 1994. This chapter briefly summarises the technical details of the CT2 CAI Standard.

17.1 Aims and Origins

The original UK CAI specification was prepared by a working group of representatives of CT2 equipment manufacturers and telecommunications operators[1]. This group set itself the following aims:

- To produce a standard that would ensure interoperability between different manufacturers' products and comply with the UK CT2 coexistence specifications in force at the time (MPT 1334 and BS 6833 [3, 4]).
- To permit manufacturers to produce cost-effective equipment across a range of product specifications, i.e. to prejudice neither simple nor complex products.
- To cater for all possible modes of operation envisaged at the time and not to restrict future enhancements.
- If necessary, to sacrifice basestation simplicity if doing so would decrease handset complexity.

When the standard became adopted by ETSI in 1990, much of it remained unchanged from a technical standpoint. (Major editorial changes were of course necessary to convert the document to I-ETS format and a number of operating parameters needed to be "internationalised", for example those concerning the analogue local loop interface.) Improvements in technology and development of the cordless telephony market meant that interoperability and quality of service were increasingly being regarded as factors of prime importance in ensuring the overall success of the system. Consequently, a decision was taken to "tighten up" a number of existing equipment specifications, such as minimum transmitter power, despite the effect that this might have on equipment complexity and cost.

Recall of existing products was not mandated however, and manufacturers were permitted to continue with their production for a limited period of time.

   Following publication of the first edition of the I-ETS, the body responsible for its maintenance, the "Radio Equipment and Systems" Sub Technical Committee of ETSI, continued working on improvements to the system, notably in the areas of call handover and handset location tracking. This work led to the publication in 1994 of the second edition of the standard. This second edition included yet more specification "tightening" with the associated "grace period" to allow for introduction of new equipment.

17.2   Structure of the Standard

The I-ETS 300 131 document is structured into a number of clauses as follows:

   (i) Clauses 1 to 3 contain references, definitions, etc. – all ETSI Technical Specification (ETS) documents begin with these.

   (ii) Clause 4 defines the radio interface by specifying details such as timing and the exact modulation method to be employed, etc. Data is exchanged between handset and basestation (referred to as CPP and CFP in the specification) in binary form using the same radio channel for both directions. This is done by a method known as time-division duplex (TDD). Each end of the radio link transmits data at $72\,\text{kb s}^{-1}$ for 1 ms, then receives data for 1 ms, then transmits, and so on. After allowing for "guard time" (when the transceivers switch from receiving to transmitting and vice versa), this system provides two-way communication at up to $34\,\text{kb s}^{-1}$. (The bandwidth of each channel is 100 kHz.)

   (iii) Clause 5 specifies the manner in which the available bandwidth is partitioned into speech, signalling and synchronisation data channels and also defines the bottom layer of a three-layer protocol for the signalling channel. Layer 1 is responsible for link initialisation and confirmation of the existence of a good link throughout the duration of a call by means of handshake exchange.

   (iv) Clause 6 specifies layer 2 of the signalling protocol, which deals with equipment identification and data error correction/retransmission such that error-free data is passed to layer 3. Layer 2 is based on two existing data transmission standards, MPT 1317 and ITU I.441 [5, 6].

   (v) Clause 7 defines the top layer of the signalling protocol, which consists of a set of message definitions covering, for example, user dialling, handset ringing, public access (telepoint) authorisation, etc. The message set is adapted from the ISDN "digital access signalling system" [7] used by many modern digital telephone exchanges.

   (vi) Clause 8 defines the contents of the speech channel, together with associated acoustic/audio specifications. As with the other parts of I-ETS 300 131, frequent reference is made to existing standards.

   (vii) Clauses 9, 10 and 11 specify equipment tests to ensure compliance with the standard.

Clauses 4 to 8 (which make up the functional description of the standard) are described in more detail below.

17.3   The Radio Interface

The radio interface specifies the conditions pertaining to the transmission of digital data across the RF link. Most of the specifications within it concern the transmitter. In fact, the only conditions placed on the receiver are that it must have a sensitivity of at least 40 dB relative to $1\,\mu V\,m^{-1}$ at a BER of less than 1 in 1000 and must meet certain blocking and intermodulation requirements, originally derived from the UK BS 6833 standard but made more stringent in the second edition.

As mentioned previously, the radio interface employs TDD transmission at $72\,kb\,s^{-1}$. The signal must be transmitted using two-level FSK with a modulation index between 0.4 and 0.7, i.e. the peak deviation from the centre frequency under all possible patterns of "ones" and "zeros" must lie between 14.4 kHz and 25.2 kHz. The FSK signal must be filtered using an approximately Gaussian filter to avoid interference to adjacent channels.

The transmitter centre frequency must be within $\pm 10\,kHz$ of the published channel centre frequency. This $\pm 10\,kHz$ frequency accuracy means that it is possible for each end of a link to be transmitting at frequencies up to 20 kHz apart. Since the FSK deviation may be as low as 14.4 kHz, this means that it is possible for a transmission never to cross over the centre frequency of the receiver. Receivers must therefore employ some form of AC-coupled data demodulation or automatic frequency control (AFC) to receive data. If AFC is used, then it is permitted for this to be linked to the handset transmitter, to bring its transmitted centre frequency closer to that of the basestation. (Clearly, transmitter AFC cannot be permitted at both handset and basestation, otherwise both ends could pull each other off frequency!)

In order to increase user density and also so that a handset close to a multiple-transceiver basestation does not desensitise the basestation receivers, thereby "drowning out" other calls, a means of transmitter power control is provided. All handsets must be capable of transmitting at two power levels, the normal power level being between 5 and 10 mW and the low power level being between 12 and 20 dB lower than the normal level. Low power transmission is only used if the handset receives a command (defined in layer 2 of the signalling protocol) from the basestation. It is not mandatory for the basestation to issue such commands, and it may only do so if the received signal strength exceeds 90 dB relative to $1\,\mu V\,m^{-1}$.

The TDD signalling scheme is illustrated in Figure 17.1 and operates as follows. Each end transmits bursts of either 66 or 68 bits every 2 ms at a rate of $72\,kb\,s^{-1}$. The use of 68-bit bursts is optional – both ends of the link use 66 bits at the start of communication and switch to 68 bits only if one end requests it and the other end indicates that it is capable of supporting the option. Whereas the use of 68-bit bursts obviously increases the signalling bandwidth, 66 bits permits simpler transceiver design or, alternatively, the possibility of accommodating propagation delays of greater than 1 bit.

Each transmitted burst is surrounded by "ramp-up" and "ramp-down" periods; in other words the transmitter power rises and falls gradually at the start and end of a burst. This is done to avoid AM splash, that is, modulation products outside the frequency bands resulting from rapid switching of the transmitter power level. In addition, a mandatory "suffix" of half a bit period must be

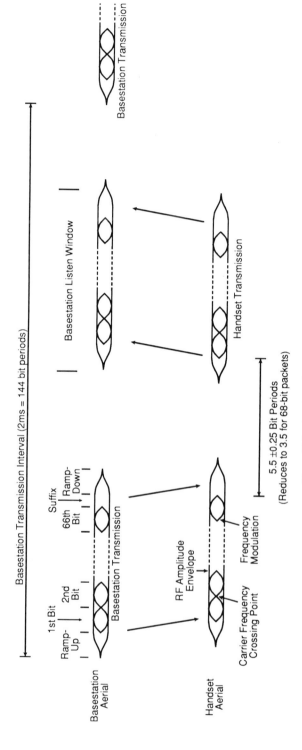

Figure 17.1 CAI time-division duplex (66-bit packets).

transmitted at the end of each burst to overcome any dispersion that may occur in receiver channel filters.

In order to synchronise the two ends of the link, so that one transmits while the other receives and vice versa, the basestation is deemed to be a "master" and the handset a "slave". The basestation is permitted to start transmitting its first burst at any time, thereafter sending bursts at 2 ms intervals. The handset must then "lock on" to these transmissions and arrange for its transmissions to coincide with the basestation's receive period. Put more precisely, the start of the first data bit of a 66-bit handset burst must be transmitted (at the aerial) 5.5 ± 0.25 bit periods after the end of the 66th bit of the basestation burst has been received (again, at the aerial). For 68-bit bursts, this "turnaround time" is reduced to 3.5 ± 0.25 bit periods.

The reason for making the basestation the master is that, in applications where a basestation contains more than one transceiver, such as telepoint, the transceivers must all transmit and receive at the same time to avoid mutual interference. It is therefore often not possible for a basestation to alter its receive window to align with a handset's transmission. This leads to a problem in that, with the approach thus far described, all communication sessions must be initiated by the basestation. Put another way, the handset is unable to initiate communication with any guarantee that its initial transmission will be received by the basestation. One solution to this would be for basestations to transmit bursts continually, purely for synchronisation purposes. An alternative solution is for the handset to transmit a special burst of longer duration than normal when it wishes to initiate communication. Such a scheme is in fact adopted in the CAI. The handset transmits for 10 ms, then listens for 4 ms, before transmitting again. No matter where the basestation's listen window is, it is thus bound to receive part of this "long" burst.

Finally, the rate of data transmission must be maintained to an accuracy of ±50 ppm. The handset must also, of course, maintain long-term synchronisation with the basestation burst rate due to the master–slave relationship explained above. The figure of ±50 ppm derives from the ITU G.711 standard for PCM coding of speech [8].

17.4 Signalling Layer 1

Clause 5 of I-ETS 300 131 defines the bottom layer of the data signalling protocol including the bandwidth allocations for the various channels at different stages of a call. Three channels are defined:

- The B channel, primarily intended for speech, although data may be transmitted over this channel (see section 17.8).
- The D channel, used for in-band signalling.
- The SYN channel, used to allow both ends of a call to obtain bit and burst synchronisation.

These channels are then allocated various bandwidths in four different allocations or "multiplexes" as follows:

(i) MUX1.2 is defined as a 66-bit burst containing 64 B-channel bits and two D-channel bits. It is used after a call has been set up and therefore there is no

SYN channel. (Each end of the link must rely on edge detection to maintain synchronisation in this multiplex.) Since these bursts are transmitted every 2 ms, the resulting data rates are $32\,\mathrm{kb\,s}^{-1}$ for the B channel and $1\,\mathrm{kb\,s}^{-1}$ for the D channel.

(ii) MUX1.4 is similar to MUX1.2 except that the burst is 68 bits long and there are four D-channel bits. Consequently the D-channel data rate is doubled ($2\,\mathrm{kb\,s}^{-1}$). This multiplex is only used if both ends of the link indicate during call set-up that they are capable of supporting 68-bit bursts.

(iii) MUX2 is a 66-bit burst consisting of 32 D-channel bits and 34 SYN-channel bits. It is used at the start of a call to allow both ends of the link to synchronise and to transfer data at a high rate ($16\,\mathrm{kb\,s}^{-1}$) prior to a change to one of the MUX1 variants and connection of the B channel. (There is provision within I-ETS 300 131 for communications to revert from MUX1 back to MUX2 for high-speed data applications but this would not normally happen in the course of an ordinary telephone call.) The SYN channel consists of 10 bits of preamble (101010. . . pattern) followed by one of three different 24-bit synchronisation patterns, referred to as CHMF, SYNCP and SYNCF. (A fourth pattern, CHMP, is also defined and is used in MUX3 – see below.) These patterns are specially chosen to have low correlation with each other and other commonly occurring bit patterns in order to reduce the probability of false detection. The use of a particular pattern out of the three depends on whether it is being transmitted by a basestation or a handset and whether or not that end is trying to initialise the link.

(iv) MUX3 is the "long" 10 ms burst transmitted by the handset when it is trying to establish a link with a basestation. In this multiplex, the 10 ms transmission is divided into five 2 ms "frames", each of which is divided into four identical "sub-frames", i.e. data is repeated four times within each frame. The first four frames contain 20 bits of D channel and 16 bits of preamble in each sub-frame, while the fifth frame contains only SYN channel (12 bits of preamble and the 24-bit synchronisation word CHMP in each sub-frame). The fourfold repetition of data ensures that, no matter where the basestation's listen window is positioned relative to the start of the handset's burst, the basestation will be able to see one of the sub-frames of each of the five frames and thus lock on to the CHMP pattern and receive 80 bits of D-channel data. The reason that preamble is transmitted in the first four frames is to ensure that the D-channel data does not accidentally mimic CHMP and thus cause incorrect synchronisation.

Layer 1 of the signalling protocol deals with link initialisation, handshaking and link re-establishment following loss of data. Link initialisation operates in two different ways depending on whether the handset or basestation wishes to set up the link. In the case of link initialisation by a handset (for example, if a user wishes to make an outgoing call), the handset starts by repeatedly transmitting a MUX3 burst, receiving for 4 ms between each 10 ms transmission. The basestation scans all 40 RF channels periodically for the CHMP pattern and, when it detects one, decodes the following 80 D-channel bits. These will contain a layer 2 message called LINK_REQUEST. If the basestation "recognises" the handset (depending on the contents of various fields in the LINK_REQUEST message), it replies in MUX2 with a message called LINK_GRANT. The handset then slaves to the MUX2 bursts, transmitting in MUX2 aligned to the basestation's receive window. After a period of data exchange in MUX2, the basestation sends a layer 3 message ("channel control") to the handset telling it to switch to MUX1. When the basestation receives a reply in MUX1, it switches to MUX1 itself.

Link initialisation by the basestation (for example, when incoming call ringing is detected by the basestation) is slightly different in that the basestation may "broadcast" messages to up to 32 different handsets. This provides a "group call" facility. Only one handset may request the link, however. (This would occur when a user decided to answer a call. At that point, the other handsets would "drop out" and stop ringing.) It is important to realise that the case of a basestation calling a single handset is in fact a special case of "group call".

When a basestation wishes to call one or more handsets, it starts transmitting MUX2 bursts using the CHMF pattern in the SYN channel. The contents of the D channel is a sequence of messages every other one of which is a "poll" message to indicate which handset(s) is (are) being called. For example, if three handsets are being called, then three different poll messages are transmitted in a cyclic fashion, one message addressing each handset. (The other intervening messages are broadcast to all the handsets being polled and might include commands to switch the ringer on and off, to flash an icon or to display a message. These intervening messages may, however, be omitted and the equivalent amount of "idle" pattern (see below) transmitted instead.) A handset scans all 40 channels periodically and, if it detects a CHMF, decodes the following D-channel data. If it is being called, it replies with a poll response message in three MUX2 bursts. (When more than one handset is being called, their responses are interleaved.) When all the polled handsets have replied, the basestation switches to using the SYNCF pattern in the SYN channel, to avoid waking up other handsets. When one of the polled handsets wishes to establish the link, it replaces its poll response with a LINK_REQUEST message. The basestation then replies with LINK_GRANT and the set-up proceeds in the same manner as for the handset-to-base process described above.

Throughout link establishment, a series of time-outs ensures that neither end repeats transmission of a message "for ever". Safeguards are also built in to resolve "deadly embrace" (i.e. both ends trying to establish a link simultaneously) and response collision situations.

Once a link is set up, both ends transmit handshake messages in the D channel at a rate of between once every 400 ms and once every second. The lower rate is chosen to ensure that losses of link can be detected reasonably quickly, while the upper rate ensures that handshakes are not "forced through" a poor-quality link by rapid retransmission.

If either the handset or the basestation detects that the link quality is poor (e.g. through loss of handshake messages), it may request that the link be "re-established". In the first edition of the standard, the method of re-establishment is similar to link set-up by a handset in that the handset starts transmitting in MUX3, the basestation replies in MUX2, and so on. Re-establishment is permitted to occur on the same RF channel not more than once every 600 ms when using MUX1.2 or once every 300 ms when using MUX1.4. The handset may also attempt to re-establish the link on a different RF channel, but this may only be done after handshake messages have been lost for at least 3 s.

The second edition of the standard makes significant changes to the rules for channel re-establishment, which are designed to improve handover performance in multiple-basestation systems such as cordless PABXs and two-way public access systems. Basestations may now command handsets to re-establish on a different channel immediately (i.e. even if handshake has not been lost) and an optional method of synchronous re-establishment is defined that avoids the use of MUX3.

17.5 Signalling Layer 2

Layer 2 of the signalling protocol defines the contents of the D channel and is responsible for link maintenance, that is:

- Providing error-free communications between the handset and basestation layer 3 processes.
- Acknowledgement and retransmission, if necessary, of messages. (Unacknowledged transmission is also provided and is used, for example, for broadcast messages from a single basestation to multiple handsets.)
- Validation of handset and/or basestation identities during link set-up.
- Monitoring of link quality and re-establishment if necessary.

It is based on a pre-existing data signalling protocol MPT 1317 [5]. "Packets", consisting of between one and six "code words", are transmitted in the D channel, each preceded by a 16-bit synchronisation word (called SYNCD). Each code word consists of 48 data bits and 16 check bits. The first code word in a packet is called an address code word (ACW) and subsequent code words are called data code words (DCWs). If there are no packets to send, an idle pattern (IDLE_D) is transmitted.

Layer 2 packets are divided into two types – fixed length and variable length. Fixed-length packets consist of a single ACW and are used during link set-up and also for handshaking. (The messages LINK_REQUEST and LINK_GRANT mentioned above are examples of this type of packet.) Note that a fixed-length packet occupies 80 D-channel bits (16-bit SYNCD + 64-bit ACW) and therefore fits into one MUX3 burst or three MUX2 bursts.

Variable-length packets consist of an ACW and optional DCWs and are further subdivided into two types called "information" and "supervisory". Information-type packets contain layer 3 messages, while supervisory types contain messages concerned with link maintenance. (Examples of the latter include the command from the basestation to the handset to switch the latter's transmitter power between the normal and low settings and the message from either end requesting link re-establishment.) Both types of variable-length packet may be sent unacknowledged or acknowledged. In the case of acknowledged transmission, packets are alternately numbered 0 and 1 and a packet is transmitted repeatedly until an acknowledgement is received. The acknowledgement takes the form of a 1-bit number indicating which packet is next expected (i.e. the inverse of the packet last successfully received) and is contained in one of the fields of a packet being transmitted in the opposite direction (a technique known as "piggybacking"). Such an acknowledgement/retransmission scheme is referred to as a "1-bit sliding window protocol" [9].

The second edition of the standard introduces a number of new features to layer 2 of the signalling protocol to provide extra system functionality in multiple-basestation applications. These include the definition of a "CFP identity and status" (CIS) code word and procedures for location tracking of handsets. The CIS code word may be broadcast by basestations to indicate information such as the type of service provided, location of the basestation, etc. (Handsets could use such information to indicate to the user whether they are within range of a particular public access service, for example.)

Location tracking of handsets allows a multiple-basestation system to select the most appropriate basestation with which to communicate with a handset at any particular time. The standard permits two methods. The first, referred to as "polling only", simply involves the basestation detecting the presence of handsets by using the polling part of the layer 1 link initialisation procedure (i.e. the "group call" procedure described earlier). The second method, "CIS assisted", requires handsets to monitor CIS transmissions and to inform the base system by means of a short call whenever they move out of the coverage area of one basestation into that of another.

17.6   Signalling Layer 3

Layer 3 of the signalling protocol defines the meaning of the messages that are passed between handset and basestation by layers 1 and 2. As mentioned previously, it is based on the ISDN DASS, although only a subset of the messages has been used. (This has also been recoded and a number of new messages added.)

The signalling system used in layer 3 is known as "stimulus mode signalling". In this type of signalling, the information contained in messages is of a fairly low-level nature. Thus, instead of the handset sending a command message to the basestation, such as "Attempt to connect to telephone number 01123 456789", the messages contain information, such as "The user has pressed the 0 key", or commands, such as "Switch the ringer on".

It is not mandatory for either the handset or the basestation to generate or respond to the entire message set. For example, handsets without a display may ignore commands to display numerals and/or icons. A minimum subset of messages must, however, be supported by all equipment and a further subset supported by equipment intended for public access use.

The message set may be considered as being divided into the following broad areas. (It is not intended to discuss individual messages in detail – readers are referred to I-ETS 300 131 for further details.)

- Transmission of dialled digits from handset to base, and display information vice versa.
- Generation of tone caller alerting signals, e.g. incoming ringing, error, etc.
- Outgoing call selection, including ordinary domestic, telepoint, PABX, intercom, emergency, etc., together with other exchange features such as recall, follow-on, hold, etc.
- Call progress indication. This might be used, for example, to control icons on a handset display.
- Connection and disconnection of the audio channel together with sidetone control.
- Registration of handsets to basestations "over the air". Basestations do not recognise handsets until they have been programmed to do so.

- Telepoint call authentication. Whereas a domestic basestation only recognises handsets that are registered on to it, telepoint basestations initially recognise all handsets and then proceed to determine whether or not the user has a valid account with the service provider.
- Indication of handset and basestation capabilities. For example, a handset may indicate to a base that its display can only indicate numerals.
- Storing and retrieval of "parameters" (information particular to an individual handset or basestation). For example, a basestation may programme handsets with different "classes of service", e.g. ability to make international calls, etc.
- Selection of alternative message sets for future enhancements.

17.7   Speech Coding and Transmission Plan

Clause 8 of I-ETS 300 131 contains the specifications for coding and decoding of analogue information transmitted in the B channel. These are subdivided into two categories:

- The specifications relating to the conversion of the analogue signals to digital form and their subsequent compression.
- The specifications for the remaining analogue parts from the mouth reference point (MRP) to the telephone line and back to the ear reference point (ERP) – the "transmission plan".

The analogue-to-digital conversion specification is straightforward, namely that the contents of the B channel is that produced by the ADPCM algorithm defined in ITU recommendation G.721 (1988) [10]. (Unfortunately, this algorithm differs sufficiently from the one published in the previous ITU recommendations (1984) to make interworking between the two impossible.)

Two concessions are permitted from the full G.721 algorithm in certain instances. Firstly, if a basestation is connected to an analogue line interface (as opposed to, say, an ISDN interface), it need not implement features such as PCM format conversion, etc. Secondly, the standard permits handsets manufactured before March 1993 to use an algorithm of reduced specification, provided that, when used with a basestation employing the full algorithm, certain speech quality tests (derived from BS 6833) are met.

The transmission plan specifies parameters such as frequency response, loudness, noise and distortion for the various analogue processes. (A high-level block diagram of the plan, with nominal values for some of these parameters at 1 kHz, is shown in Figure 17.2.) It also specifies a maximum handset "loop" delay, i.e. the time taken for a signal to travel from the handset MRP to the basestation line interface and back to the ERP. This has been specified as 5 ms and has been chosen to maximise the permissible processing time for the ADPCM algorithm whilst avoiding the need for echo cancellation.

The transmission plan has been devised to comply with different countries' specifications for connection to analogue line interfaces, together with ETS 300 085, the candidate NET33 standard for connection to digital exchanges [11]. One important consequence of making the CAI compatible with digital interfaces is that handsets must be capable of generating sidetone locally. (With an analogue connection, sidetone is generated by the hybrid circuit in the basestation.)

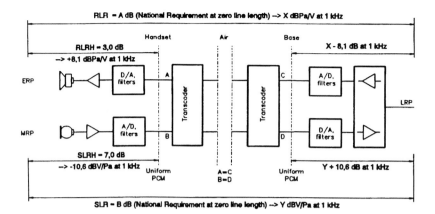

Figure 17.2   CAI voice transmission plan. (*Source*: ETSI, from I-ETS 300-131, used by permission.)

17.8   Data Services

Although the CAI is primarily intended for voice telephony, circuit-switched data services are permitted and protocols are defined in an annex to the standard. In order to allay concerns that a large number of data calls (which in some applications are set up for much longer periods that voice calls) may occupy an unfair proportion of the overall RF bandwidth, data calls are classified as a "secondary service" and subject to special call set-up rules. These rules may prevent a data call being set up if a large number of RF channels are busy.

Two classes of data protocol are specified – synchronous and asynchronous. The synchronous protocol includes forward error correction (FEC) by means of Reed–Solomon codes [12] but no retransmission of errored data and hence does not guarantee data integrity. The asynchronous protocol includes both FEC and automatic repeat request (ARQ) error control. The latter is a modified form of the X.25 LAPB protocol [13] and operates in a similar manner to that of CAI layer 2 acknowledged code word transmission.

References

[1]   "Radio Equipment and Systems; Common Air Interface Specification to be Used for the Interworking Between Cordless Telephone Apparatus in the Frequency Band 864.1 MHz to

868.1 MHz, Including Public Access Services", I-ETS 300 131, 2nd edn, ETSI, Sophia Antipolis, November 1994

[2]   "Common Air Interface Specification", MPT 1375, UK Department of Trade and Industry, London, May 1989; amended November 1989 and February 1990

[3]   "Performance Specification for Radio Equipment for Use at Fixed and Portable Stations in the Cordless Telephone Service Operating in the Band 864 to 868 MHz", MPT 1334, UK Department of Trade and Industry, London, 1987

[4]   "British Standard: Apparatus Using Cordless Attachments (Excluding Cellular Radio Apparatus) for Connection to Analogue Interfaces of Public Switched Telephone Networks", BS 6833: 1987, British Standards Institution, London, 1987

[5]   "Transmission of Digital Information Over Land Mobile Radio Systems", MPT 1317, UK Department of Trade and Industry, London, 1981

[6]   "Integrated Services Digital Network (ISDN)", ITU Red Book, Vol. III, Fasc. III.5, Recommendation I.441, 1984

[7]   "Digital Access Signalling System", ITU Red Book, Vol. VI, Fasc. VI.9, Recommendation Q.931, 1984

[8]   "Pulse Code Modulation (PCM) of Voice Frequencies", ITU Red Book, Vol. III, Fasc. III.3, Recommendation G.711, 1984

[9]   "Computer Networks", AS Tanenbaum, Prentice-Hall, Englewood Cliffs, NJ, 1981

[10]   "32 kbits/s Adaptive Differential Pulse Code Modulation (ADPCM)", ITU Blue Book, Recommendation G.721, 1988

[11]   "Integrated Services Digital Network: 3.1 kHz Telephony Teleservice; Attachment Requirements for Handset Terminals", ETS 300 085, ETSI, Sophia Antipolis, 1990

[12]   "Error Control Coding", S Lin and DJ Costello Jr, Prentice-Hall, Englewood Cliffs, NJ, 1983

[13]   ITU Blue Book, Vol. VIII, Recommendation X.25, 1988

Note

1   The author of this chapter was one of the participants in the working group.

18 The Personal Communications Interface, PCI

Gary Boudreau

The Personal Communications Interface (PCI) standard [1] is designed to ensure interoperability and interworking between second-generation (digital) cordless handsets and basestations from different manufacturers operating in the North American unlicensed frequency band (1920 to 1930 MHz) allocated for personal communications services (UPCS). The equipment is intended to convey digitally encoded speech and data with associated digital signalling, via a radio channel, to and from a nearby fixed station, or network of fixed stations. The PCI standard is derived from the CT2 Common Air Interface [2] to allow operation of CT2-derived equipment in the UPCS band according to the US Federal Communications Commission (FCC) "Spectrum Etiquette" rules defined in [3] and summarised in Chapter 8. In addition, a number of CT2 capabilities have been enhanced in PCI: automatic handoff, location tracking, roaming, rapid call set-up and battery life. This chapter briefly describes the technical differences between the PCI and CT2 standards and may be best understood with reference to the previous chapter.

18.1 Aims and Origins

The PCI standard has been prepared under the auspices of the TIA TR41.6 Subcommittee. This Subcommittee has developed standards for two-way interworking between fixed and portable radio devices operating in the isochronous UPCS band. Applications of the equipment include wireless access to business and residential services. The standard is intended to allow a user to migrate from one personal communications environment (domestic, private branch exchange, key system, Centrex) to another with a single portable device. Owing to the additional constraints required to follow the FCC Part 15 Subpart D "Etiquette" rules for sharing the spectrum in the unlicensed band, compatibility or interoperability is not maintained between equipment operating according to the PCI standard and equipment following I-ETS 300 131.

18.2 Structure of the Standard

The PCI standard follows the same basic structure as the I-ETS 300 131 document [2], which defines the CT2 standard.

(i) Chapters 1 to 3 of the standard define the scope and organisation of the document, as well as providing lists of references and definitions.

(ii) Chapter 4 provides the details of the physical radio interface including the channel frequencies and requirements for modulation, channel monitoring, channel selection and interference rejection.

(iii) Chapter 5 defines the layer 1 communications interface; it includes the definition of time-division duplexing, data multiplexing, link initiation and handshaking. This layer allows systems to obtain mutual synchronisation over a digital channel and provides bi-directional channels over the Common Air Interface (CAI) for digital data and digital speech data.

(iv) Chapter 6 defines the layer 2 communications interface. Layer 2 covers the channel protocols, message formats, error detection, error correction and message acknowledgement. This layer allows systems to communicate over an established link using data and channels that are established and maintained free from interference.

(v) Chapter 7 defines the layer 3 communications interface. Layer 3 defines the structure of and attaches meanings to messages. Part of the message space is open in order to accommodate future expansion of services and facilities.

(vi) Finally, chapter 8 defines the speech coding and telephony requirements.

18.3   Principal Changes from CT2

This section defines the major changes added in PCI in order to ensure adherence with the spectrum etiquette defined in the FCC Part 15 Subpart D rules [3] (hereafter FCC 15D). Additional details are defined in the following sections of this chapter. The main changes are as follows.

(i) To facilitate sharing of the spectrum among different systems, FCC 15D [3] defines the following requirements:

- Both the cordless fixed part (CFP) and cordless portable part (CPP) must always monitor the intended transmit channel during the intended transmit interval prior to transmitting to ensure the channel is free. This process is referred to as channel monitoring or "listen before transmit" (LBT).
- The CFP or CPP must not transmit on a channel if foreign signals are detected above a threshold.
- There are a number of RF parameter requirements dealing with power level, power spectral density and allowed levels of in-band and out-of-band emissions that require changes from CT2.

(ii) The spectrum assignment for PCI spans the frequencies of 1920 to 1930 MHz. This defines 99 possible frequency channels. Of these 99 channels, not all can be employed, since some of them fall on the 1.25 MHz segment boundaries set by FCC rules [3] and are thus excluded from use. The expanded channel range introduces a number of operational and protocol changes to accommodate the larger number.

(iii) The CFP (or group of CFPs) will, at all times when it is capable of providing service, maintain a marker channel (MC) stream to indicate to nearby mobile devices (CPPs) that they are allowed to operate in the area. The CPP is forbidden from operating its transmitter unless it can properly receive and decode a marker channel provided by the CFP or the CPP is paged on a traffic channel. Marker channels operate using the MUX2 transmissions. The marker channels are intermixed with the traffic channels and may change their channel assignments according to local traffic and interference conditions.

(iv) Operation of equipment in the UPCS band can only take place in areas coordinated (coordinatable areas) by UTAM Inc. (Unlicensed PCS *Ad Hoc* Committee for 2 GHz Microwave Transition and Management). This requirement originates from the fact that incumbent microwave users still employ the UPCS band in certain geographic locations in the USA as part of the fixed radio service. The radio equipment operated in the UPCS band is forbidden from causing interference to equipment operating as part of the fixed radio service. The fixed radio services may be operating as receivers or transmitters in the same portion of the spectrum as the PCS equipment. To ensure that there is no interference between the two services, the PCS equipment is only to be operated in (geographic) areas where its operation is known not to cause interference to the fixed service. It should be noted that UTAM is an interim organisation put in place until all of the incumbent microwave users have been cleared from the UPCS band. To ensure that CPP devices do not operate outside of UTAM coordinated areas, PCI employs beacon or marker channels (MCs). All CPPs are forbidden from transmitting unless they have detected a valid transmission on a PCI marker channel. Furthermore, owing to this requirement, MUX3 is not supported in PCI. Marker channels are also employed to track the CPP location and to initiate call set-up.

18.4   The Radio Interface

There are several changes in the PCI radio interface, as compared to the CT2 standard.

(i) The maximum RF transmission power has been increased to 15 dBm, as allowed by the FCC etiquette rules. The normal level of operation is between 12 dBm and 15 dBm.

(ii) The second major difference in the radio interface of PCI as compared to CT2 is the FCC requirement for channel monitoring. Before using a channel, each device (both CFP and CPP) must monitor the channel during at least the intended transmit intervals. Monitoring consists of measuring the signal strength over the RF emission bandwidth centred on the channel for an interval of not less than 10 ms, i.e. five transmit time slots. The channel is considered free if the signal meets *one* of the following threshold criteria:

- The monitored level is below a lower threshold of −94 dBm.
- If all of the channels that the CFP is capable of accessing are monitored above the lower threshold and at least 40 of these have been monitored within the preceding 10 s, then the channel that has the lowest strength of all channels monitored and does not exceed −74 dBm (the upper threshold) can be considered to be free.

The monitoring function reports the peak signal strength for the monitoring interval and must be capable of reliably reporting a pulse signal as short as 175 μs at the lower threshold and as short as 120 μs at a level 6 dB above the lower threshold. The channel may be used only if the monitored level is below the applicable threshold. If the level is above the threshold, then another channel may be selected and monitored or the call may be blocked. An additional constraint placed on the selection of a channel is that the lowest numbered channel that is free according to the above definition must be chosen in order to satisfy the channel packing rules of the FCC spectrum etiquette [3]; for PCI, channels must be packed from the bottom end of the available spectrum (i.e. 1920 MHz and up).

(iii) In each cell, one radio fixed part (RFP) of a CFP is designated for signalling operation and the designated RFP selects a marker channel subject to the above channel monitoring requirements. Once a marker channel has been selected, the designated RFP establishes a signalling stream on a free marker channel by transmitting a code word, typically a CFP_INFO code word, on the marker channel (see section 18.6). If a CFP signalling on a marker channel does not receive an acknowledgement from a CPP within 30 s of establishing the signalling stream or receiving a previous acknowledgement from a CPP, the CFP shall stop transmission and remonitor the channel to verify that it is free before continuing transmission.

(iv) In addition to the above major changes from CT2, there are several additional RF changes for PCI in terms of channel emission requirements, spurious response, power spectral density and interference rejection capabilities [1]. These changes are due to a combination of FCC spectrum etiquette requirements and the increase in transmit power of PCI with respect to CT2.

18.5   Signalling Layer 1

There are several changes in PCI layer 1 compared with CT2.

(i) As mentioned earlier, the major change in signalling layer 1 is that there is no MUX3 in PCI. Since there is no MUX3 in PCI, the CHMP synchronisation pattern employed in the SYN channel is also not defined.
(ii) CPPs and CFPs support MUX1.4 only, which is referred to as MUX1; MUX1.2 is not employed in PCI.
(iii) In addition, the call set-up procedures have been modified to accommodate and exploit the requirement for marker channels.

As described earlier, the need to coordinate usage of the spectrum with existing users necessitates the use of marker channels for all call set-up activity. Within each cell, the CFP maintains its marker channel at all times when it is capable of handling calls by transmitting the CFP_INFO code word. The CPP is expected to find the strongest marker channel and locally register to it in order to facilitate location tracking. Furthermore, the CPP will regularly monitor the marker channel for network-originated calls. Only the CFP selects traffic channels; the CPP is restricted to responding on the active marker channels or the traffic channels on which it is polled. All transmissions over the marker channel take place in MUX2.

PCI supports link re-establishment for both mobility handover and interference handover in a manner similar to CT2. Mobility handover allows the CPP to re-establish its link with a different CFP in a different cell. In PCI handover is provided that is synchronous, such that the CFP transmits first in MUX2. If a synchronous handover fails, the CPP tries to re-establish the call using a procedure similar to initial call set-up through use of the marker channel.

18.6   Signalling Layer 2

The following changes have been made in the PCI layer 2 operation with respect to CT2:

(i) For location tracking, the concept of making the CPP responsible for local registration is emphasised. Location registration via marker channels is a mandatory automatic background task for each CPP. Four additional fixed-length packet types have been added to PCI in order to accommodate local registration at layer 2. A handshake protocol has been defined using these messages to allow CPPs to locally register in the cell that they are located. This is a layer 2 activity in PCI as compared to CT2, for which local registration is a layer 3 activity.

(ii) Link identification codes (LIDs) have been redefined in PCI in terms of link references and system access LIDs in order to clarify LID usage.

(iii) Both the fixed- and variable-length messages have a marker channel field added, which indicates whether the channel in use is a marker channel or a traffic channel. This aids the CPP in detecting the presence of a marker channel.

(iv) The CFP_INFO code word has been modified to include a CFP_INFO identifier field (CID), which includes a local-area identifier (LAI) and local CFP identifier (LCI). The LAI typically is employed to group CFPs into common areas, typically cells. An LAI can also be a cluster of cells. The CFP_INFO code words are employed as beacon messages on the marker channels and are transmitted at least every 12 ms. After having detected a valid marker channel beacon and locally registered, a CPP will periodically monitor the MC at least every 1.5 s to ensure that the marker channel signal can still be detected.

(v) For CFP-originated calls, the CFP sends a paging signal sequence on the marker channel consisting of a LINK_GRANT code word, SCAN_MAP code word and CFP_INFO code word. This may possibly be from more than one active marker channel. The LINK_GRANT code word is identical to that employed in CT2. The SCAN_MAP code word is a variable-length link supervisory code word that provides channel information to the CPP concerning which channel will be employed for the traffic channel poll during call set-up. The CPP will be polled concurrently on the selected traffic channel. The CPP may respond to the page on its selected marker channel and establish the link on the indicated traffic channel.

(vi) For CPP-originated calls, the CPP sends a call request message (LINK_REQUEST) over its selected marker channel to alert the CFP. The CFP responds to this on the marker channel and will begin polling on its selected traffic channel. Call set-up proceeds in an identical manner to CFP-originated calls. This method allows the CFP to set up the call faster than in CT2 and does not use MUX3 operation since this would violate the requirements for coordinatable operation.

(vii) PCI does not support group calling in the same manner as supported in CT2. Rather than each CPP in a group call being polled on a single traffic channel, as in CT2, multiple CPP calling is supported over the MC through the use of a POLL IN PROGRESS message. Each CPP in the group call is polled on a separate traffic channel.

18.7   Signalling Layer 3

PCI supports all of the layer 3 messaging of CT2 with the exception of the following changes and additions:

(i) When answering non-group CFP-originated calls, the CPP automatically begins layer 3 by issuing a NULL feature activation (FA) message. This allows acknowledged information transfer to take place prior to the CPP taking any action. Features such as caller ID can thereafter be offered by the CFP (see below).

(ii) If, during a link, the CPP is deliberately placed in a state in which it cannot continue communications, such as being powered off, the CPP will then issue a CLEAR FA message to terminate the link.

(iii) Minimum mandatory feature sets are defined for most of the layer 3 information elements.

(iv) Message waiting indication has been added as a layer 3 feature.

(v) Calling line identification has been added as a layer 3 feature. This parameter provides the CPP with information about the far-end party's name and number. This parameter may be sent during alerting or during the call to provide the CPP with data about the caller.

(vi) The majority of the layer 3 packets are acknowledged at layer 2.

(vii) The local registration parameter information elements are no longer supported at layer 3 as in CT2 since this functionality is now supported as part of layer 2 in PCI.

(viii) Layer 3 preferred channels are not supported in PCI since this would violate the FCC spectrum etiquette channel packing requirements.

18.8   Speech Coding and Telephony

The speech coding and telephony requirements are almost identical to those of CT2 except for several minor modifications to the requirements for frequency response, loudness rating, sidetone and noise. These changes were made in order to harmonise PCI with the requirements for operation in North America.

References

[1] "The Personal Communications Interface (PCI) Standard", TIA Standard Proposal SP-3244, 19 June 1995

[2] "Radio Equipment and Systems: Common Air Interface Specification to be Used for the Interworking Between Cordless Telephone Apparatus in the Frequency Band 864.1 MHz to 868.1 MHz Including Public Access Services", I-ETS 300 131, ETSI, November 1994

[3] "Radio Frequency Devices, Subpart D – Unlicensed Personal Communications Service Devices", FCC 47 CFR 15

19 The DECT Specifications

Heinz Ochsner

The Digital Enhanced Cordless Telecommunications (DECT) specifications developed by ETSI Subtechnical Committee RES03 provide an air interface specification for cordless technology able to support a wide variety of services. DECT is essentially a common interface specification; a subset of it, however, may serve as a coexistence specification in those cases where no adequate profile exists. The DECT standardisation process began within ETSI in 1988 and still continues in 1996, as its power is recognised and new extensions and applications are developed. Many of the DECT specification documents have been approved and published by ETSI and have been available since 1992. By 1995 there were at least seven manufacturers offering commercial DECT products from all the application categories, with basic handset products, for example, in very high-volume manufacture. In this chapter we outline the structure and content of the DECT specification documents, describing various aspects of the operation of the DECT protocols, also pointing the reader to the relevant original sources for more specific information should this be needed.

19.1 The DECT Services

The main objective of the DECT standard was to provide a specification able to support many applications, such as residential cordless telephones, business systems or public access networks, and evolutionary applications, such as radio local-area networks (radio LAN) or radio local loop (RLL) systems. To understand the DECT standard, we need to realise how the ETSI standardisation committees were seeing cordless telecommunications during the late 1980s.

The main service principles adopted were to provide a system specification for both voice and non-voice applications. DECT equipment should function as an equivalent replacement for a wired telephone connected directly or indirectly, e.g. via a PABX, to a public wired network such as a PSTN or an ISDN. In the next subsections, these four areas of applications are described in further detail.

19.1.1 Residential Cordless Telephony

This is the ordinary domestic cordless telephone, a product that first emerged in the early 1980s. In its simplest form it consists of a cordless handset and a basestation that also acts as a charging unit for the handset's batteries. Current DECT residential products include the use of several handsets with intercom facility. More evolutionary products allow connection to more than one PSTN line or to an ISDN basic-rate interface.

19.1.2   Cordless Business Communication Systems

Cordless business communications systems basically provide the same functions as today's business telephone networks, i.e. key telephone systems or private automatic branch exchange (PABX) systems, but work with cordless handsets. To provide the user's premises with enough capacity whilst using the spectrum efficiently, the system concept envisages use of microcellular coverage. As illustrated in Figure 19.1, a cordless business communication system consists of a wired switching system with a unit called a cordless extension. Cordless basestations attach to this extension normally using wireline connections. The extension routes calls to basestations in one or several cells via which cordless handsets or portable radio data terminals can communicate with the network. Portable-to-portable traffic always goes via the network, even though the radio conditions could allow direct communication. Of course, the cordless extension in this case needs to provide mobility functions (as in cellular systems), in particular handover or, in the case of very large systems, location registration. Many terminals of the system may still remain wired such as high-comfort telephones, workstations, or access lines to mainframe computers.

19.1.3   Public Access

Public access DECT services potentially encompass telepoint, wireless local loop and integration with public cellular networks.

Telepoint was the original cordless public access service introduced using not DECT but CT2 technology in the late 1980s. The telepoint concept provides cordless access to the public telephone network via public basestations, offering two-way calling but with more limited range and mobility than a cellular telephone system. A handset, which can also be used at home and in the office, can access the telephone network and the owner can be billed to his or her home address. An authentication and billing centre is needed (Figure 19.2).

Telepoint systems cannot provide nationwide coverage but may be installed at locations where high call traffic is to be expected, e.g. railway stations, airports, shopping centres, restaurants. The main difference between a telepoint service and a cellular mobile telephony service is that telepoint makes use of the existing telephone network while cellular systems require dedicated networks with sophisticated intelligence including mobile switching centres, location registration centres, and so on. It is these differences which allow telepoint to be a cheaper service, as well as requiring cheaper equipment compared to cellular systems.

Service providers may further get access to a large population of owners of cordless handsets by using cordless technologies; in some countries the

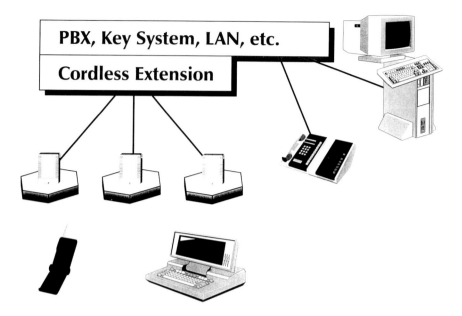

Figure 19.1   Cordless business communication system.

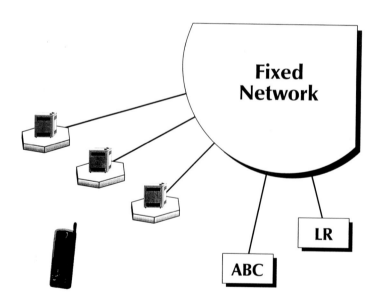

Figure 19.2   Telepoint system; ABC = authentication and billing centre, LR = location register (in some networks).

penetration of cordless telephones already has reached 80% of all households. In return, the limited capabilities of many of today's telephone networks do not yet allow the outgoing calls from the network to the handset. This drawback, however, is being overcome with the evolution of the wired network into ISDN (offering supplementary services such as "follow me") and by the use of complementary services such as wide-area paging.

Most of the early European telepoint services were commercial misfortunes and closed down after only a few years. There are still telepoint networks being installed and opened outside Europe. It is not in the scope of this chapter to discuss their likely commercial success because the telepoint service needs a wider vision in the context of mobile communications. Issues like the availability of cellular services (in terms of coverage, available capacity and affordability) require attention as well. One of the European telepoint installations – the French Bi-Bop system – has solved one of the major disadvantages of the original telepoint idea by introducing two-way calling. This indicates that the original concept was just too simple, at least for the European markets.

For such reasons, modern approaches to cordless public access using DECT – while avoiding any reference to telepoint – combine a two-way service of that type, potentially covering large areas with clusters of cells, with other features, such as offering two-way services to the home or business (wireless local loop – see below) or combining DECT access with cellular infrastructure, as discussed in earlier chapters.

19.1.4   Evolutionary Applications

DECT allows integration of voice and non-voice services into one radio transmission scheme. In particular, cordless data applications such as the radio LAN have been implemented. Powerful dynamic allocation of radio spectrum allows efficient use of the available capacity and as such provides high data rates – a single DECT data link can support throughput in excess of $250\,\mathrm{kb\,s^{-1}}$ as described in Chapter 14. Appropriate radio access techniques guarantee efficient use of radio spectrum in the case of bursty or variable-throughput data transmission.

Another important application for which DECT has been specifically designed is the so-called radio local loop (RLL) or wireless local loop (WLL). RLL provides the telephone link to the subscriber premises by means of radio rather than copper. Radio in many cases is a cheaper medium than buried copper, as outlined in Chapter 6. What is even more important, radio links allow far quicker installation than wireline connections. This is particularly important in situations where telephony service to the premises is generally offered by competing operators. It is, therefore, no surprise that, as soon DECT became available, manufacturers were offering DECT-based RLL systems.

19.1.5   Basic Features and Capabilities

To allow all these applications, the designers of the DECT standard included many unique basic features, notably:

- Very high voice call capacity, up to $10\,000\,\mathrm{E}\,\mathrm{km}^{-2}$ per floor in indoor environments (E = Erlang), with speech quality similar to wired networks.

- Very high data capacity ranging from $24.6\,\mathrm{kb}\,\mathrm{s}^{-1}$ for individual connections up to several $\mathrm{Mb}\,\mathrm{s}^{-1}$ total network capacity at bit error rates below 10^{-8}.

- High-capacity basestations, starting from 5 Erlang (12 simultaneous calls), with flexible, yet simple, radios.

- Flexible and adaptive use of the time–frequency space (240 slots), avoiding the need for frequency planning.

- Support of ISDN services including Group IV fax, charging services and others.

19.2   Structure of the Specification Documents

The DECT standard is applicable to many different products. Many such products are already available on the market today – e.g. cordless phones, radio local loop, radio LAN, plug-in PC cards, etc. Other innovative applications of DECT have still to be discovered but will be possible. The challenge of providing a standard that at the same time is applicable to a cordless telephone as well as to a product we do not even yet know has led to the concept of the DECT access profiles.

Figure 19.3 shows the overall structure of the DECT standards. The "bedrock" is the system specification. It contains all the building blocks needed to design any DECT-based cordless product. Of course, most products only require a subset of the whole collection. These subsets are defined in profiles. Such a profile should be no more than just a list of which elements from the base specification have to be combined in order to make a product according to that profile. In reality this is almost true; in practice there was sometimes a need to extend the base specification a little bit. To use again the example of the building blocks – there are a few cases where the building blocks are defined in the system specification but not the way that they have to be put together. Here, the profiles may go beyond just a list of items. Section 19.5 lists very briefly some of the profiles available or being produced; Chapter 20 goes into more specific details.

As we have seen in Chapter 7, there are a number of essential steps needed to move from the system specification to a mandatory type approval specification. Firstly, there is a basic type approval applicable for any DECT equipment, irrespective of the availability of a profile; this is basically the coexistence specification introduced earlier. Then, in the case of DECT, the next step for ETSI was to create a test-case library. This is (more or less) a definition of a type approval test for each individual building block. Consequently, there exist collections of tests available as profile test specifications. Processes outside ETSI convert these specifications to binding documents according to the procedures outlined in Chapter 7.

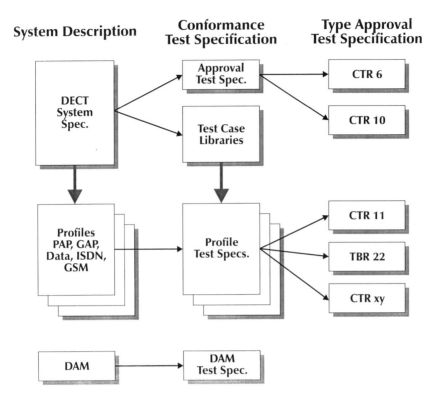

Figure 19.3 The structure of the DECT standards

Similar to many other modern radio technologies, e.g. GSM, the DECT concept treats the communications terminal responsible for the data transmission and the module that contains the user-relevant data (e.g. the telephone number) as two separate network elements. In the DECT case the module containing the subscriber personal information is called the DECT authentication module (DAM). A comparable structure of specifications exists or is being developed for this module.

Finally, there also exists an EMC standard for DECT equipment (not shown in Figure 19.3), published as ETS 300 329.

19.3 The System Specifications

This section discusses the base DECT system specifications contained in ETS 300 175. These specifications were developed by ETSI RES03, largely during the period 1988–92 and were published in 1992; the summaries below refer to this first edition. A second edition was issued by ETSI for public enquiry (PE) in early 1996, which essentially reflects additions and enhancements, rather than changes, to the standard. At the time of writing, the public enquiry is ongoing and changes to the PE version are still likely – the interested reader may obtain the final version of the second edition specification from ETSI when it is published.

19.3.1   ETS 300 175-1: Overview

The introduction to the DECT specification, besides containing editorial information such as scope of the document, list of references, etc., introduces the basic concepts of the DECT radio communication. It is noted that the DECT system specification only specifies the radio communication link between a DECT terminal and the network (which may contain many basestations, common control modules, possibly even location registers). Although the standard is based on certain assumptions about the network architecture, it does not deal explicitly with the functions within the network. However, some of the profile standards specify interconnection and interworking with partner networks (see Chapter 20).

Figure 19.4 shows the layered structure employed for the communication protocols. Two layers are common to both user data and signalling data: the physical layer (PHL), which transmits the data over the radio link, and the

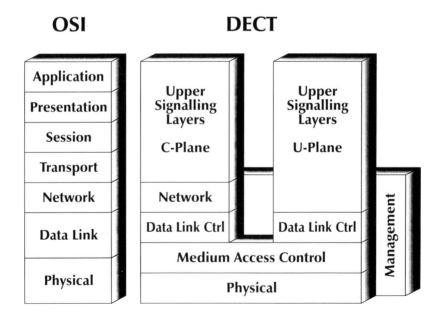

Figure 19.4   The layered structure of the DECT radio interface protocols.

medium access control layer (MAC), which is responsible for effective allocation of radio resources and for multiplexing user and signalling data onto one bit stream. PHL and MAC layers together provide the radio transmission. The communication peers are, therefore, the portable handsets and the basestations in the cells.

The upper layers are separate for user data and signalling. The data link control layer (DLC) maintains a secure data link for signalling and user data even when the cell has to be changed during a call, i.e. when a handover is executed. The network layer (NWK) is responsible for routing the calls from a portable wishing to make a call to the fixed network and vice versa. In contrast to PHL and MAC, the DLC and NWK layers' tasks are to "hide" the presence of non-ideal and possibly changing radio transmission links from the upper communication layers. Their procedures thus involve the DECT network control and the portables, but not the basestations.

Not directly involved in the peer-to-peer communication is the lower layer management entity (LLME), which controls the four lower layers and takes the necessary decisions. A typical decision might arise if the quality of the radio connection were no longer adequate, in which case the LLME would decide to initiate a handover.

For reference, Figure 19.4 also shows the standard OSI layered structure. Two points need to be mentioned here. Firstly, in the OSI model anything above the network layer concerns end-to-end communication, while the lower three layers belong to the communication between nodes. The same applies to DECT, that is the top of the NWK layer corresponds to the top of OSI layer 3. As a consequence, the fact that there is a radio link is known only up to the network layer. This layer and those below it provide all the functions needed to manage the difficult radio communications environment. Secondly, the DECT structure uses four layers for the node-to-node communication, that is DECT terminal to DECT network, while OSI would suggest the use of only three. The OSI model, however, does not adequately consider multiple access to one transmission medium. Most specifications dealing with multiple access systems do, in fact, have the MAC layer, or an equivalent of it. Thus, for example, in the CT2 CAI, as specified in I-ETS 300 131, the radio interface specification is equivalent to the DECT physical layer, the CAI signalling layer 1 corresponds to the DECT medium access control, while CAI's signalling layers 2 and 3 have, with a few exceptions, their counterpart in the DECT data link control and network layers.

19.3.2   ETS 300 175-2: The Physical Layer (PHL)

The PHL is responsible for segmenting the radio transmission medium into physical channels. This is done using a time-division multiple access (TDMA) scheme on multiple carriers. Ten carriers are allocated in the frequency band between 1880 MHz and 1900 MHz. The centre frequencies are defined as follows:

$$f_c = 1897.344 \, \text{MHz} - c \times 1728 \, \text{kHz} \qquad \text{where } c = 0, 1, \ldots, 9$$

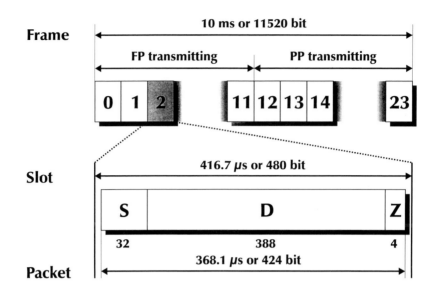

Figure 19.5   DECT physical layer.

Each carrier contains the TDMA structure. A time frame of 10 ms is formed and each frame is divided into 24 time slots (Figure 19.5), resulting in time slots of approximately 416.7 μs. During these time slots, data can be sent in packets. As seen from Figure 19.5, the burst contains three fields, a synchronisation field S of 32 bits, a data field D of 388 bits, and a collision detection field Z of 4 bits. The synchronisation data allows the receiver to demodulate a single packet immediately. This is necessary to enable very fast call set-up and unnoticeable handovers to be performed. The data field D is received from the MAC layer. The Z field just repeats the last four bits of the D field and allows early detection of collision with TDMA bursts from unsynchronised systems. The Z field is optional in the system specification but is mandatory in some profiles. The packet shown in Figure 19.5 is the one used for voice telephony. There are alternatives with D fields ranging from 64 up to 868 bits for a variety of services.

The packet bits are modulated onto the carrier using a two-level FSK modulation with Gaussian prefiltering at a relative bandwidth (BT product) of 0.5. The modulated data rate is 1152 kb s^{-1}. Therefore, the complete packet has a length of 424 bits modulated at a rate of 1152 kb s^{-1}, i.e. a duration of approximately 368.1 μs (including Z bits). Modulation deviation may lie between 202 kHz and 403 kHz. If the modulation deviation were exactly 288 kHz (that is, one-quarter of the modulation rate), the modulation method would be GMSK; nominal deviation is, however, not necessary. The chosen modulation scheme permits the use of a simple receiver with non-coherent demodulation, bit-by-bit decision, easily implementable IF filters, and so on.

When these packets are transmitted within a time slot, there remains a guard space of 48.6 μs. This guard space is needed to allow for propagation delays and for smooth ramp-up and ramp-down of the transmitter and synthesiser switching between packets. Maximum peak transmit power is nominally 250 mW. Taking into account the TDMA structure, this results in an equivalent average transmit power of slightly less than 10 mW.

A physical channel is created by transmitting one packet every frame during a particular time slot on a particular carrier. The throughput of a physical channel available to the MAC layer is therefore 388 bit/10 ms, that is 38.8 kb s^{-1}. The PHL does not itself choose which time slot and carrier on which it is to transmit but is instructed in this by the MAC layer. As indicated in Figure 19.5, the first 12 slots are normally used for transmission to the portable and the second 12 slots are used for transmission in the opposite direction. The PHL itself, in fact, does not know that there is a time-division duplex (TDD) transmission mode, but is simply told by the MAC to transmit on time slot 5 and to receive on time slot 17, for example.

19.3.3   ETS 300 175-3: The Medium Access Control Layer (MAC)

The MAC layer performs three main functions. Firstly, the MAC allocates and releases physical resources according to the requests of the upper layers. Secondly, it multiplexes the logical channels from both the upper layers of signalling and user information onto the physical channels offered by the PHL. Finally, it guarantees the secure transmission of both the signalling and user information by appropriate error control.

Allocation and Release of Physical Resources

Three special features comprise the DECT radio resource management: the dynamic channel selection (DCS) algorithm, the use of a so-called beacon at each basestation, and the possibility of allocating multiple physical channels to a connection.

By using a DCS algorithm, the full capacity offered by DECT is available to all DECT users, no matter whether they belong to one installation or to different ones. Whoever wants to activate a physical channel (this is usually the portable) looks for a channel that is free locally and then may use it. There are no preassigned channels or frequencies for a particular cell site, as is the case in a cellular radio network. In this way, the algorithm may adapt to changing, or unknown, propagation and traffic conditions. DCS is a prime requirement for effective use of radio shared spectrum in cordless applications. It is, therefore, not surprising that DCS had already previously been applied in CEPT/CT1 and in CT2 CAI.

Another special feature of DECT is the use of a so-called beacon at each basestation. Every basestation transmits on at least one channel. The portable then scans the channels until it finds the transmissions of a close basestation and locks on to this channel. Call set-up requests (so-called paging calls) from the network are then transmitted using this channel. The portable receives these requests and may immediately respond to the request. This procedure has the main advantage that the portable does not need to scan all the channels continuously but may stay on an appropriate one. By doing so, very fast channel set-up times can be realised (which is important particularly for non-voice services) and the portable's power consumption is considerably reduced since there is no need for continuous scanning.

Of course, it would not be wise to dedicate a particular channel only for the purpose of calling portables. If a dedicated channel is subject to interference, the basestation could then no longer be used. Furthermore, the relatively moderate load of paging calls does not justify one complete channel to be used at each base site. The concept chosen for DECT rather allows any base-to-portable downlink, which is in use for a specific terminal, to act *simultaneously* as a beacon. If there are, say six conversations going on at one base site, this site would be equipped, momentarily, with six beacons! Only in the case where there is no conversation going on at all, would one dummy channel (usually the last one that was in use) be kept active as the beacon.

When the portable is in idle mode, i.e. not currently involved in a conversation, it scans the environment for the beacon signals of a nearby basestation. Once found, its receiver stays locked on to this signal. Should the radio environment change, or should the locked-on channel disappear, the portable looks for a new beacon.

It is noted that the type of call set-up described in this section is suitable for circuit-oriented connections as required for telephony. Other mechanisms for packet-oriented transmissions similar to those supported by CT2 CAI are provided by DECT as well.

Multiplexing of Logical Channels

The second task of the MAC layer is to multiplex the information coming from the upper layers onto one physical channel. The specifications define several multiplex schemes. The one suitable for most voice applications is shown in Figure 19.6. Information coming from the upper layers is grouped into logical channels I, C, M, N, P and Q. Other services would employ different multiplex schemes and involve control channels not shown in Figure 19.6.

Table 19.1 illustrates the type of information these channels bear. These channels have similar meaning and functionality to those existing in other ETSI standards with different names, e.g. GSM or TETRA.

Information from the upper layers is provided as a continuous bit stream, as is the case for speech transmission, or in complete data frames, as the data link control layer would provide them, in the case of signalling information. In both cases, the MAC first has the task to create short data segments that fit into the packets of 388 bits that the PHL can transport.

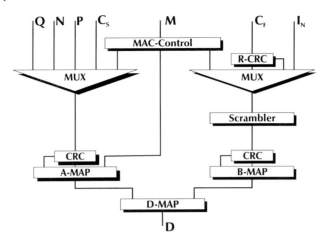

Figure 19.6 Multiplex scheme for DECT voice applications.

Table 19.1
Contents of the different DECT logical channels

Channel	Type of information
I	Information
	I_N: Non-protected information
	I_P: Protected information
C	Control channel for signalling
	C_S: Slow signalling (concurrent with information channel)
	C_F: Fast signalling (stealing information channel capacity)
M	MAC internal control channel
N	Identities channel
P	Paging channel
Q	System information broadcast channel

The information channel I bears the user information. As Figure 19.7 shows, 320 bits are available in each packet, hence the I channel throughput is $32\,\text{kb s}^{-1}$. In the case of voice this would be a bit stream of ADPCM-coded speech. An I channel exists in both fixed-to-portable and portable-to-fixed directions.

The control channel C is used for the actual signalling data. "Signalling" embraces everything that is needed to set up, maintain and release a call with a particular terminal. The C channel is divided into two subchannels, C_S for the data that has to be transmitted continuously and in parallel to the information, and C_F for signalling needing high throughput and hence stealing information channel capacity. Again a C channel exists in both fixed-to-portable and portable-to-fixed directions. The M channel is very similar to the C channel; its signalling information, however, is truly MAC internal and not visible to the upper layers. On the N channel, identities of the portable and the basestation are exchanged at regular intervals (handshake).

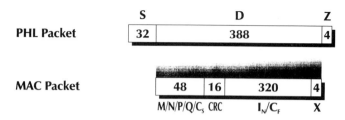

Figure 19.7   DECT medium access control (MAC) layer packet.

As explained earlier, network-originated call set-up is carried out by paging the portable, which then, in return, builds up the link. Any active channel can be used to page a portable. The paging information is carried by the P channel and is therefore broadcast to all portables locked on to the channel, whilst the signalling data carried by the C channel is intended only for the one receiver in conversation. Obviously a P channel exists in the fixed-to-portable direction only.

Before a portable can lock on to a basestation, it has to verify that the basestation belongs to a system with which it wants to communicate. The system identification, together with some other important parameters, is broadcast in the Q channel. The Q channel again exists in the downlink direction only and is a logical channel within the MAC layer only.

The C, P and Q channels are multiplexed onto 48 bits of each burst; together they may occupy a signalling rate of $4.8\,\text{kb s}^{-1}$. The allocation to the individual channels, however, is dynamic and depends on the actual needs. The algorithm of this statistical multiplexing scheme, however, is such that for each logical channel a minimum throughput is guaranteed.

Error Control

The multiplexed C, P and Q channels are protected by a cyclic redundancy check (CRC) with 16 bits. The CRC allows recognition of transmission errors. In the case of the C channel, an automatic retransmission request (ARQ) protocol is used to recover from transmission errors. The data of the other channels is not retransmitted. The CRC only allows verification of the correctness of the received data.

As Figure 19.7 illustrates, the I field is followed by four further bits, denoted X. These are calculated from the bits in the I field. They are not used to protect the information data; a 4-bit parity check can never reasonably protect 320 bits of data. Rather, the X bits are used to recognise the presence of interference that only affects a part of the burst. This is the case, for example, if the interferer is a signal from an unsynchronised system drifting slowly into the wanted signal. Recognising the interference by constantly observing the X bits allows counteraction to be taken, for example by handing the connection over to another channel. Note also that these bits are repeated by the PHL into the optional Z field, hence enhancing the collision detection capabilities.

The multiplex scheme outlined in Figures 19.6 and 19.7 is called the U-multiplex, U indicating that the multiplex contains user data. In DECT there exists also an E-multiplex, which does not have an I channel but has increased capacity for an extended C channel. This multiplex is used in particular during the channel set-up phase.

19.3.4   ETS 300 175-4: The Data Link Control Layer (DLC)

As mentioned earlier, the radio link becomes more and more invisible at higher layers. This is mainly due to the DLC layer, whose main task is to guarantee error-free transmission to the network layer. Once a data link is set up by the DLC layer it has the task of maintaining this link even if physical resources have to be changed during the conversation. This is the case where, for example, a handover must be executed, or when the data throughput needs are changed during a call and the allocated resources need to be adapted accordingly.

Another consequence of the invisibility of the radio communication to the higher layer is that more and more existing protocols may be used. These may be adapted to fit the particular needs of DECT but they do not need to be completely newly designed. Thus, the DLC entity uses a protocol derived from the ISDN LAPD (link access protocol D-channel), ITU-T I.441.

19.3.5   ETS 300 175-5: The Network Layer (NWK)

The highest protocol layer covered within the DECT specification is the network layer (NWK). As for the DLC layer, there is no need to define a completely new protocol, but an existing specification can be used. In the case of a DECT circuit-oriented transmission, a network protocol similar to the ISDN I.451 protocol is used.

One task of the NWK that is unique to a microcellular network application such as DECT is that it has to route calls to a handset, the exact location of which is not *a priori* known. Therefore, the protocol needs to be expanded to include such mobility management procedures. A second unique task is the provision of a transport medium for security-related functions, in particular authentication.

19.3.6   ETS 300 175-6: Identities and Addressing

This part of the specification deals with the identities used in the DECT system. In a system like DECT, identities have to be treated differently from other systems like GSM or TETRA for the reasons outlined below.

Firstly, DECT is a radio access specification. As such, DECT alone does not provide a service but rather simply supplies the radio connection to the system providing the service. Such a system may be one of many different types of network – telephony networks like the PSTN, a PBX, data networks like a X.25 or an IEEE 803 type network, other networks like cellular networks or the ISDN. The signalling protocols must therefore be capable of supporting all the specific identities used in these varied networks.

Secondly, identities must be organised in such a manner that a terminal can access more than one network. If needed, a cordless PBX terminal must be able to access all wireless PBX systems of a corporation, the headquarters as well as all the subsidiaries. The system must allow calling a terminal on one system using several telephone numbers as well as calling it on different systems using the same number.

We cannot describe here the full detail of the organisation of identities and addresses – the reader must study the specification for this. However, here we describe the basic concept, which is the one of "access rights". An access rights identity (ARI) uniquely identifies a system or a service provider. The ARI (there may be more than one per system) are regularly broadcast by the system. Each portable has stored at least one portable access rights key (PARK), which fits to the ARI of those systems granting access rights. As a side issue of this concept, the distribution of unique ARI had to be organised. For private systems, the ARI is linked to the equipment manufacturer; public systems have an ARI tied to the service provider or operator.

19.3.7   ETS 300 175-7: Security Features

Similar to most other recently developed radio systems, the DECT development has seen (and is still seeing) an extensive discussion about security. Security here firstly means authentication, i.e. checking whether access to the system is permitted or not; the second issue, of course, is encryption. We do not go into details of the algorithms, mainly because they are not publicly available; they need to be ordered from an ETSI custodian by signing a confidentiality agreement.

The basic principle is similar to those used in other systems; authentication is granted by using a challenge/response technique. The network sends a random number to the terminal, which then calculates a response using the security algorithm and secret information. Sending back the response and comparing it with the response calculated by the (home) network may or may not grant access. A challenge/response pair is only used once. During this process, also the session keys for the encryption are derived.

The DECT security features differ in two different points from those used in GSM. Firstly, there is a higher degree of standardisation needed in the DECT case as the DECT authentication module (DAM) is not necessarily a separate piece of equipment as in the GSM case. Secondly, there might be many DECT networks or basestations around. For this case the DECT security features allow the authentication of the *network* by the portable. Here, the portable can verify whether the network is really the one it pretends to be. This reverse authentication is not foreseen in GSM.

19.3.8   ETS 300 175-8: Speech Coding and Transmission

The problems arising from delay in cordless telephony systems have already been discussed earlier in this book. In DECT a system-inherent one-way delay of 10 ms is introduced, arising from the TDMA framing. Other delay sources, for example digital signal processing, may add further delay to total no more than 15 ms. Part 8 of the DECT specification defines all the necessary means to be taken to guarantee high-quality telephone conversations with DECT systems. These means include the complete end-to-end transmission planning with all the echo control requirements, necessary loudness ratings, etc.

The specification also defines the digital speech coding to be used in DECT. The requirement is that the resulting speech quality perceived by the user must be equal to the quality of a fixed digital telephone. Therefore, the quantisation distortion of the speech coding should be acceptable even in normal fixed telephony networks. The adaptive differential pulse-code modulation (ADPCM) coding algorithm defined in ITU-T Rec. G.721 fulfils this requirement with a data throughput need of $32 \, \text{kb s}^{-1}$. Since no codec with lower bit rate but fulfilling the quality requirement currently exists, the G.721 algorithm has been chosen for the DECT standard. Nevertheless, the standard is open to allow the use of other speech coding algorithms should they become available in the future.

19.4 Other Standards

There exist many other DECT standards as well as the base system specification. Some of these make up the basis for type approval specification, including test-case libraries. Again, others relate to profiles and interworking with other networks, and these are dealt with in the next chapter. There remain a few other important standards, which should be described here.

19.4.1 ETS 300 329: EMC Standards for DECT Equipment

This standard defines the generic EMC requirements imposed upon DECT equipment to guarantee coexistence with electronic and telecommunications equipment other than DECT. This document was published in 1994.

19.4.2 ETS 300 331: DECT Authentication Module (DAM)

The DECT system uses a function called DAM, which is capable of holding all the identities and authentication algorithms. It is therefore similar to the GSM SIM, but with some differences.

Firstly, GSM assumes a SIM card bearing some of the security algorithm, which is always a different piece of equipment. DECT uses a similar concept like the SIM, called DAM (DECT authentication module). The DAM, however, may or may not be a separate piece of equipment (for example a chip card). For this reason, it is not just the interface that is being standardised but also the minimum functions within the DAM; this includes the organisation of the data in the DAM.

Secondly, the DAM must hold many identities (i.e. PARKs) as well as the security- and authentication-related information. The specification requires the DAM to hold many subscription records, each of them allowing several (up to eight) authentication keys. The actual number of subscriptions is not defined but is limited to the type and length of data that is stored in the DAM as well as by the addressing capabilities of the file system within the DAM.

For the cases of external DAM cards, the specification defines the physical sizes: Standard ISO (credit-card size), Mini ISO (66 mm × 33 mm) and Plug-In (25 mm × 15 mm).

The DAM specification ETS 300 331 was adopted in March 1996.

19.4.3   ETS 300 700: Wireless Relay Stations (WRS)

During 1994 the issue of using something like radio repeaters has been taken on board by ETSI. The result is a specification of a wireless relay station (WRS) recently finalised. Standard repeaters receive a radio signal with one antenna and retransmit it at the same time via another antenna (on the same frequency). Such an apparatus cannot be used for DECT systems because of the time-division duplex employed. Rather, signals need to be demodulated and retransmitted during different time slots. The specification defines two options for doing so. One minimises the additional spectrum capacity that is needed by repeating the DECT signal; the alternative minimises the speech delay.

The WRS specification ETS 300 700 was approved for public enquiry in March 1996.

19.5   Access Profile Specifications

As explained earlier, certain DECT products are defined in terms of profiles, such as the public access profile (PAP), the generic access profile (GAP), the data services profile (DSP), the radio local loop access profile (RAP), etc. A detailed list of these profiles is given in the next chapter, where a number of them are more fully described.

19.6   Type Approval Specifications

Chapter 7 outlined the creation of mandatory type approval specifications. Normally, ETSI takes a system standard it has created (that is an ETS or an I-ETS) and creates the type approval specification based upon this as a "Technical Basis for Regulation" (TBR). This document would then be "rubber-stamped" to become a "Common Technical Regulation" (CTR). In the case of the generic access profile (GAP), this is TBR 22, which is being converted into CTR 22. Hence, the type approval specification does not bear an ETS number.

Some years ago, these procedures had not yet been strictly established. As a consequence, there exists a type approval specification, called I-ETS 300 176. This ETS was later converted to CTR 6 and CTR 10, but still exists as an I-ETS. For the same reason, the type approval specification for the public access profile (PAP) is available as ETS 300 323-1 to 300 323-7. The test-case library will be published as an ETSI Technical Report (ETR).

19.6.1   CTR 6

CTR 6 is virtually the coexistence specification of DECT. It describes all the mandatory radio features and requirements that all DECT equipment has to follow, even in the absence of a profile.

19.6.2   CTR 10

CTR 10 adds to CTR 6 those mandatory requirements for DECT equipment intended for voice telephony and which will be connected directly or indirectly (that is through a PBX) to the PSTN.

19.7   Reports

Apart from the "standards", ETSI creates so-called "reports" – ETSI Technical Reports (ETRs). These documents are not standards – that is, they have no normative character – but they give additional explanation and guidance on the application of the standards. We do not go into details of these reports; they are available from ETSI and most of them read fairly simply. The ETRs shown in Table 19.2 apply specifically to or have at least some influence on DECT.

Table 19.2
ETSI Technical Reports (ETRs) relating to DECT

Report	Name	Status
ETR 015	DECT Reference Document	Published 1991
ETR 041	DECT Transmission Aspects, 3.1 kHz Telephony	Published 1992
ETR 042	Guide to Features That Influence Traffic Capacity	Published 1992
ETR 043	DECT Services and Facilities Requirements Specification	Published 1992
ETR 056	DECT System Description Document	Published 1993
ETR 139	Radio in the Local Loop (RLL)	Published 1994
ETR 159	DECT Wide Area Mobility Services Using GSM	Published 1995
ETR 178	A High Level Guide to the DECT Standardisation	Published 1995
ETR 183	Conformance Test Specification for DECT	Completed 1995
ETR 285	Data Services Profiles Overview	Completed 1995
ETR 246	Wireless Relay Stations (WRS)	Completed 1995
ETR xxx	DECT/GSM Interworking Profile (GIP)	In progress 1996
ETR xxx	Radio Local Loop Access Profile (RAP)	In progress 1996
ETR xxx	Traffic Capacity and Spectrum Requirements for Multisystem and Multiservice Applications	In progress 1996

20 The DECT Access Profiles

Peter Olanders

The core concept of DECT has been to replace, or extend, the last pieces of telephone wire, between the user and the networks, with radio – "no strings attached". Thus, DECT is an access technology, with the DECT standards focused on the radio interface and extending beyond this only when necessary to achieve the required functionality. This means that DECT has been specified with no particular background network of its own, in contrast to, for example, some cellular telephone systems (e.g. GSM) – this is a fundamental difference between DECT and such systems.

As an access technology to other networks and the services supported therein, DECT access must be transparent for these networks and services. Consequently DECT has to be adapted to these networks – it is this adaptation which is described in the so-called "profiles". The specific networks that have been studied within the ETSI standardisation committees to date for DECT access include the following: PSTN, PTN (PBX), GSM, ISDN, X.25 and the Internet, a formidable selection.

In OSI terms a profile is defined as:

> a combination of one or more base standards, and where applicable, the identified chosen classes, subsets, options and parameters of those base standards, necessary for accomplishing a particular function (for such purposes as interoperability).

Each of the DECT profiles describes a particular way of using DECT, with most of the profiles aiming at *interworking* and *interoperability*. Interworking is usually regarded as the ability to use two or more systems together, whereas interoperability usually refers to the possibility to use equipment from different manufacturers together. Some profiles even have the words "interworking profile" attached to the name, abbreviated as IWP, e.g. the DECT/GSM IWP. Most of the profiles have interoperability as a declared aim. For example, in the DECT/ISDN IWP it is stated:

> One of the main objectives is to describe how the ISDN services are mapped across the DECT air interface in a formal way, so that interoperability of different manufacturers' equipment can be achieved. This is achieved by describing the Inter-working Unit (IWU) procedures and mappings loosely following CCITT Recommendations and by describing air interface protocols.

The DECT profiles may be extensive in their standards. The actual testing and type approval of the profiled equipment may as well be extensive. However, the actual implementation usually only requires some added software rather than hardware changes; thus DECT equipment with sufficient data memory capacity may easily host all of the profiles. In theory it is even possible for the user to load the profiles wanted, although from a practical point of view there may be some obstacles in doing this and manufacturers may prefer to handle this process themselves.

This chapter thus provides an overview of the various DECT profiles and describes some of the more important ones that have already been developed within the ETSI standardisation forums.

20.1   The DECT Profiles – Overview

By early 1995 the following DECT profiles had been defined in ETSI:

- Public access profile (PAP)
- Generic access profile (GAP)
- DECT/ISDN interworking profile
- DECT/GSM interworking profile
- DECT data profile(s)

The PAP and the GAP are profiles describing a particular use of DECT as an access to a network, for a given set of services. Backbone networks for PAP and GAP may be PSTN, ISDN or GSM, for example, but the PAP and GAP profiles are focused on supporting speech services, as explained more fully below.

The DECT/ISDN and DECT/GSM IWPs, on the other hand, describe how ISDN and GSM services respectively shall be offered via a DECT radio link.

The DECT data profiles are a set of standards (and then profiles) for data transport applications over the DECT air interface. The data profiles specify interworking with data networks (with LANs, X.25, etc.), for interoperability (to be able to use the same infrastructure as other DECT applications) and for particular types of data services.

Profiles for other networks and purposes have not yet been developed. In 1995 work began on the RAP (radio local loop access profile, i.e. DECT applications in the local loop) and the CAP (CTM access profile). Most of the information and standardisation elements for these actually already exist in the base standards, and in DECT GAP, the profiles are for defining a common set of elements optimised for these purposes.

20.2   DECT Goes Public – PAP and GAP

When the concept of DECT was first defined, one of the major applications envisaged for DECT was its use in public environments, as an access to public networks. The DECT public access profile was thus developed in parallel with the basic DECT standard (the DECT common interface) – the PAP is in fact the last part of the DECT common interface standard, ETS 300 175.

The part of the PAP that has attracted most interest is that related to the "telepoint" public access application. In this concept DECT is used in a superficially similar way to a cellular network – a user in an area of DECT radio coverage may call or be called using his or her DECT portable telephone. Such DECT coverage areas are of course smaller than the corresponding areas in a (higher-transmission-power) cellular system. On the other hand, provision of DECT infrastructure does not require the same high degree of investment as cellular systems; basestation costs are vastly lower.

A similar concept was launched some years ago using CT2 technology, with some important differences, however. Firstly, in the original CT2 concept incoming calls were not supported; secondly, the user could not move from one cell to another during a call – handover was not supported. The original CT2 telepoint concept was almost "a telephone booth in your pocket"[1].

The development of the PAP was influenced by a philosophy of "not to overstandardise", which was dominating in ETSI at the end of the 1980s. Most of the features in the PAP are consequently not mandatory, and some of the features may be implemented in more than one way. As all processes and procedures in the PAP are taken from the DECT common interface, interoperability (the capability for equipment from different manufacturers to work together) is ensured for all the basic features (such as speech). As a few PAP services may be implemented in more than one way, then these few services might not always be accessible by all DECT handsets (depending on implementation).

In 1993 the ACTE and TRAC regulatory bodies in Europe decided, in general terms, to introduce a new profile to ensure *full* interoperability. When this concept was accepted in ETSI, it was agreed that this profile should be a minimal standard, which should be interpreted as either a subset of the PAP (from the service viewpoint) or as an increase of the basic DECT threshold. The PAP was regarded as too extensive for a mandatory interoperability profile, and it was thus proposed to develop a new "minimal" profile.

At about the same time the generic access profile (GAP) was discussed in ETSI within the context of Cordless Terminal Mobility (CTM) – see Chapter 6 – where it became apparent that the mobility management functionality in networks could be utilised with DECT to offer new services.

Thus today's DECT generic access profile provides for the basic requirements both of general mobility and of interoperability. The profile is likely to become mandatory for all equipment supporting 3.1 kHz speech, i.e. an additional requirement on speech equipment. From a regulatory point of view, there will be a TBR (to be converted to a CTR) for the DECT GAP, which is expected to be used together with CTR 10.

The DECT GAP standard was finally adopted by ETSI in June 1995. The GAP has been designed to be compatible with the PAP, in order to give users and manufacturers the possibility to use PAP until the GAP has reached its final official status.

It has also been proposed by some either to extend the scope of the DECT GAP in its current form or to develop an extended version (E-GAP). Such an extended version would then cover all PAP services as well (in fact, GAP already covers all the important areas of the PAP) and some data services. If data services are to be included in an extended GAP, this will be a fairly simple operation, merely a formality – the DECT data profiles (see below) could just to be referred to in the extended GAP profile.

Other extensions of the GAP that are under discussion at the time of writing include "external handover", "synchronisation" and "emergency calls". External handover is the capability to move from one DECT system (e.g. a cluster of cells) to another completely separate DECT system. One requirement for doing this may be to have the systems synchronised – with synchronised systems, the DECT portable will have no problem with being connected to both systems at the same time. The process of external handover is actually indicated in the DECT common interface standard, but it needs some elaboration and further standardisation if it is to be used for handover between completely separate systems. It should be pointed out that both external handover and synchronisation are features that are not included in, for example, GSM.

"Emergency calls" is a feature taken from GSM and other similar systems. In GSM it is possible to make emergency calls from every mobile (i.e. handset) without any subscription, SIM card, etc. It is disputable if the same feature should be implemented in DECT, as it will never have the same wide-area coverage as the cellular systems. Furthermore, the way DECT is usually applied, connected to backbone (fixed) networks, emergency calls in DECT will be based on the performance and qualities of these backbone networks.

20.3   Wireless ISDN – The DECT/ISDN Interworking Profile

The aim of the DECT/ISDN profile is to facilitate ISDN services over DECT access – *wireless ISDN* or the *ISDN radio*. As full ISDN requires rather high capacity, two different reference configurations have been defined within the DECT/ISDN interworking profile:

- DECT end system
- DECT intermediate system

In the DECT end system the ISDN is terminated in the DECT fixed system (DFS), sometimes called the DECT fixed part, which is similar to a basestation. The DFS and the DPS (DECT portable system) may be seen as an ISDN terminal equipment (TE1). The DFS may be connected to an ISDN S, S/T or P interface, as illustrated in Figure 20.1.

In contrast, in the DECT intermediate system the ISDN is terminated in the DECT portable part; thus, this variant is fully ISDN transparent. It goes beyond the end system configuration, such that the (ISDN) S interface is even regenerated in the DECT portable part. This means that the ISDN is regenerated in the DECT portable, which may then work as an ISDN distribution point, to which other ISDN terminals may be connected.

The standards for these configurations cover the following ISDN bearer services (a "bearer service" is a defined capability of signalling):

- Speech
- 3.1 kHz audio
- Unrestricted digital information, 64 kb s^{-1}

The DFS will support basic-rate access (BRA), one or more, as well as primary-rate access (PRA).

The DECT/ISDN IWPs will, at least, include:

- 3.1 kHz telephony, i.e. standard telephony
- 7 kHz telephony, i.e. high-quality audio
- Video telephony

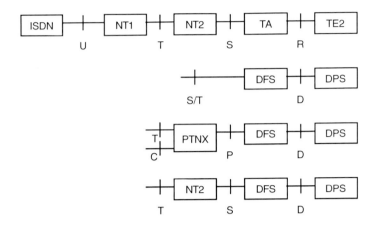

Figure 20.1   DECT/ISDN interworking: end system reference configurations.

- Voice-band data, e.g. Group III fax and modems
- Digital data transmission, e.g. X.25 over ISDN
- Telematic services, e.g. Group IV fax, telex, teletex, videotex

The ISDN supplementary services are also expected to be supported.

For the end system configuration, it should be noted that neither the bearer service "packet mode" nor any mobility management, such as location registration and subscription registration, are covered by the standard. The end system configuration standard describes four major issues:

- *Interworking requirements*, which include the reference configurations and the protocol architecture model (Figure 20.2) and a description of the main service requirements.
- *IWU mappings*, which show the important mappings. The signalling mappings are described in terms of IWU procedures with informative data flow diagrams, followed by detailed descriptions.
- *Support of ISDN bearer services*, which identifies the main DECT connection types at the air interface supporting optimised groups of services, from the IWU mappings.
- *ISDN access profile* (IAP) for the end system and description of the air interface protocol in detail.

The DECT/ISDN profile is compatible with the PAP (and GAP) in such a way that a PAP portable can be used together with a DECT/ISDN fixed part, and an ISDN portable can be used with a PAP fixed part.

The end system configuration is likely to be used in (large) companies, where there is a need for mobile ISDN services, together with an ISDN PBX. The portables in such environments will be small and for use with specific ISDN services (e.g. videophone via a $64\,\text{kb s}^{-1}$ data connection), i.e. there is no need to regenerate the ISDN S interface.

The intermediate system configuration is expected to be of interest in wireless local loop applications, as well as for connections to small businesses. The ability

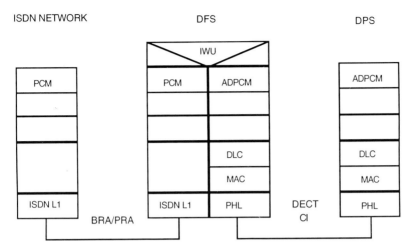

Figure 20.2   An example of the protocol stacks involved in a basic speech connection in DECT/ISDN end system configuration. Shown here is the U-plane, i.e. an emphasis on the end-to-end handling.

to regenerate the S interface may be very interesting, as this will open up new possibilities for the users.

20.4   Data Interworking Profile

One of the very first DECT products to be launched on the market in Europe, as described briefly in Chapter 14, was a system for data transmission, i.e. without any voice services, designed to work as a wireless LAN, offering medium data transmission rates, up to some few megabytes per second (depending on configuration). Such equipment is obviously likely to be used in offices and other locations where there may be an interest in using cordless telephones at the same time. To have two independent systems would be inefficient and waste radio spectrum.

20.4.1   Purpose of the Profile

Thus the data interworking profile enables cordless portable telephones to be used efficiently in the same system environment as cordless data terminals, even if the equipment stems from different manufacturers. The data interworking profile describes one way in which data services may be implemented in DECT, such that products complying with this profile will be able to interoperate. The profile is thus a support for both customers and manufacturers. The data profile actually comprises a number of subsets and demonstrates well the flexibility of the DECT standards; for this reason, we describe these in some detail below.

Data transmission was traditionally by wire to fixed locations. There were many reasons for this, one being the lack of an adequate wireless technology offering both a reasonable data rate to the user (at least in the order of 10–

$100\,\mathrm{kb\,s^{-1}}$) and at the same time a mobility better than offered by a long wire. The cost of providing reasonable radio coverage must also be limited.

The need for mobility for data transmission has increased just as the way computers are used has changed dramatically in recent years and as the computers themselves have become readily portable. Laptops dominate the PC market and personal intelligent communicators (PICs) or personal digital assistants (PDAs) are predicted to become a mass market in a few years.

DECT has been designed to offer a very high data capacity – with a single radio unit a data rate of about $565\,\mathrm{kb\,s^{-1}}$ (with BER of 10^{-8}) can be achieved, with a radio range of about 50–200 m, and with mobility functions giving coverage in predefined areas. From these viewpoints, DECT is very well suited for wireless data transmission.

Roaming between wireless LANs and PBXs allows users' speech and data services to be available and automatically re-routed to them as they move between different LAN/PABX sites either within one company or between several companies. It is even possible to extend the mobility to public areas such as airports, etc.

20.4.2 Service Types and Mobility Classes

The DECT data profiles have been structured into "services types" and "mobility classes", here outlined. The service types are:

A Low-speed frame relay, data rate up to $24.6\,\mathrm{kb\,s^{-1}}$. For low power, bursty data and low complexity.

B High-performance frame relay, throughput up to $552\,\mathrm{kb\,s^{-1}}$.

C For data streams requiring link access protocol (LAP) services. Provision for packet assembly/disassembly function for asynchronous data streams is also included.

D Transparent and isochronous connections.

E Short message transfer, like paging, which may, or may not, be acknowledged.

F Support of teleservices, such as fax.

Two mobility classes are specified:

1 Local-area applications, for which terminals are pre-registered.

2 Roaming applications, both public and private, for which terminals may move between fixed parts within a given domain.

With two separate mobility classes, it is possible to optimise the equipment for the use for which it is intended, e.g. equipment for restricted mobility only needs to have a very thin network layer.

The services types and mobility classes may then be mixed freely, thus offering a range of opportunities. Some combinations are of course more interesting than others, and the first standard to be developed was the combination AB/1, followed by C/2 and AB/2. Other DECT data standards that have been developed are the C/1, F/2 and E/2. With the given structure of types and classes, it is unlikely that there will be any F/1, as such a profile will anyhow require a rather extensive network layer; the difference between F/1 and F/2 will be too small to motivate any development of the F/1 profile (all services provided by F/1 will of course be covered by F/2). Among these AB/1 is fully adopted within ETSI, and all the other mentioned standards have started the ETSI approval procedure (as at late 1995). Work on the profiles D/1 and D/2 (isochronous connections) have started, and they are foreseen to be finalised in 1996 and 1997.

20.5 DECT/GSM Interworking Profile

The interworking with the Global System for Mobile (GSM) has attracted considerable interest, partly stimulated by the worldwide adoption of GSM as a standard for digital cellular radio. It was, in fact, decided at a very early stage of DECT standardisation activity to develop an interworking capability between DECT and GSM, long before the success of GSM outside of Europe was foreseen. GSM, being the emerging dominant digital network (or set of networks), however, is very attractive for interconnection to an access system like DECT; the mobility functions in GSM of course further increase the attraction of GSM for DECT users.

20.5.1 DECT/GSM Interconnection

Generally, the advantages with the combination of DECT and GSM may be expressed as follows: The user will be able to access the GSM fixed network (GSM PLMN) with DECT using a higher data rate, thus offering higher speech quality (compared to GSM) and possibilities for intensive data or multimedia communication (such as supported by the DECT data profiles). The GSM operator will be able to offer services to DECT users, thereby re-using his investments in the GSM PLMN. In Europe today the bottleneck in GSM in high-user-density urban centres is the availability of spectrum (at least in the 900 MHz band); DECT radio access offers a way of providing very high capacity to users, overcoming this problem.

In the context of using DECT and GSM together, the DECT/GSM *dual-mode handset* is a very interesting concept. The dual-mode handset consists of two parts: one part for GSM radio access and the other part for DECT radio access (these could be more or less integrated). The GSM radio access is primarily to the GSM PLMN, whereas DECT radio may access any network, including GSM PLMN. Integrating these two radio accesses into one handset offers the user a single handset for most of his or her telecommunications, fixed and portable. As much of the telephone is common – e.g. the keypad, display, battery and encapsulation – such a device will cost less to produce than two separate handsets. It has been estimated that a dual-mode GSM/DECT handset will cost only some 10–15% more than a normal GSM handset. Some manufacturers have already publicly declared their intention to deliver dual-mode terminals to the market by late 1995/early 1996.

DECT may be interconnected with GSM PLMN in principally two different ways:

- DECT (fixed part) connected to the GSM infrastructure like DECT access is provided to any other fixed network, i.e. the normal (GAP) DECT radio

interface is used. The DECT user will then benefit from a limited set of services, like speech and maybe mobility (roaming, handover, etc.).
- DECT connected to GSM in a "GSM-specific" way, where most of the GSM services are handled (transferred or terminated) in the DECT equipment. The DECT radio interface will in this case be optimised in order to accommodate the GSM information. This case is the approach adopted for the DECT/GSM interworking profile (the same reasoning may be applied as was presented in the DECT/ISDN case).

These are two completely different concepts, which need to be investigated in further detail in order to understand their respective benefits and limitations.

The obvious advantage of the first case is the lack of any further standardisation and corresponding implementation in handsets. From a technical viewpoint any GSM operator can easily start to operate DECT, under the same conditions as all other DECT operators (as e.g. PSTN-based operators). It may be possible to use the mobility functionality of the GSM PLMN, but it may require a separate DECT subscription (i.e. the user may need to have a GSM identity and a separate DECT identity). The drawback is the lack of "GSM services" – most likely speech will be the only successful service used with such an approach.

The second case offers an access not only to the GSM physical network, but also to the services therein. One major feature is then the "GSM identity" (i.e. the user's telephone number, subscription, etc.), which may be used also via the DECT radio access in this case. Thus, this option gives a truly integrated DECT/GSM offering to the user. In principle, it is even possible to offer handover between the two systems (at least in the direction DECT to GSM), which indicates the very close interworking that the DECT/GSM IWP provides. As noted earlier, the difference between the DECT profiles (for speech) is surprisingly small in hardware in the portable. A DECT (speech) profile could be said to define the DECT elements to be used, and in which order. Thus, it is likely that these profiles are implemented in software, using the same hardware. A portable may then contain more than one profile; it may also be able to change from one profile to another very easily and fast.

20.5.2   DECT/GSM Services and Features

As pointed out in the introduction to this chapter, the major objective in the DECT/GSM IWP is the access to the *services* offered by the GSM PLMN. The services are handled in some different parts of the connection link and are sometimes directly correlated to certain objects, e.g. the handset itself. This section is subdivided into the following aspects: services, architecture, portable features, smart cards and finally "out of scope", i.e. issues not covered by the DECT/GSM IWP standards.

Services

The DECT/GSM interworking profile supports the GSM phase 2 services:

- Teleservices

- Bearer services
- Supplementary services

Teleservices

Teleservices are services such as:

- Telephony (GSM teleservice 11)
- Emergency calls (GSM teleservice 12)
- Short message service mobile-terminated point-to-point (GSM teleservice 21)
- Short message service mobile-originated point-to-point (GSM teleservice 22)
- Short message service cell broadcast (GSM teleservice 23)
- Alternate speech and facsimile Group III (T/NT) (GSM teleservice 61)
- Automatic facsimile Group III (T/NT) (GSM teleservice 62)

Bearer services

Bearer services are "a type of telecommunication service that provides a defined capability for the transmission of signals between user–network interfaces" (usual definition in DECT standards).

Supplementary services

In GSM phase 2 supplementary services include the following.

Number identification supplementary services:

- Calling line identification presentation (CLIP)
- Calling line identification restriction (CLIR)
- Connected line identification presentation (CoLP)
- Connected line identification restriction (CoLR)

Call offering supplementary services:

- Call forwarding unconditional (CFU)
- Call forwarding on mobile subscriber busy (CFB)
- Call forwarding on no reply (CFNRy)
- Call forwarding on mobile subscriber not reachable (CFNRc)

Call completion supplementary services:

- Call waiting (CW)
- Call hold (HOLD)

Multi-party supplementary services:

- Multi-party service (MPTY)

Community of interest supplementary services:

- Closed user group (CUG)

Charging supplementary services:

- Advice of charge information (AOCI)
- Advice of charge charging (AUCC)

Call restriction supplementary services:

- Barring of all outgoing calls (BAOC)
- Barring of outgoing international calls (BOIC)
- Barring of outgoing international calls except those directed to the home PLMN country (BOIC-exHC)
- Barring of all incoming calls (BAIC)
- Barring of incoming calls when roaming outside the home PLMN country (BIC-Roam)

Architecture

The DECT/GSM IWP interworking requirements are based upon mobile switching centre (MSC) attachment for the DECT fixed part (FP). The DECT FP uses the A interface towards the GSM MSC, in the sense that the FP emulates a GSM basestation controller (BSC) with regards to the relevant GSM messages. The complete interface used between the DECT FP and the GSM mobile switching centre (MSC) is specified in one specific standard (see section 20.5.4). The DECT/GSM IWP does not cover other interfaces for attachment to GSM networks (however, section 20.5.6 contains some indications of future development of the profile).

Portable Features

The portable, usually identical to the handset, will have to support some basic features necessary for handling of the services provided by the GSM PLMN. In addition to these, there are also the supplementary features. Most of these features – basic as well as supplementary – are mandatory. The basic features in the portable are thus:

- Display of called number
- Indication of call progress signals
- Country/PLMN indication and PLMN selection
- Keypad – in itself this is not mandatory, but the portable must in some way (e.g. voice-activated) be able to generate keypad signals
- IMEI/IPEI – international mobile station equipment identity (GSM feature) and international portable equipment identity (DECT feature) – is correlated to the identity of the portable
- International access function ("+")
- Auto-calling restriction capabilities
- Emergency calls capabilities
- Subscription identity management
- Support of DECT encryption
- Support of the ERMES[2] alphabet

The supplementary portable features are:

- Control of supplementary services – the only mandatory one among the supplementary features
- Abbreviated dialling
- Fixed number dialling
- Selection of directory number in short messages
- Last numbers dialled (LND)

Smart Cards

The smart card used in the GSM system, the SIM (subscriber identity module), is by now well known. DECT provides for use of a similar smart card, called the DAM (DECT authentication module). The DECT and GSM smart cards are in reality very similar. They have the same physical size and mechanical and electrical interfaces. However, because the SIM card is directly connected to the GSM system (operator and, indirectly, services), they are not identical; different rules apply to the operation of these systems, for example. Furthermore, the SIM is mandatory for use in GSM, whereas DAM is optional: the DAM does not even have to be implemented in a portable. However, a DAM smart card may include SIM information, if a GSM operator wishes to distribute such multifunctional smart cards. It is expected that these smart cards (as well as others such as UPT, etc.) will be totally aligned in the future.

With this background, it is natural to request for DECT/GSM IWP interworking that the DECT portable should accept the GSM SIM, as well as the DECT DAM, with a GSM application. The functional separation of the DECT/GSM interworking among DECT fixed parts and portables allows for terminal portability. All roaming scenarios based on SIM roaming as described in GSM specifications are then applicable. Additionally, subscribers using a SIM or DAM with a GSM application in conjunction with a DECT portable, capable of DECT GSM interworking, may also roam among DECT fixed parts connected to different GSM networks (assuming, of course, that the user has the necessary subscriptions).

Out of Scope

Handover among DECT radio access systems and GSM radio access systems as well as handover among DECT cells connected to different common control fixed parts (CCFPs), which are individually linked to an MSC, is not covered by the standard.

20.5.3 DECT/GSM IWP Radio Interface

The radio interface part of the DECT/GSM IWP is central in the profile. The DECT radio interface must accommodate the GSM services (described above), which places some specific requirements on the interface. It is also central from the viewpoint that it is really this part which will have important consequences for the DECT portables and thereby the users.

One standard in the profile focuses on the radio part, expressed as "access and mapping", a phrase which is also included in the title of the standard. The standard also contains the procedures and implementation descriptions of GSM authentication and derivation of the DECT ciphering key from the respective GSM cipher key, the GSM international mobile subscriber identity (IMSI) and temporary mobile subscriber identity (TMSI) and the GSM location area identity (LAI). The main parts of the standard focus on the following three aspects.

Interworking requirements

These include reference configurations and the protocol architecture models. The context of the interworking profile is also required, e.g. backwards compatible with GAP, etc. The interworking requirements contain the mandatory and optional requirements for the DECT fixed and portable parts, as well as the requirements on the DECT radio interface. The reference configuration, shown in Figure 20.3, is based on the GSM A interface (reference point "a" in the figure refers to the interface that supports the functional structure of the A interface). The GSM MAP (mobile application part) is assumed for the mobility management and call control.

Interworking Unit (IWU) Mapping

This part describes the mapping for the DECT fixed part to GSM PLMN attachment. IWUs are described for the DECT fixed part, which is expected to be most used, although an IWU attached to the DECT portable part is also described. Interworking models (one is shown in Figure 20.4) are used to describe the protocol interactions at the control plane and in the DECT fixed and portable parts. The protocol architecture model also shows the location of the IWUs. In Figure 20.4 the IWU in the FP (DECT fixed part) provides the mapping of the GSM MM (mobility management) sublayer (a subset of the GSM layer 3) to the respective DECT layer 3 protocols (NWL/MM) and vice versa. The call control entity is composed of a similar interworking model. The IWU in the portable provides the mapping of a subset of the DECT layer 3 protocols to the DAM GSM AP/DECT AP (most of the other abbreviations are explained in Chapter 19).

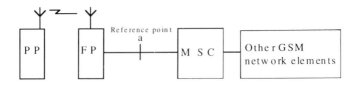

Figure 20.3   DECT/GSM IWP reference configuration.

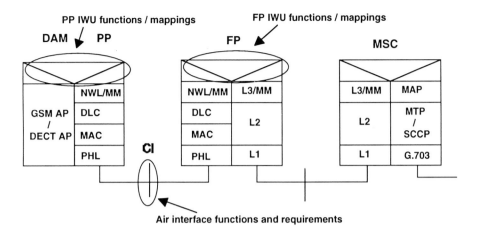

Figure 20.4 The interworking model for mobility management for the attachment of a DECT fixed part to the GSM PLMN.

Connection Types

The main DECT connection types at the air interface support optimised groups of services, from the IWU mappings for different configurations/models. The standard only covers one such connection type, which then is equivalent to the GAP connection type.

20.5.4 DECT/GSM Fixed Interconnection

As mentioned in the previous sections, the DECT/GSM IWP is based upon the GSM A interface. One of the standards in the DECT/GSM IWP addresses simply this aspect. This is the last link in the DECT/GSM chain to be standardised in the profile, since this is the link between the DECT fixed part and GSM PLMN. Some aspects are of particular interest in this standard, as follows.

(i) For *call control* (CC) and *mobility management* (MM), most messages are not interpreted, but relayed to the underlying DECT radio subsystem. Some messages may, however, be interpreted and handled at the interconnection. The DECT access protocols and a particular set of FP interworking functions/ mappings (to ensure that the GSM services can be provided over DECT) are specified in other DECT standards.

(ii) The DECT fixed part uses the A interface in such a way that it (the fixed part) emulates a GSM BSC (basestation controller). The standard then describes the procedures that are applicable on the A interface in order to ensure an appropriate emulation. At the interface are the so-called error-handling (mostly terminated messages), timers and message headers.

(iii) The functional split between the FP and MSC is similar to the functional split between a BSC and MSC. Since DECT has a completely different radio management, the number of functions handled by the FP will be less than for a GSM BSC. In particular, the DECT fixed part will handle the following functions:

- Terrestrial channel management
- User data encryption
- Signalling element encryption

20.5.5   Other Standards in the DECT/GSM IWP

As the GSM phase 2 supplementary services are so numerous, there is a separate standard covering these in the DECT/GSM IWP. The standard has two main sections:

- Supplementary services support procedures
- Interworking mappings for supplementary services

Supplementary Services Support Procedures

This section describes the general approach for the handling of call-related and call-independent GSM supplementary services over the DECT air interface. The procedural support is based upon the general approach where the DECT keypad protocol acts as transport mechanism and where the functional protocol of GSM is implemented in the DECT FP. The signalling mappings are described in terms of IWU procedures. The handling of individual services is described, when necessary, as well as error-handling and recovery procedures. Procedures described in detail include:

- Unstructured supplementary service data
- Closed user group
- User-determined user busy
- Call-related procedures not using facility information element
- Call waiting
- Call hold
- Call retrieve

Interworking Mappings for Supplementary Services

These show the mappings for GSM to DECT and DECT to GSM in respective order. One interworking unit (IWU) is considered – the FP/IWU. The detailed information element mappings between the stimulus DECT keypad protocol and the functional GSM protocol is the responsibility of the IWU and is described merely as guidance for implementation.

Apart from the standard describing the GSM phase 2 supplementary services, three additional standards are to be developed describing:

- Implementation of short message services, point-to-point and cell broadcast
- Implementation of bearer services
- Implementation of facsimile Group III

20.5.6 Future Development of DECT/GSM

The standards of the DECT/GSM IWP are at the time of writing at a very advanced stage. The basic standard, "Access and Mappings" (ETS 300 370), is approved, others have reached the approval procedure and the remainder are well developed. Are any further developments of the profile then to be expected? There are two major arguments for future development of the profile:

• Firstly, the global success of GSM will most probably lead to further development of GSM itself. In that case the interworking profile will itself need to be updated accordingly.
• Secondly, the existing profile is totally based on the GSM A interface. Lately, it has been suggested to consider interworking of DECT with the GSM PLMN using the so-called DSS1+ interface.

Considering this second proposal, the DSS1+ interface is an ISDN interface, yet to be developed (the DSS1 interface is standardised; the "+" sign indicates the addition of a channel for mobility management). To use the DSS1+ interface for interworking will offer several benefits. Firstly, this interface is the only one being considered for CTM (Cordless Terminal Mobility; see Chapter 6), which means that mappings between DSS1+ and DECT fixed part will be developed, i.e. DSS1+ will be an interface for which DECT will be standardised. Secondly, the DSS1+ interface could then become the general standardised interface for mobility (and other services). The development of corresponding standards for DSS1+ has started in ETSI, but it is premature to give any forecasts of timescale, etc., for the complete set of standards needed.

20.5.7 Other Aspects of the DECT/GSM IWP

There are of course a number of aspects not covered by standards, and in some cases there does not even exist at present any common understanding on how to proceed. One such aspect is roaming, i.e. the possibility to use (access) networks other than the "home" network (or subnetwork). This is a general unresolved issue in DECT, which as a short-range system will not have the possibility to offer the same wide-area coverage as GSM. Regulators may generally require roaming between different DECT operators.

In interworking with GSM it is to be expected that the same rules will be applied when using DECT/GSM IWP as when using GSM access. In particular, it is not possible to roam between the GSM operators in one's "home" country, i.e. the country of the GSM operator that issued the GSM subscription. If this rule is applied in the same way on DECT/GSM interworking, it will not be possible to roam between different GSM operators. One solution of this could be to have DAM cards with more than one GSM subscription.

Furthermore, non-GSM operators may also be interested in public operation of DECT using the DECT/GSM IWP, thereby offering GSM (or GSM-like) services to the users.

20.6   Profile Type Approval

To ensure that a piece of equipment does comply to a standard, usually a type approval process is used. The most important kind of type approval is the regulatory type approval, which usually is limited to terminal equipment. Equipment declared to comply with DECT PAP shall undergo such approval, described in a corresponding CTR. The same process is valid for GAP equipment.

It has not yet been decided to have separate type approval for DECT/GSM and DECT/ISDN terminal equipment, but it is likely that this will be necessary as in these cases the radio interface is specific for the application.

DECT data equipment will presumably not have to be regulatory approved, but it may be worth while to define a type approval for such equipment in order to inform users/buyers, operators, etc., of the qualities of a piece of equipment.

20.7   Profile Standardisation Status

The development of standards describing the DECT profiles started already at the time when the DECT basic standards were developed (DECT common interface). This development gained speed when the basic standards and regulatory documents were finalised. Table 20.1 lists the standards in the DECT profiles that have been finalised, or are in progress at the time of writing. The full names of the standards are usually very long, and thus the names have been shortened; preliminary names are used in some cases.

Some technical reports have been developed in ETSI on the profiles. These reports are "ETSI Technical Reports" (ETRs) and are used to explain the general outline of their respective subjects. The ETRs related to the DECT profiles have been listed in the previous chapter.

20.8   Further Profiles?

As mentioned above, the need for profiles for other applications has been discussed. The major benefit of such profiles is the agreed and standardised implementation of important features.Theoretically, a drawback of new profiles could be the time delay connected to the development of profiles. This is today hardly a problem as there is a very good and stable basis in the DECT standards – indeed, a new profile for a well-known area can now be developed within a timescale of a few months.

One area already discussed for a possible new profile is the wireless local loop application. DECT technology can be used in many ways in the local loop and an ETR including this topic was produced in 1993. It has been agreed to produce a (voluntary) profile defining how DECT may be used in an optimal, future-proof, way for this application.

A further interesting area for a DECT profile could be Cordless Terminal Mobility (CTM). The CTM project in ETSI aims to develop standards for network support for terminal mobility using the fixed public networks (primarily ISDN, but also PSTN is to be considered). Thus, CTM will offer general mobility

Table 20.1
Standards in the DECT profiles

Profile	Name	ETS	Approval	Notes
PAP	DECT Common Interface, part 9	300 175-9	1992[a]	
	Profile Test Specification, parts 1–7	300 323	1994	Approved 1995, 2nd edn
	TBR 11	–	1994	Approved 1995, 2nd edn
GAP	DECT GAP	300 444	1995	Approved 1995
	Profile ICS[e]	300 474	PE 95/6[b]	Approved
	Profile Test Specification	300 494	PE 95/6	Approved
	TBR 22	–	PE 95/6	On public enquiry (PE)
DECT/	3.1 kHz Speech	300 370	1994	Published, 2nd edn stable
GSM	3.1 kHz Speech, Profile ICS	300 xxx	–[d]	Approved for PE
	3.1 kHz Speech, Test Specification	300 xxx	–	Approved for PE
	General Description of Service Requirements	300 466	PE 94/12	Approved for vote
	Fixed Part to MSC Interconnection	300 499	PE 95/6	Approved for PE
	GSM Phase 2 Supplementary Services	300 703	PE 95/x[c]	Approved for PE
	GSM Short Message Services	300 xxx	PE 95/x	Started
	Bearer Services	300 xxx	PE 95/x	Stable
	Facsimile Services, Group III	300 xxx	PE 95/x	Started
	Enhanced Bearer Services	300 xxx	–	–
	Profile ICS	300 704	PE 95/x	On public enquiry
	Profile Test Specification	300 702	PE 95/x	On public enquiry
	TBR 35	–	–	EC mandate given
DECT/	End System Configuration	300 434	PE 95/12	Approved for vote
ISDN	Intermediate System Configuration	300 xxx	–	Started
	End System Profile ICS	300 705	PE 95/x	Approved for PE
	Intermediate System Profile ICS	300 xxx	–	
	ESC, Profile Test Specification	300 xxx	–	Started
	ISC, Profile Test Specification	300 xxx	–	
	ESC, TBR xx	–	–	EC mandate given
	ISC, TBR xx	–	–	EC mandate given
DECT	Frame Relay Services: A & B, Class 1	300 435	Vote 95/6[c]	Approved Nov. 1996
data	Roaming Data Stream: C, Class 2	300 651	PE 95/8	
	Roaming Frame Relay: A & B, Class 2	300 701	PE 95/x	
	Data Stream Services: C, Class 1	300 xxx	PE 95/x	
	Roaming Low Rate Messaging: E, Class 2	300 xxx	PE 95/x	
	Multimedia Messaging Service With Specific Provision for Facsimile Services: F, Class 2	300 xxx	PE 95/x	
	Isochronous Services: D, Class 1	300 xxx		Expected spring 1997
	Roaming Isochronous Services: D, Class 2	300 xxx		Expected spring 1996
	Conformance Test Specification	300 xxx	–	Expected spring 1996

[a]The year, 1992, for example, means the year of final approval in ETSI.
[b]PE 95/6, for example, means that the standard exists as a stable draft, and is expected to be approved for public enquiry in the year and month shown.
[c]Vote 95/6 or PE 95/x mean the year and month of start of vote or public enquiry in ETSI. These dates are not fully stable, but the standard has been approved for vote or PE.
[d]A dash under "Approval" means that the standard is expected.
[e]Implementation Conformance Statement.

comparable with the cellular networks (e.g. GSM), based, however, on using the fixed networks and cordless terminals. The DECT generic access profile fulfils all existing requirements from CTM phase 1. In order to meet the requirements of future phases of CTM, it has been decided to develop a specific DECT CTM profile, in which CTM-related standards, or elements thereof, are gathered or listed.

Notes

1   Of course, later implementations of CT2 telepoint, such as those in Asia described in the early chapters of this book, have to some extent incorporated such features.
2   ERMES = European Radio Messaging System, the new ETSI digital paging standard.

21 Personal Wireless Telecommunications, PWT

Walter Tuttlebee

The Personal Wireless Telecommunications (PWT) standard, TIA/EIA 662 [1], is based upon the now well established Digital Enhanced Cordless Telecommunications (DECT) standard, with minimum modifications necessary to allow operation in the North American unlicensed frequency band (1910 to 1930 MHz) allocated for personal communications services (UPCS). Originally designated WCPE (for Wireless Customer Premises Equipment), the nomenclature was changed to PWT in 1995 to reflect the intention of several manufacturers to develop equipment suitable also for outdoor and wireless local loop operation in the licensed PCS regime – a variant known as PWT-E being standardised initially within the JTC TAG6, with responsibility transferred during 1995 to the TIA TR 41.6 Subcommittee.

The PWT standard is designed to allow operation of DECT-derived equipment in the UPCS band in accordance with the US Federal Communications Commission (FCC) "Spectrum Etiquette" rules defined in [2] and summarised in Chapter 8. This chapter briefly summarises the technical differences between the PWT/PWT-E and DECT standards.

21.1 Aims and Origins

The PWT standard has been prepared under the auspices of the TIA TR 41.6 Subcommittee. This subcommittee has developed an interoperability standard that is essentially based upon the DECT GAP standard, with the minimum essential changes to meet the FCC rules for the use of the unlicensed PCS spectrum. The aim of the standard has been to facilitate the availability of low-cost equipment capable of supporting a range of services, voice and data, and applications, residential and office, in the US marketplace. A wider range of applications can be supported by PWT [1] and PWT-E [3] – as with most of the US UPCS standards, the concept is that a single dual-mode handset should be economically capable of supporting both licensed and unlicensed modes of operation. PWT-E is a high-capacity picocellular technology, for cell radii up to about 1 km, with range extension possible using wireless basestations, known as cordless radio fixed parts (CRFPs).

The PWT standard was initiated by Motorola and AT&T and received early backing from most of the major international telecommunications manufacturers, including Ericsson, Nokia, Nortel, Philips and Siemens (Rolm).

21.2   Structure of the Standard

The PWT standardisation documents [1] mirror essentially the same basic structure as the ETSI documents ETS 300 175 [4], which define the DECT standard. Specifically, the content of the various documents is summarised below:

- TIA/EIA 662-1 provides an overview of the standard, describing the scope and organisation of the document, as well as providing lists of references and definitions.
- TIA/EIA 662-2 provides the details of the physical radio interface, including the channel frequencies and requirements for modulation, channel monitoring, channel selection and interference rejection.
- TIA/EIA 662-3 describes the medium access control layer.
- TIA/EIA 662-4 describes the data link control layer.
- TIA/EIA 662-5 describes the network layer.
- TIA/EIA 662-6 describes PWT identities and addressing aspects.
- TIA/EIA 662-7 describes security features of PWT.
- TIA/EIA 662-8 describes the speech coding and transmission aspects.
- TIA/EIA 662-9 describes the customer premises access profile, essentially equivalent to the DECT generic access profile.

The PWT-E standard has been developed as standards project PN-3614. The draft standard document for PWT-E comprises a delta document from the above PWT specifications and was issued for ballot in early 1996.

21.3   Basic Parameters and Rationale

This section describes the basic parameters of PWT/PWT-E and the major differences between PWT and DECT that have been incorporated in order to ensure adherence with the spectrum etiquette defined in the FCC Part 15 Subpart D rules [2].

The basic PWT parameters are summarised in Table 21.1. The need to make changes from the DECT standard reflects the FCC stipulation on coexistence of multiple PCS technologies in the unlicensed spectrum and the resultant implications arising from the spectrum etiquette rules. These changes are detailed further in section 21.4.

Parameters are also shown below for the PWT-E standard. The PWT-E standard is an application of the PWT standard in the licensed frequency bands 1850–1910 MHz and 1930–1990 MHz with enhanced capabilities. A PWT-E system can provide TDD operation in (parts or the whole of) one or both duplex bands of a licensed FDD block of spectrum. The offered services are the same as those supported by DECT. Wireless relay stations are described in the PWT-E standard, as a means of providing cost-effective coverage enhancement, in a similar manner as for range extension for DECT-based wireless local loop in Europe [5].

Table 21.1
Main parameters of the PWT and PWT-E standards

Parameter	PWT	PWT-E
Modulation	$\pi/4$ DQPSK	$\pi/4$ DQPSK
Bit rate	1.152 Mb s^{-1}	1.152 Mb s^{-1}
Symbol rate	576 kb s^{-1}	576 kb s^{-1}
Frame duration	10 ms	10 ms
Duplexing	TDD	TDD
No. of half slots (8 kb s^{-1} user data)	48	48
No. of full slots (16 kb s^{-1} user data)	24	24
No. of double slots (32 kb s^{-1} user data)	12	12
Transmit power level 1 (2 mW)	Optional	Optional
Transmit power level 2 (90 mW)	Mandatory	Mandatory for handsets
Transmit power level 3 (200 mW)	Not allowed	Optional
Transmit power level 4 (500 mW)	Not allowed	Optional for basestations
Receiver sensitivity (at 0.1% BER)	−90 dBm	−90 dBm for handsets −92 dBm for basestations
Antenna gain	3 dBi maximum	Allowed
Tolerance to time dispersion	200 ns (with selection diversity)	200 ns (with selection diversity)
Extended preamble transmission	Optional	Mandatory
Operating frequency band	1910–1930 MHz	1850–1910 MHz 1930–1990 MHz
Number of carriers	16 carriers	120 carriers
Carrier frequency spacing	1.25 MHz	1 MHz

Source: from [3, 6].

The rationale for introducing the PWT-E standard alongside PWT has been that the FCC unlicensed PCS allocation alone cannot serve the public with the wide range of reliable residential, office, public outdoor and wireless local loop services that is available elsewhere using DECT and its 20 MHz initial allocation. The reason that the traffic capacity is lower is because:

- Only 10 MHz is allocated for periodic transmissions (speech and ISDN).
- FCC etiquette rule power restrictions preclude antenna gain.
- Use of the spectrum by a range of *different* unlicensed PCS technologies, i.e. multiple system coexistence, decreases the overall efficiency of the spectrum usage.

The FCC unlicensed PCS allocation is well suited to office and residential applications, including private outdoor sites, such as campuses, but not for outdoor public services. The full range of services supported by DECT (public as well as private) can, however, be offered by using a part of a licensed PCS block, preferably in combination with the unlicensed PCS band. Thus, for example, a handset can operate in the whole of the licensed and unlicensed bands, whilst a PWT-E basestation will be programmed to operate in the appropriate specific part or parts of the licensed PCS band and may also operate in the unlicensed PCS band (although when doing so it must conform to the PWT base standard, e.g. 1.25 MHz channelling, lower power, etc.).

Illustrations of such usage are given in [3, 6], which also provide traffic capacity results for usage of PWT/PWT-E in a variety of scenarios, including office environments, pedestrian street usage and wireless local loop. Thus, for example, if a licensed PCS operator were to allocate just 2 MHz of his up- and downlink FDD licensed blocks for PWT-E, these 48 channels would support an average traffic per single basestation of 2.8 Erlang in outdoor pedestrian usage and up to 4.1 Erlang in large office applications. Further consideration of traffic capacity and spectrum efficiency is presented in section 21.4.3.

21.4   Principal Changes from DECT

The main changes from DECT for the PWT standard are as follows:

- Issues relating to the spectrum allocation and channelling – impinging primarily upon the physical layer.
- Issues relating to the etiquette rules and implications – necessitating a minor change to the MAC layer.

Other small changes are also proposed, some of which have been incorporated in the second edition of the base DECT specification.

21.4.1   Physical Layer Changes

The primary change in the PWT physical layer, as compared to the DECT standard, relates to the modulation scheme and channel spacing.

The division of the 1920–1930 MHz isochronous band into 1.25 MHz channels precluded the adoption of the original DECT physical layer and has resulted in the adoption of a narrower-band, $\pi/4$ DQPSK, modulation, with $\alpha = 0.5$, resulting in the required 1.25 MHz channel spacing. Compared with DECT equipment, this change will require modifications to the radio circuitry, implying the need for new radio chips. A linear RF power amplifier with higher peak power requirement (by 3 dB) is also required, implying a slightly higher power consumption (circa $\times 1.3$, for similar output power) for the handset.

The robustness of PWT to delay spread is similar to DECT. Although the symbol duration is doubled, tolerance to delay spread also depends upon receiver co-channel rejection performance C/I. For $\pi/4$ DQPSK the C/I performance is in fact some 3–4 dB worse than for GFSK, counterbalancing the improvement at first sight offered by the increased symbol duration. An extended preamble is optional for PWT, but mandatory for PWT-E, facilitating as it does improved selection diversity performance.

21.4.2   MAC Layer Changes

Minor changes to the DECT MAC layer has been necessary, in order to accommodate the FCC channel selection (spectrum etiquette) rules. These changes are reflected in small changes in the MAC software; otherwise, however, the implications of these changes are trivial. Existing DECT baseband chipsets (burst mode controllers) may thus be employed in PWT terminals, offering potential scale economies, arising from the already significant volumes of DECT product sales in Europe.

21.4.3 Traffic Capacity Performance

Simulations developed in the course of standardisation and reported in [3, 6] have indicated that the PWT/PWT-E standard is spectrally efficient, in both absolute and relative terms. Note, however, the differences expressed in section 21.3 between the use of PWT and PWT-E. In relative terms, the utilisation of PWT in the unlicensed PCS band 1910–1930 MHz is less efficient than the use of a similar amount of spectrum purely for DECT or PWT-E. The reasons for this are the frequency band split and segmentation, coexistence of different technologies and the fact that the channel selection procedures mandated by the FCC spectrum etiquette rules have been optimised with the need to take into account other issues as well as spectrum efficiency.

Disregarding the reduction due to coexistence of different technologies, to compare like with like, a 10 MHz PWT system (eight carriers) will support the same traffic capacity as will a 10 MHz DECT system (five carriers) in an office environment.

The traffic capacity per megahertz is larger for PWT-E than for PWT, owing to the narrower channel spacing (1 MHz as compared to 1.25 MHz) and the use of different (DECT) channel selection rules. Simulations [6] indicate that for most realistic scenarios PWT-E provides about 30% more traffic per MHz than DECT; theoretically no improvement in capacity is anticipated for a propagation law of $n = 2$ (line of sight), but about 30% improvement is indicated for a propagation law of $n = 4$, which will reflect many realistic deployments.

The second edition of the DECT standard ETS 300 175 has defined extended carrier frequencies up to 1937.088 MHz, i.e. beyond the original 1880–1900 MHz. Thus, both PWT-E and DECT are defined for the frequency band 1880–1937 MHz. PWT/PWT-E and DECT provide very efficient coexistence on a common spectrum allocation, because they share a common time-domain slot structure and channel selection procedures. The difference in carrier frequencies has only marginal influence on the coexistence efficiency. This means that in technology-neutral markets the two standards will be able to coexist without significant loss of capacity within a common spectrum allocation.

21.4.4 Interoperability Profiles

The fact that differences between DECT and PWT are limited as they are means that great potential for re-use is offered, as well as for baseband chipsets (and elements of the radio design), also for software, including software for a variety of profiles defined already for DECT. Thus, a wide-ranging set of applications that have already been developed, proven and implemented can be readily supported by PWT. These could potentially include the GSM interworking profile (GIP), which would allow interworking between PWT/PWT-E and PCS1900 public PCS systems now deployed in North America for speech, short message, bearer and fax services. This must enhance the competitive position of PWT/PWT-E as a low-tier cordless type standard in the North American market.

Acknowledgements

The assistance of Dag Åkerberg, pioneer of DECT and a key architect of its adaption as PWT for the North American marketplace, is gratefully acknowledged, as is the help of Peter Bligh, Chairman of the TR 41.6 Subcommittee standardising PWT. Any errors, however, remain the author's own.

References

[1] "The Personal Wireless Telecommunications (PWT) Interoperability Standard", TIA/EIA Standard Proposal 662, 1996

[2] "Radio Frequency Devices, Subpart D – Unlicensed Personal Communications Service Devices", FCC 47 CFR 15

[3] "The Personal Wireless Telecommunications – Enhanced (PWT-E) Interoperability Standard", Standards Project PN3614, TIA TR 41.6, 1996

[4] "DECT Common Interface (CI), parts 1 to 9", ETS 300 175, 2nd edn, European Telecommunications Standards Institute, Sophia Antipolis, Nice

[5] "Application of DECT Wireless Relay Station", ETR 246, European Telecommunications Standards Institute, Sophia Antipolis, Nice

[6] "DECT Standards in North America", D Åkerberg, DECT '96 Conference, organised by IBC, London, January 1996

22 The PACS-UB Standard

Tony Noerpel

This chapter details the technical operation of the North American Personal Access Communications System, Unlicensed B (PACS-UB) system, which was presented in overview earlier in this book. PACS-UB [1–3] is a time-division duplex (TDD) version of the PACS [4] Common Air Interface standard for low-tier PCS, with modifications to conform to the FCC etiquette rules for the unlicensed spectrum. Both standards are based on the Bellcore WACS air interface specification for low-tier service [5, 6].

The PACS-UB protocol was designed with the following guidelines in mind:

- A dual-mode subscriber unit terminal capable of both PACS and PACS-UB operation should be only slightly more complex than a single-mode PACS terminal.
- PACS-UB should use the PACS higher-layer protocols, to facilitate service interoperability (e.g. automatic registration between licensed and unlicensed spectrum access modes).
- Channel scanning, access rights determination and channel access for PACS-UB should be fast to facilitate service interoperability.
- PACS-UB should retain from PACS the system design philosophy that emphasises inexpensive, highly reliable and simple infrastructure.
- PACS-UB should be robust in the presence of interference from unlike and/or unsynchronised systems that share the same spectrum.
- The protocol underpinnings of PACS-UB should scale gracefully with system size and teletraffic capacity requirements, from large office wireless Centrex systems, to multi-line key sets in small business environments, down to residential use of cordless home ports.

The FCC has encouraged technologies that can interoperate in both the licensed and unlicensed spectrum so as to give future personal communications services (PCS) customers opportunity for economic, high-quality and flexible communications. Notably, Julius Knapp of the FCC, in an address at the WINForum conference in Dallas, Texas [7], stated that we should "expect dual-mode devices that operate in either the licensed or unlicensed PCS spectrum", a view shared by the developers of the PACS standard [4]. Thus, the PACS air interface standards have an FDD mode to enable operation in the licensed frequency allocation and a TDD mode to enable unlicensed isochronous operation in the 1920–1930 MHz spectrum [8]. The licensed and unlicensed mode standards were defined to facilitate manufacturing equipments that place both air interfaces in the same low-cost package [9]. The PACS air interface standards are designed to bridge the gap between indoor and outdoor coverage. Outdoor applications dictate the use of frequency-division duplexing (FDD) because of the capacity advantage over time-division duplex (TDD) methods when basestation signals are unsynchronised [10]. However, that advantage is mitigated for indoor applications. Additionally, the PACS protocol is easily extensible to "high-tier" applications either in combination with existing cellular air interfaces such as PCS-1900 (GSM) [11] or IS-136, or by simple modifications to the PACS protocol itself [12–16].

Motorola and Hughes Network Systems introduced the Bellcore WACS standard [17] to the Joint Technical Committee (JTC) of TIA and T1 [18] in late 1993 and, with Bellcore, played the lead role in forming the WACS Technology Advocacy Group (TAG). The WACS standards proposal was endorsed within the JTC by potential PCS service providers including Bell Atlantic, Time-Warner, Sprint and US WEST.

NEC, Panasonic, Hitachi and PCSI originally proposed a low-tier air interface standard to the JTC based on the Japanese Personal Handyphone System (PHS). They withdrew this proposal in favour of modifying the proposed WACS standard to incorporate aspects of PHS. The merger of the WACS and PHS air interface standards proposals was renamed PACS. The PACS TAG produced a single air interface standard for operation in the licensed bands and two PACS-compatible air interface standards for operation in the unlicensed PCS (UPCS) spectrum [19]. One unlicensed standards proposal is based on PHS and is called PACS-UA [20]; the other unlicensed proposal, which is based more closely on the original WACS, is called PACS-UB [1] and is described in this chapter.

PACS allows for low-complexity implementations of both portables and basestations in order to reduce power consumption. The portable peak transmit power is 200 mW and the average power is 25 mW. The radio ports (RPs) function largely as RF modems, depending on the centrally located radio port controller units (RPCUs) for most of the functionality traditionally associated with RP electronics. For example, PACS RPs can be powered to 3.5 km using local exchange company supply voltage of 130 V on HDSL (high-density subscriber line) deployed on 24 or 26 gauge copper pairs. Line powering eliminates both the need for batteries at the RP and the need to derive local power at the RP site. Another advantage of locating most of the electronics in the RPCU is that system upgrades to support new services or improve speech codecs do not require visits to RP sites.

Each PACS burst carries 120 bits of information, including 80 bits of payload or user information on the fast channel (FC) and 40 bits or 20 symbols of overhead. The 80-bit FC, used once per 2.5 ms frame, provides a raw data rate of 32 kb s^{-1}, adequate for good-quality speech coders. PACS also supports sub-rate channels of 16 kb s^{-1} and 8 kb s^{-1}, achieved by using one burst per two frames, or one burst per four frames, respectively.

The advantages of a short frame structure cannot be overstated. Varma et al. [21] have shown that frame erasure lengths have an impact on the recovery time of speech quality in the presence of radio link errors. Longer system frame lengths lead to longer speech frame erasures, which in turn lead to slower recovery of speech quality for the ITU standard ADPCM decoders. For example, for 2.5 ms erasures, the decoded speech signal recovers to within 3 dB of the error-free signal (in terms of signal energy) within 5 ms after restoring correct transmission for 80% of burst errors. For 10 ms erasures, the 80th percentile delay for recovery to within 3 dB of error-free speech is 35 ms. Therefore, systems with shorter frame lengths can provide more robust speech quality in the presence of link errors.

RP operating frequencies are assigned automatically and autonomously, eliminating the need for manual frequency planning. The automatic frequency assignment is called quasi-static autonomous frequency assignment (QSAFA). QSAFA is a self-regulating means of selecting individual RP frequency channel pairs that functions without a centralised frequency coordination between different RPs. PACS uses a version of QSAFA called "least interference algorithm" (LIA) and PACS-UB uses a version called "lowest frequency below threshold algorithm" (LFA) [22]. QSAFA combines the principal advantage of dynamic channel allocation (DCA), in that pre-engineering of a frequency plan is unnecessary, with the performance advantages of a fixed frequency assignment, i.e. elimination of blind time slots for channel assignment, elimination of resource blocking, and faster call set-up and handoff times [23]. The resource blocking probability for DCA technologies is higher than Erlang-B blocking because of the blind time slot problem [24].

Several features of PACS-UB make it particularly amenable to operation in 1.9 GHz unlicensed PCS spectrum:

- Its short frame structure and low end-to-end delay allow fast frequency scanning by subscriber unit terminals, eliminate the need for echo control processing in full-rate 32 kb s^{-1} traffic channels, and allow sub-rating down to 8 kb s^{-1} traffic channels.
- Its narrow-band transmission format creates a relatively large number of frequency channels in the 10 MHz allocations, which should be advantageous in conditions of high intersystem interference.
- Finally, its emphasis on low-cost, low-complexity hardware is highly amenable to a radio technology that could span both residential and business uses.

The sections that follow provide a general overview of the PACS-UB radio system for application to wireless business and residential telephony. The rest of this chapter is organised so as to reflect the PACS-UB air interface standards document as published by ANSI.

22.1 Objectives and Document Organisation

Section 1 contains the objectives of the air interface standards document and describes the document's general organisation. The overall objective is to provide a standardised layered air interface that is optimised to allow residential and business customers to gain wireless access to a typical wireline exchange using the radio spectrum allocated by the FCC for unlicensed operation.

This document specifies the bandwidth, frame structure, elements of procedure, format of fields and procedures for the proper operation of the PACS-UB air interface.

22.2 Definitions

Section 2 contains definitions of the acronyms used in the standards documents. A glossary of terms is included, as are definitions of timers and parameters used in PACS-UB components and required by the air interface.

22.3 Overview of the PACS-UB Concept

Section 3 provides an operational overview of PACS-UB components, their functions, and the system's operating environment. The functional representation of the system architecture is shown in Figure 22.1. This representative architecture is similar to that given for PACS. The architecture consists of portable terminals or subscriber units (SUs) communicating through radio ports (RPs) that access via a radio port control unit (RPCU) and an access manager (AM) to a switching function. It should be recognised that, though the RP, RPCU and AM are often discussed in this document as separate logical entities, some or all of their functions may be combined as a stand-alone unit for some applications.

The AM, in conjunction with the RPCU, facilitates aspects of radio access, such as automatic link transfer (ALT) or handoff for a call in progress moving from one RPCU to the next. The AM function may be implemented in a stand-alone RP, an RPCU, a switch adjunct, or the switch.

The switching network provides a variety of functions, including provisioning, performance management, capacity management, etc. The AM and switching functions are shown for completeness. The air interface specification does not encompass the AM, the switching function, or the configuration of the radio system or network.

The following interfaces are defined for the system:

- Interface A (the "air interface") connects the SU and the RP. The air interface consists of layers 1, 2 and 3 that are described in sections 5, 6 and 7 of the PACS-UB standard document.

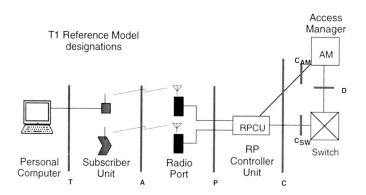

Figure 22.1 Functional reference architecture.

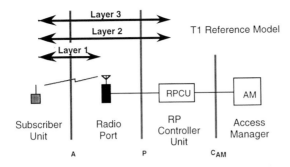

Figure 22.2 PACS-UB signalling protocol connections.

- Interface P provides connectivity between the RPCU and its RPs. The protocols provided by layers 2 and 3 provide for interaction between the SU and the RPCU that, therefore, traverse interface P. Interface P also carries an embedded operations channel (EOC) that supports control functions between the RPCU and its RPs. The physical interconnection (layer 1) between the RPCU and its RPs may be implemented in any of several means and may be an internal interface. As such, layer 1 of interface P is not specified in the document; however, any implementation must be capable of supporting the protocols as defined in the document for layers 2 and 3.
- The C, D and T interfaces are depicted in Figure 22.1 but are not specified; they are shown for reference purposes only.

Figure 22.2 illustrates the protocol layers defined in the standards document. Layer 3 messages shown in the call flows of the document make use of the acknowledge mode transfer service of the underlying layer 2 process. Thus layer 3 messages arrive at their destination error-free and in the order they were sent. The PACS-UB "layers 1, 2 and 3" are defined specifically for the air interface and should not be interpreted as the layers 1, 2 and 3 of any other protocols such as OSI, ISDN or Signalling System Number 7 protocols.

22.4 Radio Parameters

Section 4 specifies the layer 1 radio parameters for the air interface. The isochronous unlicensed operating band, in the UPCS spectrum as allocated by the FCC, is divided into 8×1.25 MHz channels. For PACS-UB these channels are considered to be further subdivided into 4×300 kHz channels. This results in 32 traffic frequencies numbered from $N = 702$ to $N = 798$, so that the RF channel centre frequency equals $1850 + N/10$ MHz as shown in Figure 22.3.

Since the uplink and the downlink for PACS-UB use the same frequency channel, the system can take advantage of reciprocity and use transmitter antenna diversity at the RP to achieve downlink diversity protection. The performance of this form of diversity is limited by the time delay between the uplink burst received at the RP and the transmission of the downlink burst using the antenna that the RP determined received the best-quality uplink signal. This

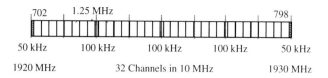

Figure 22.3   Operating channel structure in the unlicensed PCS band.

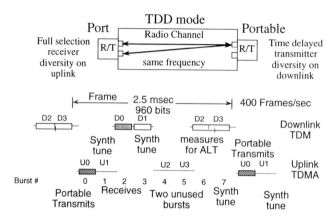

Figure 22.4   PACS-UB diversity operation and TDMA/TDD frame structure with portable activity.

latency is equal to 625 µs and therefore the diversity protection for PACS-UB should be good for user speeds up to about 20 mph (~9 m s^{-1}). Additionally, the portable uses selection diversity when in simplex mode of monitoring port signals for alerts or system information, or when measuring port signals to find proximate and accessible PACS-UB ports for registration or other access or as handoff candidates during calls. The diversity protection method and the frame structure of PACS-UB are shown in Figure 22.4. The diversity implementation in the PACS-UB system, together with the excellent receiver sensitivity specification of −101 dBm, both contribute to the excellent range and coverage area performance.

This section also covers in-band and out-of-band emissions, adaptive power control, carrier-off condition, ramp-up and ramp-down of the transmitters, transmit power, frequency stability, modulation, modulation accuracy, pulse shaping, symbol alignment and other physical requirements for PACS-UB portables and basestations. The fast uplink power control of the PACS-UB radio system enhances its capacity advantage [25].

22.5   Layer 1 Interface Specification

Section 5 details the layer 1 air interface specification. The PACS-UB TDD frame structure is shown in more detail in Figure 22.5. The bursts are paired to form

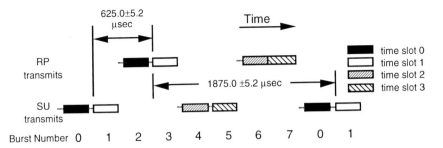

Figure 22.5 Burst pairing forming time slots.

duplex channels as follows: (0,2), (1,3), (4,6) and (5,7). Thus PACS-UB has four servers per frequency channel and 32 frequency channels in the 10 MHz isochronous unlicensed band, giving a total of 128 traffic channels in 10 MHz.

The channel assigned for broadcasting of system information is designated the system broadcast channel (SBC). The SBC is asymmetric in that it is primarily used in the downlink. Its uplink is used by portables for access requests.

At any time while a time slot is available (not in use as a traffic channel) at an RP, it is used to broadcast system and alerting information and to handle access requests from SUs; when assigned in this manner, this channel is known as the system broadcast channel. There is a superframe structure associated with the SBC that is referred to as the SBC-SF. The SBC-SF is composed of a group of 400 consecutive frames, i.e. 1 s in duration. The SBC-SF is used to allow the SBC to carry a number of logical channels (the alerting channel, the system information channel and the access request channel).

In performance of the channel access etiquette, RPs employ two different structures of the SBC-SF known as the "basic SBC-SF" and the "access SBC-SF"; their structure is illustrated in Figure 22.6.

Consistent with PACS, the beginning of the basic SBC-SF is marked by the transmission of the first alert phase or group. The basic SBC-SF is further divided into two intervals designated as interval A and interval B. During each of these intervals, the system broadcast channel is used as follows:

- Interval A (200 ms) – the SBC downlink carries the alerting channel (AC) and the access request channel (ARC).
- Interval B (800 ms) – the SBC downlink carries the system information channel (SIC) and the access request channel (ARC).

The basic SBC-SF structure is employed by the RP for most transmissions. However, the access SBC-SF structure is utilised by the RP during execution of the RF channel access procedure. This access SBC-SF structure reserves an interval for the RP to perform the required RF channel search process. However, this access SBC-SF structure in divided into four intervals used as follows:

- Interval A (200 ms) – the downlink SBC carries the AC and the ARC.
- Interval B (200 ms) – the downlink SBC carries the SIC and the ARC.
- Interval C (400 ms) – the RP searches for an acceptable RF channel and, upon identification of an acceptable channel, the SBC carries the ARC.

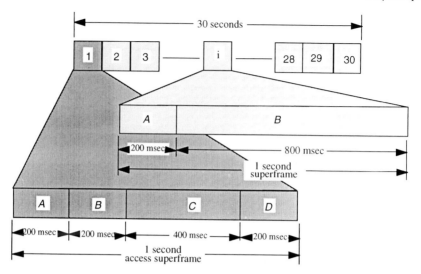

Figure 22.6 The basic and access system broadcast channel superframe structures.

- Interval D (200 ms) – the downlink SBC carries the ARC.

In order to conform to the access etiquette, an RP must initially employ the access SBC-SF to select a suitable RF channel for transmission. Following successful identification of an RF channel, an RP uses the basic SBC-SF format for further transmissions for up to 30 s of unacknowledged transmission. Without receipt of a transmission from an SU, the RP must employ the access SBC-SF structure at least once every 30 s (i.e. a single access SBC-SF followed by, at most, 29 basic SBC-SFs). This process (as shown in Figure 22.6) ensures that the RP relinquishes its selected RF channel each 30 s unless the channel has been used for SU activity.

TDM downlink avoids blind time slots, resource blocking and other problems normally associated with DCA schemes, releasing more bandwidth for system information, alerting and system ID. This allows handsets to rapidly determine access rights to private networks and basestations and thus conserves battery charge life.

The RP searches for an available RF channel in a specific manner described by the etiquette state diagram shown in Figure 22.7. Upon initial power-up, the RP begins in the "idle" state. When it reaches interval C in its first SBC-SF, it moves to the "measurement" state and begins to measure received signals on channels within the frequency band. Its measurements must begin at any channel 727 or below and proceed sequentially through each channel (e.g. 727, 730, 733 and so on) until an acceptable channel is identified.

When an acceptable channel has been found, the RP moves to the "transmit" state where it remains (using the basic SBC-SF) until it is required to perform the access procedure again or until it receives an access request from an SU. When the RPCU determines that an RP must relinquish the channel, it sends a broadcast directive in interval A of the access SBC-SF to inform all SUs that the RP is relinquishing the channel.

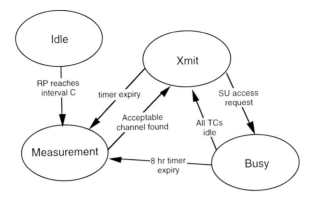

Figure 22.7 Etiquette states.

Upon receipt of an access request from an SU, the RP moves to the "busy" state. The RP remains in the busy state as long as a call is present on any of its traffic channels. When the RP no longer has a call present, it returns to the transmit state. The RP must return to the measurement state at least once every 8 h even if it is carrying an active traffic channel.

PACS-UB employs two burst formats. These are identified as format A and format B. Burst format A (Figure 22.8) is used for bursts 0–6 at all times. Burst format A is also used for burst 7 (the burst that normally carries the downlink SBC) when it is carrying a traffic channel. Both downlink and uplink bursts carry a 10-bit slow channel (SC), an 80-bit fast channel (FC) and a 15-bit cyclic redundancy check (CRC). The final bit of the downlink burst is the power control channel (PCC), and the final bit of the uplink is reserved for future use.

Burst format B (Figure 22.9) is used for burst 7 when it is carrying SBC. The downlink SBC burst (format B burst) contains 14 bits of synchronising information in the sync channel (SYNC) to facilitate frame initialisation (initial synchronisation) by the SU.

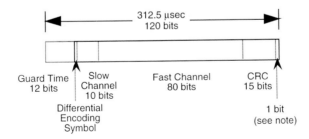

Note: The last bit in the burst is used differently in the uplink and the downlink. This bit is reserved in the uplink and is the PCC in the downlink.

Figure 22.8 Format A burst.

Downlink Burst Format B for the SBC

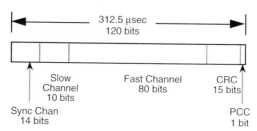

Figure 22.9   Format B burst.

22.6   Layer 2 Interface Specification

Section 6 specifies the layer 2 protocol operation for the core elements of the air interface. This protocol includes link establishment and release, an acknowledged mode transfer protocol to provide error and flow control for higher-layer messages, encipherment of user information and signalling for privacy, handoff or automatic link transfer control, portable synchronisation with the basestation transmission, and channel coding for joint error detection and burst synchronisation.

Section 6 also describes the format and use of the dedicated system broadcast channel (SBC). Because of the relatively large alerting channel bandwidth available in the SBC, PACS can support up to 200 000 users and PACS-UB can support up to 80 000 users per alerting/registration area (ARA) with approximately zero probability of alert blocking [26]. ARAs can be made arbitrarily large, which considerably reduces registration traffic. An effective polling procedure is also supported, which can be implemented as required by a service provider.

While in licensed mode (i.e. not currently registered to an unlicensed private system), an SU can scan periodically for an unlicensed system or RP to which it has access rights. Unlike scanning for channels in the licensed mode, where the presence of signals in a given band conveys both air interface compatibility and access rights, the portable RF channel search process in the unlicensed band is a threefold process. Firstly, the portable must detect the presence of energy or the proximity of a basestation. Secondly, the portable must be able to demodulate the transmitted signal (i.e. the detected transmission must be from a PACS-UB basestation and not from a basestation using some other air interface protocol). Thirdly, the portable must determine access rights to the basestation (i.e. it must be the residential cordless basestation associated with the portable or a basestation in the wireless PBX or Centrex system to which the portable belongs or some public or telepoint basestation to which the portable has subscribed).

For PACS-UB a portable can measure all 32 frequencies and determine access rights in about 200 ms; if no suitable ports are found, then the SU can return to standby mode. This performance is facilitated by two factors: the frequent broadcasting of access rights and system identities, and the 2.5 ms frame structure. Channel acquisition for initial access takes less than 20 ms and channel access for the intra-RPCU handover process takes less than 30 ms.

Both PACS and PACS-UB support a low-duty-cycle standby mode of operation. The SU need only activate its receiver for a listening period ranging from 312 µs to 12.5 ms out of every 1 s alerting superframe period to listen for alerts or system directives. Since this "listening" operation consumes low power, the battery life of the SU can last longer. The average duty cycle for PACS and PACS-UB is only 0.7%.

22.7 Layer 3 Interface Specification

Section 7 specifies the layer 3 protocol operation for the core elements of the air interface. Layer 3 constitutes the messages and procedures necessary to establish and maintain an end-to-end connection through the network. The defined procedures support the establishment of $32\,\text{kb s}^{-1}$ circuit-switched voice-band information calls and circuit-switched data calls. Layer 3 also includes certain mobility management functions (such as terminal registration) that are not in themselves end-to-end connections but facilitate network services. This section includes the message formats, information element formats, specification description language (SDL) diagrams and message flow diagrams for successful registration, unsuccessful registration, call origination, call delivery with answer, call delivery with no answer, far-end disconnect, portable disconnect, three-way calling, call waiting, service rejection, and rejection of individual message service during a call.

This section also describes the privacy and authentication (P&A) technique specified for the air interface, which is based on an enhanced version of the secret-key P&A methods developed for TIA Interim Standard IS-54, DAMPS [27]. Migration to the use of public key techniques is also supported.

22.8 Speech Service Option

Speech service is considered to be an optional service in the PACS-UB system, i.e. not all equipment is required to support speech services. However, if equipment is designed to support speech services, manufacturers must adhere to the specifications presented in section 8 of the standard.

The first version of the PACS-UB air interface specification provides only for $32\,\text{kb s}^{-1}$ speech to be encoded using ADPCM, the algorithm specified in CCITT G.726 [28].

22.9 Messaging Service Option

Section 9 specifies those supplementary elements of the air interface that are required in order to implement messaging services. The PACS-UB individual messaging protocol (IMP) supports a private, reliable, two-way messaging service. IMP supports very long messages (up to 16 Mbyte) in a wide variety of message formats. The protocol provides for two-way service, offering users the ability both to send and to receive messages. All messages may be enciphered using the same procedures described in sections 6 and 7 of the air interface document. IMP supports a wide variety of message types, including unstructured text, structured textual messages, images, audio, video, and application-specific formats. IMP messages may be transferred (either sent or received) during a call. Examples might include: e-mail, a 10-digit string received by an alphanumeric pager, a voice-mail message, a multi-page facsimile transmission, an image taken by a surveillance camera, or a service order dispatch to a field service agent.

22.10   Circuit Mode Data Service Option

Section 10 specifies those supplementary elements of the air interface that are required in order to implement circuit mode data services. A component of circuit mode data services is link access procedures for radio (LAPR). LAPR is an automatic repeat request protocol, whereby user data is fragmented into numbered frames. The peer LAPR process sends a numbered, positive acknowledgement when specific frames are received with no error detected. LAPR supports ITU-T V.42bis [29] compression. LAPR provides a full error correction mode of operation; in this mode, LAPR continues to attempt to deliver every octet of user data until it is acknowledged by its peer as received without error. The latency that a given packet experiences varies with the instantaneous radio link error performance. A call using LAPR uses the same procedures for link access, authentication, call set-up, and link encipherment as speech calls. Handoff during circuit mode data calls is supported.

22.11   Interleaved Speech/Data Option

Section 11 specifies those supplementary elements of the air interface that are required in order to implement interleaved speech/data services. In the interleaved speech/data service, a single UIC (user information channel) is used to transmit speech and data information. To assign the UIC for speech or data transmission, the user information mode messages are sent in the synchronous directive channel (SDC). These directives set the use of the fast channel as primary user information or the secondary user information transmission channel. Normally, the primary mode is associated with speech codec-processed voice-band information and the secondary mode is associated with non-codec-processed (raw) data, but it will be set otherwise by using the higher-layer call control procedures. The user information mode is not symmetrical in that the uplink and downlink maintain separate states.

22.12   In-Service Testing Option

Section 12 specifies those supplementary elements of the air interface that are required in order to implement in-service testing services. In-service testing service is considered to be an optional service in the PACS-UB system. That is, not all equipment is required to support in-service testing services. However, if equipment is designed to support in-service testing services, manufacturers must adhere to the specifications presented in this section.

Acknowledgements

The author acknowledges the assistance of many individuals who participated in the PACS development and standardisation and in particular Ken Felix, Chairman of the JTC PACS TAG, Steve Hawkins, editor of the PACS Standards documents, and Enrique Laborde, editor of the PACS Minimum Performance Standards documents and Chairman of the T1P1.3 PACS Working Group. However, the author accepts the responsibility for any errors and inaccuracies.

References

[1]   "Personal Access Communications System Unlicensed (Version B)", ANSI J-STD-014 supplement B
[2]   "PACS-UB, A Protocol for the Unlicensed Spectrum", AR Noerpel, LF Chang and RA Ziegler, IEEE ICC '95, Seattle, June 1995
[3]   "PACS-UB for Use in Unlicensed Spectrum", AR Noerpel, E Laborde and K Felix, First Annual WINForum User PCS Workshop, Dallas, October 1994
[4]   "Personal Access Communications System", ANSI J-STD-014, 1995
[5]   "Generic Criteria for Version 0.1 Wireless Access Communications Systems (WACS)", TR-INS-001313, Issue 1, Bellcore, October 1993
[6]   "A Flexible Low-Delay TDMA Frame Structure", VK Varma, NR Sollenberger, LF Chang and HW Arnold, IEEE ICC '91, Denver, June 1991
[7]   "Making User PCS a Reality – A Journey on the Information Highway", J Knapp, First Annual WINForum User PCS Workshop, Dallas, October 1994
[8]   "A United States Perspective for a Flexible PCS Standard", AR Noerpel, YB Lin and H Sherry, ITU Telecom '95, Geneva, October 1995
[9]   "Low Complexity Hardware Implementation for Interoperable Licensed and Unlicensed Personal Communications Services: the PACS-UB System", RA Ziegler, AR Noerpel, LF Chang and NR Sollenberger, WINLAB Workshop, East Brunswick, NJ, April 1995
[10] "Performance Limitations TDD Wireless Personal Communications With Asynchronous Radio Ports", JC-I Chang, Electronics Letters, Vol. 28, No. 2, pp. 532–3, 1992
[11] "An Interoperable PACS and DCS1900 Subscriber Unit Radio Architecture", R Malkemes, P Lukander and P Harrison, IEEE PIMRC '95 Conference Proceedings, Toronto, September 1995
[12] "Performance of Unequalized Frequency-Hopped TDMA on Dispersive Fading Channels", LF Chang and S Ariyavisitakul, IEEE Global Telecommunications Conference, Texas, 1993
[13] "Hybrid Slow Frequency-Hop/CDMA-TDMA as a Solution for High-Mobility, Wide-Area Personal Communications", PD Rasky, GM Chiasson and DE Borth, WINLAB Conference Proceedings, New Brunswick, 1994
[14] "An Experimental Slow Frequency-Hopped Personal Communications System for the Proposed US 1850–1990 MHz Band", PD Rasky, GM Chiasson and DE Borth, IEEE ICUPC Conference Proceedings, Ottawa, 1993
[15] "Performance of Slow Frequency-Hopped TDMA With a Hard Limited Receiver", LF Chang and S Ariyavisitakul, International Journal of Wireless Information Networks, Vol. 2, No. 2, April 1995
[16] "Slow Frequency-Hopped CDMA for High Mobility Personal Communications Systems", PD Rasky, GM Chiasson and DE Borth, Proceedings of the 31st Annual Allerton Conference on Communication, Control and Computing, September 1993
[17] "Generic Criteria for Version 0.1 Wireless Access Communications Systems (WACS)", TR-INS-001313, Issue 1, Bellcore, October 1993, Revision 1, June 1994
[18] "Development of Air Interface Standards for PCS", CI Cook, IEEE Personal Communications Magazine, Vol. 1, No. 4, pp. 30–4, 1994
[19] "Coexistence and Access Etiquette in the United States Unlicensed PCS Band", DG Steer, IEEE Personal Communications Magazine, Vol. 1, No. 4, pp. 36–43, 1994
[20] "Personal Access Communications System Unlicensed (Version A)", ANSI J-STD-014 supplement A

[21] "Performance of 32 kb s^{-1} ADPCM in Frame Erasures", VK Varma, M Thomas, S Konish, L Seltzer and D Goodman, IEEE VTC '94, Conference Proceedings, Stockholm, June 1994

[22] "Distributed Measurement-Based Quasi-Fixed Frequency Assignment for Personal Communications", M Cheng and JC-I Chuang, IEEE ICC '95, Seattle, June 1995

[23] "Digital European Cordless Telecommunications System Blind Spot Algorithm Evaluation Results", S McCann and AP Croft, IEEE Globecom '90 Conference Proceedings, p. 604.5.1, 1990

[24] "A Mathematical Model for Dynamic Channel Selection in Conformity With the Digital European Cordless Telecommunications Standard", Bout, Sparreboom, Brouwer and Prasad, IEEE PIMRC '93, September 1993

[25] "Performance of PACS-UB for Unlicensed Operation With Uplink Power Control", LF Chang, AR Noerpel and A Ranade, IEEE PIMRC '95, Toronto, September 1995

[26] "Multiplexing Protocol for the WACS System Broadcast Channel", DJ Harasty and AR Noerpel, IEEE-ICUPC '93 Conference Proceedings, Ottawa, October 1993

[27] "Cellular System Dual-Mode Mobile Station–Basestation Compatibility Standard", EIA/TIA Interim Standard IS-54, Rev. B, 1992

[28] "40, 32, 24, 16 kbit/s Adaptive Differential Pulse Code Modulation (ADPCM)", ITU-T Recommendation G.726, Geneva, December 1990

[29] "Error-Correcting Procedures for DCEs Using Asynchronous-to-Synchronous Conversion", CCITT Blue Book, Vol. VIII, Fasc. VIII.1, Recommendation CCITT V.42, Melbourne, 1992

23 The PHS Standard

Yuichiro Takagawa

The development process of the Personal Handyphone System in Japan has been described earlier in Chapter 9 of this book and initial products in Chapter 10. In this chapter we present the technical details of the PHS base technology and standard [1, 2], explaining the air interface structure and parameters, signalling structures and briefly summarising the network interface standards.

23.1 System Overview

The PHS system has been designed to allow all of the normal public and private network functionality associated with cordless systems. In addition, PHS supports direct "personal station–personal station" (PS–PS) communications for handsets registered to the same home basestation. In PHS terminology, the handset is referred to as a personal station (PS). The term "cell station" (CS) is used to describe what is often in other systems referred to as a public basestation (BS) or radio fixed part (RFP).

PHS has been standardised as a common air interface for public applications, whereas for private applications the standard is in effect a coexistence specification, with manufacturers allowed to introduce different proprietary features. Thus, for example, an NTT handset can be re-registered to operate on public PHS networks of other operators, but such a handset will not operate with the other PHS service provider's home basestations.

The main system parameters for the PHS air interface standard are presented in Table 23.1. In the sections below we discuss several of these parameters, as well as broader aspects of system standardisation and operation.

23.2 Radio Aspects

23.2.1 Modulation

PHS employs $\pi/4$-shifted QPSK modulation with Nyquist roll-off factor $\alpha = 0.5$. This modulation scheme permits a variety of demodulation techniques to be used, such as delay detection, coherent detection and frequency discrimination

Table 23.1
PHS system parameters

Parameter	Specification
Radio access and duplex method	TDMA/TDD
Number of TDMA multiplexed circuits (i.e. speech traffic channels per RF carrier)	Four (when full-rate codec is used)
Carrier spacing	300 kHz
Modulation	$\pi/4$-shift QPSK with $\alpha = 0.5$
Bit rate	384 kb s^{-1}
Speech coding	32 kb s^{-1} ADPCM G.726
Diversity	Provided in cell station: Post-detection selection for uplink Transmitter antenna selection for downlink
Frame duration	5 ms
Cell station RF power average (peak): Outdoor CS – high power Outdoor CS – standard power Indoor CS – low power	100–500 mW (4 W max) 20 mW (160 mW) 10 mW (80 mW)
Maximum antenna gain	2.14 dBi (private basestation or PS) 10 dBi (public cell station)
Average (peak) RF power (handset PS)	10 mW (80 mW)
Receiver sensitivity	–95 dBm (for a 10^{-3} BER)
Channel planning	Quasi-static autonomous frequency assignment for common control channel, CCCH, carrier Dynamic channel assignment per call for traffic channel, TCH

detection. Furthermore, the use of the QPSK modulation method makes it possible to use the allocated bandwidth more efficiently than with GMSK modulation – i.e. good spectrum efficiency.

23.2.2 Multiple Access/Duplexing Method

Time-division multiple access with time-division duplexing (TDMA/TDD) is employed, avoiding the need for paired frequency bands. The use of TDMA multiplexing also makes it possible to modify service bit rates with relative ease, making the system a promising one with room for expansion to include radio multimedia services in the future.

23.2.3 Diversity

For both uplink and downlink, diversity is provided in the cell station, avoiding the need for circuitry for diversity in the handset. For the downlink, transmitting antenna diversity at the cell station is utilised, taking advantage of the time-division duplex scheme with short frame duration; this is effective for small handset equipment. For the uplink, post-detection selection diversity is used in the cell station.

23.2.4 Speech Coding

The full-rate speech codec uses the ITU-T G.726 standard $32\,\text{kb s}^{-1}$ ADPCM algorithm. This provides good speech quality and small path delay, as well as efficient integrated circuit implementation. A super-frame configuration allows multiple codecs with lower rates, down to $8\,\text{kb s}^{-1}$, to be incorporated in the system when they become available.

23.2.5 Radio Frequency Band

The radio frequency band allocated for PHS service is some 23 MHz in the 1.9 GHz region (1.8950–1.9181 GHz) in Japan. The report from the Radio Regulatory Council recommends allocation of 11 MHz, 1895–1906 MHz, for public and private use, including direct communications between handsets, and an additional 12 MHz, 1906.1–1918.1 MHz, exclusively for public use; these recommendations are based upon considerations of anticipated traffic density and spectrum requirements. There are 77 communication channels in this band (control channels and traffic channels are included); the channel numbering plan is shown in Table 23.2.

Table 23.2
Channel numbering plan

Carrier	Frequency (MHz)	Carrier	Frequency (MHz)	Carrier	Frequency (MHz)	Carrier	Frequency (MHz)
1	1895.15	21	1901.15	41	1907.15	61	1913.15
2	.45	22	.45	42	.45	62	.45
3	.75	23	.75	43	.75	63	.75
4	1896.05	24	1902.05	44	1908.05	64	1914.05
5	.35	25	.35	45	.35	65	.35
6	.65	26	.65	46	.65	66	.65
7	.95	27	.95	47	.95	67	.95
8	1897.25	28	1903.25	48	1909.25	68	1915.25
9	.55	29	.55	49	.55	69	.55
10	.85	30	.85	50	.85	70	.85
11	1898.15	31	1904.15	51	1910.15	71	1916.15
12	.45	32	.45	52	.45	72	.45
13	.75	33	.75	53	.75	73	.75
14	1899.05	34	1905.05	54	1911.05	74	1917.05
15	.35	35	.35	55	.35	75	.35
16	.65	36	.65	56	.65	76	.65
17	.95	37	.95	57	.95	77	.95
18	1900.25	38	1906.25	58	1912.25		
19	.55	39	.55	59	.55		
20	.85	40	.85	60	.85		

Source: from [2].

Two fixed-frequency control carrier frequencies are assigned for private use: 1898.45 MHz and 1900.25 MHz in Japan, and 1903.85 MHz and 1905.65 MHz for other countries. Four fixed-frequency control carriers are assigned for public service in Japan, one for each public PHS service operator and one spare. For public usage, guard channels are assigned around the control channels. The channel usage plan is shown in Table 23.3.

Table 23.3
Channel usage plan

Channel number	Usage
1–10	Direct communication between PSs *and* Private use (can also be shared with public use)

11–37	Private use (can also be shared with public use)
	    Channels 12 and 18 assigned as control channels in Japan
	    Channels 30 and 36 assigned as control channels outside Japan

38–77	Public use
	    Channels 38 to 69 assigned as traffic channels
	    Channels 73, 75 and 77 assigned as control channels
	    Channels 71 assigned as a spare control channel
	    Channels 70, 72, 74 and 76 assigned as guard channels

Source: from [2].

23.2.6   Microcellular Architecture

Microcellular architecture, which permits efficient spectrum utilisation and the use of low-power handset terminals, is a key technology for the PHS. The ability to be able to deploy large numbers of cell stations without the need for coordination of frequency planning is indispensable for microcellular systems.

23.2.7   Cell Layout

Cell stations with 10 mW of average power are deployed at about 200 m spacing in the downtown area. Wider separations between cell stations is possible using cell stations with the higher-power transmitter and extremely low-noise amplifiers.

23.2.8   Handover

PHS has two handover types – "recalling type" and "traffic channel switching type" [2]. In the recalling type, the change of communication channel is done in such a manner that the handset establishes a connection to the new cell station in the same way as a new call originating in that cell. In the traffic channel switching type, the handset establishes a connection in the new cell by synchronising burst signals, and a new channel must be assigned from the new cell before handover occurs. There is a short time intermission when handover occurs in the recalling-type handover, for which the handover process is longer than for the traffic-switching-type handover.

Basically, measurement of the transmission quality is conducted in both the cell station and the handset during communication. The handover is triggered by both stations. The candidate cell and channel are determined by the handset without concern as to whether this will involve inter-cell or intra-cell handover. When agile frequency synthesisers and frame synchronisation between the cell stations become available, seamless handover will be supported.

23.2.9 Adaptive Channel Assignment

The channel assignment is conducted by a quasi-static autonomous algorithm for the common control channel and by an adaptive algorithm per call for the traffic channel. A distributed control architecture is introduced for radio channel management as well as handover.

23.2.10 Spectrum Sharing

The adaptive channel assignment used in PHS is based on slot-unit interference detection. Before transmitting, both stations sense their local interference, then automatically allot channels where there is little interference. This channel management will allow coexisting independent operators without the need for coordination of the spectrum sharing. This is also effective for coexistence with fixed wireless systems. This idea, based on frequency and slots assignment, is effective for not only outdoor applications but also indoor applications for office and residential use as a digital cordless telephone system.

23.3 Radio Channel Structures

The functional channel structure of the air interface is shown in Figure 23.1 and the composition and usage of these channels is described below. The detailed bit structures supporting these channels are shown in Figure 23.2. The way in which these map onto the PHS time slot structure is described in section 23.4.

23.3.1 Control Channel (CCH)

The control channel (CCH) is composed of a combination of a common control channel (CCCH) and an associated control channel (ACCH), which together make up a single dedicated channel, giving it improved performance in conditions of intermittent communications reception.

23.3.2 Broadcast Channel (BCCH)

This is a one-way downlink channel to report control information from the cell station to the mobile station. It transmits information related to channel structure, system information, etc.

common channels
CCH
(control channel)
BCCH (broadcast channel)
CCH (common control channel)
UPCH (user packet channel)
PCH (paging channel)
SCCH (individual cell channel)
individual channels
ACCH (auxiliary control channel)
SACCH (slow ACCH)
FACCH (fast ACCH)
TCH
(information channel)

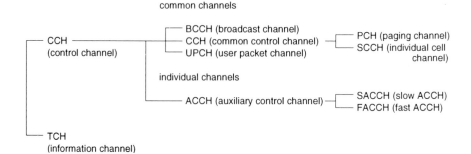

Figure 23.1 Function channel structure.

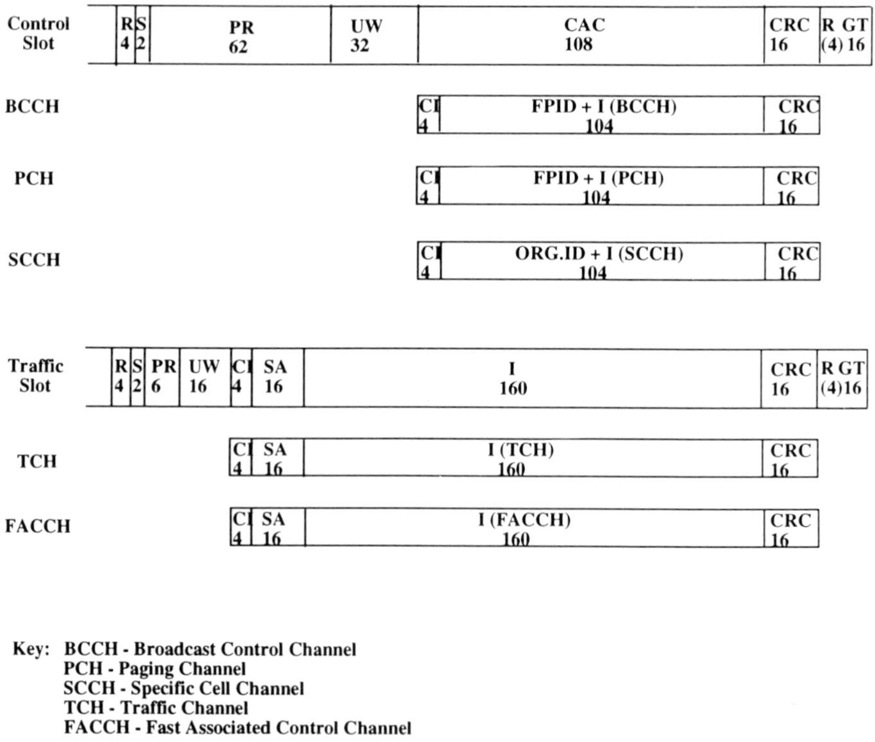

Control Slot
R 4
S 2
PR 62
UW 32
CAC 108
CRC 16
R (4)
GT 16
BCCH
CI 4
FPID + I (BCCH) 104
CRC 16
PCH
CI 4
FPID + I (PCH) 104
CRC 16
SCCH
CI 4
ORG.ID + I (SCCH) 104
CRC 16
Traffic Slot
R 4
S 2
PR 6
UW 16
CI 4
SA 16
I 160
CRC 16
R (4)
GT 16
TCH
CI 4
SA 16
I (TCH) 160
CRC 16
FACCH
CI 4
SA 16
I (FACCH) 160
CRC 16
Key: BCCH - Broadcast Control Channel
PCH - Paging Channel
SCCH - Specific Cell Channel
TCH - Traffic Channel
FACCH - Fast Associated Control Channel

Figure 23.2 PHS channel structures.

23.3.3 Common Control Channel (CCCH)

This channel carries out control information transmission necessary for call connection. The common control channel performs in an inherently random access manner. In the CCCH, a preamble of 64 bits and unique word of 32 bits are allotted at the starting position of the frame to assure frame synchronisation and permit low standby power consumption. This also provides the packet-type data transmission capability in PHS service.

23.3.4 Paging Channel (PCH)

The paging channel (PCH) subset of the CCCH is a one-way downlink, point-to-multipoint, channel that simultaneously transmits identical information to individual cells or a wide area of multiple cells (the paging area) from a cell station to the mobile station.

23.3.5 Specific Cell Channel (SCCH)

The specific cell channel (SCCH), again a subset of the CCCH, is a bi-directional, point-to-multipoint, channel that transmits information needed for call connection between the cell station and the mobile station, and it transmits independent information to each cell. The uplink channels are random access.

23.3.6 User Packet Channel (UPCH)

This is a bi-directional point-to-multipoint channel. It carries control signal information and user packet data transmission.

23.3.7 Associated Control Channel (ACCH)

This is a bi-directional channel that is associated to the traffic channel (TCH). It carries out transmission of control information and user packet data needed for call connection. The channel that is ordinarily auxiliary to TCH is the slow ACCH (SACCH), and the channel that primarily steals TCH and carries out high-speed data transmission is designated the fast ACCH (FACCH).

23.3.8 Traffic Channel (TCH)

This channel transmits user traffic information. It is a point-to-multipoint bi-directional channel.

23.4 Radio Circuit Control

23.4.1 Control procedures

Control procedures are defined as those which connect in receiving and sending to a mobile station, register mobile station location and switch channels during communication, and identify services and so forth. It is necessary for these procedures to be performed reliably by common and individually assigned slots.

23.4.2   Slot structure

Figure 23.3 shows the slot arrangement considering appropriate sending/
receiving slot separation in the TDD transmission. A three-phase link set-up
process is used in PHS. This process provides signal compactness for spectrum
efficiency in the radio channel access phase and sufficient signalling capacity,
with flexibility to support high-level protocols and expanded services.

23.4.3   Radio Channel Access Phase (Phase 1)

The radio channel resource necessary for phase 2 is assigned in this handshake
sequence using a non-layered signal structure.

23.4.4   Link Connection Phase (Phase 2)

Ordinary call control (CC) and higher-level mobility management (MM) and
radio transmission management (RT) functions are performed in this phase.
Various user protocols can be provided depending on service demand and
network conditions such as service validation, authentication sequences, etc. A
layered structure is most suitable for this phase.

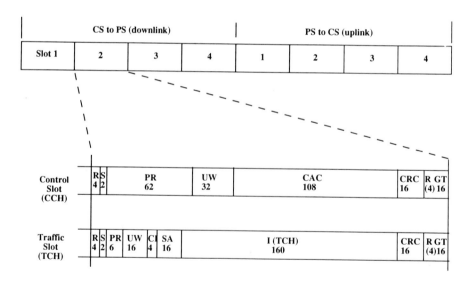

Figure 23.3   PHS time slot structure.

23.4.5   Communication Phase (Phase 3)

User information such as voice and non-voice data are transmitted in this phase. Layered control signals are transmitted through an associated control channel (FACCH and SACCH – see above).

23.5   Network Interfaces

The PHS system has been defined to allow operators a variety of different network architecture options (see [2]) in the sense of the way in which cell stations are interconnected and managed. Table 23.4 summarises the functions of the main network interface standards and their relationship to existing international (ITU) standards.

Table 23.4
PHS network interface standards

Standard	Description	Features
TTC JT-Q 921b	Limited version of ITU-T Q.921	Layer 2 of CS–switch interface
TTC JT-Q 931b	ITU-T Q.931 plus TTC supplements	Layer 3 of CS–switch interface: paging
TTC JT-Q 932a	ITU-T Q.932 plus TTC supplements	Layer 3 of CS–switch interface: location registration, handover, authentication
TTC JT-Q 761, 762, 763	ITU-T Q.761, 762, 763 plus TTC supplements	Internetworking: functions to connect a PHS call between different PHS carrier networks
TTC JT-Q 1218	ITU-T Q.1218 plus TTC supplements	Basic IN for switch–switch interface
TTC JT-Q 1218a	TTC supplement to ITU-T Q.1218	PHS capability of IN for switch–switch interface: roaming (conforms to IN CS-1 architecture)

Source: from [3].

The CS–network interface, the I′ interface (or modified I interface), is based upon the ISDN interface (I interface) with supplements. The layer 1 (physical layer), layer 2 (data link layer) and call control function of layer 3 (network layer) are identical to the ISDN protocol; however, application parts have been modified to support PHS-specific functions. The enhancements to the ITU-T Q.931 specification include:

- *PS number notification* – a function to send and receive the PS number in order to identify a PS in an outgoing or incoming call.
- *Multiple interface paging* – a function to enable a group of interfaces to support handset paging in support of an incoming call.
- *D-channel sharing* – a function to allow multiple basic rate (2B+D) interfaces connected to a single CS to share a common D channel.
- *Incoming call when busy* – a function to allow reception of an incoming call even when all channels are busy.

The enhancements to the ITU-T Q.932 specification include:

- *Recalling type handover* – a function to allow the call to be maintained whilst the traffic channel is switched between CSs or interfaces as the personal station moves during the call.
- *PB tone transmission* – a function to generate the PB signal by the network on request through the D channel.
- *Location registration* – a function to enable the network to be informed of the location of a PS, or through which interfaces a PS is to be paged as the PS moves to another area.
- *Authentication* – a function to allow the network to authenticate a PS whenever it attempts to originate or receive a call, or to make a location registration.

References

[1]   A full copy of the PHS Standard may be obtained from the Research and Development Centre for Radio Systems, RCR, 6–8 Hirakawacho 1-Chome, Chiyoda-ku, Tokyo 103, Japan

[2]   "Personal Handy-Phone System (PHS) Guidebook", Ministry of Posts and Telecommunications, Japan, 1995

[3]   "PHS – A Promising Solution for Digital Cordless Systems", Y Ito and M Ohashi, presented at the Digital Cordless Telephone Conference, September 1995

[4]   "Toward the Personal Communication Era – A Proposal of the Radio Access Concept from Japan", K Ogawa, K Kohiyama and T Kobayashi, International Journal of Wireless Information Networks, Vol. 1, No. 1, pp. 17–27, 1994

[5]   "Standardisation of the Personal Handy Phone (PHP)", T Habuka and H Sekiguchi, NTT Review, Vol. 5, No. 5, pp. 101–5, 1993

24 The PACS-UA Standard

William H. Scales Jr

The Personal Access Communication System – Unlicensed version A (PACS-UA) is an air interface standard designed for private unlicensed PCS applications in North America [1]. The PACS-UA standard has been prepared under the auspices of the TIA TR 46.3/T1 P1.4 Joint Technical Committee (JTC).

PACS-UA will provide many new services and products to meet a wide range of wireless voice and data applications. It can be expected to offer many new products such as wireless PBX systems, enhanced cordless phone systems, and wireless communications for portable and desktop computers. PACS-UA can provide radio connectivity for these systems in North America without the need for individual user or station licences. The ability of a system to operate on an unlicensed basis will enable these products to be introduced more rapidly than would be possible on a licensed basis.

PACS-UA is compatible with the PACS licensed band standard [2], which makes possible interoperability of licensed and unlicensed services. PACS-UA is derived from the Japanese Personal Handyphone air interface standard [3]. The modifications to PHS are to allow operation in the unlicensed PCS band in North America. Since there is such commonality between PACS-UA and PHS, this chapter only describes the major differences between the two systems.

24.1 System Characteristics and Structure

24.1.1 Characteristics

The following items list the key capabilities of PACS-UA:

- A single handset can be used for home cordless telephones, for wireless PBX and for access to the PSTN.
- Wireline-quality voice transmission is achieved by using $32\,\mathrm{kb\,s^{-1}}$ ADPCM codecs. A symbol rate of 192×10^3 symbols per second combined with a short frame duration does not introduce a harmful delay or distortion. Therefore, complex equalisers and echo cancellers are not required.
- A wide range of features and applications are available, covering voice, messaging, fax and data services, from indoors and outdoors, and for private, business, or public use.

24.1.2 Architecture

The PACS-UA system is composed of subscriber units (SUs) and radio ports (RPs). Groups of RPs may be interconnected and controlled, in larger systems, by a radio port control unit (RPCU). An SU is a mobile communications terminal and is used to establish communications with a fixed station (radio port). A RP is a fixed communications terminal and is used to establish communications with an SU. The RP provides interconnection to the external wired network (PSTN, PBX, etc.). Typically, diversity is provided in the RP. Terminal equipment (TE) of various functions may be connected to the SU as shown in Figure 24.1, which shows the PACS-UA architecture model. There are three interface points, the T, A and P:

- A *interface* – the interface between the SU and the RP (the "air" interface).
- T *interface* – the interface between the SU and external terminal equipment.
- P *interface* – the interface between the RP and the communication network.

24.1.3 Spectrum Etiquette

PACS-UA conforms well to the "Spectrum Etiquette" rules [4] required by US regulations, to allow independent systems to share the unlicensed PCS band (1920–1930 MHz). This etiquette has also been adopted in Canada. The etiquette rules restrict the peak transmit power to be proportional to the emission bandwidth, as described in Chapter 8. Therefore, narrow-band systems are allowed lower transmit power compared to wider-band systems. The main foundation of the etiquette is a listen-before-transmit mechanism that requires every transmitting device to monitor its intended time and spectrum window to confirm that the window is vacant before transmitting. This etiquette is in fact similar to the method used in PHS but with different monitoring thresholds; hence PHS has adapted easily to the US environment. Also, the etiquette rules necessitate that the broadcast control channel cannot be at a fixed frequency as it is in PHS.

24.2 Main Differences from PHS

In the PHS specification, RCR STD-28 [3], certain functions are somewhat different for private and public use. In many cases they are defined as a reference only for private use. So it has allowed manufacturers to develop their own proprietary specifications. It has, however, been recognised (just as in Europe) that many different private solutions are not good – indeed, in Japan the effort is now to update the PHS standard to standardise some of the functions for interoperability among private systems. These are reflected in Version 2 of the PHS standard.

Thus, with PACS-UA, an attempt has been made to "plug some of these holes", to develop a true interoperability standard. Thus all of the public functions have been included, but, where there has been a difference between the public and private specifications, the private version has been adopted. Although PACS-UA could be used in a public network, its primary application is for private systems.

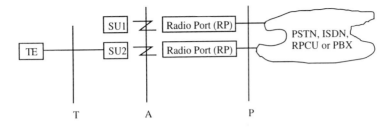

Figure 24.1   PACS-UA architecture model.

Table 24.1
Parameters of PACS-UA, compared with PHS and PACS

Parameter	PHS	PACS-UA	PACS
Operating frequency	1895.0–1906.1 MHz	1920–1930 MHz	1850–1895 MHz 1930–1975 MHz
Duplexing	TDMA/TDMA/TDD	TDMA/TDMA/TDD	TDMA/TDMA/FDD
Channel spacing	300 kHz	300 kHz	300 kHz
Transmit power	80 mW handset 80 mW private basestation 160 mW–4 W max. public basestation	53 mW handset and basestation	200 mW handset 800 mW basestation
Modulation	$\pi/4$ QPSK	$\pi/4$ QPSK	$\pi/4$ QPSK
Channel bit rate	384 kb s^{-1}	384 kb s^{-1}	384 kb s^{-1}
Voice channels per basestation	4	4	8
Frame duration	5 ms	5 ms	2.5 ms
Channel assignment	DCA[a]	Etiquette rules	Fixed or QSAFA[b]
Voice coding	G.726 ADPCM	G.726 ADPCM	G.726 ADPCM

[a]Dynamic channel assignment.
[b]Quasi-static automatic frequency assignment.

Table 24.1 shows a comparison of PACS-UA and PHS; the PACS licensed band system is also shown for comparison. The basic channel structure and slot structure is the same for PACS-UA and PHS. This section describes the differences between PACS-UA and PHS, mainly due to the adaptation for operation in the unlicensed PCS band in North America. The main system differences between PACS-UA and PHS are as follows:

• *Private systems* The PHS standard supports both public and private operation. PACS-UA adopts only the private system portion of the PHS standard.

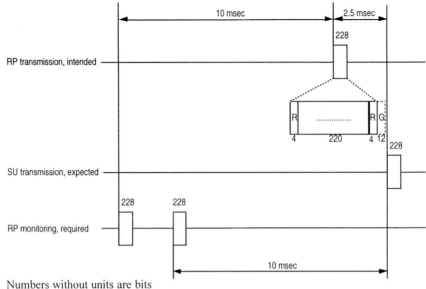

Figure 24.2   PACS-UA CCH monitoring protocol.

- *Operating frequency*   PACS-UA is specified to operate within the 1920–1930 MHz band, i.e. slightly higher than the PHS allocation.
- *Transmit power*   The peak transmit power that is allowed for both the RP and SU is 53 mW, which is a slight reduction from the values adopted by PHS of 80 mW (SU and private RP).
- *Control channel*   The primary difference between PACS-UA and PHS is in the control channel (CCH) architecture. For PACS-UA, for both traffic channels and control, the monitored time and spectrum window can be used only after it has been monitored for two or more frames and confirmed to be idle as shown in Figure 24.2. The search for an idle channel must begin within a 3 MHz window at the 1920 MHz band edge, as determined by the etiquette rules, set by Government regulation. This method differs from PHS in which there is a fixed assignment for the CCH. This results in fact in PACS-UA having a greater system robustness, compared to the basic PHS standard, in which interference on a signalling channel could preclude its use, whereas in PACS-UA the equipment simply searches for a free channel to use for signalling purposes.

References

[1]   "Personal Access Communication System – Unlicensed Version A", ANSI J-STD-014, Supplement A, 1996
[2]   "Personal Access Communication System", ANSI J-STD-014, 1996
[3]   "Personal Handyphone System, Version 2", RCR STD-28, 1995. An English version of this standard is obtainable from the Association of Radio Industries and Businesses (ARIB)
[4]   FCC Code of Federal Regulations, Title 47, Part 15, Subpart D

25 The Orthogonal CDMA Wireless Telephone System

Herman Bustamante and Horen Chen

The Orthogonal CDMA Wireless User Premises Equipment (OCDMA WUPE) system has been developed for PCS and other applications in the USA by Stanford Telecom. Unlike many of the early cordless telephony systems, this one utilises spread-spectrum technology. The system has, however, many characteristics shared with other systems in one form or another. The unique feature of OCDMA is the particular manner in which these characteristics are combined and, especially, the fact that it operates in a truly synchronous orthogonal manner in both the outbound and inbound paths – this leads to an inherently high system capacity.

The material presented in this chapter reflects the structure and relative maturity of the OCDMA WUPE standard document [1]. Section 1 of the standard document is an introductory chapter outlining the key system features, providing definitions and key references. Section 2 provides a system description, whilst sections 3, 4, 5 and 6 describe the physical, MAC, data link and network level layers respectively. The final sections, sections 7 and 8, describe privacy/authentication and audio aspects.

In this chapter, considerable emphasis is given to a description and explanation of the lower-level air interface aspects, the more mature aspects of the system and a key area of innovation and difference from some of the other cordless standards. For more details of the higher-layer specifications, the reader should consult the latest version of the full OCDMA WUPE standard document [1].

25.1   System Overview

25.1.1   OCDMA WUPE System Parameters and Concept

The OCDMA WUPE system has a number of key features, some common to other cordless systems and some unique. These are:

- Low-cost handsets and basestations
- Low power consumption
- High system user capacity
- Robustness against narrow-band interference
- Robustness against multipath fading

- Priority access and channel assignment capability
- User location capability (in the case of emergency calls)
- Suitable for operation in the ISM bands, as well as the unlicensed PCS spectrum
- Open network interfaces, for public and private networks

This system is essentially a time-division duplex (TDD) orthogonal CDMA system; the use of orthogonality offers a significant increase in system capacity, of order three times, compared to conventional CDMA, since only adjacent cell interference contributes to the perceived interference levels. It also offers reduced sensitivity to power control errors, compared to other CDMA systems.

System Parameters

Key parameters are summarised in Table 25.1; after elaborating these issues further in the text, we then proceed to describe the system concept within its three levels of complexity.

Table 25.1
OCDMA system specifications

Parameter	System value
Frequency	1910–1930 MHz
Duplexing	Time-division duplexing (TDD)
Voice processing and baseband data rate	Compressed voice data rate 16 kb s^{-1}
FEC coding	None
Security coding	4096 chip PN sequence and 32 chip RW function
Transmitted data rate	41.6 kb s^{-1}, or 20.8×10^3 symbols per second, in TDD format
Chip rate	665.6×10^3 chips per second
Spreading gain	32 chips/symbol, i.e. 15 dB
Subgroup bandwidth and number of subgroups (carrier spacing and number of carriers)	1.25 MHz, 16 subgroups
Number of simultaneous users	62/group, activity factor = 100% 310/group, activity factor = 20% 3720 users full system, activity factor = 20%
Modulation	QPSK data, BPSK spreading code
Spreading sequence	(RW) mod-2 (PN)[a]
Diversity	Cross-polarisation antenna diversity at handset
Order-wire (OW) channel	Acquisition and control functions
Peak transmit power	100 mW total per carrier
Composite system bit rate	1.33 Mb s^{-1}
Receiver sensitivity	$\leq$ –83 dBm
Minimum received signal operating range	–33 dBm to –93 dBm
Communication range	From 2 m up to 100–500 m, typically less than 50 m (see below)

[a]Rademacher–Walsh code modulo-2 combined with a pseudo-noise sequence.

Operating frequency

The system will operate in the licensed and unlicensed frequency band extending from 1910 to 1930 MHz. A 20 MHz total system bandwidth is assumed. Each subgroup signal is allocated a 1.25 MHz bandwidth such that a total of 16 subgroups can be accommodated. The system frequency plan is illustrated in Figure 25.1.

Time-division duplexing

TDD of the signals will be employed such that the receive and transmit frequencies are the same. A 10 ms frame will be used, providing equal-length 5 ms subframes for transmit and receive operations.

Voice processing and FEC

The system will employ a voice compression algorithm providing a $16 \, \text{kb s}^{-1}$ baseband data rate. No FEC code will be used.

Security

This is provided by the use of Rademacher–Walsh code modulo-2 combined with a pseudo-noise (PN) sequence; different PN sequences can be employed in different systems. If a greater degree of security is required, such as scrambling or encryption, it can be added.

Transmission and chip rates

A 10 ms baseband data segment at $16 \, \text{kb s}^{-1}$ will be transmitted at approximately 20.8 ksps QPSK data rate within a 5 ms subframe interval of the TDD format. A chip rate of 665.6 kcps is used. The use of 32 chips/symbol provides a processing gain of 15 dB.

Grouping of users

A subgroup consisting of 31 users (31 voice traffic channels) occupies 1.25 MHz. The minimum basic basestation can support one user group (group) consisting of 62 users, or two subgroups. A total of 16 subgroups or eight groups can be accommodated with a 20 MHz total system bandwidth. In addition to a $16 \, \text{kb s}^{-1}$ voice capability, each voice traffic channel provides a full-duplex $1.6 \, \text{kb s}^{-1}$ control data channel used for real-time control operations when a call is in process.

Number of simultaneous users and activity factor

A basic basestation provides 62 voice traffic channels, which can support 62 users 100% of the time, providing an activity factor of 100%. A user community

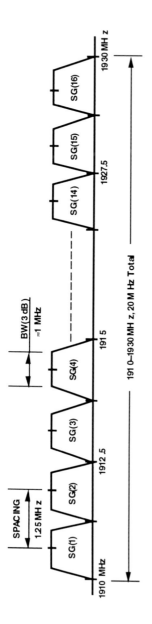

Figure 25.1   System RF spectrum characteristics.

of 310 users can also be supported 20% of the time (activity factor or usage = 20%). This activity factor is found to be generally more than adequate for most commercial applications. The baseline design provides for operations with up to 128 users with an activity factor of 48%. The number of users to be supported, and thereby the activity factor, is a programmable feature.

Modulation

The data will be QPSK modulated and the spreading code will be BPSK.

Spreading sequence

The spreading sequence will consist of the modulo-2 combination of a Rademacher–Walsh code with a PN sequence. The Rademacher–Walsh codes will be 32 chips long and will be applied one code length spanning one data symbol of a voice traffic channel. The PN sequence will be 4096 chips long and will be at the same chip rate as the Rademacher–Walsh code. Two different PN sequences will be used within a given basestation, one for all inbound signals and one for all outbound signals. Different PN sequences will be used in nearby basestations if necessary, to provide a frequency re-use capability of unity. Thirty-two different Rademacher–Walsh codes will be used in each subgroup. A unique code will be assigned to each voice traffic channel and the all-ones code is assigned to the order-wire (OW) channel.

Diversity

Diversity will be provided by using a single antenna at the basestation and a pair of linearly cross-polarised antennas at the handset. The signal quality received on both antennas is sampled every TDD frame. The antenna with the best signal quality is selected for use on the next TDD frame.

Order-wire channel

An OW channel is provided for each subgroup. It will be used for handset network entry, call establishment operations, power control, signal timing control and some telephone system (TELCO) required data transfer operations. The OW will have all the same characteristics as a voice traffic channel except for the data format and data rate characteristic.

Communication range

The communication range of the system will depend on the building or city characteristics within which the system is to operate. Indoor operation ranges will typically be restricted to less than 50 m, although ranges exceeding 100 m may be possible in outdoor line-of-sight environments. Propagation and coverage issues are discussed further below.

This system has requirements imposed on it from three levels. Firstly, it must provide the capability to function as a self-contained system supporting a user group population providing each user with a voice traffic channel when required. Secondly, it must provide the appropriate interface capability to communicate to and from a TELCO system for the establishment of new calls and for the continuing support of ongoing calls. Thirdly, it must provide a handover capability for a user roaming from cell to cell; this third capability must be provided for both conditions of no ongoing call and with a call in progress. The system concept is described below, addressed at these three levels of complexity.

Basic Single Basestation System Configuration

Figure 25.2 illustrates the hardware configuration for one 62-user system hardware set. Each hardware set consists of one basestation and up to 62 handsets with cradle. The system defines a star network configuration with the basestation as the centre of the star. The basestation contains one transceiver for each individual user handset in the operating system.

TELCO System Considerations

To provide voice traffic capability to the user community, a telephone system (TELCO) support and interface capability must be provided without which the voice traffic operations cannot exist. These TELCO support functions consist of:

• Call establishment operations support
• User information database support and update
• Multicall programming operations capability
• Peripheral support functions

Call establishment operations

These consist of interfacing with the TELCO, providing and interpreting all signalling operations required to establish incoming and outgoing calls. Such functions include dialling, busy signal and phone ringing operations and are handled by the OW channel.

User information database

A typical multiple-basestation system configuration is illustrated in Figure 25.3, where a system of n basestations each with 62 voice traffic channel capability is shown. Also shown is that each basestation may be required to support up to 128 users part-time. For these assumed conditions the TELCO[1] basestation system must have the capability to recognise and properly route calls to $128n$ different phone numbers (different users). From this we see that there are a minimum of four pieces of data required for each user, as follows:

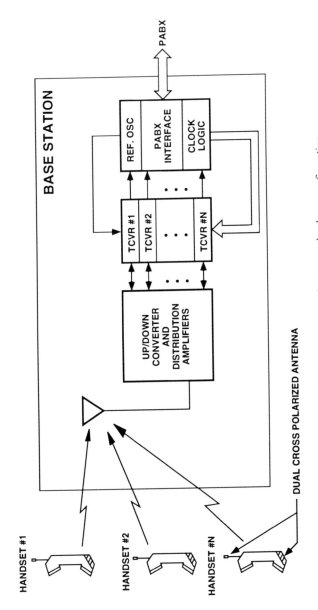

Figure 25.2 Single basestation system hardware configuration.

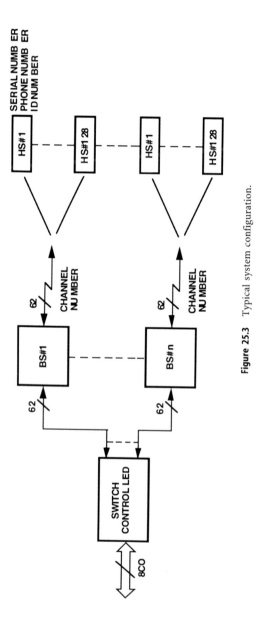

Figure 25.3 Typical system configuration.

- A serial number unique to a particular handset must be known. The serial number is a fixed, manufacturer-assigned number and identifies the handset as an authorised system user.
- An identification number identifies a handset as one of the 128 members of a particular user community associated with a basestation. This number is arbitrarily assigned by the basestation when a handset becomes a member of its user community.
- The channel number identifies one of the 62 voice traffic channels that are assigned for use arbitrarily each time a call is established.
- The set of phone numbers are the phone numbers assigned uniquely by the TELCO to the set of system users.

Figure 25.4 illustrates the relationship of the various data. Note that a phone number and handset serial number are uniquely related so long as the user maintains his or her TELCO service. Even though an identification number is arbitrarily assigned, once assigned, it is recognised as being associated with a particular handset serial number. Similarly, once a call established and a voice traffic channel is assigned for use, the identification number, handset serial number and channel number are uniquely associated for the duration of the call. It is clear that this data is necessary not only for establishing and maintaining full knowledge of the users in the system but for the proper identification of all call tariffs and billing operations.

Multicall programming

A number of operations such as "three-way calling" and "call waiting" require the processing of multiple calls simultaneously while a call is in process. This demands the existence of a two-way control channel within the voice traffic channel.

Peripheral support functions

There are a number of support functions that may be provided, which are not critical to the basic system but which make the telephone more convenient to the user. This includes such things as "speed calling" or speed dialling, which permits the dialling of frequently called numbers by pushing only two buttons on the handset. Such functions can be supported.

Multiple Basestation and Roaming Capability

So long as users are confined to operate through only one particular basestation, operations are well defined and the equipment need concern itself only with maintaining signal timing and appropriate transmitter power level. If the system is expanded to consist of many basestations over an extended geographical area, or covering multiple floors in a multi-floor building, system behaviour becomes more complex, since the user must be able to roam, executing handover operations from one basestation to another. In a multiple-basestation system, it

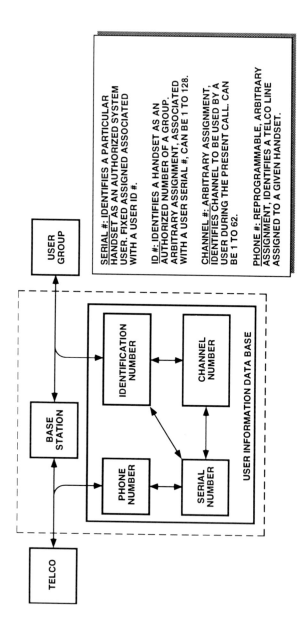

Figure 25.4 Basestation user information database.

is assumed that any user can roam from the cell area serviced by his or her original basestation to the cell area covered by any other compatible basestation.

As a user roams about his or her cell, he or she will at times reach the boundary of good coverage. As the handset realises it is reaching the limits of its operating range, it must identify the cell area the user is about to enter. The handset will constantly search for signals from other adjacent user groups that are members of the total system but outside its present cell. This will be done by searching for other OW signals than the OW of its own cell group. In order to minimise the search time and minimise the likelihood of losing the existing voice channel before it can establish a new one with the next basestation, a handset maintains a database defining relative timing between all adjacent basestations.

Once the OW of the "next" cell is contacted, the handset must now request admission to the cell as a new user. If admitted, the handset is assigned an identification number as an authorised user of the group. At this time all pertinent data on the handset, i.e. handset serial number, identification number and telephone number, must be relayed to and stored in the basestation database (see Figure 25.4). The local TELCO database must also be updated so that it knows where, i.e. to which basestation, to direct calls intended for that particular telephone number. If a call is in progress, handover now involves the local TELCO intimately. The local TELCO must now not only have its database updated, it must re-route a call in progress from one basestation to another in real time.

25.1.2   System Range, Coverage and Capacity

The propagation loss characteristics for operation at 1920 MHz in both indoor and outdoor conditions have been characterised by many research teams [2, 3]. All these extensive studies have concluded that the range exponent n in the propagation loss equation term $(1/d)^n$ may sometimes be as low 2, as in free space, but generally is at least 3, is often 4 and sometimes as large as 5. Thus, the OCDMA system must be designed to operate with a range exponent of at least 3 and preferably 4. Based on this data, Figure 25.5 can be used to estimate the propagation loss for different range exponents. The assumption is made that the basestation antenna is positioned on the ceiling, for indoor operation, and on a utility pole or the side of a building, for outdoor operations. For indoor operation, the assumption is that the signal has at least a 2 m propagation path from the basestation before it begins to experience any obstructions.

Power Budget and Operational Range

For the OCDMA system the operating range for a given value of required received signal level was computed, assuming the following worst-case conditions:

- Transmitter power, P_t = 3.2 mW, 5 dBm
- Receiver noise figure, NF = 6 dB
- Transmit antenna gain, G_t = 0 dBic
- Receive antenna gain, G_r = 0 dBic

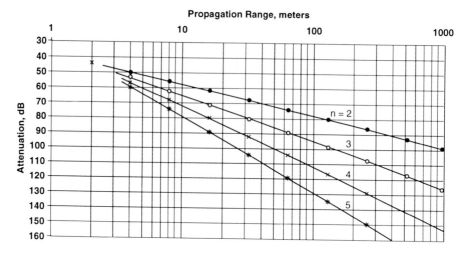

Figure 25.5 Propagation loss indoors, reference range 2 m.

- Required energy per bit/noise spectral density $E_b/N_0 = 10.3$ dB
- Transmitted bit rate, $R_b = 43$ kb s^{-1}
- Background noise temperature, $T_b = 3000$ K (worse case)

Without polarisation diversity, nulls in the received signal level in excess of 50 dB may be expected. Use of polarisation diversity and switching to the antenna with the higher signal level on a per-frame basis reduces the signal fades to approximately 15 dB. It is also estimated that the maximum degradation of the handset antenna gain due to the proximity of the user's head will be 10 dB. Taking these two factors into account indicates an effective worst-case transmitter power level, $P_t(\text{eff}) = -20$ dBm.

The received power level may thus be computed as follows:

$$P_r = P_t(\text{eff}) + G_t + G_r - L_s \quad (\text{dB})$$

Thus

$$E_b/N_0 = P_r/(N_0 R_b)$$
$$= P_r - kT - R_b$$
$$= P_t(\text{eff}) + G_t + G_r - L_s - kT - R_b \quad (\text{dB})$$

$$L_s = P_t(\text{eff}) + G_t + G_r - kT - R_b - E_b/N_0 \quad (\text{dB})$$
$$= -20\,\text{dBm} + 0\,\text{dB} + 0\,\text{dB} + 198.6\,\text{dBmHz}^{-1} - 35.6\,\text{dBK} - 46\,\text{dBbs}^{-1} - 10.3\,\text{dB}$$
$$= 86.7\,\text{dB}$$

This value of propagation loss is the maximum that can be experienced and still allow the required E_b/N_0. The achievable communication range for one carrier transmitting 3.2 mW as a function of range exponent is shown in Table 25.2. All studies have shown that the range exponent will exceed 3 in most indoor

Table 25.2
Communication range versus range exponent

Range exponent	Achievable range (m)
2	240
3	52
4	23
5	13

facilities. This being the case, most facilities will be limited to operation with a maximum communication range on the order of 50 m.

Capacity and Basestation Requirements

A typical office size can vary from a minimum of approximately 80 sq ft to 120 sq ft (~ 7.4–11.2 m^2); it is estimated that very densely populated buildings can have individual person areas as small as 48 sq ft (~ 4.5 m^2). Based on this, we can determine the number of users and number of subgroups required for a given user density as shown in Figure 25.6, where dense, typical and sparse population office areas are marked.

Considering an example building of 200 ft × 200 ft (~ 60 m × 60 m), three subgroups or two basestations would cover the entire floor for the sparsely populated case and require an operating range on the order of the length of the building, 60 m, at most. At the other extreme, for the densely populated office area, 19 subgroups or 10 basestations are required to service one floor with each basestation operating at a short range.

Note that Figure 25.6 shows the number of subgroups required for both a 100% activity factor (or usage) and a 20% activity factor. For an activity factor of 20% and the maximum density situation, an entire floor can be serviced with only four subgroups, or two basic basestations.

Power Control and AGC

With multi-user CDMA systems, power control is employed to ensure similar received power levels at the basestations, to minimise mutual interference between signals from different users, and hence maximise capacity.

- *Inbound links* Each basestation incorporates a fixed reference signal level against which all estimates of received handset signal levels are compared. The transmit power bias term in each handset is adjusted on the basis of these comparisons. This slow bias control at the base function in each handset is sufficient to maintain the power received at the basestation from each handset to within the accuracy of several dB without the need for any AGC circuitry in the basestation.
- *Outbound links* The outbound link transmit power level is held fixed at the maximum power setting at the basestation. As a handset is transported throughout the cell, its received signal level will vary over a maximum dynamic range of about 45 dB. In order to maintain the input voltage to the main signal-

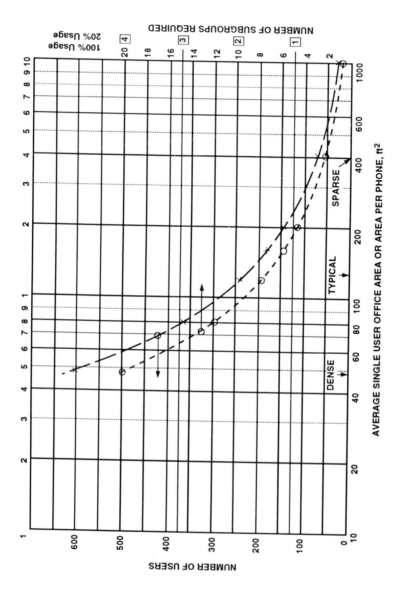

Figure 25.6   Number of users and number of subgroups as a function of office area.

path A/D converter at nominally half of full scale, and thereby avoid clipping and loss-of-resolution problems, an AGC function is implemented upstream of the A/D.

25.2 Physical Layer Description

Section 3 of the OCDMA WUPE specification details the air interface physical layer characteristics. WUPE physical channels are radio communication paths between two radio endpoints, a radio endpoint being either part of the fixed infrastructure, a radio fixed part (RFP), or a portable part (PP), typically a handset.

The physical layer interfaces with the medium access control layer and with the lower-layer management entity. On the other side of the physical layer is the radio transmission medium, which has to be shared extensively with other OCDMA WUPE users and with a wide variety of other radio services. The tasks of the physical layer can be grouped into five categories:

- Modulate and demodulate radio carriers with a bit stream of a defined rate to create a radio frequency channel
- Acquire and maintain bit and slot synchronisation between transmitters and receivers
- Transmit or receive a defined number of bits at a requested time and on a particular frequency
- Transmit the synchronisation field
- Observe the radio environment to report signal strengths

25.2.1 Physical Layer Characteristics

A physical channel provides a simplex bit pipe between two radio endpoints. To establish, for example, a duplex MAC [3] connection, two physical channels must be established between the endpoints.

The radio spectrum space for the OCDMA system has four dimensions: geometric (geographic) space, frequency, time and orthogonal code. Spectrum is assigned to physical channels by sharing it in these four dimensions. The OCDMA approach to handover allows release of a physical channel and establishment of another one in any or all of these four dimensions without releasing the end-to-end connection.

RF Channels (Access in Frequency)

The radio frequency band allocated to OCDMA WUPE unlicensed PCS equipment (UPCS) is 1910–1930 MHz. As described earlier in Figure 25.1, 16 RF carriers shall be placed in this band with centre frequencies F_c given by

$$F_c = F_0 + c \times 1.25\,\text{MHz}$$

where[2]

$$F_0 = 1910.625 \, \text{MHz} \qquad \text{and} \qquad c = 0, 1, \ldots, 15$$

The frequency band between $F_c - 625 \, \text{kHz}$ and $F_c + 625 \, \text{kHz}$ shall be designated RF channel c. All OCDMA WUPE equipment shall be capable of working on all 16 RF channels.

TDD OCDMA Structure (Access in Time and Orthogonal CDMA)

The serial bit stream input is encoded as a symbol sequence by the following mapping:

$$s_0 = p_0, p_1$$
$$s_1 = p_2, p_3$$
$$\vdots$$
$$s_n = p_{2n}, p_{2n+1}$$

To access the medium in time and orthogonal code, the following TDD and orthogonal CDMA structure is used. The TDD structure is of 10 ms length, consisting of two 5 ms half-frames. Each half-frame contains a total of 103 data symbols, or 206 data bit time intervals, for an equivalent simplex channel capacity of $20.6 \, \text{kb s}^{-1}$, plus some additional overhead time allocated to guard spaces. For use as effective useful data there are 80 data symbols (with two spares), or 160 data bits per half-frame (with four spares), for a simplex channel at a $16 \, \text{kb s}^{-1}$ rate (or $16.4 \, \text{kb s}^{-1}$ including the spares). Each half-frame interval is used to communicate one half of the data for a full-duplex channel. Sixteen such full-duplex channels are multiplexed in a CDMA fashion on the same RF carrier by assigning to, and spreading, each channel signal with a different orthogonal Rademacher–Walsh (RW) code. One such set of 16 OCDMA channels is a subgroup. Two of these OCDMA subgroups spaced at a frequency spacing of at least two frequency spaces is a group. A subgroup consists of 32 full-duplex channels at $16 \, \text{kb s}^{-1}$, and a group consists of 64 full-duplex channels at $16 \, \text{kb s}^{-1}$. One channel of each subgroup is assigned to operate as an order-wire (OW) channel for the subgroup and the remaining 31 as data channels.

Figure 25.7 shows the half-slot, and full-slot, frame format. A 10 ms TDD frame is divided into two equal-length 5 ms halves. The first half contains the normal RFP transmission, and the second half contains the normal PP transmission. Each half contains an effective 160 data bit transmission for an effective data throughput of $16 \, \text{kb s}^{-1}$ each way, thereby providing a $16 \, \text{kb s}^{-1}$ full-duplex link. A half-slot channel operates on a single RF carrier operating with an assigned OCDMA code. A full-slot channel operates on a single RF carrier and two orthogonal CDMA channels each with a different RW orthogonal code. The composite capacity of the two OCDMA channels is $32 \, \text{kb s}^{-1}$ each way, thereby providing a $32 \, \text{kb s}^{-1}$ full-duplex link.

A full-duplex slot is identified by the slot and OCDMA code assignment, and identified as full slot (i, n, k). Half-slots are labelled $i = 0$ or 1. The OCDMA channels are labelled $c = 0$ to 15, with the $c = 0$ channel always assigned to be used for the OW. The other OCDMA channels used in a full-slot configuration are labelled n, k, where $n = c$ and $k = c$ but $n \neq k$.

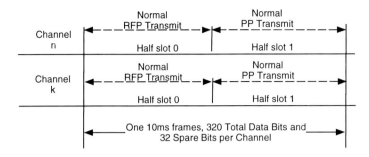

Figure 25.7 Half-slot, full-slot data format (synchronisation signals, gaps and other data not shown).

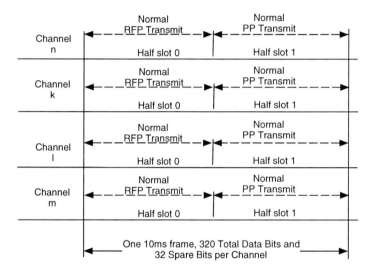

Figure 25.8 Double-slot data format (synchronisation signals, gaps and other data not shown).

A double slot is made up of two full slots. Four OCDMA channels are now utilised such that the full slot is identified as (i, n, k, l, m). As before, the i denotes the slot location, $i = 0$ or 1, and the n, k, l, m are all the four different OCDMA code channels assigned (see Figure 25.8).

Symbol intervals within any transmission slot are labelled f_0 to f_{79} (or f_{81} including the spare bits) where interval f_0 occurs earlier than interval f_1 (see Figure 25.9).

Reference timing

When an RFP and one or more PPs are intended to be used as an independent set of equipment units by themselves, as in a private home or small office environments (performing as a single cell), without the possibility of handover,

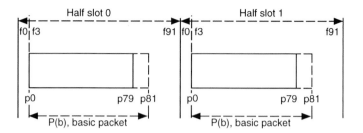

Figure 25.9 Basic packet placement within half-slot structure.

the reference timer is a notional clock to which the timing parameters of the TDD frame structure are related.

When an RFP and PP are intended to be used as a part of a larger system, permitting handover and multi-cell operation, the reference timer is an "absolute" timing reference slaved to the Global Positioning System (GPS).

A PP shall take its reference timer parameters, including half-slot, full-slot, frame, multi-frame and receiver scan, from the OW channel of the subgroup and RFP to which it is assigned and to which it is locked. It is allowed (but not required) to have more than one PP reference timer. The reference timer used for a PP transmission to an RFP shall be synchronised to packets received from that RFP or from an RFP to which handover is allowed. A reference timer is nominally (by this definition) synchronised to the time when the last packet used for synchronisation occurred at the PP antenna. Connections to different RFPs are allowed (but not required) to have different reference timers.

System synchronisation

RFPs on the same FP shall be in half-slot, full-slot and frame synchronism. If handover is provided, receiver scan and multi-frame synchronism is also required. Reference timers of RFPs of the same FP shall all be synchronised to GPS and have a mutual timing error of less than $\pm 0.1\,\mu s$. Synchronisation between FPs of different systems is provided via a GPS capability.

To obtain system and intersystem synchronisation, an RFP or PP may alter the length of a single frame by any amount, or it may alter the length of successive frames by up to two symbols.

Cells (Access in Space)

The fourth dimension to divide spectrum space is the geographical volume. Propagation losses and use of different codes in adjacent cells allow time–frequency–OCDMA code combinations to be re-used continuously.

Physical Channels

Data is transmitted within the frequency, time, OCDMA and space dimensions using physical packets. Physical packets are all of the same size. The different levels of channel capacity, i.e. throughput capability, are varied by transmitting packets related to the same data on the same RF carrier frequency, but on one, two or four OCDMA channels simultaneously, as described above for half-slot, full-slot and double-slot data transmission structures.

The basic physical packet $P_{(b)}$ consists of 80 symbols or 160 bits, with eight spare symbols or 16 spare bits. The data symbols are denoted p_0 to p_{79} (or up to p_{87} if the spare bits are used). When the packet is transmitted, the beginning of symbol p_0 coincides with the beginning of symbol interval f_3 of the half-slot being used (see Figure 25.9).

Physical channels shall be created between a PP and RFP by transmitting modulated physical packets on a particular RF channel, during a particular time in successive frames, the signal being spread by a particular OCDMA code or codes, at a particular location. It should be noted that one physical channel provides a connectionless, simplex service; a pair of physical channels must be used for a full-duplex link capability.

Physical channel notation

Physical channels shall be denoted as $R_b(i, n, k, l, m, C, N)$. The parameters are shown in Table 25.3.

Table 25.3
Parameters of physical channel notation $R_b(i, n, k, l, m, C, N)$

Parameter	Meaning
$b = 80$	Basic packet length in use
$b = \{81, 82\}$	Packet length should any spare symbols be used
$i = 0$	Packet transmission starts at symbol interval f_3 of half-slot 0
$i = 1$	Packet transmission starts at symbol interval f_3 of half-slot 1
$n = \{1, \ldots, 31\}$	OCDMA code used for the first orthogonal channel used
$k = \{1, \ldots, 31\}$	OCDMA code used for the second orthogonal channel used for a full-slot transmission channel, where k is different from n
$l = \{1, \ldots, 31\}$ and for $m = \{1, \ldots, 31\}$	OCDMA codes used for the third and fourth orthogonal channels used in a double-slot transmission channel, where n, k, l and m are all different
$C = \{0, \ldots, 15\}$	Number of the RF channel used to transmit the physical packets
N	Number (radio fixed part number $RPN = N$) of the RFP using the physical channel. This parameter will depend on the individual system and may be meaningless in many cases. It is, however, particularly helpful in describing handover algorithms

Minimum-rate channel

The minimum-rate physical channel, illustrated in Figure 25.7, shall be created by transmitting a physical packet $P_{(b)}$ during the first or second half-slot of the TDD frame, using OCDMA code n on RF carrier C in cell N, where:

$b = 80; i = 0$ or $1; n = 1, \ldots, 31; C = 0, \ldots, 15;$ and N is arbitrary

Medium-rate channel

The medium-rate physical channel, illustrated in Figure 25.7, shall be created by transmitting a physical packet $P_{(b)}$ during the first or second half-slot of the TDD frame, using two different OCDMA codes n and k on RF carrier C in cell N, where:

$b = 80; i = 0$ or $1; n = 1, \ldots, 31; k = 1, \ldots, 31$ but $k \neq n;$

$C = 0, \ldots, 15;$ and N is arbitrary

High-rate channel

The high-rate physical channel, illustrated in Figure 25.8, shall be created by transmitting a physical packet $P_{(b)}$ during the first or second half-slot of the TDD frame, using four different OCDMA codes n, k, l and m on RF carrier C in cell N, where:

$b = 80; i = 0$ or $1; n = 1, \ldots, 31; k = 1, \ldots, 31$ but $k \neq n;$

$l = 1, \ldots, 31$ but $l \neq n$ or $k; m = 1, \ldots, 31$ but $m \neq n$ or k or $l;$

$C = 0, \ldots, 15;$ and N is arbitrary

Signal Structure, Data Content, Protocols and Signal Processing

An overall description of the detailed timing waveform and data format structure will be given before detailing the various fields and their content. The signal structure for the system is predicated on two underlying objectives:

- To operate synchronously with the typically 20 ms frames of a 16 kb s^{-1} voice coder.
- To keep added signal path delays to under 10 ms.

Overall and voice channel signal structure

Accordingly, the signal structure is a sequence of 10 ms frames, as shown in Figure 25.10, each consisting of four distinct periods, two for inbound and two for outbound signalling, and each being one of 64 frames composing a 640 ms master frame as shown in Figure 25.11. The inbound signals are spread with a different PN code than the outbound signals (same code length and chipping rate, however).

The data consists of 16 kb s^{-1} bi-directional digital voice, plus a 1600 bit s^{-1} bi-directional control link. The data modulation is differentially encoded QPSK,

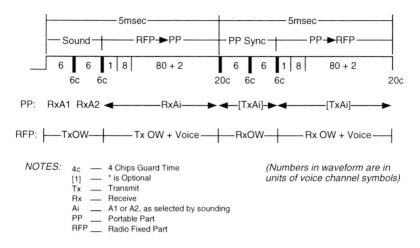

Figure 25.10 Overall and voice channel signal structure.

transmitted at a burst rate of 20.8 ksps. The data signal is bi-phase-modulated with a spreading code at 32 times the burst symbol rate (665.6 kHz). The spreading code is the modulo-2 sum of a length-256 PN sequence and a length-32 Rademacher–Walsh (RW) function. The all-ones RW function is used as the OW channel within each 32-channel subgroup; the remaining 31 functions are each associated with a different voice channel in that subgroup.

From the perspective of a PP already associated with a particular RFP, the four periods of each frame may be viewed as follows[3].

(i) *Sound* The RFP transmits a 12-3/8 symbol all-ones sounding pattern (i.e. no data transitions) on each OW channel, at a level 15 dB higher than for individual RFP → PP voice channels; each PP receives the first six symbols on antenna A_1, switches to antenna A_2 during the next 3/16 symbol, receives the next six symbols on A_2, compares the power between A_1 and A_2, chooses the antenna with the higher power, and switches to the chosen antenna during the next 3/16 symbol. The power from the chosen antenna is also used by the PP to determine its transmit power (if applicable) during the following PP Sync and PP → RFP portions of the signal, and as a code sync error measure to be input to its delay lock code tracking loop.

(ii) *RFP → PP* On each active voice channel, the RFP transmits a voice data burst of 91 symbols, followed by a guard time of 20 chips. The PP receives this data on the antenna selected during the sounding period. The voice channel data is constructed as follows:

- One phase reference symbol
- Eight channel control symbols
- 80 encoded voice data symbols
- Two spare symbols (reserved for future use)

(iii) *PP Sync* On an automatic cyclic TDMA basis, one member PP in each 32-member subcommunity (i.e. one per OW channel) transmits a continuous all-

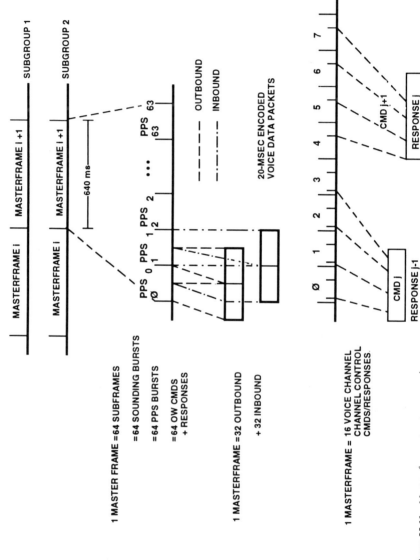

Figure 25.11   Master frame structure overview.

ones ranging signal (i.e. no data transitions) to the RFP on its associated OW channel for a duration of 12-3/16 symbols, followed by a 3/16-symbol guard time. The RFP OW channel performs a delay lock loop error measurement on this signal, and prepares and queues a timing correction command, if required, to be sent to that PP at the next opportunity. Each transmitting PP transmits on the antenna it selected during the sounding period.

(iv) *PP → RFP*   On each active voice channel, the PP transmits a voice data burst of 91 symbols, followed by a guard time of 20 chips, on the antenna selected during the sounding period. This inbound burst is of the same format as the RFP → PP burst of period (ii).

Thus the time-division duplex signal is symmetrical, with respect to format and content, its inbound and outbound portions being essentially identical to each other, and of the total time available, 76.9% is used for voice data, 10.6% for related overhead and spare capacity, 5.8% for channel sounding, 5.8% for PP timing synchronisation, and 1% for various switching and guard times. The advantages of this signal structure are summarised in Table 25.4.

Table 25.4
Advantages of selected signal structure

One dedicated bi-directional OW channel (for link control) for each 31 voice channels

No voice channel activity during sounding burst (at 15 dB higher than individual voice channels, allows very accurate measurements of received power, time offset and frequency offset)

Dedicated PP sync period per channel allows accurate measurement of PP power and time offset with no interference due to timing errors in other channels

Bi-directional 1600 bit s^{-1} control link incorporated into each voice channel (for PP power and timing control, as well as link control)

Order-wire channel signal structure

Looking at the OW signal by itself, we see the structure shown in Figure 25.12, wherein we note that the four periods of the overall time-division duplex structure are superimposed on an OW signal structure consisting of (in each direction) two OW symbol periods followed by 10 actual OW symbols plus a seven-voice-channel symbol frame sync/parity check signal and a 32-chip guard time. Each half-subframe is exactly 13 OW symbol periods in duration.

The OW signal structure has been designed so as to maximise signal search effectiveness, i.e. to minimise expected search times. Each OW symbol period has a length of 256 PN chips, i.e. one PN code sequence length. Thus by taking energy measurements over one OW symbol period, we are integrating over one PN code sequence length and taking full advantage of the PN code's autocorrelation properties.

Also, the choice of an exact integer number of PN sequence lengths per half-frame was made for two reasons:

• It greatly simplifies the PN coder design and the search algorithm.

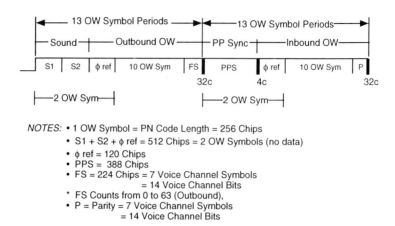

NOTES: • 1 OW Symbol = PN Code Length = 256 Chips
        • S1 + S2 + φ ref = 512 Chips = 2 OW Symbols (no data)
        • φ ref = 120 Chips
        • PPS = 388 Chips
        • FS = 224 Chips = 7 Voice Channel Symbols
                       = 14 Voice Channel Bits
        * FS Counts from 0 to 63 (Outbound),
        • P = Parity = 7 Voice Channel Symbols
                     = 14 Voice Channel Bits

Figure 25.12   Order-wire channel signal structure.

- It is critical for avoiding code phase ambiguities, which would increase typical and worst-case initial search times by more than 10-fold.

During the two sounding periods, the switching times allotted at the end of each and the reference phase period – i.e. for a total of $(192 + 4) \times 2 + 120 = 512$ chips = 2 OW symbol periods – the basestation is transmitting a continuous (spread) tone corresponding to an all-ones data modulation (i.e. no data transitions). The next 10 OW symbols contain OW data as described earlier.

The outbound OW channel frame sync field contains seven voice channel symbols (14 bits) organised as 6 bits parity check on the 20 OW bits, 6 bits frame number within master frame (0–63) and 2 bits parity check on the frame number. Thus 12/13 = 92.3% of the RFP OW channel transmit time (i.e. 46.1% of the total time) is available to PPs for signal acquisition purposes.

The inbound OW signal format consists of two segments. During the first, on a cyclic TDMA basis, one PP out of each community of 32 transmits a continuous (spread) tone corresponding to an all-ones data modulation (i.e. no data transitions), for a duration of 388 chips, for the purpose of allowing the RFP to measure that PP's transmit code synchronisation, power and quality during a period wherein there is guaranteed to be no interference from other PPs on the same channel.

Four chips guard time later, if the current OW time slot is assigned, the PP assigned to this slot transmits first a 120-chip phase reference symbol, then 10 OW symbols, and finally a seven-voice-channel symbol (14-bit) field containing a parity check of the 20 OW bits; the last 32 chips of the inbound OW signal segment are merely guard time.

If the current OW time slot is not assigned, it may be accessed on a collision sense multiple access (CSMA) basis by roaming PPs seeking membership in a new RFP community, or by PPs that have just been switched from "standby" to "active" mode and are seeking a voice channel assignment. The signal structure for such accesses is identical to that for assigned accesses.

Collision sense multiple access (CSMA) issues

PPs seeking entry to a cell are unknown entities to that cell's RFP, and thus some means other than assigned TDMA must be provided for the PP to access the RFP. Also, in order to accommodate other asynchronous events (e.g. PP transition from standby to active mode and requesting allocation of a voice channel) and to avoid the delays inherent in a purely cyclical TDMA or polling approach, again, some other means is desirable. A CSMA approach is well suited to supporting these relatively infrequent demands, but it brings with it the requirement to manage the CSMA resources intelligently. Several design features have been incorporated in this regard.

Firstly, the number of slots available for CSMA use will be kept to at least five in a continuous cluster at the end of the master frame. (Any slot that is not an assigned TDMA slot is available for CSMA use. This essentially means all broadcast slots, i.e. all slots that follow an OW broadcast by 15 ms.) This level of CSMA resource availability provides an excellent probability of no collision on the first access attempt.

Secondly, the RFP will maintain statistics of the use of available CSMA slots and will broadcast these statistics to the PPs for use in making intelligent choices of initial access and backoff strategies.

Thirdly, the extensive parity check code included in inbound OW transmissions minimises the possibility that when collisions do occur they would not be recognised as such; thus the likelihood of the RFP erroneously interpreting the demodulated results of collided transmissions is extremely low.

Any CSMA access attempt that is not acknowledged within 35 ms will be considered to have failed, the appropriate backoff strategy will be selected, and a retry will be scheduled accordingly.

Voice channel control data structure and protocol

Each voice channel burst contains an eight-symbol field allocated for channel control, i.e. in-band OW functions such as PP transmit power control, PP transmit code phase control and other functions to be identified. This provides a bandwidth of

$$800 \text{ symbols/second} = 512 \text{ symbols/frame}$$
$$1600 \text{ bits/second} = 1024 \text{ bits/frame}$$

in each direction, inbound and outbound, for these purposes, so that PPs with calls in progress still have access to full OW functionality.

Outbound channel control data is organised into 16-bit commands and acknowledgments formatted as shown in Figure 25.13 and frame synchronised to provide 64 such commands per master frame (100 per second) per voice channel. Each command is composed of a 6-bit function field and a 10-bit data field. Unlike the OW channel, no PP ID field is required since the PP being addressed is implicit in the voice channel assignment.

Inbound channel control data is organised into 16-bit requests and acknowledgements formatted identically to outbound commands and synchronised with them but offset by half a frame. Inbound responses to outbound commands

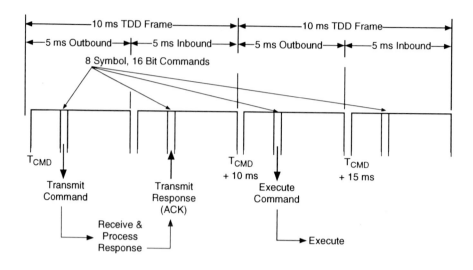

Figure 25.13 Voice channel control command format and extraction from voice channel signal.

commence one half-frame after the command transmission is complete, and outbound responses to inbound requests commence in the TDD master frame immediately following completion of the request.

The protocol philosophy for channel control dialogues between PP and RFP is essentially the same as that for OW dialogues (but even simpler, since there are no CSMA operations to support), namely that commands and requests be verified via acknowledgement prior to being acted on. The example shown in Figure 25.14 and described below in steps 1–4 serves to illustrate this:

1. The RFP determines that a PP's transmit code phase should be adjusted. The PP has a call in progress, so the indicated command is issued via the channel control field of the voice channel rather than via the OW channel. It is transmitted during one outbound 5 ms TDD burst. This step extends from $T = T_{CMD}$ to T_{CMD} + 5 ms in Figure 25.13.
2. The PP receives the command and acknowledges it by echoing it during the following response period, i.e., during the period $T = T_{CMD}$ + 5 ms to T_{CMD} + 10 ms in Figure 25.13.
3. The RFP receives the acknowledgement and issues an "execute" command during the period $T = T_{CMD}$ + 10 ms to T_{CMD} + 15 ms in Figure 25.13.
4. The PP receives the "execute" command, and some time during the next 5 ms executes the previously received "adjust transmit code phase" command.

As indicated earlier, and specifically in Figure 25.10, there is an excess capacity of two symbols per frame (400 bit s^{-1}), which can be allocated for transmission of control data. This can be used to reduce the control channel response time if found necessary. This can be especially useful for maintaining transmit power control requirements in a multipath environment.

STEP	OUTBOUND	INBOUND
1	ADJUST TRANSMIT CODE PHASE	
2		ACK ADJUST TRANSMIT CODE PHASE (ECHO)
3	EXECUTE	
4		<PORTABLE PART ADJUSTS TRANSMIT CODE PHASE >

Figure 25.14   Example voice channel control dialogue.

25.2.2   Transmission of Physical Packets

The transmission requirements are graphically represented in Figure 25.15. Modulation, differential encoding and signal shaping are involved. The modulation method shall be differential quadrature phase shift keying (DQPSK). It is necessary to provide signal filtering so as to minimise the effects of adjacent channel interference; at the same time excessive filtering increases the effects of mutual channels interference of RW signals on the same carrier. An optimum operating condition is found to exist with signal filtering using a fourth-order Chebychev filter with 0.1 dB ripple and with a cutoff frequency at about 0.7 of the signal chipping rate. This provides in excess of 24 dB attenuation for a frequency offset of 1.1 MHz from the centre frequency of the transmitted signal, and an operating signal-to-noise condition in excess of 15 dB for mutual channel interference.

The binary data stream p_n is converted into two separate binary data streams with all odd-numbered bits forming stream X_k and all even-numbered bits forming stream Y_k. The in-phase and quadrature components of an unfiltered carrier are given by:

$$I_k = I_{k-1} \cos[\Delta\phi(X_k, Y_k)] - Q_{k-1} \sin[\Delta\phi(X_k, Y_k)]$$

$$Q_k = Q_{k-1} \sin[\Delta\phi(X_k, Y_k)] - Q_{k-1} \cos[\Delta\phi(X_k, Y_k)]$$

The phase change shall be Gray encoded as determined from Figure 25.16.

25.2.3   Reception of Physical Packets

In this section, sensitivity requirements are given in terms of power and interference levels at the receiver input; steady-state, non-fading conditions are assumed for both wanted and unwanted signals. Equipment without an external antenna connection may be taken into account by assuming a 0 dBi gain antenna and converting these power level requirements into field strength requirements.

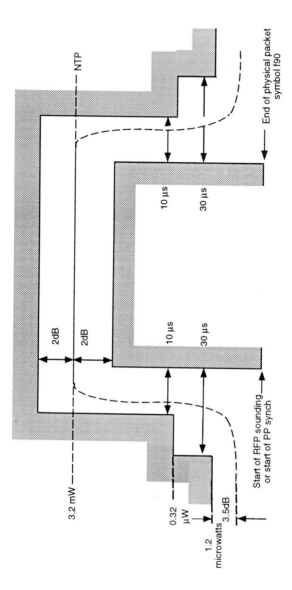

Figure 25.15 Physical packet power–time template.

X_k	Y_k	$\Delta\phi$
0	0	0
0	1	$\pi/2$
1	1	π
1	0	$3\pi/2$

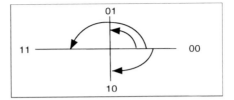

Figure 25.16 Differential data encoding.

This means that the tests on equipment without an external antenna will consider field strengths (E) related to the power levels (P), as specified by the following formula:

$$E\,(\mathrm{dB}\,\mu V\,m^{-1}) = P\,(\mathrm{dBm}) + 142.9$$

derived from

$$E = P + 20\,\log F\,(\mathrm{MHz}) + 77.2 \qquad \text{assuming} \qquad F = 1920\,\mathrm{MHz}$$

Radio Receiver Sensitivity

The radio receiver sensitivity is defined as the power level at the receiver input at which the bit error rate (BER) is 10^{-3} in the D field. The radio receiver sensitivity shall be –99 dBm (i.e. 43.9 dB $\mu V\,m^{-1}$), or better.

Before using an OCDMA WUPE physical channel for transmission or reception, the receiver shall be able to measure the strength of signals on that physical channel that are received stronger than –114 dBm (i.e. 29 dB V m^{-1}) and weaker than –54 dBm (i.e. 89 dB V m^{-1}) with a resolution better than 6 dB. Signals that are received weaker than –114 dBm shall produce a result equal to, or less than, that produced by a signal of –114 dBm. Signals that are received stronger than –54 dBm shall produce a result equal to, or greater than, that produced by a signal of –54 dBm.

Radio Receiver Reference Bit Error Rate

The radio receiver reference bit error rate is the maximum allowed bit error rate for a power level at the receiver input of –92.6 dBm or greater (i.e. 50.3 dB $\mu V\,m^{-1}$); the reference bit error rate is 10^{-5} in the D field.

Radio Receiver Interference Performance

With a received signal strength of –92.6 dB (i.e. 50.3 dB $\mu V\,m^{-1}$) on RF channel M, the BER in the D field shall be maintained better than 10^{-3} when a modulated, reference OCDMA WUPE interferer of the indicated strength is introduced on the other RF channels as shown in Table 25.5.

Table 25.5
Performance in the presence of interferers

Interferer on RF channel Y	Interferer signal strength	
	(dB μV m^{-1})	(dBm)
$Y = M$	37.3	−105.6
$Y = M \pm 1$	64.3	−78.6
$Y = M \pm 2$	84.3	−58.6
$Y =$ any other channel	89.3	−53.6

Radio Receiver Blocking

With the desired signal set at 10 dB greater than the sensitivity limit, the BER shall be maintained below 10^{-3} in the D field in the presence of any one of the signals shown in Table 25.6.

Table 25.6
Blocking performance

Frequency (f)	Continuous sine wave carrier level (dB μV m^{-1})
25 MHz $\leq f <$ 1490 MHz	96.1
1490 MHz $\leq f <$ 1885 MHz	81.4
$|f - f_c| >$ 6 MHz	76.1
1935 MHz $< f \leq$ 2000 MHz	81.4
2000 MHz $< f \leq$ 12.75 GHz	96.1

25.2.4   Enhanced Emergency 911 System Operation

When a user makes a telephone call to the emergency services – police, fire, ambulance – with a conventional wired telephone, the emergency services operator can immediately locate the geographical location of the caller. With a wireless phone, however, the caller can only be located to within the coverage area of the RFP. With the OCDMA system, an enhanced emergency 911 (ENH911) handset location capability can optionally be provided, as follows.

In the event that an ENH911 call is initiated by the user of a PP of the system, the PP automatically changes mode of operation. All of the preceding system description remains valid with respect to the transmission of data or voice. The operational procedures are unchanged, including all MAC, DLC and network-related functions. What changes is the inclusion, once per master frame, of a 100 ms burst of a sequence of half-slot transmissions of unmodulated overlay PN sequence and RW OCDMA code emanating from the PP only. This is illustrated in Figure 25.17, where HS0 and HS1 represent half-slot 0 and half-slot 1 in the standard transmission format. This pattern is repeated once per master frame. The 540 ms interval between ENH911 bursts will be normal voice communications.

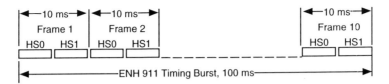

All HS 0 Transmissions contain normal RFP to PP data/voice modulated signals.

All HS 1 Transmissions from PP to RFP contain a data free RF carrier modulated only by RW + PN for 100 ms out of each frame.

Figure 25.17 Enhanced emergency 911 timing burst format.

A message will be contained as part of the 1600 bit s^{-1} control channel data to the RFP from the PP alerting the RFP that an ENH911 call is in progress. Upon reception of this information by the RFP, it alerts the other RFPs in the immediate area surrounding the receiving RFP.

All RFPs will be instrumented with matched filter receivers, which can process the received OCDMA and overlay PN code spread signals so as to derive accurate time-of-arrival (TOA) data from them. At least three of the RFPs within range of the transmitting PP will be assigned to derive TOA data from the ENH911 signal. Accurate initial frequency, timing and OCDMA code and overlay PN code data are provided to the matched filter receiver by the standard data/voice receiver. By performing smoothing of the derived measurements over an extended time interval, and by use of triangulation of measurements from all participating RFPs, very accurate position determination can be performed on the PP transmitting the ENH911 signal.

25.3   Media Access Control (MAC) and Higher Layers

25.3.1   MAC Layer Description

Section 4 of the OCDMA WUPE document presents the media access control (MAC) layer protocol architecture, utilised by the lower layers. The CDMA, order-wire (OW) and communication channel structures are defined and the alternate slot structures identified together with the required overhead signalling requirements needed for the establishment and transmission of the various data types and data rates. A complete description of the protocol and signalling structure required in all cases is provided. Other issues described in this section include:

- Order-wire channel data formats and communication access protocol and procedures
- Communication channel access data formats
- Definition and description of alternate types of channels

25.3.2 Data Link Layer Description

Section 5 will encompass the data link layer description. As such, it provides all the necessary information elements that are intrinsic to the data link layer protocol architecture. These elements are described in detail, including the definition of the message, the required format, numbers of bits or field size and application of the message. This section describes the functions that are one level above the frame and slot structure but include the critical components that provide differentiation between types of traffic. Call flow diagrams are dependent on the message set described in this section and are provided. Other material included in this section includes:

- Information elements
- Multiple mode traffic

25.3.3 Network Layer Description

Included in section 6 of the document is a description of network layer signalling for the physical layer elements, i.e. a network layer description. A control channel is provided between the basestation (BS) and the basestation controller (BSC) for the purpose of transporting the requisite network layer signalling information. This channel will be a 64 kb s^{-1} DSO link supporting the CCITT ISDN data link layer protocol. This channel provides network layer signalling as required for the MAC layer elements, network layer signalling as required for the data link layer elements, and network layer signalling as required for the overall network layer operations. Other issues described include:

- Procedures for mobility management
- Procedures for paging operations
- Procedures for handover

25.3.4 Privacy and Authentication Requirements

Section 7 provides the details of the privacy and authentication techniques employed in the system; this includes both a discussion of the DCS1900 and IS-54 methodologies as well as privacy and authentication requirements peculiar to OCDMA.

25.3.5 Audio System Performance Requirements

Section 8 provides a description of the protocol and digital data processing techniques to be employed by the system for speech and data coding, as well as the description of echo canceller characteristics and requirements.

Patent/Copyright

References

[1]   "Proposed Orthogonal CDMA (OCDMA) Wireless User Premises Equipment (WUPE) Air Interface Compatibility Standard", TR 41.6.4
[2]   "Indoor Radio Communications for Factories of the Future", TS Rappaport, IEEE Communications Magazine, May 1989
[3]   "Broadband CDMA for Personal Communications Systems", DL Schilling et al., IEEE Communications Magazine, November 1991

Notes

1   This TELCO unit is sometimes referred to as a Mobile Telecommunications Switching Office (MTSO).
2   To satisfy the requirements of FCC 47 CFR 15 Subpart D 15.323, the spectrum window intended for transmission must be monitored for at least $50\,\mu s$ prior to transmission.
3   Throughout this discussion, the term "symbol" is used to mean "voice channel symbol duration", i.e. 32 chip times, even when the activity is on the order-wire channel, and the term "voice channel" means one frequency channel/non-unity Rademacher–Walsh code combination.

26 The Composite CDMA/TDMA Standard, CCT

J. Neal Smith

This chapter details the technical operation of the North American composite CDMA/TDMA (CCT) system, which was awarded the US Federal Communications Commission (FCC) "Pioneer Preference Award" for PCS technology development in 1994. This technology has subsequently been standardised in the US ANSI technical committees in 1995 as IS-661, with initial deployment in the New York Metropolitan Trading Area (MTA) in the mid-1996 timeframe. CCT offers short-range cordless-style operation in the private environment, as well as long-range wireless loop and cellular-type operation.

26.1 Design Objectives

The Omnipoint composite CDMA/TDMA (CCT) system was designed to address the seven most important areas of concern to prospective PCS operators:

1. The flexibility to deploy a single radio technology to serve multiple applications economically, including high-speed vehicular mobility, outdoor public access, wireless local loop, indoor fixed wireless access and indoor cordless mobility.
2. Wireline-quality voice, wireless ISDN and a wide variety of data rates up 256 kb s^{-1} to support existing and future wireless services such as digitised video and multimedia applications.
3. Single handset and common air interface (CAI) for public systems and private applications using either licensed or unlicensed radio spectrum.
4. Efficient provision of full-service, full-mobility, cordless terminal mobility, and wireless local loop PCS at much lower radio infrastructure and cell site costs than conventional cellular systems and other prospective mobile PCS technologies, while still providing cell sizes ranging from several hundred feet (a few hundred metres) to as large as 20 miles (~32 km) in diameter.
5. Minimal deployment costs due to miniaturised basestations (as small as 27 inch × 14.5 inch × 13 inch (~ 69 cm × 37 cm × 33 cm) for 16, 32 or 64 simultaneous voice users.
6. Efficient use of existing network infrastructures such as GSM, AIN and IS-41-based architectures.

7. Enormous increases in system capacity through the use of high-speed, mobile-directed handoff (in 4–200 ms), which allows one network both to manage large cells efficiently and yet also to deploy very low-cost microcells wherever capacity is needed due to localised peak demands.

As a result of these design objectives, the following five unique characteristics have been incorporated into the CCT system design: flexible cell sizes, private and public operation, minimal interference to other systems, mobile-directed mobility management, and improved capacity.

26.1.1 Flexible Public Cell Sizes

CCT systems can be configured to provide indoor cells, typically 200–400 ft (~ 60–120 m) in diameter, as well as outdoor cells with diameter of up to 4 miles (~ 6.5 km) in urban areas, 8 miles (~ 13 km) in suburban areas and 20 miles (~ 32 km) in more open areas (with proper antenna configurations and small reductions in capacity) as shown in Figure 26.1.

26.1.2 Public and Private System Operation

The CCT system has been designed to serve both public and private indoor environments. The indoor version of CCT systems has been tested in more than 100 buildings of virtually every size and type of construction.

Indoor use will require the use of unlicensed frequencies (versus in-building re-use of licensed bands) in many locations where large market penetrations are

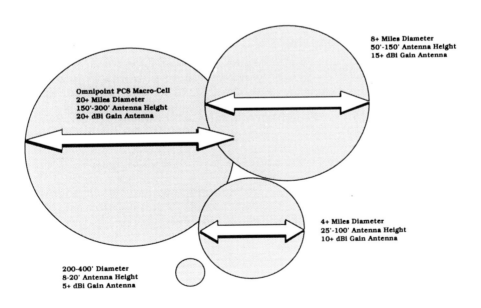

Figure 26.1 CCT system cell sizes.

to be achieved. The CCT system is able to meet private premises' needs in either unlicensed or licensed frequencies while allowing customers to use the same handsets in the outdoor mobile environment. Handsets can also operate in the unlicensed 2.4 MHz ISM (industrial, scientific and medical) band.

Wireline-quality voice and data services are available with the CCT system for both private and public users, with higher-speed digital transport being used for "premium" services, such as ISDN rates, supported by the system's capacity to deliver over 144 kb s^{-1} full-duplex (over 256 kb s^{-1} broadcast) for data applications.

26.1.3   Minimal Interference to Other Systems and Users

Interference to other systems and users is kept to a minimum with the CCT system for two reasons: the more Gaussian-noise-like characteristics of CCT's implementation of spread spectrum; and the use of high-speed TDMA, which ensures that no more than one transmitter is operating at a given instant in time within an entire cell. Because there is only one user transmitting at a time, there is no additive noise effect as other users are added, unlike CDMA-only or FDMA-only systems. CCT's approach yields significantly better C/I ratios for frequency re-use planning because of the ability to offset nearby users in different co-channel cells into different time slots. With its unique spread-spectrum TDMA architecture, an entire CCT-based system operating with the maximum output power being generated by each handset and the maximum number of handsets operating simultaneously produces less interference than a single traditional narrow-band analogue (FDMA) cellular handset channel.

26.1.4   Mobile-Directed Mobility Management

The CCT system utilises a "mobile-directed" handoff scheme. In this approach, the terminal makes the decision to hand off between cells and directs the basestation controller or network to switch once an alternative basestation is acquired. The entire handoff process takes from 4 to 200 ms, regardless of the size of the cell or the speed of the mobile unit. Thus, with CCT systems, even with a cell radius of 250 ft (~75 m), mobile handoffs are undetectable.

26.1.5   Flexible System Capacity

Capacity in a wireless system is a geographic phenomenon. The minimum cell radius sets the limit on the maximum capacity of a wireless system on a geographic basis (i.e. either localised cell area or total city area). If a PCS operator using traditional technologies is limited in capacity by a minimum of a half-mile (~800 m) cell radius while a PCS operator using CCT technology can easily install cells with 500 ft (~150 m) radii[1] in areas where capacity is needed, then the CCT system has 25 times the capacity of the other PCS system over that area. By using 250 ft (~75 m) cell radii to serve a particular "hot spot", the Omnipoint IS-661-based system would have 100 times the capacity of the cells from other systems in that coverage area (all other factors being equal).

26.2   Standardisation (IS-661)

The CCT-based system that is now an official industry standard was first proposed to the standards bodies on 1 November 1993 as a PCS Standard in the Joint Technical Committee (JTC) on Wireless Access (see Chapter 8 for details of the JTC). Currently, the IS-661-based system is one of only six PCS technologies (originally there were 17 proposals) that have passed by ballot as an official mobile standard for the US PCS industry.

The PCS standardisation process has granted the CCT proposal the standards designation IS-661. The Technical Ad Hoc Group (TAG) that proposed the CCT system included technical input from various industry participants including wireless manufacturers, current cellular service providers, network and switch manufacturers, the semiconductor industry and the US Government.

26.3   Applications Flexibility

PCS requires full coverage and vehicular mobility. However, if strategically beneficial to the PCS operator, the CCT system can support services with reduced mobility, such as cordless terminal mobility and fixed-loop bypass. Because of the CCT system's flexibility, it can support new products such as PDAs and video phones or new services such as software, database, or multimedia downloads as they are supported by the network infrastructure connected to the CCT system. In some cases, ancillary equipment must be deployed with the CCT system and switching infrastructure to handle billing, teleconferencing, voice messaging and other services. This equipment may be provided by the PCS operator or a third party.

Figure 26.2 provides an overall picture of the CCT system as deployed in a complete network environment. Database access for HLR/SCP functionality is provided via D-channel messaging from the basestation and over SS7 in the network. These functions can be performed by the LEC network or by the PCS operator in a stand-alone network.

26.3.1   Mobility

Mobility has three aspects: terminal, personal and network. The CCT system supports all three forms of intra- and internetwork mobility, including the capabilities provided by GSM- and AIN-based solutions.

Terminal mobility involves handoff and is most often associated with a terminal travelling from the coverage area of one cell site to another. With a handoff, the terminal is re-registered on the network on a different cell site.

Personal mobility allows users to place and receive calls from any wireless or wired phone and carry along his or her service profile. This process entails either a registration process that essentially downloads or enables the user service profile from the network, or a smart card that allows the user to carry his or her profile for use with any device.

Network mobility allows the user to use any handset or wired phone and maintain a profile of features separate from a specific device. This network

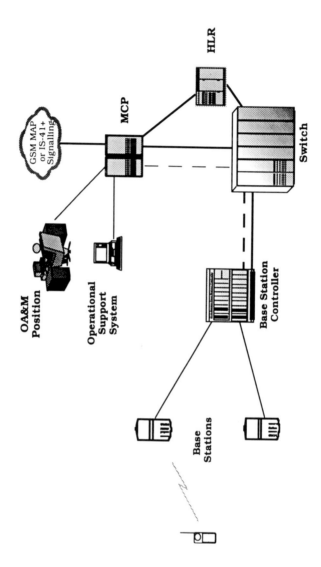

Figure 26.2   CCT network architecture.

capability allows the user to be "found" by the network and receive calls at any location based on a single dialled telephone number. The CCT system supports network mobility in conjunction with the network provider.

26.3.2 Wireless Local Loop

For wireless local loop, a variation of the mobility system is required: a fixed radio (transceiver) unit is mounted in the subscriber's residence or business, which can deliver from one to several dozen subscriber lines or trunks.

Subscriber Unit

The subscriber unit for wireless local loop provides the radio interface from the subscriber's location to the CCT basestation. In general, the subscriber unit must provide the following capabilities:

- Provision of dial tone instead of mobility "send" operation, as well as RF communications (including CAI, RF performance and O&M) with the basestation and the PCS provider's system.
- RJ-11, E1, or T1 trunk access from the indoor telephone outlets or PBX to the radio for wireless transmission to the basestation.
- Environmental enclosure for outdoor locations.
- Antenna interconnection for communication with the basestation.
- Ability to aggregate slots for higher-speed data services.
- Potential for multiple line terminations that can be used simultaneously.
- AC power from the site.
- Back-up power capabilities (i.e. battery).

In addition to the above, considerations must be given to required landline services such as E911 (emergency calls).

Wireless Loop Basestation

The basestation for wireless loop is essentially the same as the mobile basestation. However, it is possible that the service provider might not want to enable the basestation for mobility and may be able to offer service with a subset of the CCT system's capabilities to the subscriber (i.e. no mobility management, no need for routing through a BSC, etc.) Therefore, a GSM switch, a Class 5 central office switch, or a PBX-like platform (similar to a BSC without mobility management) can be the access point for wireless loop basestations. The switch can then provide for billing services and other landline-like services without the need for BSC handoff capabilities and database services.

26.4 Services, Features and Functions Supported

The CCT system provides a full array of voice and data services through its radio access and transport system. Specific features supported depend on the network backbone selected, GSM or AIN. All GSM features are supported transparently to the handset. With AIN Class 5 networks, the system supports all LEC and CLASS features as well as mobile-specific features. This section describes services supported by the CCT system for all network architectures.

26.4.1   Voice Services

The CCT system can provide landline-quality voice services to the end-user. The system provides transparent operation for network functionality including DTMF tones for interactive services and music on hold. Active call features remain intact even after a handoff between cell sites. Standard vocoder options include 32 kb s^{-1} ADPCM and a very high-quality (MOS greater than 3.9) 8 kb s^{-1} algorithm. Forward error correction (FEC) is then added to these rates for additional protection.

Users place calls on the CCT system in a manner similar to placing calls on wireline telephones. CCT provides two handset interfaces that allow for either cellular-like calling (entering the number then pressing a call button to commence call set-up) or PSTN-like calling (off-hook dial tone and DTMF tones for each digit). The exact implementation of the user interface is programmable and can be modified for the service operator.

System features are inherently robust because of CCT's transparency to the network switching platform. In the case of LEC applications, the Omnipoint system supports delivery of most business telephone network features to the wireless user. These features include custom calling, caller ID and other CLASS/AIN features, PVN, conference calling, do not disturb, custom billing and others. The CCT system also supports most of the advanced features proposed by various industry (PCIA) and standards development (JTC) organisations.

26.4.2   Data Services

The CCT system enables the user to request any bandwidth (up to the maximum of the RF channel at the time of call origination) that is consistent with the current application being used. For example, a mobile voice subscriber may require only a 9.6 kb s^{-1} voice channel for a vehicular telephone call, where an office user may request 32 kb s^{-1} voice for enhanced voice quality or to support a data modem or fax. A data user may also request either a small number of time slots for bursty packet data or multiple 9.6 kb s^{-1} channels for higher data traffic. The CCT system supports over 144 kb s^{-1} full-duplex or over 256 kb s^{-1} simplex rates in the 1.875 MHz version and twice that in the 3.75 MHz version.

In addition to the user voice/data channel, the CCT design provides for a D channel that allows for continuous 400 bit s^{-1} of data traffic to every user. This channel is separate from (but simultaneous with) the bearer channel for messaging even when the handset is in use. The D channel can carry information for applications such as paging, voice-mail notification and short message service.

The CCT system provides bearer rates to the handset that fully support fax transmission through the CAI. Connection to a fax device can be made via a stand-alone module or a data port on the handset. Data rates for transparent, voice-circuit fax service can be provided, which allows the existing installed base of faxes and modems direct, transparent access to the network. Higher-speed fax transmission (such as $56/64\,\text{kb s}^{-1}$) can be provided through the direct data support of the CCT system.

The CCT system provides transparent error detection/correction for certain data applications through a CRC on both the control information and the entire packet. The protocol provides for ARQ at the application level for data error correction, an approach that is more efficient than FEC within the packet structure.

26.4.3   Features Supported

The CCT system provides radio access and transport facilities for PCS services. When coupled with a full-featured switching platform, the CCT system will enable the features listed below:

- Basic POTS
- Residential and business services
- CLASS
- Paging, voice-mail and short messaging services
- Central office features, including Centrex
- Data services
- 911 emergency or safety services
- Attendant, line restriction, hunting, abbreviated dialling, call waiting/forward/ transfer, conference calling, call screening/routing
- International Direct Distance Dialling (IDD)
- Private Virtual Network (PVN)
- Access features, private facility and automatic or flexible routing
- Authorisation and account codes, message detail recording

26.5   Technology Overview

The following sections provide an overview of technical operations of the critical components of the CCT system. Included are descriptions of the RF specifications, operating frequencies, channelisation plans, frequency re-use plans, spread-spectrum and multiple access techniques, traffic and control channel use, the available traffic modes, associated link budgets and the use of vocoders.

26.5.1   RF Specifications

The main RF specifications are summarised in Table 26.1.

26.5.2   Operating Frequency

The CCT system is frequency-agile in the entire 1850–1990 MHz band, including the unlicensed band (1910–1930 MHz) recently allocated by the FCC. Because CCT mobiles are frequency-agile across the entire 140 MHz band, PCS operators

Table 26.1
IS-661 RF specifications

RF channelisation	1.875 or 3.75 MHz RF channels
TDMA	16 full-duplex 9.6 kb s^{-1} slots 32 full-duplex 4 kb s^{-1} slots
Mobile output power	Maximum of 2 W EIRP (per FCC rules) Maximum of 1 W EIRP in handhelds
Basestation output power	Maximum of 1640 W EIRP (per FCC rules)
Antenna gains (basestation)	Options to 26 dBi gain
Max cell site coverage	Depends on antenna heights, and propagation environment: approximately 2–4 miles (~3–6 km) diameter in urban/suburban areas and up to approximately 14 miles (~22 km) diameter in rural areas with only 20% reduction in the number of users per RF channel

can set or change cell frequencies as required. In Version 3.0X, the CCT system will provide for remote frequency selection via a data port on the basestation. The system does not require tuning when the basestations are changed to transmit at a different frequency.

Frequency planning with the CCT system is relatively easy because most configurations only require planning within the $N = 3$ re-use structure. Because all mobiles in a cell operate on the same frequency separated by time, frequency planning consists of simply planning for RF channels (up to three per cell) in an appropriate configuration.

26.5.3 Channelisation

Nominal RF channel spacing for the CCT system is 1.875 MHz or 3.75 MHz, depending on which option is selected by the service operator. The system also provides lower chip rates with closer channel spacing for applications that use lower bandwidth spread.

26.5.4 Frequency Re-use

The CCT system uses a patented combination of CDMA, TDMA and FDMA for separation of users and cells. In general, cells are separated by frequencies (F) and code sets (CS) as shown in Figure 26.3. Adjacent cells are separated by frequency; CDMA is used for additional orthogonality when frequencies are repeated.

26.5.5 Spread-Spectrum TDMA Approach Within a Cell

Within a cell, direct-sequence spreading codes are used to achieve significantly higher data rates. Users are separated within a cell by a unique, high-speed TDMA approach as shown in Figure 26.4.

The very high data rates created by CCT's direct sequence spread spectrum allow many more TDMA users per RF channel than is generally possible with

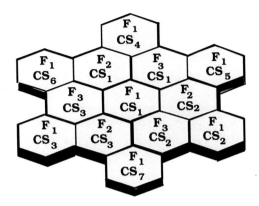

Figure 26.3 $N = 3$ frequency re-use using CDMA.

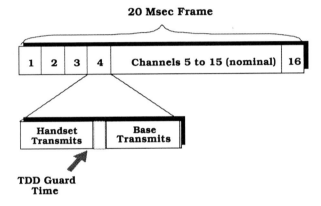

Figure 26.4   TDMA frame structure.

traditional, fully mobile TDMA systems (without equalisers). CCT's TDMA design permits over 256 kb s^{-1} of bearer channel information (after all overheads) simplex or over 144 kb s^{-1} of bearer channel information in full-duplex mode per 1.875 MHz of RF bandwidth. The actual instantaneous burst data rate is significantly higher to accommodate addressing, control channel information, other overheads and guard time. The most significant benefit is that the cost per user channel declines as the number of channels increases per basestation.

By utilising a TDMA approach within a cell, and not relying solely on CDMA for separating multiple handset signals at the basestation, self-interference at the base receiver is greatly reduced, permitting greater coverage area for a given handset transmitter power output level. Also, TDMA within a cell eliminates the "near–far" problem that plagues CDMA-only systems, which require expensive and complex power-control algorithms to ensure that the same signal levels reach the base receiver from all users in a cell.

In the CCT system, CDMA is used for two reasons. Firstly, the code sets mitigate multipath, maintaining low symbol rates while providing very high data rates; and secondly, to achieve better C/I ratios and frequency re-use. Without the use of these spread-spectrum techniques, high data rates are difficult to maintain in fully mobile channels.

26.5.6   Multiple Access

CCT Version 3.0X uses a TDD and TDMA structure based on a 20 ms polling loop for handset access to the basestation. For the 1.875 MHz version, the 20 ms TDMA frame supports 32×9.6 kb s^{-1} simplex time slots after all overheads and is equivalent to 16 paired 9.6 kb s^{-1} full-duplex time slots. Provisions are made for each handset to aggregate multiple time slots, giving a user more bandwidth for higher data rates as required. Asymmetric data rates can be supported on a frame-by-frame basis, permitting high-speed data transport and data broadcasting by the basestation. An FDD structure is also under development to ensure better coexistence with existing FDD mobility systems in similar frequency bands.

26.5.7   Traffic/Control Channels

The standard CCT basestation provides 32 simultaneous simplex 9.6 kb s^{-1} bearer rate traffic channels per 1.875 MHz channel. If used in full-duplex mode, this results in 16 full-duplex 9.6 kb s^{-1} traffic channels. Each of these channels in any of the above configurations also provides all the overhead beyond the bearer channel for addressing, CRCs, internal control messages, plus a "slow-speed" D channel (at 500 bit s^{-1}). Other control messages use the bearer channel during call set-up, handoff, or for fast message services, paging, etc. Any time slot can be used for either control or traffic information. All types of functional "channels" for other systems, such as GSM, can be logically mapped to CCT's control channels.

26.5.8   Traffic Mode

The traffic mode in the CCT system occurs after a basestation and handset have established a transmission path with the appropriate allocation of air-slot resources for the service supported. Two types of messages are supported in traffic mode:

- *Normal or bearer traffic* – the actual information carried by the wireless link, either voice or a form of digital information.
- *Control traffic* – link-specific data messaging or call control information.

26.5.9   Link Budgets

The CCT system is designed for a maximum path loss of between 160 dB and 137 dB, depending on basestation configuration. For most RF modelling, a maximum of 19 dBd-gain, three-sector, cells are assumed, for a maximum path loss of 149.4 dB.

Modelling for CCT cell radii is based on hundreds of real-world measurements. For computer modelling predictions in urban areas, Omnipoint uses the Hata [1] large-city median propagation path-loss models to predict median propagation path loss, adjusting them to reflect real-world measurements. Large-scale variations about the median are characterised by a log-normal distribution [2] representative of the statistical effects of RF shadowing and obstruction. Small-scale and indoor [3] multipath fading effects are accounted for separately in the CCT analysis based on frame length, frame error rate and associated minimum E_b/N_0.

In the 1850–1990 MHz frequency range, log-normal standard deviations are usually between 6 dB and 10 dB. In urban areas, Omnipoint assumes a standard deviation of 10 dB, while in suburban and rural areas Omnipoint uses a 7 dB standard deviation. With a 100 ft (~30 m) cell antenna height, 6 ft (~1.8 m) mobile height and a centre frequency of 1850 MHz, our propagation path-loss models are as shown in Table 26.2.

Table 26.2
Propagation path-loss models[a]

Urban area	Median path loss (dB) = $133.80 + 35.18 \log_{10}(R)$ Log-normal standard deviation = 10 dB
Suburban area	Median path loss (dB) = $121.78 + 35.18 \log_{10}(R)$ Log-normal standard deviation = 7 dB
Rural area	Median path loss (dB) = $101.72 + 35.18 \log_{10}(R)$ Log-normal standard deviation = 7 dB

[a]R is the cell radius in kiiometres.

26.5.10   Vocoders

The Version 3.0X CCT system can use virtually any vocoder. Version 3.0X comes standard with a vocoder that provides 9.6 kb s^{-1} voice (after FEC) as well as options for 32 kb s^{-1} ADPCM and 40 kb s^{-1} data ADPCM (Omnipoint proprietary) applications and provides for later migration to 4 kb s^{-1}, or lower, as future vocoders become available without requiring any change to the CAI or the basestations. This flexibility allows the user to select different voice data rates for each application (e.g. 9.6 kb s^{-1} for outdoor mobile, and higher voice rates such as 32 kb s^{-1} ADPCM for in-building office use).

Forward error correction on the voice link is implemented within the vocoder for the 9.6 kb s^{-1} voice application, also allowing for interpolation of the speech in error conditions. Error detection is also used as a parameter in determining the quality of the link on the entire frame.

26.6   Layer 1 Air Interface Specification

26.6.1   Modulation

To produce the direct sequence spread-spectrum (DSSS) characteristic of the system RF signal, a form of quadrature modulation called root raised cosine (RRC) modulation is used. This modulation requires linear RF power amplification to prevent spectral regrowth of modulation sidelobes. DSSS conveyance of information is accomplished by using multiple DSSS PN chip sequences to encode the baseband data. The PN sequence modulates the carrier to a 1.875 MHz bandwidth. By shaping the PN chip waveforms before modulation, all modulation sidelobes at frequencies more than 0.9 MHz from the centre frequency of the DSSS RF signal are greatly attenuated.

26.6.2   TDMA, FDMA and CDMA Multiple Access Methods

The system employs a unique combination of time-division multiple access (TDMA), frequency-division multiple access (FDMA) and code-division multiple access (CDMA) for multiple user access to the PCS network. Within a PCS cell, TDMA is used to separate users. To provide greater area of coverage, or to provide greater capacity for densely populated regions, multiple cells or sectored cells are deployed using FDMA, thus separating cells by frequency. Also, to permit multi-cell deployments in a given region, the DSSS form of CDMA is used for each RF link to reduce co-channel interference between cells re-using the same RF carrier frequency. DSSS also improves system response to RF channel impairments.

TDMA Frame and Time Slot Structure

The TDMA frame and time slot (channel) structure is based on a 20 ms polling loop for user access to the RF link – see Figure 26.5. Utilising TDD, the 20 ms frame is equally divided between 16 full-duplex channels within the frame. Each resulting time slot (channel) is capable of supporting a 9.6 kb s^{-1} full-duplex user.

At the basestation, the first half of the TDMA/TDD time slot is allocated for the MS transmit function. During the second half, the BS transmits to the MS assigned to that particular time slot.

The BS receives during the first half of the time slot and transmits during the last half. After each TDMA transmission from either the base or mobile unit, a small portion of each time slot (designated guard time) is allocated to allow the transmitted signal to propagate from a mobile transmitter at the maximum specified distance from the BS (maximum cell radius), and back again. This is necessary to prevent received and transmitted signals from overlapping in time at the base and mobile terminals.

The signal received from the MS serves as a channel sounding signal to determine link propagation loss and to serve as a measurement of link quality for the power control subsystem. This is also used to determine which of the multiple antennas to use for the spatial diversity scheme and permits spatial diversity control to be updated during each TDMA time slot period.

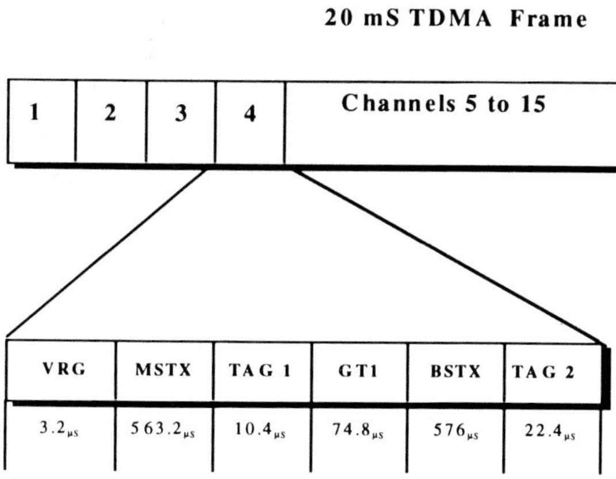

20 mS TDMA Frame

1	2	3	4	Channels 5 to 15	16

VRG	MSTX	TAG 1	GT1	BSTX	TAG 2
3.2 μs	563.2 μs	10.4 μs	74.8 μs	576 μs	22.4 μs

Figure 26.5   TDMA frame and channel time slot structure.

TDMA Burst Data Rates

The TDMA data burst rate during transmission is determined by the requirement to provide a 9.6 kb s^{-1} digital capability for 16 full-duplex channels per frame during the 20 ms frame period. This requires a TDMA burst data rate of 307.2 kb s^{-1} in TDD mode. Owing to the guard time requirement of the TDD mode of operation, and because of additional overhead bits required for link protocols and control functions, the total TDMA burst rate is 468.75 kb s^{-1}.

TDMA Channel (Time Slot) Assignment

Multiple slots in the polling loop may be negotiated for and assigned to an individual MS. The negotiation may take place at any time via signalling traffic. The slots, if available, are assigned by the BS to the MS. Slot synchronisation is maintained for each assigned slot. For example, by employing two channels (time slots), the user terminal can operate at a 19.2 kb s^{-1} data rate, versus 9.6 kb s^{-1} for one channel (time slot). The maximum data rate supported per user is 153.6 kb s^{-1} full-duplex or 307.2 kb s^{-1} half-duplex (see Figure 26.6).

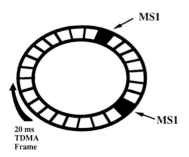

20 ms
TDMA
Frame

Figure 26.6 Multiple TDMA channels (time slots) per user.

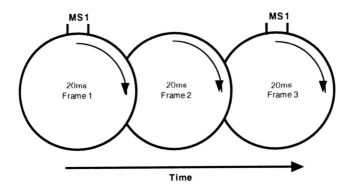

Figure 26.7 Submultiple TDMA channels (time slots) per user.

Submultiple TDMA Channels (Time Slots) Per User

A MS need not be granted a channel (time slot) in every frame. Slots may be granted in frames separated by an integral number of intermediate frames to support user data rates of less than $9.6 \, \text{kb s}^{-1}$. The maximum limit on the separation of slots allocated to a single MS is 0.5 s. This equates to a per-user data rate of $384 \, \text{bit s}^{-1}$ (see Figure 26.7).

Basestation-to-Network and Basestation-to-Basestation Synchronisation

The BS provides the basic TDMA loop timing structure for its cell or sector. To maximise PCS system throughput capacity, the TDMA frame times for all BSs within the same geographical area should be synchronised. For example, one method of obtaining the required timing is to utilise a GPS receiver at the BSC (and optionally at the BS) to generate the primary reference timing marker for the TDMA frame timing. This marker is captured at the BSC every second and transmitted down the backhaul lines to the attached BSs.

This synchronisation of BSs within a given multi-cell PCS deployment allows a BSC temporarily to turn off any TDMA time slot of a given cell that may be interfering with a neighbouring cell. It also facilitates time slot interchange (TSI), i.e. switching a MS to a different time slot if a current time slot is being interfered with by an adjacent cell using the same time slot.

Basestation-to-Network Synchronisation

The primary data timing standard in a digital network backhaul system, such as T1 or ISDN BRI or PRI, is the PSTN timing standard. To prevent data precession into over-run or under-run, the BSC and its BSs are synchronised to the PSTN timing standard. The actual data movement clock, generated by the PSTN and rendered to an 8 kHz timing marker, is used by the system to get the data rate throughput.

Mobile Station-to-Basestation Synchronisation

The MS can synchronise to a new BS within one channel (time slot) and is capable of synchronising with multiple BSs when those BSs are synchronised to a common digital network. The system allows non-coherent detection to be used by the BS and MS receivers, and they do not have to be phase-locked. However, the transmit and receive local oscillator frequencies of the BS and MS are automatically controlled to prevent data precession between the BS and MS.

Link Establishment: Polling

A polling channel (time slot) is used by an MS to acquire a connection to a BS. Any channel (time slot) that is not already seized by an MS contains a general poll command message in the BS-to-MS transmit interval. To acquire a channel (time slot), the MS responds to the general poll message with a general poll response message. The BS, upon receiving the general poll response message, sends a specific poll message, which includes a time slot (or time slots) assignment for the MS – see Figure 26.8. Following these signalling data exchanges, the BS and MS begin bearer traffic exchanges.

Bearer Traffic Exchange

Once the MS has seized a channel (time slot) and been validated, the exchange between the MS and BS leads to bearer traffic exchange. Bearer traffic can be voice, error-controlled data, or non-error-controlled (raw) data. From the BS software point of view, the information in the bearer traffic header includes the absence of a bit signifying that this is poll-type message, with no transmission error bit set, and no bit signifying signalling data. Each TDD channel (time slot) can be configured with the majority of the slot time given to the BS-to-MS direction, the majority of the slot time given to the MS-to-BS direction, or with symmetric slot time distribution giving 50% of the channel to both BS and the MS.

Typically, voice traffic utilises the bearer channel symmetrically. In a data exchange, more data is usually sent in one direction and less in the other. For instance, if fax data was being sent to an MS, a higher data rate would be used in the BS-to-MS direction. In this case the majority of the TDD channel (time slot) is given to the BS-to-MS link.

Packet Structure

Each time slot is divided into subframes that are normally used for bi-directional base-to-handset communication. Each subframe is packetised as shown in Figure

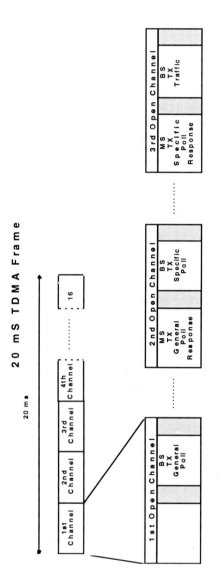

Figure 26.8 Mobile access link set-up.

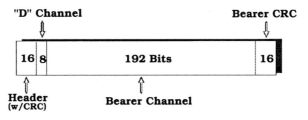

Figure 26.9   CCT packet structure.

26.9. All data rates shown are after all overheads because only the 192 bits of the bearer channel is used in the 9.6 kb s^{-1} per-user capacity calculation. Polls are used by the basestation for establishing and maintaining communications with all handsets in the cell coverage area.

26.6.3   Signal Acquisition and Synchronisation

At the bit level, CCT handsets can synchronise and acquire an air channel in less than 10 μs. Omnipoint's proprietary synchronisation techniques provide much of the spectrum efficiency of the CCT system.

At the protocol level, channel acquisition occurs whenever a handset needs to access or resynchronise with a base slot. During channel acquisition, the handset searches for an available basestation in its general vicinity by listening for a common signalling poll on any time slot from any possible frequency. Once a base is located with sufficient signal strength and adequate load available, the handset instantaneously seizes an air slot in response to the basestation's general poll and provides the base with ID and user information. The basestation then transmits a specific poll for that handset to commence communications. At this point, the time slot is removed from common signalling mode and the handset and basestation enter traffic mode and begin communications on that time slot. (Rare collisions during time slot acquisitions are resolved by a CSMA/CD IEEE 802.3-like backoff procedure.)

Although not required for unlicensed band use, synchronisation of the basestations is used in the CCT system for public networks to increase capacity and is derived from the BSCs. GPS receivers in the BSCs will provide this timing initially, although network-based DS1 level timing can also be used by the system if this becomes ubiquitous for wide-scale applications.

26.6.4   RF Stability Requirements

For the basestation, CCT requires a guaranteed transmitter carrier frequency stability equal to or less than ±1 ppm over the operating temperature range. For the mobile stations and handsets, CCT requires a guaranteed transmitter carrier frequency stability equal to or less than ± 10 ppm over the operating temperature range. The basestation operating frequency may be synchronised to the digital network or other external sources for more precise timing in the digital network interface. Mobile units may also acquire synchronisation from the basestation for similarly precise frequency control.

26.6.5   Transmit Power

Mobile Station (MS)

Although the FCC permits up to 2 W EIRP for the handset, the nominal peak power output for the CCT handset is 600 mW into a dipole antenna, or 1 W EIRP. With CCT's 1.875 MHz version, the average handset power output in the 9.6 kb s^{-1} time slot mode is only 18.7 mW, permitting long durations between battery recharges. Even with this low value of handset transmit power, large PCS cell sizes can be accommodated because of the spread-spectrum modulation technique and the unique, bi-directional CCT diversity technique.

Basestation (BS)

Additionally, high-gain directional antennas can be incorporated economically into CCT's basestation design. The FCC rules permit up to 1640 W peak EIRP per RF channel for PCS basestations. Since the BS peak power output to its antenna is 2 W, the maximum permissible BS antenna gain (ignoring feed losses) is therefore limited by the FCC rule to 29.15 dB.

With CCT's TDMA radio system, only one electronic steering mechanism is required to address all users in one cell or cell sector, significantly reducing system complexity and cost. This capability contrasts with other systems that require as many electronic steering mechanisms as there are simultaneous users within each cell. With TDD, each user link within the cell can take advantage of the gain of the directional antenna at the basestation for both the reverse and forward links.

26.6.6   Power Control

Power control is not used within a cell to separate users as in a CDMA-only system. While the CCT method of TDMA does not require the strict-precision, adjustable control of transmitter RF power output necessary to resolve the "near–far" problem experienced by CDMA-only PCS systems, the CCT system does use a proprietary method of controlling transmitted RF power levels. However, dynamic power control is provided in simple 3 dB steps in the CCT system.

The BS transmits a power control command (PCC) at the beginning of its transmit period to each MS. The PCC provides a power control signal to adjust the MS power output level to a value just large enough to provide the required signal-to-noise plus interference ratio at the BS, as determined by the quality of the received signal from the MS. This is done in the CCT system for each time slot independently of other time slots in a frame, in order to satisfy the following requirements:

- Minimise interference to other cells, which may be operating on the same or adjacent RF channels
- Improve *C/I* ratios between cells
- Increase link quality
- Minimise interference to nearby OFS users
- Conserve battery life

26.6.7   Receiver Selectivity

The CCT systems use multiple techniques to mitigate multipath effects. For example, the wide spread-spectrum bandwidth coupled with CCT's unique antenna diversity techniques can maintain the received signal strength within a very tight range around the mean signal strength. As a result, significantly fewer excess link margins are required to maintain proper RF link performance.

The required receiver IF bandwidth is 1.875 MHz. This bandwidth is maintained through down-conversion to the intermediate frequency (IF) until despreading occurs and processing of the information content of the received signal begins. At that time, the receiving channel bandwidth is reduced to the level required for the digital information signal. By providing the proper RF, IF and baseband signal bandwidths, the receiver is optimised for both thermal noise and potential interferers. The same selectivity considerations apply to both the handset and basestation receiver designs.

26.6.8   Receiver Sensitivity

Each RF receiver in the CCT basestation system is designed for -104 dBm receiver sensitivity at 10^{-3} BER, and -105 dBm at 10^{-3} with FEC thermal noise. The mobile station and handset are designed for -102 dBm at 10^{-3} BER receiver sensitivity and -103 dBm at 10^{-3} with FEC thermal noise. CCT's system modelling is based on a $C/I = 8$ dB for co-channel interference. However, when coupled with CCT's unique antenna diversity capability, diversity is provided in both directions. Even with each RF receiver in a diversity configuration operating at 10^{-3} BER, the basestation can operate error-free in most circumstances.

26.6.9   Signal Quality Measurement of the Physical Interface

To reduce the required interference dynamic range rejection of the BS receivers, CCT uses an adaptive power-control algorithm (not to be confused with the precision power adjustment needed for IS-95 CDMA-only-based systems). Implementation of this power control feature uses measurement of the signal quality received from each handset.

Because the BS transmits sequentially to all handsets within its cell in different TDMA time slots, each BS-to-MS link may vary in path loss. Control of BS transmitter power is possible on a slot-by-slot basis in traffic mode. Each MS measures received BS signal quality from its own cell and uses this information to determine when to measure signal quality from adjacent cells and, further, when to request handoff to an adjacent cell when the received signal quality from its present BS falls below a preset threshold level.

26.7   Layer 2 Air Interface Specification

26.7.1   Channel Acquisition, Etiquette and States

An MS attempting to communicate with a BS shall seize at least one channel on that BS. This is accomplished by responding to a base general poll with an MS general response. The general response shall contain the unique personal identification number (PID) of the MS. If the BS receives the general response for this MS, the BS shall respond with a specific poll, which contains the PID of this MS. On reception of such a specific poll, the MS may transition into the traffic mode.

Until the MS receives a specific poll containing its PID, the MS shall not seize the channel and must wait a pseudo-random time based upon the PID and then try again in a manner similar to the backoff procedure of ANSI/IEEE 802.3. When the BS is ready to assign a channel to the MS and initiate communications with the MS, the BS shall issue a specific poll packet containing the PID of the MS.

26.7.2   Multiple Associated Signalling Time Slots/Frame

Normal time slot synchronisation shall be accomplished by timing. Both the BS and MS shall know which time slots have been assigned for communication. The MS shall send its signalling information to the MS in the first half of the TDD time slot; the BS shall send its signalling information in the second half of the TDD time slot. The MS shall synchronise and shall maintain timing synchronisation on the BS transmissions. The MS shall maintain timing synchronisation with the BS for up to 1 s in the absence of received BS transmissions.

If available, multiple time slots per frame shall be used for polling and signalling traffic. To accommodate this, the BS shall assign a temporary address, known as the correlative ID, to the MS on the first specific poll. This correlative ID shall then be carried in further signalling traffic from the BS to the MS. The MS shall search for this ID in all traffic. The MS can then respond to any signalling traffic time slot containing this correlative ID. Unused correlative IDs shall be maintained in a pool by the BS. When communication has ended between the BS and MS, the correlative ID shall be returned for re-use. Any available time slot may be used by the BS to continue signalling communications with the MS.

The last time slot used by the BS for signalling traffic will become the first time slot for bearer traffic use unless otherwise specified by slot mapping information given to the MS by the BS. If the BS returns to signalling traffic at a later time on the current channel, the correlative ID will still be effective, and the BS may use any available time slot for further control traffic

26.7.3 Superframe Structures

Asymmetric Channels

Traffic flow between the BS and MS may be either symmetric or asymmetric. The total number of bits per TDD time slot shall remain constant in either case. The flow shall be controlled by the BS acting upon the bandwidth request bit in the MS-to-BS traffic header. The normal flow is symmetric, with an equal number of bits (except for the header bits) assigned in each direction. The bandwidth grant bits in the header of the BS-to-MS traffic channel shall establish the actual number of bits to be used in the next time slot of the channel.

The BS shall assign TDD time slot bandwidth (number of bits) using the following procedure:

1. If only the BS requires additional bandwidth, then the BS shall be granted the additional bandwidth for the next time slot assigned to that MS.
2. If only the MS requires additional bandwidth, then the MS shall be granted the additional bandwidth for the next time slot assigned to that MS.
3. In all other cases, symmetric bandwidth shall be granted for the next available time slot assigned to that MS.

Broadcast Channels

The asymmetry of the channel may be taken to its logical extreme by granting the entire bandwidth of each time slot to the BS to produce a broadcast channel. The nature of this channel shall be indicated by the bandwidth grant bits in the BS time slot header (they apply to the next time slot in the channel). Multiple simultaneous broadcast channels shall be supported. During broadcast, the bits normally used for the D channel shall be used as a broadcast identifier. Since this occurs in the same position as the correlative ID, the difference in usage is signalled by the bandwidth grant bits.

Superchannels

The ability to assign multiple time slots in the frame may be negotiated for and assigned to an individual MS. The negotiation may take place at any time via signalling traffic. The assigned TDD time slot, if available, shall be communicated by the BS to the MS via the OTA map type and OTA map information elements. Channel synchronisation shall be maintained by the MS based on frame timing.

The handover procedure shall account for the multiplicity of time slots per channel assigned to the transferring MS. A BS shall have the appropriate number of time slots available to become a candidate as a terminating BS for handover. The time slots need not be available in the same positions in the frame as those in the originating BS.

Logical Subchannels

An MS need not be granted a time slot in every frame. Time slots may be granted in frames separated by an integral number of intervening frames. The maximum limit on the separation of frames allocated to a single MS is one time slot every 25 frames or every 0.5 s. This would yield a channel with a raw full-duplex rate of $384 \, \text{bit s}^{-1}$.

26.7.4   Multiple Mode Traffic

A single MS may have multiple connections established through the basestation to the network via multiple channels. One channel may, for example, be assigned to audio traffic while other channels are devoted to data traffic.

26.8   Layer 3 Air Interface Specification

26.8.1   Handoff

In the CCT handoff procedure, the system can perform either a seamless "make before break" handoff or a "break before make" handoff in situations where all communications with a basestation are lost before a new connection is established. In the CCT "make before break" approach, the terminal can listen for other basestations during idle time slots and determine which basestations are within range and are candidates for handoff.

Intra-Cluster Handoff

When the handset determines that a handoff is appropriate, the terminal acquires an air channel on the new basestation and notifies the BSC to switch the incoming phone line from the old base to the new base.

The entire handoff process typically requires only 4–32 ms and, given proper RF coverage, can be achieved in most circumstances without missing a single packet. Even "break before make" handoff can be accomplished typically in 16–250 ms, which is barely audible to the user.

In the CCT system, the speed of handoff is the same regardless of the size of the cell or the speed of the mobile unit. This enables operators to deploy miniaturised cell sites (e.g. with 250 ft (~75 m) radii) for greater capacity without impact on mobile service. The handoff functionality is extremely low-cost because the processing in the BSC is simple and results from a mobile-directed command.

Handoffs required by a lost or high-interference air channel are handled very quickly by the CCT system through the terminal's ability to reacquire the original or an alternative basestation in 16–250 ms, even if no information is available to it. In contrast, losing a link even briefly in a traditional wireless architecture such as AMPS, IS-136 digital cellular, or DCS-1900 becomes a "dropped call."

To determine link quality (typically measured by the receive signal quality index (RSQI)), the CCT system uses the receive signal strength index (RSSI) to measure channel quality, and measurements on both frame errors and error types. When the RSQI scanning threshold is reached, the handset first scans for an alternative time slot in the same RF channel, then performs a time slot interchange (TSI). This approach provides a dramatic improvement in the cell's average C/I ratio, which is valuable for re-use planning.

In cases where TSI cannot improve the channel's performance (such as when nearing the limit of the cell's range), the handset scans for the frequencies in adjacent basestations. From the common signalling information transmitted by each basestation, the handset can determine its link quality and available capacity. When the handoff threshold is reached, the handset will change frequencies and direct the new basestation to notify the BSC to bridge the call to the new basestation channel. When appropriate, the handset further directs the BSC to shut down the bridge, completing the call switch.

Inter-Cluster Handoff

Inter-cluster handoffs between basestations managed by different BSCs are managed in the CCT system via the network platform chosen by the PCS operator. This may be through the GSM or AIN switching infrastructure in the MSC approach or the PSTN in the intelligent basestation approach.

Over the air, intra- and inter-cluster handoffs are essentially identical. The only difference is the signalling from the BSC to the main switch (MSC or PSTN). The signalling from the handset to the BSC passes through to the MSC, which in turn signals the new target BSC and the appropriate BS for that handoff. Although this signalling can add as much as 100 ms to the handoff, they will occur very infrequently, even on a large network, because a single BSC can handle up to 32 basestations.

Time Slot Interchange (TSI)

The CCT system provides for the rare instances when two users in different cells on the same frequency may experience degradation of their radio links. In these cases, the CCT system performs a time slot interchange (TSI) where one or both of the conflicting handsets are assigned different time slots within its basestation to eliminate the collisions between the two users. This is the time-domain (TDMA) equivalent of dynamic channel allocation and allows for much lower C/I figures than are normally expected. The C/I ratio required for frequency re-use is based on the average user rather than the worst-case user.

26.8.2 Registration

Registration is the process of informing the wireless network of the presence of a subscriber in a specific geographic area. For quick and efficient processing, the CCT system offers multiple levels for registering mobiles: at the basestation, the basestation controller and the VLR/HLR in the network.

26.8.3   Privacy and Authentication

The CCT system provides a "pipeline" and the necessary transport mechanisms for various P&A-related applications. GSM-based P&A is currently developed to be used over the CCT system. Both the GSM and an Omnipoint proprietary registration procedure are currently supported. This Omnipoint proprietary system is used to reduce the paging loads on the system.

26.8.4   Voice and Data

Voice, circuit-switched data and packet-switched data network layer protocols are supported.

26.8.5   Zones, Information Elements, Procedures and Algorithms

Over three zones, 80 different information elements, as well as scores of specific procedures and algorithms have been developed within the Omnipoint "Notes" protocol to support advanced features and capabilities.

26.8.6   Additional Standards Documentation

The IS-661 standard also includes sections covering message formats, information element formats and message flow diagrams for numerous features.

26.9   Network Options

CCT basestations connect to a basestation controller (BSC) for network interface implementations, including those based on GSM and AIN. CCT basestations can also connect directly to the PSTN in intelligent basestation deployments while performing all BSC functions. In the typical configuration, the basestation controllers provide for local registration, database management and switching for fast handoff. The basestation controller connects to either the PSTN or a PCS switch connects to the BSC for call delivery and outgoing calls.

26.9.1   Public/Private System Operation

Users of CCT handsets can operate on private systems as well as on either GSM-based network or AIN networks – see Figure 26.10. The CCT system can also support various service implementations through interconnection with an LEC network, CATV network, new build (PCS operator, PCS architecture) or as an extension to a cellular system (with different RF equipment, but some co-located equipment). In all cases the CCT system provides for full PCS functionality. While CCT can accommodate these limited PCS applications, the system is designed to implement a fuller vision of PCS services.

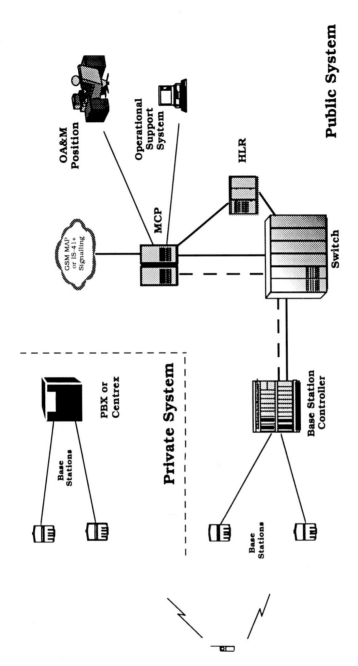

Figure 26.10   Public–private network operations.

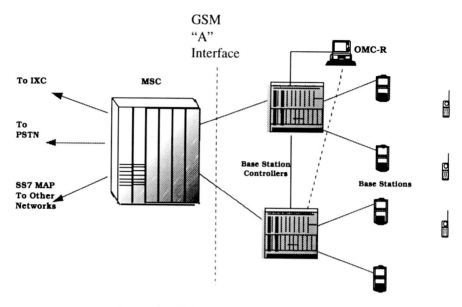

Figure 26.11 CCT connection with a GSM network.

CCT private systems will operate in unlicensed bands and can use either the local, private equivalent of a basestation controller for larger systems or connect the basestations directly to a PBX or Centrex system.

26.9.2 GSM Networks

A similar architecture is used for GSM, except that the features and functionality provided by the LEC network are provided by a GSM network infrastructure owned by the PCS operator – see Figure 26.11. In this implementation, the GSM architecture interconnects through the defined "A" interface. Features and functionality of GSM are passed to and from the CCT system over this network interface in a manner that is transparent to the end-user.

26.9.3 AIN-Based Networks

Direct interconnection to an AIN network using LEC, cable access provider (CAP), or independently owned switches is possible from the CCT system (see Figure 26.12). The specific services provided by these networks are determined by the service operator. Configurations that require no service support from IXC or LEC networks are possible with the CCT system via an appropriate switching/services platform. Database interfaces for AIN services, roaming and network management may be used, if available, from either the LEC or a third-party provider.

The LEC or CAP central office can provide connectivity from the basestations to the basestation controllers via standard lines or ISDN. This gives the PCS

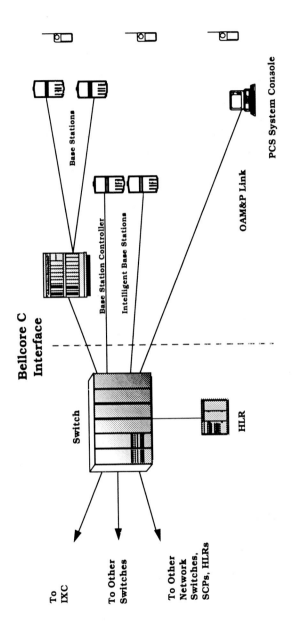

Figure 26.12 CCT connection with an AIN network.

operator the option of using LEC network resources for a variety of features, call routing and subscriber record management capabilities. In all except fully contained network configurations, the CCT system can interconnect to the LEC at the local CO (central office) level, providing a level of integration that allows the PCS operator to leverage the embedded LEC network for many of the PCS network requirements.

CCT supports AIN 0.1 (initially) through AIN 0.2 (or higher as AIN evolves) for full ISDN interconnectivity, including network database support and handoffs via the LEC central office. Signalling between network elements will consist primarily of standard signalling protocols, including SS7.

Omnipoint has already tested ISDN basestation interfaces, including direct interconnection to an AIN-based CO switch. Support for ISDN upgrades will be provided by Omnipoint as supported by NISDN-3 for interconnection to the PSTN (as this becomes available from the switch vendors). Omnipoint is also working with several vendors to provide SS7 interconnectivity to ensure SCP (service control point) involvement in the network architecture for PCS, while also allowing for enhanced network services (i.e. CLASS, other AIN features) to be provided to end-users.

26.9.4   Cable TV Networks

The CCT system offers PCS basestations that are miniaturised for installation inside standard cable TV amplifier boxes. Omnipoint has already developed and tested several different approaches including interface to analogue RAD/RASP systems and the use of digital transport mechanisms. One of the tests used T1 and FT1 digital multiplexer outputs from the CATV network, and basic rate (BRI) ISDN links to transport digital channels. This basic ISDN transport capability is currently in testing as a prototype and will be available to Omnipoint CCT customers in 1996. Omnipoint is providing the basic digital link transport system (DLTS) interface for customer testing.

Omnipoint's CCT-based cable antenna transceiver (CAT) extends coverage from a basestation in cable TV deployment. This device was successfully tested and shown to provide cost-effective deployment of PCS to cable subscribers, while providing full functionality including handover and vehicular mobility. For example, one Version 1.0 Omnipoint CCT basestation and four CAT modules covered 153 homes in a residential neighbourhood. Using Version 2.0, one Omnipoint CCT basestation can cover an area of roughly 10 square miles ($\sim$30 km^2) with CAT modules used as localised enhancers within that area. Additionally, Omnipoint's CCT basestations and/or CATs can be mounted at nodes.

The CCT system does not require the modification of existing PSTN network equipment. Omnipoint is working with potential PCS providers, equipment suppliers and Bellcore to refine further the specifications for the "C" generic interface operating in AIN 0.2 and NISDN-3. No changes are required to the CATV networks specifically for an Omnipoint CCT system. Obviously, any CATV network operator must introduce a switch architecture, two-way amplifiers, redundancy and other features to provide full PCS service. Additionally, some form of a CAT-like system must be deployed to use the coax portion of the CATV network to extend coverage from a basestation.

References

[1]   "Empirical Formula for Propagation Loss in Land Mobile Radio Services", M Hata, IEEE
       Transactions on Vehicular Technology, August 1980
[2]   "Field Strength and its Variability in VHF and UHF Land Mobile Service", Okumura, 1968.
       Reprinted in "Land Mobile Communications Engineering", IEEE Press, 1984
[3]   "Indoor Radio Communications for Factories of the Future", T Rappaport, IEEE Communica-
       tions Magazine, May 1989

Note

1   CCT cell sizes can be as small as 100 ft (~30 m) or as large as 20 miles (~30 km) in diameter.

Glossary of Acronyms

This glossary provides short definitions of a range of abbreviations and acronyms in use within the cordless telecommunications field; many of the terms are defined in greater detail within this volume.

ACCH	associated control channel
ACELP	algebraic code-excited linear prediction, vocoder
ACK	acknowledgement protocol
ACTE	Approval Committee for Telecommunication Equipment
ACW	address code word
ADM	adaptive delta modulation
ADPCM	adaptive differential pulse-code modulation
AGC	automatic gain control
AIN	advanced intelligent network
ALT	automatic link transfer
AM	access manager
AMPS	American Mobile Phone System – US cellular standard
API	application programming interface
ARA	alerting/registration area
ARI	access rights identifier
ARIB	Association of Radio Industries and Businesses (Japan)
ARQ	automatic repeat request
ATIS	Alliance for Telecommunications Industry Solutions (USA)
AWGN	additive white Gaussian noise

B	echo balance return loss
B channel	user information bearer channel, $64\,\text{kb s}^{-1}$, in ISDN
BABT	British Approvals Board for Telecommunications
BCCH	broadcast channel
BCT	business cordless telephone
BER	bit error ratio
BMC/BMD	burst mode controller/device
BPSK	binary phase shift keying, modulation
BRA	ISDN basic rate access
BS	basestation – the fixed radio component of a cordless link, single-channel or multichannel; term also used in cellular radio

BS6833	a standard for digital cordless telephones allowing for proprietary air interfaces (mainly specifying telephony-related aspects) (UK)
BSC	basestation controller
BT	bandwidth × time product for GMSK modulation
BTR	bit timing recovery, used to permit synchronisation across a radio link
CAI	Common Air Interface standard, as in the CT2 standard
CAS	cordless access service(s)
CATV	cable television
CBCS	cordless business communication system, supporting wireless PABX and data applications, for example, in the business environment
CCCH	common control channel
CCF	cluster control functions
CCFP	common control fixed part
CCH	control channel
CCIR	Consultative Committee for Radio – part of the previous ITU structure (now ITU-R)
CCITT	Consultative Committee for Telecommunications – part of the previous ITU structure (now ITU-T)
CCT	Combined CDMA/TDMA (IS-661) standard (USA)
CDC	cordless data controller
CDMA	code-division multiple access
CEC	Commission of the European Communities
CELP	code-excited linear prediction, vocoder
CEN	Comité Européen de Normalisation
CENELEC	Comité Européen de Normalisation Electrotechnique
CEPT	Conference of European Posts and Telecommunications
CFP	cordless fixed part, e.g. a cordless telephone basestation
CHMF	synchronisation pattern used in CT2
CHMP	synchronisation pattern used in CT2
CI	channel identity sequence, used for over-the-air identification of a radio transmission
CI	common interface, type of specification
CISPR	Comité International Spécial Perturbations Radio
CLR	connection loudness rating
COST	Cooperation in Science and Technology Programme (Europe)
COST 231	COST committee dealing with future mobile systems
CPBX	cordless PBX – same as a wireless PBX
CPE	customer premises equipment
CPFSK	continuous phase frequency shift keying, modulation
CPP	cordless portable part – the cordless telephone handset carried by the user
CR	carrier recovery
CRC	cyclic redundancy check
CRFP	cordless radio fixed part
CS	cell station, a PHS basestation

CS-ACELP	conjugate structure algebraic code-excited linear prediction, vocoder
CS2, CS3	intelligent network, capability set 2, 3
CSF	cell site function
CSMA	carrier sense multiple access
CSMA/CA	carrier sense multiple access with collision avoidance
CT0	the original analogue VHF/LF cordless telephone technology, as used in the UK, France and elsewhere
CT1	Cordless Telephone Generation One – in Europe 900 MHz analogue FM; also known as CEPT/CT1; sometimes in the UK the term was and occasionally still is used to refer to the CT0 technology
CT1+	later version of the CEPT CT1 standard, with additional channels
CT2	Second Generation Cordless Telephone–Digital – originally developed in the UK and subsequently accepted by ETSI as an I-ETS
CT2plus	Canadian variant of CT2
CT3	early Swedish digital cordless standard
CTM	cordless terminal mobility, an application concept and an ETSI project
CTN	cordless telecommunication network
CTR	common technical regulation
CVSDM	continuously variable slope delta modulation
D channel	control and information data channel, 16 kb s^{-1}, in ISDN
DAM	DECT authentication module
DASS	digital access signalling system
DCA	dynamic channel assignment
DCE	data communications equipment
DCS	digital cellular system (see DCS1800)
DCS	dynamic channel selection
DCS1800	Digital Cellular System 1800 MHz – the 1800 MHz version of the GSM digital cellular radio standard
DCT	digital cordless telephone
DCT900	early CT3 standard cordless telephone product from Ericsson
DCW	data code word
DECT	Digital Enhanced Cordless Telecommunications – the ETSI standard (originally, prior to 1995, Digital European Cordless Telephone)
DFS	DECT fixed system
DLC	data link control layer, protocol layer
downlink	transmission link from the fixed part to the portable part
DPCM	differential pulse-code modulation
DPNSS	inter-PABX signalling system
DPS	DECT portable system
DQPSK	differential QPSK modulation
DRT	diagnostic rhyme test, used in speech quality measurements
DSP	digital signal processing

DSS1	ISDN signalling protocol
DTAAB	DECT Type Approval Advisory Board
DTI	Department of Trade and Industry (UK)
DTMF	dual tone multiple frequency, audio tone signalling system
duplex	simultaneous two-way communication
DVSDM	digitally variable slope delta modulation

EC	European Commission
ECMA	European association for standardising information and communication systems
ECTEL	European association of the telecommunications manufacturing industry
EFR	enhanced full-rate codec, developed for US PCS1900
EMC	electromagnetic compatibility
EOC	embedded operations channel
ERC	European Radiocommunications Committee
ERMES	European Radio Messaging Standard – digital paging standard specified by ETSI
ERO	European Radiocommunications Office
ERP	earpiece reference point
ERP	effective radiated power
ESPA	European Selective Paging Association
ETR	ETSI Technical Report
ETS	European Telecommunications Standard
ETSI	European Telecommunications Standards Institute

FACCH	fast associated control channel
FCC	Federal Communications Commission (USA)
FCS	frame check sequence
FCS	free channel search
FDD	frequency-division duplex
FDMA	frequency-division multiple access
FEC	forward error correction
FH	frequency hopping
FHMA	frequency hopping multiple access
FPLMTS	Future Public Land Mobile Telecommunications System
FSK	frequency shift keying
FT	fixed termination

GAP	generic access profile of DECT
GFSK	Gaussian-filtered FSK modulation
GIP	GSM interworking profile of DECT
GMSK	Gaussian-filtered minimum shift keying
GoS	grade of service
GPRS	generalised packet radio service, of GSM
GPS	Global Positioning System

GSM	Global System for Mobile – the digital cellular radio system specified by ETSI and now adopted in many countries worldwide
handoff	procedure whereby communications between a cordless terminal and a basestation are automatically routed via an alternative basestation (or possibly via an alternative channel to the same basestation) when this is necessary and appropriate to maintain or improve communications
handover	another term for handoff
HDLC	high-level data link control – type of protocol
HDSL	high-density subscriber link
HIPERLAN	High Performance Radio Local Area Network
HLR	home location register, used in cellular systems and which might be used in future enhanced public cordless systems
I-ETS	interim ETS
IAP	ISDN access profile, of DECT
IEC	International Electrotechnical Commission
IEEE 802.11	a spread-spectrum data communication standard (USA)
IF	intermediate frequency
IMEI	international mobile station equipment identity (GSM feature)
IN	intelligent network
INAP	intelligent network application part
IPEI	international personal equipment identity (DECT feature)
IS54	a digital AMPS standard (USA)
IS95	a CDMA digital cellular standard (USA)
ISDN	integrated services digital network
ISM	industrial, scientific and medical frequency band
ISO	International Standards Organisation
ISPABX	integrated services PABX
ITSTC	Information Technology Steering Committee
ITU	International Telecommunications Union
IWP	interworking profile
IWU	interworking unit
JLR	junction loudness rating
JTC	TIA/T1 Joint Technical Committee – US PCS standardisation committee
LAI	local area identifier
LAN	local area network
LAPB	link access protocol–bearer channel
LAPD	link access protocol–data channel
LAPR	link access protocol radio

LBT	listen before talk – part of the US UPCS spectrum etiquette, FCC Part 15 Subpart D rules
LCD	liquid-crystal display
LCI	local CFP identifier
LD-CELP	low-delay code-excited linear prediction, vocoder
LDB	location database
LDM	linear delta modulation
LE	local exchange
LEC	local exchange carrier
LFA	lowest frequency below threshold algorithm
LIA	least interference algorithm
LID	Link identification code
LLME	lower layer management entity
LNA	low-noise amplifier
LPC	linear predictive coding
LR	location register (as in cellular radio)
LR	loudness rating

MAC	medium access control, protocol layer
MAHO	mobile assisted handover
MAP	mobile access part, of the SS7 protocol
MC	marker channel
MCHO	mobile controlled handover
MDT	mobile data terminal
microdiversity	diversity between antennas with relatively small spacing (of order quarter wavelength) to counter multipath fading effects
MKK	radio equipment inspection and certification institute (Japan)
MMI	man–machine interface
MOS	mean opinion score, used in speech quality measurements
MoU	Memorandum of Understanding
MPEG	Motion Pictures Expert Group – developing video codec standards
MPT1317	a standard for data signalling over a radio path (UK)
MPT1334	a standard for digital cordless telephones allowing for proprietary air interfaces (mainly specifying radio-related aspects) (UK)
MPT1375	a standard for Common Air Interface (CAI) digital cordless telephones (UK)
MRP	mouthpiece reference point
MSC	mobile services switching centre, as in cellular radio
MSK	minimum shift keying modulation
MUX1	traffic channel multiplex in PCI; same as MUX1.4 in CT2
MUX1.2	multiplex within the CT2 standard containing 64 B-channel bits and 2 D-channel bits
MUX1.4	multiplex within the CT2 standard containing 64 B-channel bits and 4 D-channel bits
MUX2	multiplex within the CT2 standard containing 32 D-channel bits and 32 SYN-channel bits

| MUX3 | "long" 10 ms burst transmitted by a CT2 handset when trying to establish a call |

NACK	not acknowledge – a form of protocol where action is taken if no acknowledgement is received
NAFTA	North American Free Trade Area – USA, Canada and Mexico
NAMPS	narrow-band AMPS cellular standard
NCHO	network controlled handover
NET	Norme Européenne de Télécommunications – a type of standard
NMT	Nordic Mobile Telephone – an analogue cellular standard
NPAG	North American PCS Advisory Group
NRZ	non-return to zero
NTIA	National Telecommunications Information Administration (USA)
NWK/NWL	network layer, protocol layer

O&M	operations and maintenance
OAM	operations, administration and maintenance
OAM&P	operations, administration, maintenance and provisioning
OCDMA	orthogonal CDMA
OELR	overall echo loudness rating
OFS	operational fixed services
OFTA	Office of the Telecommunications Authority (Hong Kong)
OFTEL	Office of Telecommunications (UK)
OLR	overall loudness rating
OSI	Open Systems Interconnection – ISO standard for communications, a seven-layer reference model for protocol specification
OTAR	over-the-air registration

P&A	privacy and authentication
PA	portable application
PA	power amplifier
PABX	private automatic branch exchange
PACS	Personal Advanced Communication System
PACS-UA	PACS variant A for the US unlicensed bands; related to PHS
PACS-UB	PACS variant B for the US unlicensed bands
PAP	public access profile, of DECT
PARK	portable access rights key
PBX	same as PABX
PCH	paging channel
PCI	Personal Communications Interface – US unlicensed standard based on CT2
PCM	pulse-code modulation
PCMCIA	Personal Computer Manufacturers' Card Industry Association – a standard for plug-in PC cards, now known also simply as PC Card

PCN	personal communications network
PCS	personal communications service
PCS1900	a cellular standard, 1900 MHz version of DCS1800 (USA)
PDA	personal digital assistant
PDC	Personal Digital Cellular standard (Japan)
PHL	physical layer, lowest protocol layer
PHS	Personal Handyphone System – Japanese cordless standard
PIC	personal intelligent communicator
PINX	private integrated services network exchange
PISN	private integrated services network
PLL	phase-lock loop
PLMN	Public Land Mobile Network
PMR	private mobile radio
POTS	plain old telephone service
PP	portable part
PRA	ISDN primary rate access
PS	personal station
PSK	phase shift keying modulation
PSTN	public switched telephone network
PT	personal telecommunications
PT	portable termination
PT	Project Team (in ETSI)
PTN	personal telephone number
PTO/PTT	term used to refer to a national telecommunication administration
PWT	Personal Wireless Telecommunications – US cordless technology, a variant of DECT
PWT-E	PWT (Enhanced) – enhanced version of the PWT technology for operation in the licensed PCS bands
QCIF	Quarter Common Interface Format
QDU	quantisation distortion unit
QPSK	quadrature phase shift keying modulation
QSAFA	quasi-static autonomous frequency assignment
QSIG	inter-PABX signalling protocol
quantisation	process of representing samples of an analogue waveform by the nearest whole number of predefined voltage steps
RACE	Research and development in Advanced Communication technologies in Europe – research initiative instigated by the CEC
RAP	radio local loop access profile of DECT
RBOC	Regional Bell Operating Company
RCR	Research and Development Centre for Radio Systems (Japan)
RES 3	Technical Subcommittee, Radio Equipment and Systems 3, of ETSI; responsible for the specification of cordless systems
RFI	radio frequency interference

RFP	radio fixed part – same as cordless fixed part
RLAN	radio local area network
RLL	radio local loop
RLR	receiving loudness rating
RP	radio port
RPCU	radio port controller unit
RPE-LTP	regular pulse excitation–long term predictor – type of LPC speech coding used in the GSM digital cellular radio system
RSSI	received signal strength indication

SACCH	slow associated control channel
SAW	surface acoustic wave
SBC	system broadcast channel
SCCH	specific cell channel
SCI	synchronisation, control and information
SCP	service control point
SDL	system description language
SDLC	synchronous data link control
sidetone	attenuated component of the speaker's voice signal fed back to his or her own earpiece to provide confidence that the equipment is functional
SIM	subscriber identification module, of GSM
simplex	one-way communication
SLR	sending loudness rating
SMS	short message service, of GSM
SMT	surface mount technology
SNR	signal-to-noise ratio
SoHo	small office/home office
SPCS	Satellite Personal Communications System
SQER	signal-to-quantising-error ratio
STMR	sidetone masking rating
SU	subscriber unit
SYN	channel used in CT2 to allow synchronisation to be established
SYNCD	synchronisation pattern used in CT2
SYNCF	synchronisation pattern used in CT2
SYNCP	synchronisation pattern used in CT2

TACS	Total Access Communication System – an analogue cellular system, based upon AMPS, the US system
TAL	telephone acoustic loss
TBR	Technical Basis for Regulation – a European standard mechanism
TCH	traffic channel
TCL	telephone coupling loss
TDD	time-division duplex
TDM	time-division multiplex
TDMA	time-division multiple access
TE	terminal equipment

TELR	talker echo loudness rating (nearly the same as OELR)
TIA	Telecommunications Industry Association (USA)
TIM	telephone identity module
TQFP	thin quad flat package
TRAC	Technical Recommendations Applications Committee
TTC	Telecommunication Technology Committee (Japan)

UIC	user information channel
UMC	universal mobile communicator
UMTS	Universal Mobile Telecommunication Service – being standardised by ETSI
UNI	user network interface
UPCH	user packet channel
UPCS	unlicensed PCS
uplink	transmission link from the portable part to the fixed part
UPT	universal personal telecommunications
USP	unique selling proposition
UW	unique word, used for synchronisation and/or identification across a radio link

VCO	voltage-controlled oscillator
VDU	video display unit
VPN	virtual private network

WACS	Wide Area Communication System – wireless access technology developed by Bellcore, which has formed the basis for PACS
WCPE	wireless customer's premises equipment
WLAN	wireless local area network
WLL	wireless local loop
WPABX	wireless private automatic branch exchange
WPBX	wireless private branch exchange
WRC	World Radio Conference – part of the ITU activity, superceding the previous WARC process
WRS	wireless relay station
WUPE	wireless user premises equipment

Contributors' Biographies

Editor

Walter Tuttlebee

Roke Manor Research, Romsey, UK

Dr Walter Tuttlebee graduated from the University of Southampton with a BSc in electronics in 1974. Following PhD studies and postdoctoral research, he joined Roke Manor in 1979, where he contributed to and led feasibility studies, project definition studies and development teams on several successful radio communication systems.

During the 1980s, as a Chief Engineer within the Radio Communications Division, Walter was responsible for a wide range of R&D, with particular interests in the field of personal communications. He has presented many invited papers at conferences on this theme, as well as undertaking consultancy assignments for the European Commission and telecommunications operators. Walter has also overseen Roke Manor's ongoing work on UMTS in RACE and ACTS, and contributed to one of the UK PCN licence applications. He is a Senior Member of the IEEE (SMIEEE) and a Fellow of the IEE (FIEE).

In 1992 Walter graduated with an MBA degree from the Cranfield School of Management, Bedford, UK. Reflecting this, his role has evolved to one of technology and business development for the growing Radio Communications Business Unit at Roke Manor, in which role he has pioneered new initiatives in low-cost digital audio broadcasting (DAB) receivers and in satellite communications.

Contributors

Dag Åkerberg

Ericsson Radio Systems, Stockholm, Sweden

Dr Dag Åkerberg has been, and continues to be, a pioneer and statesman of cordless technology. He joined Ericsson in 1971, after obtaining his Dr Sc (Tekn Lic) from the Royal Institute of Technology in Stockholm, and, till 1984, was Technical Manager for Paging Systems development. During the 1980s his main activities were studies, specifications and international standardisation work related to wireless office communications, in particular relating to DECT. Between 1989 and 1991 he was seconded to ETSI as part of the Project Team 10 supporting the specification of the Digital Enhanced Cordless Telecommunications (DECT) system. During the 1990s, in response to the PCS initiative in North America, he has helped formulate the DECT-based PWT standards, in addition to continuing to play an important role in the further development of the DECT standards, as chairman of the ETSI RES 03 Radio and Speech Experts' Group.

Gary Boudreau

Northern Telecom, Ottawa, Canada

Gary Boudreau obtained a BASc in electrical engineering from the University of Ottawa in 1982, an MSc in electrical engineering from Queen's University in 1984 and a PhD in electrical engineering from Carleton University in 1989. He has over 10 years of industrial experience in satellite and mobile communications systems. He has worked on the development of indoor wireless PBX systems and most recently was the Editor for the development of the PCI standard. Currently he is with the wireless division of Northern Telecom in Ottawa, Canada, and leads an advanced technology development team. His current interests include modulation, coding, digital signal processing and CDMA.

Andrew Bud

Olivetti Telemedia, Ivrea, Italy

Andrew Bud is responsible for network technology planning in Olivetti Telemedia, Olivetti's public fixed operator. Before that, he founded and ran Olivetti's DECT product business, based on its NET[3] wireless LAN. He also authored the technical chapters of Omnitel's winning bid for the second GSM licence in Italy. He joined Olivetti in 1988 from PA Technology in Cambridge, England, where he had been centrally involved in the invention and development of the Ferranti Zonephone CT2 system. A Member of the IEE, he has a first-class honours degree in engineering from the University of Cambridge. He holds a number of patents in the area of cordless telephones.

Andrew was a pioneer of DECT. From 1988 to 1992 he chaired the ETSI RES 03 Network Subcommittee, which designed DECT's network layer. Since 1993 he has been Chairman of the Data Working Party, creating standards for multimedia services.

Herman Bustamante

Stanford Telecom, Sunnyvale, CA, USA

Herman Bustamante is a Technical Director at Stanford Telecom's Wireless and Cable Products Division in California. He is one of the key inventors of Stanford Telecom's Orthogonal CDMA (O-CDMA) patent and several other related patents pending. He has also served as the Chair of the O-CDMA Standard Subgroup of TIA TR 41.6, Wireless User Premises Equipment. Herman has close to 40 years of experience in design and development of wireless and satellite communications systems. His recent effort has focused on PCS, wireless local loop, wireless LAN and wireless cable. He joined Stanford Telecom in 1974 and received his BSEE (1956) from the University of California at Berkeley and MSEE (1964) from the University of Santa Clara, California.

Ed Candy

Hutchison, Hertford, UK

Ed Candy is currently the Director of Technology for Orange, the UK PCN operation of Hutchison. In his previous role as Technical and Operations Director of BYPS, he oversaw the conception, implementation and conclusion of the Rabbit telepoint business in the UK. Previously Corporate Technology Manager for Philips Radio Communication Systems, he was responsible for radio system development and, while in that position, established the original CEC collaborative research programme into UMTS, RACE Mobile. Prior to coming to England in 1987, Ed was State Manager for Philips Communications Systems in New South Wales, Australia, and responsible for radio, computer and telecommunication systems.

Horen Chen

Stanford Telecom, Sunnyvale, CA, USA

Dr Horen Chen is Vice President of Wireless Broadband Products at Stanford Telecom, California, responsible for planning, design, development, production and sales of advanced wireless broad-band products. He has also recently served as Chairperson of DAVIC's Wireless Delivery Systems Subgroup. He joined Stanford Telecom in 1975 and has held design and system engineering responsibilities for Unlicensed PCS, Cordless Telephony, Wireless PBX and Wireless Local Loop systems using Stanford Telecom-patented Orthogonal CDMA technology. Dr Chen received his B.S. (1970) in telecommunications engineering from National Chiao Tung University, Taiwan, MSEE (1972) from Rensselaer Polytechnic Institute, and PhD (1976) in electrical engineering from Stanford University.

Dominic Clancy

Philips Semiconductors, Zürich, Switzerland

Dominic Clancy is currently International Product Marketing Manager with Philips Semiconductors, having previously held several senior-level positions in both consulting and manufacturing industry in the mobile telecommunications field, especially cordless. He wrote the first study into cordless telephony markets, both digital and analogue, in 1987. He has worked with the European Commission and has been actively involved in various watershed developments such as the ETSI strategic review of mobile communications and the PCN licence applications in the UK. He has an MBA from Bradford University and a BA in philosophy from the University of Liverpool.

Graham Crisp

GPT, Nottingham, UK

Graham Crisp has held a variety of posts concerned with the development and evolution of public, private and mobile telecommunication networks and systems' architectures. Since 1988, he has worked for GPT's Central (Corporate) Engineering Research function. As a senior consultant, he has a special responsibility for telecommunication network evolution and mobile services. Among other tasks during this period with GPT, he has been responsible for a number of research projects related to second- and third-generation mobile telecommunication systems. He chaired the ETSI committee responsible for the standardisation of data and telematic services for the GSM system. As part of Graham's current responsibilities, his services are provided by GPT to manage the ETSI Cordless Terminal Mobility (CTM) Project.

Phil Crookes

AT Kearney, London, UK

Dr Phil Crookes is a Consultant with AT Kearney who specialises in the marketing of telecommunications solutions, particularly in mobile and cordless markets. His 10 years' experience includes positions in PABX sales and marketing and consultancy to users considering the implementation of cordless PABX systems. He previously worked for PA Consulting where he undertook many marketing consultancy assignments in the cordless and mobile market, including a pathfinder study on DECT for the Commission of the European Community.

Robert Harrison

PA Consulting, London, UK

Robert Harrison is a Managing Consultant with PA Consulting Group, which he joined from the IT and telecommunications industry. His work in the industry included sales, marketing and engineering in the PBX sector, and since joining PA he has worked with many of the major players in the telecommunications industry advising on market developments, and developing market and business strategies. He led the study on DECT for the European Commission, and he was the author of the report "Study of the International Competitiveness of the UK Telecommunications Infrastructure", published in 1995 by the UK Department of Trade and Industry.

Anthony Peter Hulbert

Roke Manor Research, Romsey, UK

Peter Hulbert, CEng, MIEE, received his BSc in electronic engineering in 1974 and joined Roke Manor Research in 1975. Currently a Consultant within the Radio Communications business unit, he has contributed over a very wide technical base to numerous research and development programmes. Specifically, he was responsible for leading the Cordless Telephone research team at Roke Manor during the period 1982 to 1988. During the 1990s he has contributed significantly to the development of Roke Manor's advanced CDMA technology, holding several patents in this field. He has published many technical papers.

Javier Magaña

Advanced Micro Devices, Austin, TX, USA

Javier Magaña is currently Member of Technical Staff in the Communication Products Division at the Austin, Texas, site of Advanced Micro Devices, where he is working on radio architectures for digital cordless products for the European and North American markets. He was previously involved in digital cordless systems silicon design and definition. He holds a BSc from Rice University, Houston, Texas.

Neil Montefiore

MobileOne, Singapore

Before his appointment as Chief Executive Officer of MobileOne Singapore, Neil Montefiore was the Director, Mobile Services, at Hong Kong Telecom CSL Ltd, the largest cellular operator in Hong Kong. Prior to this he was Managing Director of Chevalier Telepoint, Hong Kong, from 1991 until 1995.

Between 1976 and 1983 Mr Montefiore held various marketing and engineering management positions with Cable and Wireless, based in Hong Kong, Bahrain, Saudi Arabia and the UK. In 1983 he joined Cable and Wireless Systems Ltd, a wholly owned subsidiary of Cable and Wireless Plc specialising in telecommunication products, projects and services in Hong Kong and the Far East region and was appointed as Chief Executive in 1987. In 1989 he returned to the UK as Managing Director of Paknet Ltd, a joint venture owned by Cable and Wireless Plc and Vodafone Plc, which developed and launched the world's first public packet radio data network.

Prior to his move to Singapore, he was President of the Mobile Services Group of the Hong Kong Telecommunications Association, a member of the Hong Kong Government Telecommunications Standard Advisory Committee and a member of the Hong Kong Government Telecommunications Numbering Advisory Committee.

Yasuaki Mori

Advanced Micro Devices, Geneva, Switzerland

Yasuaki Mori is currently Marketing Manager for Wireless Products for Advanced Micro Devices in Geneva, Switzerland. In this role he was, among others, responsible for establishing various partnerships with major players in the digital cordless arena in Europe and Asia. He has held various strategic marketing and sales posts in the USA, Japan and Europe. He holds a BSc in electrical engineering and Masters in international affairs/business from Columbia University, New York, USA.

Anthony Noerpel

Hughes Network Systems, Germantown, MD, USA

Anthony R. Noerpel is an Advisory Engineer at Hughes Network Systems in Germantown, Maryland, USA. He currently works on PCS and mobile satellite projects involving antennas, propagation and system design. He formerly worked at Bellcore as a Senior Research Scientist, where he was instrumental in the PACS and PACS-UB standards development and where he led the development of the higher-layer protocols for PACS. He has several patents and publications in areas ranging from antenna structures and multi-mode couplers to system design issues such as handover and multi-system interoperability. He has an MS degree in electrical engineering from New Jersey Institute of Technology and a BA in mathematics from Rutgers University.

Heinz Ochsner

Ochsner MTC, Solothurn, Switzerland

Dr Heinz Ochsner received his Dr Sc Techn (Doctor of Technical Science) degree from the Swiss Federal Institute of Technology in 1987. His thesis covered aspects of the use of spread-spectrum techniques in mobile radio. In 1987 he joined Ascom Autophon AG in Solothurn, where he was responsible for system studies for mobile radio and cordless telecommunication systems. In this capacity he was the Swiss representative in the CEPT/GSM Permanent Nucleus during 1987–88.

During the late 1980s and early 1990s he undertook similar responsibilities in the field of cordless telecommunications, being responsible for contacts with potential cooperation partners, competitors, EC authorities, European industry associations, etc. This included a period chairing the ETSI RES 03 R Committee, which developed the DECT standard. Between 1992 and 1993 he was a consultant and project manager with Ascom Systems for Information Technology and Communications, prior to establishing his own business offering consultancy services in the specific field of mobile communications.

Peter Olanders

Telia Research, Malmoe, Sweden

Dr Peter Olanders joined Swedish Televerket in 1988 from Ericsson Radar Systems. At this point in time he headed a group in development of land mobile radio, which focused on cordless communications, including CT2, CT3, DECT and later on radio-based LANs. The group developed into a section, and was transferred to Telia Research at its creation in 1991. Dr Olanders was appointed as Chairman of the ETSI RES 03 PAV (Public Access Validation) Working Party in 1991, as the Chairman of RES 03 in 1992, a Chair he is still holding. When ETSI decided to create a specific DECT project, Dr Olanders was appointed as the Project Manager. In 1994 he initiated the DECT Operators' Group, of which he became the first Chairman.

Annette Ottolini

Greenpoint, The Hague, The Netherlands

Annette Ottolini works for Netherlands PTT Telecom where she has responsibility for marketing the Dutch Greenpoint CT2-based telepoint network.

Frank Owen

Philips, Wien, Austria

Frank Owen has been active for the last 10 years in the design and product development of digital cordless telephony and associated systems. From a period of initial research and pre-development on the DECT system, he was appointed International Project Manager for the range of private and public DECT products available from Philips, including residential systems, cordless PBX and wireless local loop installations. He was responsible for organising the technical contribution to one of the winning telepoint bids in the UK. During 1991 and 1992 he served as Chairman of the ETSI Working Committee responsible for the DECT radio interface specification under RES 3. Frank is now responsible for product creation and marketing of facsimile products with Philips Business Electronics. Part of the product portfolio includes facsimile machines for personal use, once again integrated with cordless telephony systems.

Marc Pauwels

Belgacom, Brussels, Belgium

Marc Pauwels studied electronics and process control at the University of Ghent, receiving his civil engineering degree in 1985. He joined RTT in 1989 as a technical product manager for terminal equipment, based at their headquarters in Brussels. In 1990 he assumed technical responsibility for the telepoint CITEL pilot project in Brussels, overseeing its implementation and operation. Since the reorganisation of Belgacom in 1995, he is currently responsible for the management systems in the payphone division.

William H. Scales Jr

Panasonic, Alpharetta, GA, USA

William Scales received MSEE and BSEE degrees from North Carolina State University in 1988 and 1984 respectively. Since 1995, he has been with Matsushita Communications (Panasonic) in Alphaetta, Georgia, USA, where he is a Staff Engineer responsible for the systems design and development of PCS products. During the past two years, he has actively participated in industry standards bodies for the PACS air interface. Previously, Mr Scales held various design and manufacturing positions with Hitachi Telecom, Texas Instruments and ITT Telecom.

J. Neal Smith

Omnipoint Technologies, Colorado Springs, CO, USA

J. Neal Smith graduated from the University of Cincinnati with a BSc in electrical engineering in 1981 and from Ashland University with an MBA degree in 1985. Neal worked for 10 years for United Telephone of Ohio across a wide range of fixed and cellular telecommunications activities. In 1991, Neal became Manager of Architecture and Strategic Planning, supporting Sprint's Local Telecommunications Division by developing central office switching, transport, AIN and PCS technology platform plans. Neal subsequently began developing PCS technology deployment plans as one of the original members of the Wireless Business Development group within Sprint's Cellular Wireless Division (which eventually became Sprint Telecommunications Venture), where he managed the trial of unlicensed in-building cordless technologies (CT2, DECT and other technologies). Neal joined Omnipoint Corporation in 1994, where his initial work involved developing and getting approval, for the US PCS standard IS-661, for Omnipoint's technology. He subsequently has been leading efforts for product and business development for PCS-based wireless local loop and cordless terminal mobility products based on IS-661 technology in the licensed and unlicensed bands.

Richard Steedman

PMC-Sierra, Burnaby, BC, Canada

Richard Steedman received a BSc (Hons) degree in computer science and electronics from Edinburgh University in 1985. After graduating, he joined the VLSI design group at Roke Manor Research, where he designed a number of semi-custom ICs for military and civil applications and undertook associated studies and feasibility demonstrations. Richard contributed to the development of a CAI-compatible CT2 product and, as part of that activity, sat on the Technical Working Group that produced the CT2 CAI standard, MPT 1375. From 1990 to 1991, he served on the ETSI RES 3 R Working Group (DECT radio interface). Richard is currently a senior product designer with PMC-Sierra Inc., a manufacturer of broad-band networking semiconductors.

Robert S. Swain

RSS Telecom, Suffolk, UK

Bob Swain, CEng, MIEE, operates his own consultancy in the field of mobile communications, where recent assignments have included supporting the European Commission in their UMTS Task Force and ACTS Mobile activities. Previously Head of the Personal Radio Systems Section of British Telecom's Martlesham Research Station, Bob was responsible for pioneering digital cordless communications systems research and development within BT throughout the 1980s. This work involved extensive propagation and system studies and contributed significantly to the UK digital CT2 Common Air Interface standard. Following this activity, he served as Chairman of the ETSI RES 03 Services Definition and Authentication Working Group for DECT. Bob was awarded the prestigious Martlesham Medal by British Telecom in recognition of his outstanding contributions in the field of cordless telecommunications.

Internationally, Bob Swain has had close involvement over many years with the personal mobile communications work of the European Telecommunications Standards Institute (ETSI), the Conference of European Posts and Telecommunications (CEPT) and the European Conference of Telecommunication Manufacturers' Associations (ECTEL).

Yuichiro Takagawa

NTT, Tokyo, Japan

Yuichiro "Tuck" Takagawa is currently Executive Manager in Networked Multimedia Services with the NTT Multimedia Business Department. He was a pioneer of Cordless Telephone System development in NTT's Customer Premises Equipment Department from 1986 to 1989, where he was responsible for development of a single-line cordless system, a home telephone cordless system and a small key telephone cordless system, as well as a special cordless system for golf course usage. His areas of expertise include ISDN, CPE products and wireless systems.

Julian Trinder

Roke Manor Research, Romsey, UK

Julian Trinder, MA, CEng, MIEE, received an honours degree in physics from Oxford in 1971. This was followed by a six-year period of voice communications research with Plessey. He then worked for the UK Medical Research Council Institute of Hearing Research in the field of speech signal processing followed by a change to radio communications research with Multitone Electronics. During this latter period Julian worked within the ESPA cordless telephone working groups, towards TDMA digital cordless telephony and developed a CDMA cordless telephone demonstrator as part of a joint venture with Mitel. After joining Roke Manor as a Senior Consultant in 1987, Julian was a significant contributor in the UK CT2 CAI Working Groups dealing with radio and speech aspects, also undertaking theoretical and hardware studies leading to the development of one of the early CT2 CAI products. Recent activities have included speech aspects of PCS1900 and CDMA wireless local loop developments. Julian is the author of numerous papers and holds several patents, including ones relating to cordless telephony.

Diane Trivett

Dataquest Europe, Egham, Surrey, UK

Diane Trivett is employed as an Industry Analyst by Dataquest Europe, where she has joint responsibility for the Telephones and PBX programmes. Diane is responsible for tracking market dynamics including developments, emerging technologies, company activities and pricing issues across Europe for both the voice programme and consulting projects. These have included research and analysis of PBX and voice terminal manufacturers and their products and a focus report on cordless telephony, including market sizing and forecasts for handsets, basestations and systems, pricing trends, and vendor strategies. Prior to joining Dataquest, Diane was employed by Datapro International as the Managing Analyst of the International Communications Equipment service, for which she covered all areas of the voice and mobile industries. Prior to this position, Diane worked as a Market Analyst for GPT.

Margareta Zanichelli

Telia Research, Malmoe, Sweden

Margareta Zanichelli is a Project Manager at Telia Research AB in Sweden, in the Communications Systems Division. She has a Master of Science degree in electronical engineering and has since 1990 been working with development of radio access systems at Telia. Since 1992 her special field has been radio in the local loop, with a focus on developing and trialling short-range communications systems. Internationally she has been working with radio in the local loop in ETSI RES 03 and in EURESCOM, where she has been the task leader of a group dealing with these issues.

Index